Current Topics in Microbiology and Immunology

Volume 431

The review series *Current Topics in Microbiology and Immunology* provides a synthesis of the latest research findings in the areas of molecular immunology, bacteriology and virology. Each timely volume contains a wealth of information on the featured subject. This review series is designed to provide access to up-to-date, often previously unpublished information.

2019 Impact Factor: 3.095., 5-Year Impact Factor: 3.895

2019 Eigenfaktor Score: 0.00081, Article Influence Score: 1.363

2019 Cite Score: 6.0, SNIP: 1.023, h5-Index: 43

More information about this series at http://www.springer.com/series/82

Steffen Backert
Editor

Fighting *Campylobacter* Infections

Towards a One Health Approach

Responsible Series Editor: Shizuo Akira

Springer

Editor
Steffen Backert
Division of Microbiology
Department of Biology
University of Erlangen-Nuremberg
Erlangen, Germany

ISSN 0070-217X ISSN 2196-9965 (electronic)
Current Topics in Microbiology and Immunology
ISBN 978-3-030-65480-1 ISBN 978-3-030-65481-8 (eBook)
https://doi.org/10.1007/978-3-030-65481-8

This Springer imprint is published by the registered company Springer Nature Switzerland AG
The registered company address is: Gewerbestrasse 11, 6330 Cham, Switzerland

Foreword

A Very Personal Foreword: *Campylobacter*, My First Love in Microbiology

I feel very much honored to be contacted by Steffen Backert for providing a foreword for the Current Topics in Microbiology and Immunology (CTMI) book on *Fighting Campylobacter Infections: Towards a One Health Approach* that he has edited. After having worked in *Campylobacter* research area for several decades, I was very impressed by the extraordinary list of chapters and the great collection of internationally recognized experts who contributed to this volume. As a matter of fact, writing the foreword offered to me a perfect opportunity to study the whole book. Thus, I am very happy to write this in a very personal way.

In the long history of the *Campylobacter* research, the pathogenicity of these remarkable bacteria was gradually enlightened. In a first description more than 134 years ago, the German medical doctor and bacteriologist Theodor Escherich described "Vibrio-like" microbes in the colonic mucus of children (Escherich 1886), a study that was not recognized by the international community until 1985 (Kist 1986a). In these early days, Theodor Escherich was working as a pediatrician at a children's hospital in Munich, where he had seen 72 infants with diarrhea (*Cholera infantum*) during the summer of 1885. Seventeen of them died from the disease, and in the postmortem examination, Theodor Escherich microscopically detected "Vibrio-like" microbes in the colonic mucosa of 15 infants. He was not able to grow the bacteria under laboratory conditions, but made drawings and described their morphology in detail. The drawings and description clearly fit with *Campylobacter* bacteria (Kist 1986a). About 45 years after this first report by Theodor Escherich, bacteria isolated from the intestines of sick calves and pigs were named *Vibrio jejuni* (Jones et al. 1931) and *Vibrio coli* (Doyle 1944), respectively. In addition, these bacteria were detected in ox bile blood-cultures as a "Vibrio of bovine origin", when a gastroenteritis outbreak was investigated in a correctional institution of the USA (Levy 1946). "Related Vibrio" species were also discovered elsewhere in patients with diarrheal disease (King 1957). Later, these "related Vibrio-like" microbes, probably all corresponding to *Campylobacter jejuni* and *Campylobacter coli* as we know them today, were included in the new genus *Campylobacter* (Sebald and Véron 1963). Subsequently, firstly in 1968, *Campylobacter* spp. could be isolated and grown from human diarrheal stools (Cooper and Slee 1971; Dekeyser et al. 1972). The development of improved culturing methods in laboratories facilitated the successful isolation and growth of these bacteria from human stool samples, which was the prerequisite that *C. jejuni* or *C. coli* could be confirmed as human enteric pathogens (Skirrow 1977). Indeed, *C. jejuni* and *C. coli* are extremely fascinating, highly successful, mysterious and also quite challenging organisms. They are elegantly coiled when they are young and becoming more coccoid in aging stages, a bit like some of us. And yes, *Campylobacter* was actually my first love in microbiology. And so, for me, this was the beginning of a "love story."

At the time when Martin Skirrow's communication "Campylobacter enteritis – a new disease" (Skirrow 1977) appeared, I started as a young medical doctor and to my knowledge firstly in Germany implementing cultural and biochemical methods for detection and identification of *Campylobacter* from stool specimens in our diagnostic laboratory at the Institute of Hygiene and Microbiology at Freiburg University. And in order to prepare a dissertation for a professorship, I launched at the end of 1970 a comprehensive study on incidence, clinics and epidemiological risk factors of infectious enteritis, which covered all causative agents known at that time, including also *Campylobacter*. In this study, we investigated more than 17,000 patients with diarrhea, and in 945 cases (5.3%) *Campylobacter* ssp. were isolated from stool specimens (Kist 1986b). In a multivariate analysis, fever and bloody–watery diarrhea were shown as significant clinical symptoms due to *Campylobacter* infection, with bloody diarrhea more common in younger age-groups and predominantly watery diarrhea in older age-groups. Vomiting, however, was not a typical symptom. Follow-up of the above-mentioned 945 cases revealed the development of a total of three patients with neurological complications, namely two Guillain–Barré syndromes and one Miller Fisher syndrome (Kist 1986b).

How I, as the first German microbiologist in this field, got access to the international "*Campylobacter* community" was then due to a fortunate accident. In 1980 at a meeting of the "Austrian Society for Hygiene, Microbiology and Preventive Medicine," which took place in Klosterneuburg, a historic monastery nearby Vienna, I presented my first *Campylobacter*-related results of the above study, and I finished my talk with the conclusions "the typical campylobacteriosis patient is a young boy, growing up in a rural area, who drinks raw milk, and likes eating chicken and swimming in natural surface waters." My presentation was the last before lunch, and some people involved me into a discussion after the end of the session. And nobody got aware that the doors were locked and we were trapped in this wonderful ancient lecture room. So we missed the lunch, but I got in contact with a colleague from the British Public Health Laboratory Service (PHLS). We had a highly stimulating conversation with the result that I was invited to the very first International Workshop on *Campylobacter*, which took place in Reading, England, in 1981. At that opportunity, I was lucky again because I met there the most prominent protagonists of modern campylobacteriology at that time, namely Jean-Paul Butzler, Martin Skirrow, Hermy Lior, and furthermore Diane Newell, Martin Blaser, Roger Feldman, and later Francis Megraud. Over the years, we became friends and the ties of friendship hold up to now. Already in 1968, Jean-Paul Butzler in collaboration with Paul Joseph Dekeyser had succeeded in Brussels as the first in culturing *Campylobacter* from a stool specimen of a 20-year-old female (Dekeyser et al. 1972). This prompted Jean-Paul Butzler in the early 1970 to a systematic study at the St. Pierre University in Brussels, where *C. jejuni* and *C. coli* were isolated from 5.3% of 3,800 diarrheic stool samples (Butzler 1974). Jean-Paul Butzler published this study in his Ph.D. thesis in Flemish language, but nevertheless it came to the attention of Martin Skirrow in England, who contacted Jean-Paul Butzler in 1976 by phone. This was the beginning of a long-lasting and fruitful collaboration as well as friendship between these headliners in campylobacteriology.

It was highly exciting to me to learn how much further the *Campylobacter* field developed in recent years. The current volume comprehensively discusses our modern knowledge of research in *Campylobacter*–human interactions, the natural reservoirs, infection routes and envisaged intervention strategies. Chapter "Human Campylobacteriosis—A Serious Infectious Threat in a One Health Perspective" by Stefan Bereswill and co-workers highlights *Campylobacters* as the major cause of foodborne bacterial gastroenteritis worldwide, eventually also affecting other organs such as the nervous system and joints, associated with enormous costs in the societies. They also emphasize the urgent need for the so-called One World—One Health concept requiring the combined endeavors of public health authorities, clinicians, veterinarians and basic researchers to better manage the burden of this zoonotic disease in the future. As next, Roswitha Merle reviews the surveillance instruments of *Campylobacter* infections in humans as well as important control strategies including multiple risk assessment tools. In Chapter "Population Biology and Comparative Genomics of *Campylobacter* Species", Torsten Semmler and co-authors discuss the population biology and comparative genomics of various *Campylobacter* species, including *C. jejuni*, *C. coli*, *C. upsaliensis*, *C. concisus and C. lari.* They provide an excellent overview of the current technologies like whole-genome sequencing that allow the fast and efficient completion of entire *Campylobacter* genomes. These approaches lead to the identification of specific genomic features and their proposed impact on host adaptation by *Campylobacter* spp. The overall management schemes for prevention of *Campylobacter* infections through the poultry food chain are then comprehensively discussed by Thomas Alter and Felix Reich. In Chapter "Emission Sources of *Campylobacter* from Agricultural Farms, Impact on Environmental Contamination and Intervention Strategies", Vanessa Szott and Anika Friese nicely illuminate the emission sources of *Campylobacter* spp. from agricultural farms, their impact on contamination of the environment and associated intervention measures. In this regard, Sophie Kittler discusses an elegant approach using bacteriophages to reduce *Campylobacter* loads in poultry. This article focuses on the exciting option as to how certain bacteriophages can be used in practice and discusses the legal biosafety regulations for approval of corresponding treatment schemes in the food industry, which may assist in resolving campylobacteriosis cases in humans. In the following chapter, Nicole Tegtmeyer and Steffen Backert review the molecular virulence properties of *C. jejuni*. They highlight the known factors and various pathogenicity-linked determinants comprising bacterial motility, chemotaxis, cellular binding, host cell entry, intracellular survival and transmigration into deeper tissues and even other organs, which are discussed in detail. Furthermore, Jörg-Dieter Schulzke and Roland Bücker update us on the diarrheal mechanisms and the role of intestinal barrier dysfunction triggered by *Campylobacter* infection. In Chapter "Murine Models for the Investigation of Colonization Resistance and Innate Immune Responses in *Campylobacter Jejuni* Infections", Markus Heimesaat and colleagues review the various mouse models of *C. jejuni* infection and the role of the microbiota in this context. They highlight the current validation and standardization procedures in the murine infection model systems, which may provide the basis for future development of more innovative treatment and prevention strategies of *C. jejuni* infection. In addition, Julia Golz and Kerstin Stingl review the fundamental concepts of natural transformation and horizontal gene transfer events by *Campylobacter* spp., which clearly contribute to genome diversity and the spread of antibiotic resistances in the bacteria, which is an enormous problem of the healthcare systems in combating many infectious diseases. Finally, the group of Greta Gölz gives an excellent update of our current knowledge on the molecular mechanisms of biofilm formation and quorum sensing by *Campylobacter* spp., which are important features that may help the bacteria to survive inside the natural environment.

In summary, the book updated me greatly on *Campylobacter* and provided me with many new ideas. For example, after reading, I saw the previous concepts in a new perspective and also learnt a lot about the most recent developments in the *Campylobacter* area. The book gave me important new advice where research in the field is standing at this time and in what directions the journey might take us in near future. Together, I highly recommend this book to advanced undergraduates, graduate students, postdocs, medical doctors and other investigators, who are interested in infection biology and zoonoses research.

Freiburg im Breisgau, Germany
October 2020

Manfred Kist

References

Butzler JP (1974) *Campylobacter*, an unrecognized enteropathogen. Ph.D. thesis. Free University of Brussels, Brussels, Belgium (in Flemish).

Cooper IA, Slee KJ (1971) Human infection by *Vibrio fetus*. Med J Aust 1:1263-1267.

Escherich T (1886) Beiträge zur Kenntnis der Darmbacterien. Münchener Medizinische Wochenschrift 33:815–835.

Dekeyser P, Gossuin-Detrain M, Butzler JP, Sternon J (1972) Acute enteritis due to related vibrio: first positive stool cultures. J Infect Dis 125:390-392. https://doi.org/10.1093/infdis/125.4.390

Doyle LP (1944) A vibrio associated with swine dysentery. Amer J Vet Res 5:3–5.

Jones FS, Orcutt M, Little RB (1931) Vibrios (*Vibrio jejuni*, n.sp.) associated with intestinal disorders of cows and calves. J Exp Med 53:853–863. https://doi.org/10.1084/jem.53.6.853

King EO (1957) Human infections with *Vibrio fetus* and a closely related vibrio. J Infect Dis 101:119-128. https://doi.org/10.1093/infdis/101.2.119

Kist M (1981) *Campylobacter* enteritis: epidemiological and data from recent isolations in the region of Freiburg, West Germany. In: Newell DG (ed) *Campylobacter* – epidemiology, pathogenesis and biochemistry. Proceedings of an International Workshop on *Campylobacter* Infections held at University of Reading, 24-26th March 1981, pp.138-143

Kist M (1986a) Who discovered *Campylobacter jejuni/coli*? A review of hitherto disregarded literature. Zentralbl Bakteriol Mikrobiol Hyg A. 261(2):177-186.

Kist M (1986b) Vergleichende Untersuchung zu Inzidenz, Klinik und epidemiologischen Risikofaktoren der infektiösen Enteritis. Habilitationsschrift zur Erlangung der *Venia legendi* der Hohen Medizinischen Fakultät der Albert-Ludwigs-Universität Freiburg im Breisgau (in German).

Levy AJ (1946) A gastro-enteritis outbreak probably due to a bovine strain of vibrio. Yale J Biol Med 18: 243-258.

Sebald M, Veron M (1963) Base DNA content and classification of Vibrios. Ann Inst Pasteur (Paris) 105:897-910.

Skirrow MB (1977) *Campylobacter* enteritis: a "new' disease. Br Med J 2:9-11. https://doi.org/10.1136/bmj.2.6078.9

Contents

Abbreviations

AC	Adenylate cyclase
AHL	Acylated homoserine lactone
AI	Autoinducer
AIP	Autoinducer peptide
AMC	Activated methyl cycle
ANI	Average nucleotide identity
AspA	Aspartate ammonia lyase A
AspB	Aspartate aminotransferase B
ATP	Adenosine triphosphate
BCCA	Brilliance CampyCount agar
BumR	Butyrate response regulator
BumS	Butyrate sensor kinase
C.	*Campylobacter*
Ca^{2+}	Calcium ion
CadF	*Campylobacter* adhesin to fibronectin
cAMP	Cyclic adenosine monophosphate
Cas	CRISPR-associated protein
CASA	*Campylobacter* selective agar
CCD	Charcoal-cefazolin-sodium deoxycholate
CCP	Critical control point
CCV	*Campylobacter*-containing vacuole
CDC2	Cyclin-dependent kinase 2
Cdc42	Cell division control protein 42, a small Rho GTPase
CDT	Cytolethal distending toxin
CetA	*Campylobacter* energy taxis protein A
CFA	CampyFood Agar
CFT	Complement fixation tests
CFTR	Cystic fibrosis transmembrane conductance regulator
CFU	Colony-forming unit
cgMLST	Core genome MLST

CheY	Chemotaxis response regulator Y
CHO	Chinese hamster ovary cell
Cia	*Campylobacter* invasion antigen
cjA	*Campylobacter jejuni* autoinducer
CJIE	*Campylobacter jejuni* integrated element
CJT	*Campylobacter jejuni* enterotoxin
Cl^-	Chloride ion
CoV	Coronavirus
COVID-19	Coronavirus disease 2019
CPS	Capsular polysaccharide
CR	Colonization resistance
CRISPR	Clustered regularly interspaced short palindromic repeats
CT	Cholera toxin
DALY	Disability-adjusted life years
DegP	HtrA ortholog in *E. coli*
DegQ	HtrA ortholog in *E. coli*
DegS	HtrA ortholog in *E. coli*
dG	Deoxyguanosine
DNA	Deoxyribonucleic acid
DOC	Desoxycholate
Dock180	Dedicator of cytokinesis, a 180 kDa GEF
dpa	Days post-phage application
dPCR	Digital PCR
DPD	4,5-Dihydroxy-2,3-Pentanedione
DRA	Downregulated in adenoma
dsDNA	Double-stranded DNA
DSS	Dextran sodium sulfate
DUS	DNA uptake sequence
E.	*Escherichia*
EC	European Commission
eDNA	Extracellular DNA
EEA-1	Early endosome antigen 1
EFSA	European Food Safety Authority
EGFR	Epidermal growth factor receptor
EIA	Enzyme immunoassays
ELISA	Enzyme-linked immunosorbent assay
ENaC	Epithelial sodium channel
EPS	Extracellular polymeric substance
ERK	Extracellular signal-regulated kinase
ESBL	Expanded-spectrum β-lactamase (ESBL)–producing *Escherichia coli*
EU	European Union
FAK	Focal adhesion kinase
Fed	Flagellar co-expressed determinants
FFEM	Freeze fracture electron microscopy

FlaA	Flagellin A
FlaB	Flagellin B
FlaC	Flagellin C
FlgE	Flagellar hook protein
FlgR	Two-component response regulator
FlgS	Two-component sensor kinase
FliS	Flagellar chaperone
FlpA	Fibronectin-like protein A
FSA	Food Standards Agency
FSIS	Food Safety and Inspection Service
FspA2	Flagellar secreted protein A2
fT3SS	Flagellar type III secretion system
GalfNAc	2-acetamido-2-deoxy-D-galactofuranose
GBS	Guillain–Barré syndrome
GEF	Guanine nucleotide exchange factor
GGT	Gamma-glutamyltranspeptidase
GS1	Genomospecies 1
GS2	Genomospecies 2
GWAS	Genome-wide association studies
HAAP	Hazard analysis and critical control points
HCF	High conjugation frequency
HGT	Horizontal gene transfer
hma	Human microbiota-associated
hpa	Hours post-phage application
HtrA	High-temperature requirement A
IBD	Inflammatory bowel disease
IBS	Irritable bowel syndrome
IFN	Interferon
Ig	Immunoglobulin
IL	Interleukin
IM	Inner membrane
IRF3	Interleukin regulatory factor 3
I_{SC}	Short-circuit current
ISO	International Organization for Standardization
ISPC	Internal sample process control
IκBα	Nuclear factor of kappa light polypeptide gene enhancer in B cells inhibitor alpha
JAM	Junctional adhesion molecule
JlpA	*Jejuni* lipoprotein A
K^+	Potassium ion
Lamp1	Lysosomal-associated membrane protein 1
LDH	Lactate dehydrogenase
LFC	Low frequency of conjugation
LGT	Lateral gene transfer
LOS	Lipooligosaccharide

LPS	Lipopolysaccharide
MALDI–TOF MS	Matrix-assisted laser desorption ionization (MALDI)–time-of-flight (TOF) mass spectroscopy (MS)
MAMPs	Microbe-associated molecular patterns
MAP	Modified atmosphere packaging
mCCDA	Modified charcoal-cefoperazone-deoxycholate agar
MD-2	Myeloid differentiation protein -2
MDRGI	Multidrug-resistant genomic island
MeOPN	O-methyl phosphoramidate
MERS	Middle East respiratory syndrome
MFS	Miller Fisher syndrome
MIC	Minimal inhibitory concentration
MLCK	Myosin light-chain kinase
MLST	Multi-locus sequence typing
MOI	Multiplicity of infection
MOMP	Major outer membrane protein
MPI	Ministry of Primary Industries, New Zealand
mRNA	Messenger ribonucleic acid
mTOR	Mammalian target of rapamycin
MyD88	Myeloid differentiation primary response 88
Na^+	Sodium ion
NF-κB	Nuclear factor kappa-light-chain-enhancer of activated B cells
NGS	Next-generation sequencing
NHE	Sodium–hydrogen antiporter
NLR	NOD-like receptor
NLRP3	NLR family pyrin domain-containing protein 3
NOD	Nucleotide–oligonucleotide domain
NOD2	Nucleotide–oligonucleotide domain 2
OM	Outer membrane
OMV	Outer membrane vesicle
ONT	Oxford Nanopore Technology
ORS	Oral rehydration solution
PacBio	Pacific Bioscience
PAMP	Pathogen-associated molecular pattern
PCR	Polymerase chain reaction
pDC	Plasmacytoid dendritic cell
PDGFR	Platelet-derived growth factor receptor
PDZ	Postsynaptic density protein
PEB1	Periplasmic-binding protein 1
PFGE	Pulsed-field gel electrophoresis
PFU	Plaque-forming units
PGE_2	Prostaglandin 2
PGL	Protein glycosylation system
PI3K	Phosphatidylinositol 3-kinase
PI3K-γ	Phosphatidylinositol 3-kinase-γ

PKC	Protein kinase C
PMA	Propidium monoazide
QALY	Quality-adjusted life years
qPCR	Quantitative real-time polymerase chain reaction
QQ	Quorum quenching
QS	Quorum sensing
RA	Reactive arthritis
Rac1	Ras-related C3 botulinum toxin substrate 1, a small Rho GTPase
R^{epi}	Epithelial resistance
R^{sub}	Subepithelial resistance
SAM	S-adenosylmethionine
SARS	Severe acute respiratory syndrome
SCFA	Short-chain fatty acid
SIGIRR	Single Ig IL-1-related receptor
Siglec	Sialic acid-binding immunoglobulin-like lectin
sm	Soluble mouse
SMRT	Single-molecule real-time sequencing
SNV	Single-nucleotide variant
SodB	Superoxide dismutase B
SPC	Sample process control
SPF	Specific pathogen free
SpoT	ppGpp synthetase/pyrophosphohydrolase
SRH	S-ribosylhomocysteine
ssDNA	Single-stranded DNA
ST	Sequence type
SVR	Short variable region
SYF	$Src^{-/-}/Yes^{-/-}/Fyn^{-/-}$ triple knockout
T3SS	Type III secretion system
T4SS	Type IV secretion system
TEM	Transmission electron microscopy
TER	Transepithelial electrical resistance
TIR	Toll-Interleukin receptor domain
TJ	Tight junction
Tlp	Transducer-like protein
TLR	Toll-like receptor
Tm	Melting temperature
TNF	Tumor necrosis factor
TracrRNA	Trans-activating CRISPR RNA
TRAM	TRIF-related adapter molecule
TRIF	TIR-domain-containing adapter-inducing interferon-β
UC	Ulcerative colitis
UI	Uncertainty interval
UK	United Kingdom
US	United States

UV	Ultraviolet
VBNC	Viable but non-culturable
VD	Vitamin D
VDR	Vitamin D receptor
wgMLST	Whole-genome sequencing-based MLST
WGS	Whole-genome sequencing
WHO	World Health Organization
YLD	Years lost due to disability
YLL	Years of life lost
YPLL	Years of potential life lost
ZO	Zonula occludens

Human Campylobacteriosis—A Serious Infectious Threat in a One Health Perspective

Markus M. Heimesaat, Steffen Backert, Thomas Alter, and Stefan Bereswill

Contents

Abstract Zoonotic *Campylobacter* species—mainly *C. jejuni* and *C. coli*—are major causes of food-borne bacterial infectious gastroenteritis worldwide. Symptoms of intestinal campylobacteriosis include abdominal pain, diarrhea and fever. The clinical course of enteritis is generally self-limiting, but some infected individuals develop severe post-infectious sequelae including autoimmune disorders affecting

M. M. Heimesaat · S. Bereswill (✉)
Institute of Microbiology, Infectious Diseases and Immunology, Gastrointestinal Microbiology Research Group, Charité—University Medicine Berlin, Corporate Member of Freie Universität Berlin, Humboldt-Universität zu Berlin, and Berlin Institute of Health, Berlin, Germany
e-mail: stefan.bereswill@charite.de

M. M. Heimesaat
e-mail: markus.heimesaat@charite.de

S. Backert
Division of Microbiology, Department of Biology, Friedrich Alexander University Erlangen/Nuremberg, Erlangen, Germany
e-mail: steffen.backert@fau.de

T. Alter
Department of Veterinary Medicine, Institute of Food Safety and Food Hygiene, Free University Berlin, Berlin, Germany
e-mail: thomas.alter@fu-berlin.de

S. Backert (ed.), *Fighting Campylobacter Infections*, Current Topics in Microbiology and Immunology 431, https://doi.org/10.1007/978-3-030-65481-8_1

the nervous system, the joints and the intestinal tract. Moreover, in immunocompromised individuals, systemic spread of the pathogens may trigger diseases of the circulatory system and septicemia. The socioeconomic costs associated with *Campylobacter* infections have been calculated to several billion dollars annually. Poultry meat products represent major sources of human infections. Thus, a "One World—One Health" approach with collective efforts of public health authorities, veterinarians, clinicians, researchers and politicians is required to reduce the burden of campylobacteriosis. Innovative intervention regimes for the prevention of *Campylobacter* contaminations along the food chain include improvements of information distribution to strengthen hygiene measures for agricultural remediation. Given that elimination of *Campylobacter* from the food production chains is not feasible, novel intervention strategies fortify both the reduction of pathogen contamination in food production and the treatment of the associated diseases in humans. This review summarizes some current trends in the combat of *Campylobacter* infections including the combination of public health and veterinary preventive approaches with consumer education. The "One World—One Health" perspective is completed by clinical aspects and molecular concepts of human campylobacteriosis offering innovative treatment options supported by novel murine infection models that are based on the essential role of innate immune activation by bacterial endotoxins.

1 Introduction

Food-borne microbial infections of the human gastrointestinal tract and resulting diseases are associated with very high degrees of morbidity and mortality in the world's population. According to the World Health Organization (WHO 2020), unsafe food products cause 600 million cases of food-borne diseases and 420,000 deaths annually, especially in children and elderly people. The WHO calculated that about 33 million years of healthy lives are lost through eating unsafe food worldwide every year, and this number is likely an underestimation (WHO 2020). Over the past two decades, *Campylobacter jejuni* has been recognized as the leading source of bacterial gastroenteritis around the globe (Wassenaar and Blaser 1999; Young et al. 2007; Altekruse 2008; Dasti et al. 2010; Burnham and Hendrixson 2018). Gut disease outcomes vary from mild, non-inflammatory, self-limiting diarrhea to severe, inflammatory, bloody diarrhea associated with severe abdominal pain, which can last for a few weeks. However, *C. jejuni-infection* is also associated with more severe neurological sequelae in some patients, such as the Guillain-Barré syndrome (GBS) and the Miller Fisher syndrome (MFS) (Talukder et al. 2011; Wakerley et al. 2014). Statistical evaluations indicate that *Campylobacter* infections are responsible for considerable costs of medication and health service. In the USA alone, it was estimated that *Campylobacter* illnesses in humans cost a burden of up to $6.2 billion annually (Forsythe 2000). In fact, in numerous studies from the USA and other developed nations, *Campylobacter* was reported to cause diarrheal disease 2–7 times more frequently than pathogenic *Salmonella*, *Escherichia* or *Shigella* species (Acheson and

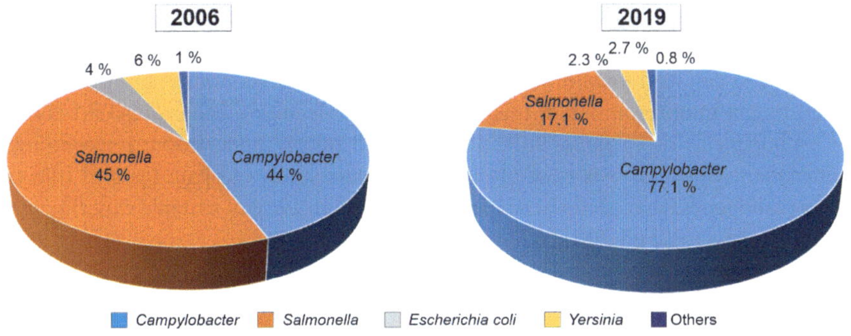

Fig. 1 Annual incidences of infections with intestinal bacterial pathogens in Germany. These data are derived from the yearly German Disease Statistics Reports for 2006 and 2019 as published by the Robert Koch Institute (RKI), Berlin (Germany). While the reported number of salmonellosis cases has slightly fallen from 19,531 to 13,281, those of campylobacteriosis rose from 22,390 to 55,959 in the shown years

Allos 2001; Tam 2001). In some countries, the fraction of reported *Campylobacter* cases has increased over the years. For example, the Robert Koch Institute's Annual Statistical Reports indicate that the incidence of reported *Campylobacter* cases in Germany constituted 45 and 77% of all reported intestinal bacterial infections in 2006 and 2019, respectively (Fig. 1). However, it was estimated that the actual numbers of campylobacteriosis cases in Germany and other countries are likely to be much larger and could exceed more than four times the published statistics (EFSA 2011; Stingl et al. 2012).

C. jejuni is a member of the ε-proteobacterial subphylum of Gram-negative bacteria. These bacteria have a relatively small circular chromosome of 1.59–1.77 million base pairs, with an average guanosine and cytosine (GC) content of about 30.3–30.6%. The high gene density of about 94–94.3% establishes it as one of the most compact bacterial genomes sequenced so far (Parkhill et al. 2000; Fouts et al. 2005; Hofreuter et al. 2006). *C. jejuni* is quite fastidious in vitro, and bacterial growth under laboratory conditions requires nutrient-rich media, yet it is well-adapted to temperatures of 40–42 °C and microaerobic conditions (Stingl et al. 2012). The catabolic capabilities of *C. jejuni* are highly restricted because several genes required for carbohydrate utilization are missing or incomplete. In this respect, the organism differs from *Salmonella enterica* serovar Typhimurium and other bacterial gut pathogens (Hofreuter 2014). Regardless of these metabolic limitations, *C. jejuni* can effectively colonize the intestines of numerous animal hosts as a commensal. In fact, *C. jejuni* can inhabit the intestinal tract of a broad variety of wild birds and agriculturally relevant poultry, cattle and pigs (Oyarzabal and Backert 2012). Consequently, the handling and consumption of contaminated poultry and other meat products, raw milk and water have been established as the most frequent sources of *C. jejuni* infection in humans (Pielsticker et al. 2012). Upon oral uptake, *C. jejuni* colonizes the distal ileum and colon of the human host. *C. jejuni* is tremendously successful in competing with the human intestinal microbiota (Masanta et al. 2013). An infectious dose of a few hundred bacteria is sufficient to result in intestinal colonization and can lead to campylobacteriosis. Despite the economic importance and clear clinical manifestations of this disease, the molecular mechanisms underlying the pathogenesis of *C. jejuni* infections are still poorly understood. Even though human campylobacteriosis is of global importance, studies to gain insights into *C. jejuni* pathogenesis have long been hampered by the absence of suitable experimental in vivo models (Newell 2001).

2 The One Health Concept: General Theory and Practical Approaches

The concept of "One Health" is based on the idea to achieve better public health outcome globally using the design and implementation of official programs as well as scientific research by multiple disciplines that need to work together. No single discipline or sector in our society possesses sufficient knowledge, skills and resources to preclude the emergence or re-occurrence of (zoonotic) diseases in the globalized world of today. Originally, this notion was stemming from the "One Medicine" concept, which demanded for an alliance between veterinary and human medicine in response to certain diseases (Schwabe 1984). This was adapted in 2004 to the "One World—One Health" concept, as conceived by the Wildlife Conservation Society (One World One Health 2020b). This initiative for the first time enunciated an interdisciplinary projection for the prevention and spread of important diseases, while at the same time maintaining the integrity of natural ecosystems. In this regard, the Wildlife Conservation Society defined 12 specific principles or practical approaches, termed the so-called Manhattan principles (Table 1), summarizing important milestones in this concept (One World One Health 2020a). Further global efforts for the establishment of official "One Health" strategies were performed by the WHO, United Nations, and various other globally operating institutions, as summarized in previous review articles (Zinsstag et al. 2011; Bardosh 2016).

Globally, we are now facing an era where the human population and degree of industrialization steadily increase, with negative effects on land use, wildlife ecology and the global climate. In addition, geopolitical conflicts can destabilize societies, and global climate changes may trigger or worsen negative developments in almost all ecosystems, while industrialization is generally associated with substantial environmental pollution, impairment of overall biodiversity by disappearance or loss of species as well as migration of millions of people due to war, social instability and natural catastrophes. These rapid global effects are ultimately associated with the emergence and re-emergence of countless infectious and non-infectious diseases (WHO 2020). Previous and very recent outbreaks of diseases caused by zoonotic viruses including Ebola fever, Zika fever, West Nile fever, MERS, avian influenza and Covid-19 clearly illustrate how animal microbes and human health are intimately

Table 1 Manhattan principles by the "wildlife conservation society" list the following 12 recommendations*

1	Recognize the essential link between human, domestic animal and wildlife health and the threat disease poses to people, their food supplies and economies, and the biodiversity essential to maintaining the healthy environments and functioning ecosystems we all require
2	Recognize that decisions regarding land and water use have real implications for health. Alterations in the resilience of ecosystems and shifts in patterns of disease emergence and spread manifest themselves when we fail to recognize this relationship
3	Include wildlife health science as an essential component of global disease prevention, surveillance, monitoring, control and mitigation
4	Recognize that human health programs can greatly contribute to conservation efforts
5	Devise adaptive, holistic and forward-looking approaches to the prevention, surveillance, monitoring, control and mitigation of emerging and resurging diseases that take the complex interconnections among species into full account
6	Seek opportunities to fully integrate biodiversity conservation perspectives and human needs (including those related to domestic animal health) when developing solutions to infectious disease threats
7	Reduce the demand for and better regulate the international live wildlife and bushmeat trade not only to protect wildlife populations, but to lessen the risks of disease movement, cross-species transmission and the development of novel pathogen-host relationships. The costs of this worldwide trade in terms of impacts on public health, agriculture and conservation are enormous, and the global community must address this trade as the real threat it is to global socioeconomic security
8	Restrict the mass culling of free-ranging wildlife species for disease control to situations where there is a multidisciplinary, international scientific consensus that a wildlife population poses an urgent, significant threat to human health, food security or wildlife health more broadly
9	Increase investment in the global human and animal health infrastructure commensurate with the serious nature of emerging and resurging disease threats to people, domestic animals and wildlife. Enhanced capacity for global human and animal health surveillance and for clear, timely information-sharing (that takes language barriers into account) can only help improve coordination of responses among governmental and nongovernmental agencies, public and animal health institutions, vaccine/pharmaceutical manufacturers and other stakeholders
10	Form collaborative relationships among governments, local people and the private and public (i.e., non-profit) sectors to meet the challenges of global health and biodiversity conservation
11	Provide adequate resources and support for global wildlife health surveillance networks that exchange disease information with the public health and agricultural animal health communities as part of early warning systems for the emergence and resurgence of disease threats
12	Invest in educating and raising awareness among the world's people and in influencing the policy process to increase recognition that we must better understand the relationships between health and ecosystem integrity to succeed in improving prospects for a healthier planet

Source https://oneworldonehealth.wcs.org/About-Us/Mission/The-Manhattan-Principles.aspx

coupled. In fact, it has been calculated that approximately 60% of all emerging infectious diseases are of zoonotic origin in nature and the majority of those (about 72%) originate from wildlife (Jones et al. 2008). Therefore, a broader knowledge about health and disease of humans, domestic animals and wildlife is urgently required. For this purpose, the "One Health" approach is particularly important in achieving better control of zoonotic infectious diseases, reducing the spread of antibiotic resistances and food safety issues (Destoumieux-Garzón et al. 2018). The last decades have also seen a significant increase in the occurrence of infectious microbes including *Campylobacter* species and many other zoonotic pathogens (Gölz et al. 2014; Iannino et al. 2019). Local animal husbandry practices, combined with international trade and traffic, have raised the risk of emergence and spread of specific pathogens, some of which can potentially cause pandemics, as the recent outbreak with SARS-CoV-2 has demonstrated. This scenario emphasizes the importance of human and animal ecosystems in the appearance and proliferation of some pathogens associated with snowballing globalization of certain health risks. Therefore, the "One Health" initiative created an important global strategic effort highlighting the need for a joint approach, which requires interdisciplinary cooperation and integrates cross-disciplinary expertise to ensure sustainable health of global flora and fauna in all ecosystems as well as mankind.

3 Human Campylobacteriosis—From Clinical Investigations to Novel Treatment Options Using Innovative Murine Models of Infection

3.1 Human Campylobacteriosis—Basic Characteristics

The fact that *Campylobacter* species that are pathogenic to humans, mainly *C. jejuni* and *C. coli*, form part of the commensal intestinal microbiota of wild and domestic animals is the basis for the transfer of respective bacterial infections to humans, which predominantly occurs by ingestion of contaminated meat products, raw milk and water. Pathogen transmission from wild birds and pets is still under debate, and actual data indicate that these additional *Campylobacter* reservoirs might be responsible for an accountable number of disease cases (Smith et al. 2020). Source attribution studies have proven that the highest risk for human *Campylobacter* infection is associated with consumption of contaminated meat from chicken and other poultry (Cody et al. 2019; Kaakoush et al. 2015). The virtual absence of clinical signs in these animals provides the basis for formation, continuous propagation and toleration of the large pathogen reservoirs in the poultry breeding industry worldwide. This constitutes a major key in understanding the epidemiology of human campylobacteriosis given that clinical manifestations in poultry would interfere with industrial meat production procedures leading to eradication of *Campylobacter* pathogens by veterinary therapeutic interventions. However, this is not the case: Actual measures to reduce human infections are rather focused on the minimization of *Campylobacter* contamination in the meat production chains (see Chapters "Management Strategies for Prevention of *Campylobacter* Infections Through the Poultry Food Chain: A European Perspective", "Emission Sources of *Campylobacter* from Agricultural Farms, Impact on Environmental Contamination and Intervention Strategies" and "Phage Biocontrol of *Campylobacter*: A One Health Approach" in this book). Since human *C. coli* infections are far less frequently responsible for campylobacteriosis cases (about 1–5%), the following considerations and discussions will be focusing on *C. jejuni* as the major pathogen causing human campylobacteriosis worldwide.

In contrast to birds, humans become severely infected by ingestion of *C. jejuni* at very low doses. Around 500 live bacteria are sufficient to effectively colonize the intestinal lumen, enter the mucus layer by motility and invade the epithelial layers to establish inflammation by activation of the innate immune system (Fig. 2). The use of proton pump inhibitors has been shown to substantially increase the risk for *C. jejuni* infection indicating that the acidic environment of the human stomach

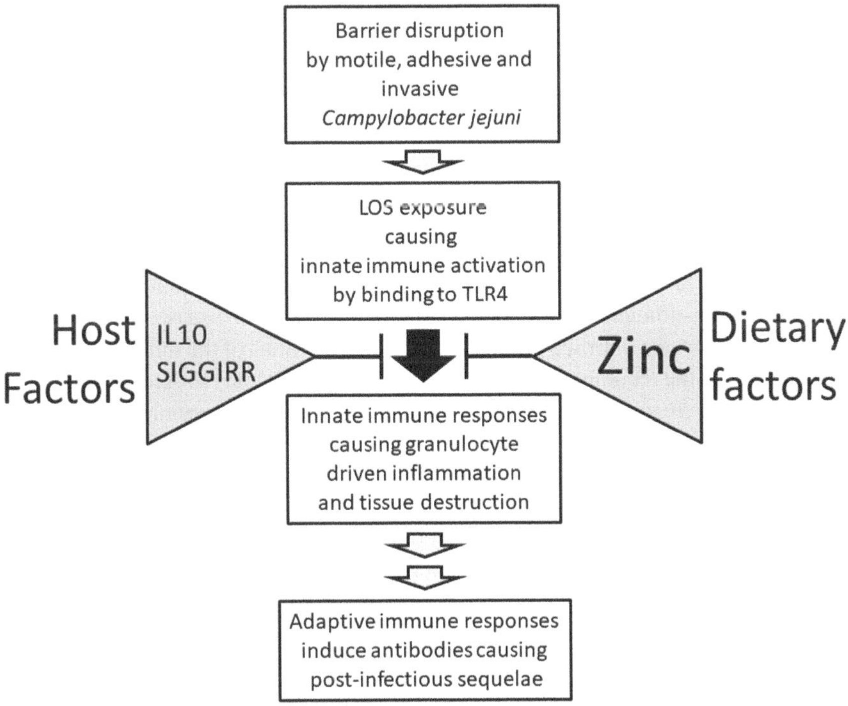

Fig. 2 Pathogenesis and modulation of *Campylobacter jejuni* infection. Both host and dietary factors such as interleukin-10 (IL-10), single IgG IL-1 related receptor (SIGIRR) and zinc, respectively, suppress innate immune responses induced by *C. jejuni* lipooligosacchgaride (LOS). As a result, the reduced inflammation levels lead to amelioration of campylobacteriosis symptoms and reduce the risk for the onset of post-infectious sequelae including RA or GBS. *Source* Mousavi et al. 2020 (adapted)

represents an effective physiological barrier directed against the pathogen (Hafiz et al. 2018). Depending on both, the immune status of the human host and the virulence factor repertoires of the infecting pathogenic strains, infected patients display a highly variable intestinal disease complex after a mean incubation period of 1–5 days (as reviewed by Skirrow 1977; Price et al. 1979; Walker et al. 1986; Blaser 1997; Awofisayo-Okuyelu et al. 2017; Facciolà et al. 2017). While some patients exhibit rather mild symptoms, others suffer from watery diarrhea or from severe campylobacteriosis characterized by purulent, bloody diarrhea, abdominal cramps and fever (Walker et al. 1986; Blaser 1997; Kist and Bereswill 2001; Janssen et al. 2008). In some instances, affected patients are even at risk for severe post-infectious autoimmune diseases such as GBS, MFS or reactive arthritis (RA) weeks or months after the initial infectious gastrointestinal manifestation (Allos 1997; Kist and Bereswill 2001; Mortensen et al. 2009). Moreover, *C. jejuni* infection is considered a potential trigger for irritable bowel syndrome (IBS), celiac disease and even inflammatory bowel diseases (IBD) which may persist lifelong (reviewed by Kaakoush et al. 2015; Keithlin et al. 2014). It is noteworthy that in immune-compromised patients spread of *C. jejuni* may cause extra-intestinal systemic manifestations affecting even the brain (Kaakoush et al. 2015).

However, in otherwise healthy individuals, *C. jejuni*-induced disease is usually self-limiting and lasts for several days up to two weeks. Even though antimicrobial treatment may reduce the duration of campylobacteriosis by 1–2 days (Ternhag et al. 2007), antibiotic application is not appropriate in general to mitigate the symptoms. One major reason is that the worldwide *C. jejuni* strain repertoire displays increasing resistance rates to macrolides and fluoroquinolones that represent first-line and second-line options for the treatment of particularly severe systemic disease manifestations, respectively (Lübbert 2016). Furthermore, the antibiotic concentrations can only insufficiently be controlled in *C. jejuni*-induced disease, particularly in the scenario of severe diarrhea with absorptive malfunctions of the inflamed mucosa. This resulted in the recommendation that antibiotic compounds should generally not be used for the treatment of campylobacteriosis with few exceptions, particularly for severely affected patients presenting with immunosuppressive comorbidities. In consequence, infected individuals do not receive causal medical treatment and rather need to sustain disease with symptomatic measures including rehydration and substitution of electrolytes to assure sufficient sodium absorption. While improvements in anti-pathogenic treatment options of campylobacteriosis are rather disappointing, the diagnostic repertoire for *C. jejuni* infection has continuously and successfully been improved, for instance by development of novel strategies in order to detect viable but not culturable (VBNC) *C. jejuni* bacteria (see also Chapter "Molecular Mechanisms of *Campylobacter* Biofilm Formation and Quorum Sensing" of this book). The search for novel alternative drugs to combat campylobacteriosis in recent preclinical studies applying novel murine infection and inflammation models revealed promising results, which await further approval and validation in clinical studies (see Chapter "Murine Models for the Investigation of Colonization Resistance and Innate Immune Responses in *Campylobacter Jejuni* Infections" of this book).

3.2 Burden of Disease

Intestinal *C. jejuni* infections and the above-mentioned post-infectious complications have been progressively rising during the last two decades worldwide (Lackner et al. 2019; Kaakoush et al. 2015). Calculated rates for the year 1997 in high-income countries were in the range of 4–5 incidents per 100,000 inhabitants. Valid global data from all reporting nations revealed that in the year 2015 incidences increased to 14–15 per 100,000 humans with a much higher estimated number of unknown cases—calculated to be in minimum 5 times higher. Thus, *Campylobacter* infections cause a high socioeconomic burden (Kaakoush et al. 2015). The increase in human campylobacteriosis cases worldwide is well in line with the rise in consumption of raw milk and other animal products including *C. jejuni* contaminated chicken and other poultry meat in high-income countries (Cody et al. 2019). Actual data from low-middle income countries show the same trend (Thomas et al. 2020) but are often too scarce to draw scientifically validated conclusions regarding worldwide incidence and prevalence rates (Kaakoush et al. 2015). By taking all these facts into account, we can summarize that campylobacteriosis is a serious inflammatory intestinal disease affecting the global human population. Mortality rate statistics indicate that newborn infants, children and immunosuppressed individuals including the elderly are at particular risk for developing severe systemic complications. Detailed data on the epidemiology of *C. jejuni* infections are summarized by Romdhane and Merle (Chapter "The Data Behind Risk Analysis of *Campylobacter Jejuni* and *Campylobacter Coli* Infections" in this book).

In recent years, more calculations on the costs of *Campylobacter* infections and the burden of *Campylobacter*-associated diseases were published. Annual costs for the USA were estimated to range from 1.2 to 4 billion $ (Eberle and Kiess 2012; Batz et al. 2014). The cost of food-borne campylobacteriosis to public health systems and to loss of individual health and productivity in the EU is estimated to be around €2.4 billion per year (EFSA 2014), with an underlying mean cost per case estimate of €267 (Pitter et al. 2018). In this context, measures of disease burden by calculating disability-adjusted life years (DALY), quality-adjusted life years (QALY), years of potential life lost (YPLL), for instance, are becoming increasingly important parameters to set priorities in health care or to assess and shape risk-based food safety policies (Fig. 3). In a recent systematic review, Lackner and colleagues analyzed studies published between 1996 and 2016 from 27 countries, with the majority of the studies focusing on Europe (Lackner et al. 2019). After adjusting study-specific DALY to 100,000 people, large differences were observed between countries, ranging from 0.4 DALY per 100,000 people in France (Van Lier and Havelaar 2007) to 109 DALY per 100,000 in Poland (Mangen et al. 2016). Differences in DALY between and even among countries were largely attributed by Lackner and co-workers to the different incidences applied in the calculations (Lackner et al. 2019). When focusing exclusively on food-borne burden of disease for *Campylobacter*, calculations ranged from 0.5 DALY per 100,000 people in Greece (Gkogka et al. 2011) to 21.2 DALY per 100,000 in New Zealand (Lake et al. 2010). In the global context, disease burden

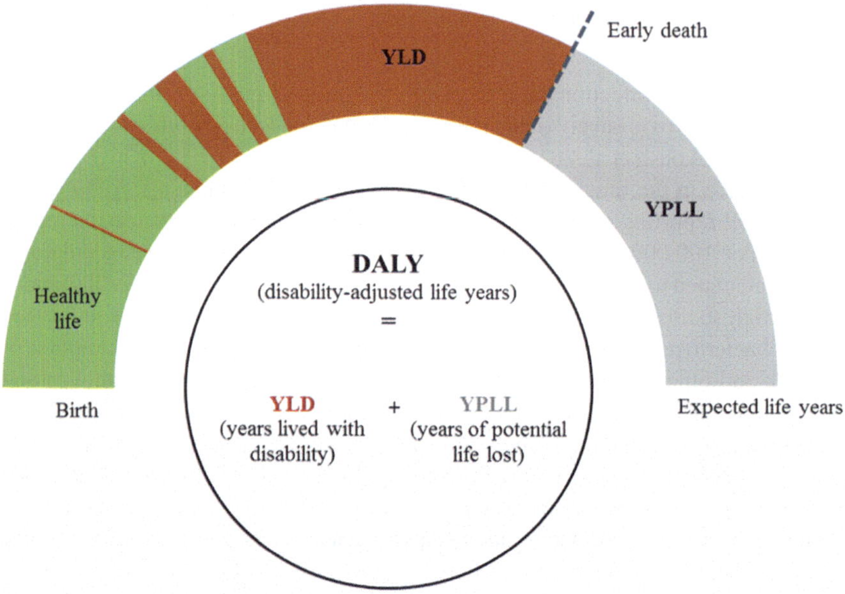

Fig. 3 Burden of disease measures. Schematic presentation of the calculation of the DALY. *Source* Public Health England (2015) (adapted)

of *Campylobacter* was generated and compared to other food-borne or diarrheal diseases (Kirk et al. 2015). The authors estimated that overall, *Campylobacter* caused 3.7 million DALY (2.9–5.3; 95% UI) in 2010 (out of estimated 78.7 million DALYs for all 22 diseases included in the study), corresponding to 31 DALY per 100,000 people (22–46; 95% UI). The ratio of DALY between the age groups <5–≥ 5 was 1.87 (1.26–2.92) compared to 0.76 for all food-borne diseases included.

Large differences were observed between regions for DALY per 100,000 people [African Region: 70 (41–112; 95% UI), Region of the Americas: 13 (8–18), Eastern Mediterranean Region: 90 (56–130) European Region: 9 (6–13), South-East Asian Region: 33 (9–83), Western Pacific Region: 10 (4–17)]. A discounted, QALY-based EU estimate resulted in 15.23 QALY loss per 1000 human *Campylobacter* cases (Pitter et al. 2018). Within that, gastroenteritis accounted for 9.96 QALY loss in 1000 human cases, *Campylobacter*-related RA accounted for 2.33 QALY loss per 1000 gastroenteritis cases, and a discounted health burden of 2.94 QALY loss due to GBS in 1000 human *Campylobacter* gastroenteritis cases. For the USA, Batz and co-workers estimated 16 QALY lost per 1000 campylobacteriosis cases, summing up to 13,256 QALY losses annually in the USA (Batz et al. 2014). Taken together, severe symptoms of campylobacteriosis lead to a significant limitation of infected individuals and cause a high socioeconomic burden worldwide.

3.3 *Molecular Concepts of* C. jejuni-*Induced Intestinal Pathogenesis*

When it comes to questioning the molecular basis of *C. jejuni*-induced intestinal pathogenesis in human patients, it is noteworthy that the bacterial virulence and pathogenicity factors mediating campylobacteriosis have been investigated to date. However, our knowledge regarding the inflammatory immune responses in the human host is still limited mainly because convenient murine infection models have not been available for a long time (see below). It is known for decades that the onset of *C. jejuni*-induced disease depends on the translocation of the highly motile pathogens from the gut lumen to the epithelial cell layer as well as on epithelial adherence and subsequent active invasion of the subepithelial tissues including the lamina propria (Backert et al. 2013 and Chapter "*Campylobacter* Virulence Factors and Molecular Host–Pathogen Interactions" in this book). Thus, essential roles of *C. jejuni* flagella, adhesins and invasins as essential pathogenicity factors in the onset, progression and clinical outcome of campylobacteriosis have been determined at the molecular level and were independently confirmed by results from a multitude of in vitro as well as in vivo studies (Cróinín and Backert 2012; Backert and Hofreuter 2013). While these bacterial factors serve as valid targets for novel treatment strategies, the inflammatory syndrome induced by *C. jejuni* in the human host is much less studied and awaits further investigation.

During the last decades of research, a substantial change in basic paradigms of immunopathogenic concepts of *C. jejuni* induced enteritis was necessary to identify bacterial molecules that are essential for the induction of intestinal inflammation during human campylobacteriosis (reviewed by Phongsisay 2016). While many scientists still follow the concept that the disease is mainly caused by a bacterial exotoxin, *Campylobacter* research was initially focused on the intensive search for a potent Cholera-like toxin (CLT), which was thought to be common to all *C. jejuni* strains (reviewed by Walker et al. 1986). This line of investigations was not successful but has led to the identification of the cytolethal distending toxin (CDT), which contributes to the virulence of *C. jejuni* but is not produced by all pathogenic strains (Facciolà et al. 2017; Pickett and Whitehouse 1999; Bang et al. 2001). In conclusion, exotoxins like CDT and CLT are not essential for the onset and progression of campylobacteriosis but may aggravate the disease when they are produced by the infecting *C. jejuni* strain(s).

Today, we follow a second concept assuming that intestinal inflammation and the post-infectious autoimmune diseases triggered by campylobacteriosis in humans are mainly caused by an intense massive innate immune response to bacterial endotoxins derived from the motile, adhesive and invasive *C. jejuni* that had translocated to the subepithelial compartment. This "endotoxin concept" was formulated very early based on results from studies of histopathological changes during intestinal campylobacteriosis in *C. jejuni*-infected humans including healthy volunteers and hospitalized patients (Black et al. 1988; Blaser et al. 1979; Price et al. 1979). In the absence of a potent exotoxin common to all *C. jejuni* strains, the accumulation of neutrophils and macrophages histologically observed at intestinal sites of hyperacute inflammation support the integrative view that *C. jejuni* endotoxins including lipooligosaccharides (LOS) trigger the pathogenesis of campylobacteriosis mainly via activation of the innate immune system (Fig. 2). In this regard, campylobacteriosis is very similar to the onset of massive innate immune activation by the LOS of *Neisseria* species such as *Neisseria meningitidis* and *N. gonorrhoeae*, primarily affecting other body compartments (Black et al. 1988; Moran et al. 1996). The revival of this second concept of *C. jejuni*-mediated gastroenteritis paved the way for the identification of bacterial agents essentially involved in intestinal inflammation during campylobacteriosis. A breakthrough in the understanding of molecular immunopathogenesis of human campylobacteriosis was based on the observation that both intestinal inflammation and the development of post-infectious sequelae are significantly associated with the production of sialylated LOS variants A, B and C by the infecting *C. jejuni* strains (Mortensen et al. 2009). The major role of this prominent endotoxin in pathogenesis was further confirmed recently by detailed analysis of intestinal barrier damage and LOS-mediated inflammatory signaling pathways in intestinal biopsies taken from *C. jejuni*-infected patients (Bücker et al. 2018). The corresponding results demonstrated that diarrhea in human campylobacteriosis results from sodium malabsorption induced by invading *C. jejuni* via a LOS-mediated cytokine storm initiated by the activated innate immune cells. Analyses of global gene expression revealed that the pathogenic LOS is the master regulator of this inflammatory scenario, which leads not only to the inhibition of sodium channels, but also to the breakdown of intestinal epithelial barrier functions, to apoptosis, and to tissue destruction (Bücker et al. 2018; reviewed by Chapters "*Campylobacter* Virulence Factors and Molecular Host–Pathogen Interactions" and "Diarrheal Mechanisms and the Role of Intestinal Barrier Dysfunction in *Campylobacter* Infections" in this book). Thus, the heterogeneity of campylobacteriosis symptoms seen in humans results, in part, from the scattered distribution, modular composition and variability of *C. jejuni* surface LOS which is due to the tremendous genetic variability of the pathogen (reviewed in Chapter "Population Biology and Comparative Genomics of *Campylobacter* Species of this book).

The roles of bacterial LOS and the innate immune system in the induction of enteritis supported the investigation of molecular mechanisms underlying campylobacteriosis (Mortensen et al. 2009). Besides their roles in human infection, LOS and other carbohydrate endotoxins of *C. jejuni* maintain the bacterial anatomic structures and protect the pathogens against environmental stress—by biofilm formation, for instance (reviewed by Chapter "Molecular Mechanisms of *Campylobacter* Biofilm Formation and Quorum Sensing" in this book). *C. jejuni* LOS is a surface glycolipid consisting of an oligosaccharide moiety and a lipid A core. Binding to its receptor, Toll-like receptor 4 (TLR4), is essential for the activation of innate immune cells (reviewed by Phongsisay 2016). Variations in LOS structures affect the inflammatory potency of *C. jejuni* and explain the variability seen in human disease outcome. Sialylation of the oligosaccharide moiety enhances bacterial invasion, binding to TLR4 and cytokine production by immune cells. Moreover, since some sialylated oligosaccharide chains in *C. jejuni* LOS are structurally related to human gangliosides, infection with respective pathogenic bacterial strains induces production of anti-ganglioside antibodies which in line with macrophage activation cause axonal destruction leading to GBS. In conclusion, although the O-antigen characteristic of bacterial lipopolysaccharide (LPS) is not present in LOS of the majority of *C. jejuni* strains (Karlyshev et al. 2005; Naito et al. 2010), the lipid A moiety of this truncated LPS molecule per se is a highly potent TLR4 agonist, and the sialylated LOS triggers severe forms of campylobacteriosis and post-infectious sequelae including GBS (Fig. 2). Thus, the functional parts of the LOS molecule provide the molecular basis for a better understanding of both, the diverse intestinal disease manifestations and the development of post-infectious sequelae in humans.

Our knowledge regarding *C. jejuni*-induced intestinal pathogenesis was further augmented by detailed investigation of the intestinal histopathology during campylobacteriosis in infected human patients. It has been known for decades that the intestinal histopathological features of campylobacteriosis are characterized by large ulcerative epithelial tissue destruction that is mostly driven by neutrophilic granulocytes and macrophages accumulating in high numbers in and under the epithelium at intestinal sites of *C. jejuni* entry (Price et al. 1979; Kaakoush et al. 2015; Backert et al. 2017). In response to LOS derived from invading *C. jejuni* bacteria, these innate immune cells produce toxic oxygen radicals including nitric oxide, peroxynitrate and superoxide in line with pro-inflammatory mediators, which *in sum* cause massive epithelial apoptosis mounting in ulcerative tissue destruction and bloody diarrhea (Walker et al. 1986; Kaakoush et al. 2015). Macrophage derived TNF-α, IL-6 and IL-8 as well as IL-1β, IL-12 and IL-23 from dendritic cells act as initiators and promoters of inflammatory responses. Activated T-cells produce IFN-γ, IL-17, IL-22 and anti-inflammatory cytokines IL-4 and IL-10 dampening the immune responses and thereby supporting self-limitation of disease (reviewed by Al-Banna et al. 2018). This inflammatory scenario was confirmed independently by artificial *C. jejuni* infection of ex vivo biopsies (Edwards et al. 2010). The resistance of *C. jejuni* to killing by phagocytosis and its ability to reside within phagocytes for up to 7 days is a very important—but often overlooked—feature of the pathogen (Kiehlbauch et al. 1985). Given that resistance to phagocytosis is an important feature of *C. jejuni* that developed during interactions with amoebae, the study of *C. jejuni* survival in those organisms is of great and stimulating importance for *Campylobacter* research (Vieira et al. 2015). While the bacterial factors mediating intracellular survival in phagosomes await further investigation, resistance to phagocytosis explains (i) the inability of macrophages and granulocytes to clear initial *C. jejuni* infection and (ii) the active transfer of live pathogens by migrating macrophages to mesenteric lymph nodes (Price et al. 1979; Walker et al. 1986; Kaakoush et al. 2015). In light of the major role of LOS in initiation of the inflammatory responses and the pronounced resistance of mice to LOS (see below), it is of note that human monocytes ingest *C. jejuni* more rapidly and vigorously than murine macrophages (Kiehlbauch et al. 1985). In conclusion, the results obtained from analyses of *C. jejuni*-induced intestinal immunopathology in human patients support the integrative view that the massive activation of the innate immune system via TLR4 signaling induced by *C. jejuni* LOS is responsible for both, the initial symptom complex of intestinal campylobacteriosis (Mortensen et al. 2009; Bücker et al. 2018) as well as the severe post-infectious sequelae such as GBS or RA (Kaakoush et al. 2015). Taken together, our knowledge on innate immune activation by *C. jejuni* endotoxins is a prerequisite for the treatment of human campylobacteriosis, which also supports the prophylaxis of post-infectious sequelae (reviewed by Mousavi et al. 2020 and by Chapter "Murine Models for the Investigation of Colonization Resistance and Innate Immune Responses in *Campylobacter Jejuni* Infections" in this book).

3.4 Novel Murine Models of C. jejuni *Infection Offering Detailed Investigations and Treatment Strategies for Campylobacteriosis and Associated Long-Term Sequelae*

The development of novel treatment options for human campylobacteriosis depends on the availability of robust and standardized animal models, which display both the symptoms and the molecular pathogenesis induced by *Campylobacter* infections in humans. Most recently, the major role of *C. jejuni* LOS in the induction and progress of campylobacteriosis was further confirmed by research groups working independently from each other on the establishment of novel animal models for the investigation of *C. jejuni*-host interactions. The development of highly convenient murine infection models for campylobacteriosis was a great challenge since conventional mice only respond very weakly to bacterial LPS/LOS and further display a pronounced intestinal colonization resistance against *C. jejuni* mediated by the murine gut microbiota. Notably, due to their low TLR4 responses, the LOS resistance of mice is approximately 10,000 fold higher (Warren et al. 2010; Munford 2010) as compared to humans (Taveira da Silva et al. 1993). Hence, detailed investigation of the LOS-driven inflammatory responses in humans was severely hampered for long time periods by the lack of appropriate murine infection models. In conclusion, the recent progress in the development of novel murine models of *C. jejuni* infection is based on the modification of both the microbiota composition as well as the LOS responses of mice (reviewed by Mousavi et al. 2020, and by Chapter "Murine Models for the Investigation of Colonization Resistance and Innate Immune Responses in *Campylobacter Jejuni* Infections" in this book). These murine models of *C. jejuni* infection were developed on the basis of the ground breaking investigations by Linda Mansfield (Mansfield et al. 2007), Christian Jobin (Lippert et al. 2009), Bruce Vallance (Stahl et al. 2014) and Richard Guerrant (Giallourou et al. 2018), who demonstrated that genetically modified mice with a reduced intestinal microbiota, sensitized to LOS by genetic manipulation or by dietary modifications inducing zinc depletion can be effectively infected by *C. jejuni* and display key symptoms of human campylobacteriosis (Poly and Guerry 2008; Heimesaat and Bereswill 2015; Stahl et al. 2017; Mousavi et al. 2020). Thus, the major role of *C. jejuni* LOS in the induction of campylobacteriosis was impressively confirmed by independent studies focused on manipulation of the murine immune system via IL-10 deficiency (Mansfield et al. 2007), single IgG IL-1 related receptor (SIGIRR) deficiency (Stahl et al. 2014, 2017) and zinc depletion (Giallourou et al. 2018), all of which resulted in abolished murine LOS resistance as a consequence of increased activation of the murine TLR4 signaling pathways (Munford, 2010; Warren et al., 2010). It is well documented that IL-10 (Emoto et al. 2003; Robertson et al. 2006, 2007), SIGIRR signaling pathways and zinc (Snyder and Walker 1976; Ohata et al. 2010; Chen et al. 2012) effectively suppress LPS/LOS-mediated inflammatory responses in mice. Moreover, oral zinc supplementation constitutes a valid measure to protect children in low and middle income countries from bacterial diarrhea including campylobacteriosis (Lazzerini and Wanzira 2016).

Taken together, the novel murine infection models represent a major breakthrough in campylobacteriosis research since the immunopathology in the murine intestines characterized by apoptosis, granulocyte and macrophage recruitment, production of pro-inflammatory cytokines such as IFN, TNF, IL-6, IL-8 as well as the activation of T and B cells is very similar to the immune and histopathological responses seen in *C. jejuni*-infected humans (Price et al. 1979; Bücker et al. 2018). Given that i) we confirmed these findings in our own investigations (Bereswill et al. 2011; Haag et al. 2012b; Masanta et al. 2013; Heimesaat and Bereswill 2015), ii) all these different animal models have in common that mice presenting with clinical signs of campylobacteriosis are sensitized to LOS by completely independent manipulations, and iii) TLR4-deficient mutants of these LOS-sensitized mice showed significantly less intestinal inflammatory responses (Haag et al. 2012b; Stahl et al. 2014, 2017), these novel insights provide final proof that *C. jejuni* LOS plays a key role in *C. jejuni*-induced inflammatory diarrhea in humans and other vertebrate hosts (reviewed by Chapter "Murine Models for the Investigation of Colonization Resistance and Innate Immune Responses in *Campylobacter Jejuni* Infections" in this book). These highly innovative murine infection models have generated substantial progress in the understanding of the molecular mechanisms underlying pathogen-host interactions during campylobacteriosis, and their standardization paves the way for the development of novel treatment strategies focused on (i) neutralization of LOS and pro-inflammatory oxygen radicals, (ii) strengthening intestinal epithelial barrier function, (iii) inactivation of the barrier-breaking *C. jejuni*-related factors including motility, tissue destruction by proteases and other invasins, as well as (iv) vaccination. While some murine models of infection could be developed to the preclinical level for validation of most of these novel intervention strategies (Stahl et al. 2017; Masanta et al. 2013; Heimesaat and Bereswill 2015), development of a potent vaccine is still most challenging given the role of sialylated LOS in the induction of post-infectious sequelae such as GBS, MFS or RA (reviewed by Chapter "Murine Models for the Investigation of Colonization Resistance and Innate Immune Responses in *Campylobacter Jejuni* Infections" in this book). In particular, the secondary abiotic IL-10-deficient mouse model of campylobacteriosis has been proven to be highly useful for the analysis of *C. jejuni* infection, mainly because disease induction in this infection model depends on motility and invasive properties of the pathogen (Schmidt et al.

2019). It is of note here that commensal *Escherichia coli*, which lack any invasive or other pathogenic properties, do not induce any immunopathology upon peroral challenge of secondary abiotic IL-10-deficient mice (Haag et al. 2012b). Recently, the secondary abiotic IL-10-deficient murine model was standardized and could be further developed to the preclinical level for pharmaceutical analysis of alternative drugs, including curcumin, resveratrol, carvacrol, ascorbate and vitamin D, which effectively suppressed inflammation in course of campylobacteriosis (reviewed by Chapter "Murine Models for the Investigation of Colonization Resistance and Innate Immune Responses in *Campylobacter Jejuni* Infections" in this book). There is good evidence that the dampening of the innate immune responses in the onset of intestinal campylobacteriosis by those interventions might reduce the risk for the development of post-infectious sequelae.

Given the central and general role of IL-10 in maintaining gut homeostasis by suppression of inflammatory responses (Neumann et al. 2019; Iyer and Cheng 2012), it is also of great impact that conventional infant mice, which do not raise a sufficient IL-10 response in their intestines, become effectively colonized by *C. jejuni* and display the typical course of self-limiting campylobacteriosis seen in humans (Haag et al. 2012a). Moreover, those mice cleared the intestinal disease and developed immune cell infiltrates at extra-intestinal sites including the liver, the kidneys and the lungs. These features highlight conventional infant mice as useful model to study systemic manifestations of campylobacteriosis including the onset of post-infectious sequelae including RA or GBS. The numbers of patients developing severe post-infectious autoimmune diseases following intestinal campylobacteriosis including RA (1–13%), GBS (0.001%), IBD such as ulcerative colitis and Crohn's disease, IBS or celiac disease are increasing to non-tolerable levels worldwide (Keithlin et al. 2014; Kaakoush et al. 2015; Facciolà et al. 2017). Particularly, immuno-compromised patients are at risk to develop extra-intestinal complications of initial intestinal campylobacteriosis entailing a multitude of disease manifestations ranging from meningitis, brain abscesses and cardiovascular complications to bacteremia and septicemia. The appearance of these secondary diseases induced by initial intestinal *C. jejuni* infection underlines the socioeconomic and individual burden of *C. jejuni*-induced inflammatory enteritis in humans. This scenario is basic to the insight that we are in urgent need for the development of novel treatment options which dampen the acute inflammation caused by the initial intestinal LOS response of the innate immune system, in order to prevent severe autoimmune reactions and extra-intestinal spread of *C. jejuni* (Keithlin et al. 2014). Valid data on the pathogenesis of GBS indicate that *C. jejuni* infection with strains producing sialylated LOS induces the production of antibodies specific for similar structured molecules decorating the myelin surface of axons in the nervous system, which leads to immune complex-mediated destruction of nervous tissues and subsequently results in GBS (Goodfellow and Willison 2016). Furthermore, similar immune-mediated post-infectious mechanisms are proposed for RA. Most recently, the innovative development of a *C. jejuni*-induced murine GBS model will allow not only for the study of molecular mechanisms underlying the GBS-inducing capacity of *C. jejuni*, but also enforces measures for prophylaxis and treatment of GBS in the near future (Brooks et al. 2017, 2019). However, the future developments will reveal whether murine models of infection may add to prevention and treatment in this regard as well.

4 Concluding Remarks

The above-outlined substantial scientific progress in understanding the molecular interactions underlying *C. jejuni* pathogenesis supports the principal view that bacterial LOS plays a major role in molecular immunopathogenesis of acute campylobacteriosis and its post-infectious sequelae. Thus, the initial intestinal inflammatory symptom complex induced by motile, adhesive and invasive *C. jejuni* in the intestinal epithelium is mainly driven by activation of the innate immune system and aggravated by bacterial exotoxins—in case these are produced by the infecting strain. This basic concept is strongly supported by the fact that the post-infectious sequelae are autoimmune diseases caused by the adaptive immune system which is initially primed by the hyper-activation of the innate immune system in response to the endotoxin LOS of the invading bacteria (Fig. 2). Therefore, the revival of the old concept of *C. jejuni*-induced inflammatory diarrhea (Moran et al. 1996; Blaser 1997) in line with the application of novel murine infection models will pave the way for preclinical evaluation of innovative prophylactic and treatment strategies to combat human campylobacteriosis in the near future. Actual therapeutic interventions for improvement of clinical symptoms during campylobacteriosis target bacterial LOS signaling pathways including anti-inflammatory approaches to dampen inflammation and tissue destruction alongside with the inactivation of bacterial pathogenicity and virulence factors such as motility, adhesins and invasins, respectively (reviewed by Chapter "Murine Models for the Investigation of Colonization Resistance and Innate Immune Responses in *Campylobacter Jejuni* Infections" in this book). Given that exclusive targeting of pathogenic structures is always accompanied by the risk of resistance development, it seems recommendable to combat both *C. jejuni* factors in line with immune responses by combined application of synergistically acting molecules.

Finally, closing the circle between asymptomatic colonization in poultry and acute disease in infected humans, the pronounced LPS/LOS tolerance of birds including chickens which is 100-fold higher as compared to mice and even 1,000,000-fold higher (Adler and DaMassa 1979) as compared to humans (da Silva et al. 1993) might provide the basis for the understanding why chickens and other poultry do not develop intestinal inflammation upon *C. jejuni* colonization and are therefore a major source for human infection (Young et al. 2007). Thus, all the novel discoveries in the active and dynamic field of campylobacteriosis research support the optimistic view that novel murine models in combination with clinical studies enable us to develop novel drugs for prophylaxis and treatment of human campylobacteriosis, which will in turn prevent or lower the risk for post-infectious sequelae in the near future.

Acknowledgements We thank Drs. Martin Blaser (Rutgers University, Piscataway, NJ/USA) and Martin Maiden (University of Oxford, UK) for helpful comments on the manuscript. This work is supported by grants from the German Federal Ministry of Education and Research (BMBF) in frame of the zoonoses research consortium PAC-Campylobacter to MMH, SBe, (project IP7/01KI1725D), SBa (project IP9/01KI1725E), TA (project IP2/01KI1725A), and from the Federal Ministry for Economic Affairs and Energy following a resolution of the German National Parliament, Deutscher Bundestag to SBe and MMH (ZIM, ZF4117908 AJ8).

References

Allos BM (2001) *Campylobacter jejuni* infections: update on emerging issues and trends. Clin Infect Dis 32(8):1201–1206. https://doi.org/10.1086/319760

Adler HE, DaMassa AJ (1979) Toxicity of endotoxin to chicks. Avian Dis 23(1):174–178

Al-Banna NA, Cyprian F, Albert MJ (2018) Cytokine responses in campylobacteriosis: linking pathogenesis to immunity. Cytokine Growth Factor Rev 41:75–87. https://doi.org/10.1016/j.cytogfr.2018.03.005

Allos BM (1997) Association between *Campylobacter* infection and Guillain-Barre syndrome. J Infect Dis 176(Suppl 2):S125–S128. https://doi.org/10.1086/513783

Altekruse SF (2008) *Campylobacter* 3rd edition. Emerg Infect Dis 14(12):1977–1977. https://doi.org/10.3201/eid1412.081165

Awofisayo-Okuyelu A, Hall I, Adak G, Hawker J, Abbott S, McCarthy N (2017) A systematic review and meta-analysis on the incubation period of Campylobacteriosis. Epidemiol Infect 145(11):2241–2253. https://doi.org/10.1017/S0950268817001303

Backert S, Boehm M, Wessler S, Tegtmeyer N (2013) Transmigration route of *Campylobacter jejuni* across polarized intestinal epithelial cells: paracellular, transcellular or both? Cell Commun Signal CCS 11:72. https://doi.org/10.1186/1478-811x-11-72

Backert S, Hofreuter D (2013) Molecular methods to investigate adhesion, transmigration, invasion and intracellular survival of the foodborne pathogen *Campylobacter jejuni*. J Microbiol Methods 95(1):8–23. https://doi.org/10.1016/j.mimet.2013.06.031

Backert S, Tegtmeyer N, Cróinín TÓ, Boehm M, Heimesaat MM (2017) Chapter 1—Human campylobacteriosis. In: Klein G (ed.) Campylobacter. Academic, pp 1–25. https://doi.org/10.1016/B978-0-12-803623-5.00001-0

Bang DD, Scheutz F, Ahrens P, Pedersen K, Blom J, Madsen M (2001) Prevalence of cytolethal distending toxin (cdt) genes and CDT production in *Campylobacter* spp. isolated from Danish broilers. J Med Microbiol 50(12):1087–1094. https://doi.org/10.1099/0022-1317-50-12-1087

Bardosh K (2016) One health: science, politics and zoonotic disease in Africa. Routledge

Batz M, Hoffmann S, Morris JG Jr (2014) Disease-outcome trees, EQ-5D scores, and estimated annual losses of quality-adjusted life years (QALYs) for 14 foodborne pathogens in the United States. Foodborne Pathog Dis 11(5):395–402. https://doi.org/10.1089/fpd.2013.1658

Bereswill S, Fischer A, Plickert R, Haag LM, Otto B, Kuhl AA, Dasti JI, Zautner AE, Munoz M, Loddenkemper C, Gross U, Gobel UB, Heimesaat MM (2011) Novel murine infection models provide deep insights into the “menage a trois” of *Campylobacter jejuni*, microbiota and host innate immunity. PLoS ONE 6(6):e20953. https://doi.org/10.1371/journal.pone.0020953

Black RE, Levine MM, Clements ML, Hughes TP, Blaser MJ (1988) Experimental *Campylobacter jejuni* infection in humans. J Infect Dis 157(3):472–479

Blaser MJ (1997) Epidemiologic and clinical features of *Campylobacter jejuni* infections. J Infect Dis 176(Supplement 2):S103–S105

Blaser MJ, Berkowitz ID, LaForce FM, Cravens J, Reller LB, Wang WL (1979) *Campylobacter* enteritis: clinical and epidemiologic features. Ann Intern Med 91(2):179–185. https://doi.org/10.7326/0003-4819-91-2-179

Brooks PT, Bell JA, Bejcek CE, Malik A, Mansfield LS (2019) An antibiotic depleted microbiome drives severe *Campylobacter jejuni*-mediated Type 1/17 colitis, Type 2 autoimmunity and neurologic sequelae in a mouse model. J Neuroimmunol 337:577048. https://doi.org/10.1016/j.jneuroim.2019.577048

Brooks PT, Brakel KA, Bell JA, Bejcek CE, Gilpin T, Brudvig JM, Mansfield LS (2017) Transplanted human fecal microbiota enhanced Guillain Barre syndrome autoantibody responses after *Campylobacter jejuni* infection in C57BL/6 mice. Microbiome 5(1):92. https://doi.org/10.1186/s40168-017-0284-4

Bücker R, Krug SM, Moos V, Bojarski C, Schweiger MR, Kerick M, Fromm A, Janssen S, Fromm M, Hering NA, Siegmund B, Schneider T, Barmeyer C, Schulzke JD (2018) *Campylobacter jejuni* impairs sodium transport and epithelial barrier function via cytokine release in human colon. Mucosal Immunol 11(2):575–577. https://doi.org/10.1038/mi.2017.78

Burnham PM, Hendrixson DR (2018) *Campylobacter jejuni*: collective components promoting a successful enteric lifestyle. Nat Rev Microbiol 16(9):551–565. https://doi.org/10.1038/s41579-018-0037-9

Chen YH, Zhao M, Chen X, Zhang Y, Wang H, Huang YY, Wang Z, Zhang ZH, Zhang C, Xu DX (2012) Zinc supplementation during pregnancy protects against lipopolysaccharide-induced fetal growth restriction and demise through its anti-inflammatory effect. J Immunol 189(1):454–463. https://doi.org/10.4049/jimmunol.1103579

Cody AJ, Maiden MC, Strachan NJ, McCarthy ND (2019) A systematic review of source attribution of human campylobacteriosis using multilocus sequence typing. Eurosurveillance 24(43):1800696. https://doi.org/10.2807/1560-7917.ES.2019.24.43.1800696

Cróinín TO, Backert S (2012) Host epithelial cell invasion by *Campylobacter jejuni:* trigger or zipper mechanism? Front Cell Infect Microbiol 2:25. https://doi.org/10.3389/fcimb.2012.00025

Dasti JI, Tareen AM, Lugert R, Zautner AE, Gross U (2010) *Campylobacter jejuni*: a brief overview on pathogenicity-associated factors and disease-mediating mechanisms. Int J Med Microbiol 300(4):205–211. https://doi.org/10.1016/j.ijmm.2009.07.002

Destoumieux-Garzón D, Mavingui P, Boëtsch G, Boissier J, Darriet F, Duboz P, Fritsch C, Giraudoux P, Le Roux F, Morand S (2018) The one health concept: 10 years old and a long road ahead. Front Vet Sci 5:14. https://doi.org/10.3389/fvets.2018.00014

Eberle K, Kiess A (2012) Phenotypic and genotypic methods for typing *Campylobacter jejuni* and *Campylobacter coli* in poultry. Poult Sci 91(1):255–264. https://doi.org/10.3382/ps.2011-01414

Edwards LA, Nistala K, Mills DC, Stephenson HN, Zilbauer M, Wren BW, Dorrell N, Lindley KJ, Wedderburn LR, Bajaj-Elliott M (2010) Delineation of the innate and adaptive T-cell immune outcome in the human host in response to *Campylobacter jejuni* infection. PLoS ONE 5(11):e15398. https://doi.org/10.1371/journal.pone.0015398

EFSA (2011) The European Union summary report on trends and sources of zoonoses, zoonotic agents and food-borne outbreaks in 2009. https://doi.org/10.2903/j.efsa.2011.2090

EFSA (2014) EFSA explains zoonotic disease-*Campylobacter*. Fact Sheet. https://doi.org/10.2805/59450

Emoto M, Emoto Y, Brinkmann V, Miyamoto M, Yoshizawa I, Staber M, van Rooijen N, Hamann A, Kaufmann SH (2003) Increased resistance of LFA-1-deficient mice to lipopolysaccharide-induced shock/liver injury in the presence of TNF-alpha and IL-12 is mediated by IL-10: a novel role for LFA-1 in the regulation of the proinflammatory and anti-inflammatory cytokine balance. J Immunol 171(2):584–593. https://doi.org/10.4049/jimmunol.171.2.584

Facciolà A, Riso R, Avventuroso E, Visalli G, Delia SA, Laganà P (2017) *Campylobacter*: from microbiology to prevention. J Prev Med Hyg 58(2):E79–E92

Forsythe SJ (2000) Food poisoning microorganisms. In: The Microbiology of Safe Foods. Blackwell Science Publishers

Fouts DE, Mongodin EF, Mandrell RE, Miller WG, Rasko DA, Ravel J, Brinkac LM, DeBoy RT, Parker CT, Daugherty SC (2005) Major structural differences and novel potential virulence mechanisms from the genomes of multiple *Campylobacter* species. PLoS Biol 3(1):e15

Giallourou N, Medlock GL, Bolick DT, Medeiros PH, Ledwaba SE, Kolling GL, Tung K, Guerry P, Swann JR, Guerrant RL (2018) A novel mouse model of *Campylobacter jejuni* enteropathy and diarrhea. PLoS Pathog 14(3):e1007083. https://doi.org/10.1371/journal.ppat.1007083

Gkogka E, Reij MW, Havelaar AH, Zwietering MH, Gorris LG (2011) Risk-based estimate of effect of foodborne diseases on public health, Greece. Emerg Infect Dis 17(9):1581. https://doi.org/10.1371/journal.pbio.0030015

Gölz G, Rosner B, Hofreuter D, Josenhans C, Kreienbrock L, Löwenstein A, Schielke A, Stark K, Suerbaum S, Wieler LH (2014) Relevance of *Campylobacter* to public health—the need for a One Health approach. Int J Med Microbiol 304(7):817–823. https://doi.org/10.1016/j.ijmm.2014.08.015

Goodfellow JA, Willison HJ (2016) Guillain-Barré syndrome: a century of progress. Nat Rev Neurol 12(12):723–731. https://doi.org/10.1038/nrneurol.2016.172

Haag LM, Fischer A, Otto B, Grundmann U, Kuhl AA, Gobel UB, Bereswill S, Heimesaat MM (2012a) *Campylobacter jejuni* infection of infant mice: acute enterocolitis is followed by asymptomatic intestinal and extra-intestinal immune responses. Eur J Microbiol Immunol (Bp) 2(1):2–11. https://doi.org/10.1556/EuJMI.2.2012.1.2

Haag LM, Fischer A, Otto B, Plickert R, Kuhl AA, Gobel UB, Bereswill S, Heimesaat MM (2012b) *Campylobacter jejuni* induces acute enterocolitis in gnotobiotic IL-10-/- mice via Toll-like-receptor-2 and -4 signaling. PLoS ONE 7(7):e40761. https://doi.org/10.1371/journal.pone.0040761

Hafiz RA, Wong C, Paynter S, David M, Peeters G (2018) The risk of community-acquired enteric infection in proton pump inhibitor therapy: systematic review and meta-analysis. Ann Pharmacother 52(7):613–622. https://doi.org/10.1177/1060028018760569

Heimesaat MM, Bereswill S (2015) Murine infection models for the investigation of Campylobacter jejuni–host interactions and pathogenicity. Berl Munch Tierarztl Wochenschr 128(3–4):98–103

Hofreuter D (2014) Defining the metabolic requirements for the growth and colonization capacity of *Campylobacter jejuni*. Front Cell Infect Microbiol 4:137. https://doi.org/10.3389/fcimb.2014.00137

Hofreuter D, Tsai J, Watson RO, Novik V, Altman B, Benitez M, Clark C, Perbost C, Jarvie T, Du L (2006) Unique features of a highly pathogenic *Campylobacter jejuni* strain. Infect Immun 74(8):4694–4707. https://doi.org/10.1128/IAI.00210-06

Iannino F, Salucci S, Di Donato G, Badagliacca P, Vincifori G, Di Giannatale E (2019) Campylobacter and antimicrobial resistance in dogs and humans: "one health" in practice. Vet Ital 55(3):203–220. https://doi.org/10.12834/VetIt.1161.6413.3

Iyer SS, Cheng G (2012) Role of interleukin 10 transcriptional regulation in inflammation and autoimmune disease. Crit Rev Immunol 32(1):23–63. https://doi.org/10.1615/critrevimmunol.v32.i1.30

Janssen R, Krogfelt KA, Cawthraw SA, van Pelt W, Wagenaar JA, Owen RJ (2008) Host-pathogen interactions in *Campylobacter* infections: the host perspective. Clin Microbiol Rev 21(3):505–518. https://doi.org/10.1128/CMR.00055-07

Jones KE, Patel NG, Levy MA, Storeygard A, Balk D, Gittleman JL, Daszak P (2008) Global trends in emerging infectious diseases. Nature 451(7181):990–993. https://doi.org/10.1038/nature06536

Kaakoush NO, Castano-Rodriguez N, Mitchell HM, Man SM (2015) Global Epidemiology of *Campylobacter* Infection. Clin Microbiol Rev 28(3):687–720. https://doi.org/10.1128/CMR.00006-15

Karlyshev AV, Ketley JM, Wren BW (2005) The *Campylobacter jejuni* glycome. FEMS Microbiol Rev 29(2):377–390. https://doi.org/10.1016/j.femsre.2005.01.003

Keithlin J, Sargeant J, Thomas MK, Fazil A (2014) Systematic review and meta-analysis of the proportion of *Campylobacter* cases that develop chronic sequelae. BMC Public Health 14:1203. https://doi.org/10.1186/1471-2458-14-1203

Kiehlbauch JA, Albach RA, Baum LL, Chang K (1985) Phagocytosis of *Campylobacter jejuni* and its intracellular survival in mononuclear phagocytes. Infect Immun 48(2):446–451. https://doi.org/10.1128/IAI.48.2.446-451.1985
Kirk MD, Pires SM, Black RE, Caipo M, Crump JA, Devleesschauwer B, Döpfer D, Fazil A, Fischer-Walker CL, Hald T (2015) World Health Organization estimates of the global and regional disease burden of 22 foodborne bacterial, protozoal, and viral diseases, 2010: a data synthesis. PLoS Med 12(12):e1001921. https://doi.org/10.1371/journal.pmed.1001921
Kist M, Bereswill S (2001) *Campylobacter Jejuni*. Contrib Microbiol 8:150–165. https://doi.org/10.1159/000060405
Lackner J, Weiss M, Müller-Graf C, Greiner M (2019) The disease burden associated with *Campylobacter* spp. in Germany, 2014. PLoS ONE 14(5):e0216867. https://doi.org/10.1371/journal.pone.0216867
Lake RJ, Cressey PJ, Campbell DM, Oakley E (2010) Risk ranking for foodborne microbial hazards in New Zealand: burden of disease estimates. Risk Anal Int J 30(5):743–752. https://doi.org/10.1111/j.1539-6924.2009.01269.x
Lazzerini M, Wanzira H (2016) Oral zinc for treating diarrhoea in children. Cochrane Database Syst Rev (12). https://doi.org/10.1002/14651858.CD005436.pub5
Lippert E, Karrasch T, Sun X, Allard B, Herfarth HH, Threadgill D, Jobin C (2009) Gnotobiotic IL-10; NF-kappaB mice develop rapid and severe colitis following *Campylobacter jejuni* infection. PLoS ONE 4(10):e7413. https://doi.org/10.1371/journal.pone.0007413
Lübbert C (2016) Antimicrobial therapy of acute diarrhoea: a clinical review. Expert Rev Anti-Infect Therapy 14(2):193–206. https://doi.org/10.1586/14787210.2016.1128824
Mangen M-J, Havelaar A, Haagsma J, Kretzschmar M (2016) The burden of *Campylobacter*-associated disease in six European countries. Microbial Risk Anal 2:48–52
Mansfield LS, Bell JA, Wilson DL, Murphy AJ, Elsheikha HM, Rathinam VA, Fierro BR, Linz JE, Young VB (2007) C57BL/6 and congenic interleukin-10-deficient mice can serve as models of *Campylobacter jejuni* colonization and enteritis. Infect Immun 75(3):1099–1115. https://doi.org/10.1128/IAI.00833-06
Masanta WO, Heimesaat MM, Bereswill S, Tareen AM, Lugert R, Groß U, Zautner AE (2013) Modification of intestinal microbiota and its consequences for innate immune response in the pathogenesis of campylobacteriosis. Clin Dev Immunol 2013:526860–526860. https://doi.org/10.1155/2013/526860
Moran AP, Prendergast MM, Appelmelk BJ (1996) Molecular mimicry of host structures by bacterial lipopolysaccharides and its contribution to disease. FEMS Immunol Med Microbiol 16(2):105–115. https://doi.org/10.1111/j.1574-695X.1996.tb00127.x
Mortensen NP, Kuijf ML, Ang CW, Schiellerup P, Krogfelt KA, Jacobs BC, van Belkum A, Endtz HP, Bergman MP (2009) Sialylation of *Campylobacter jejuni* lipo-oligosaccharides is associated with severe gastro-enteritis and reactive arthritis. Microbes Infect 11(12):988–994. https://doi.org/10.1016/j.micinf.2009.07.004
Mousavi S, Bereswill S, Heimesaat MM (2020) Novel clinical *Campylobacter jejuni* infection models based on sensitization of mice to lipooligosaccharide, a major bacterial factor triggering innate immune responses in human Campylobacteriosis. Microorganisms 8 (4). https://doi.org/10.3390/microorganisms8040482
Munford RS (2010) Murine responses to endotoxin: another dirty little secret? J Infect Dis 201(2):175–177. https://doi.org/10.1086/649558
Naito M, Frirdich E, Fields JA, Pryjma M, Li J, Cameron A, Gilbert M, Thompson SA, Gaynor EC (2010) Effects of sequential *Campylobacter jejuni* 81–176 lipooligosaccharide core truncations on biofilm formation, stress survival, and pathogenesis. J Bacteriol 192(8):2182–2192. https://doi.org/10.1128/JB.01222-09
Neumann C, Scheffold A, Rutz S (2019) Functions and regulation of T cell-derived interleukin-10. Semin Immunol 44:101344. https://doi.org/10.1016/j.smim.2019.101344

Newell DG (2001) Animal models of *Campylobacter jejuni* colonization and disease and the lessons to be learned from similar *Helicobacter pylori* models. Symp Ser (Soc Appl Microbiol) 30:57s–67s. https://doi.org/10.1046/j.1365-2672.2001.01354.x

Ohata S, Moriyama C, Yamashita A, Nishida T, Kusumoto C, Mochida S, Minami Y, Nakada J, Shomori K, Inagaki Y, Ohta Y, Matsura T (2010) Polaprezinc protects Mice against endotoxin shock. J Clin Biochem Nutr 46(3):234–243. https://doi.org/10.3164/jcbn.09-125

One World One Health (2020a) The Manhattan principles. https://oneworldonehealth.wcs.org/About-Us/Mission/The-Manhattan-Principles.aspx. Accessed 25.06.2020

One World One Health (2020b) One World—One Health™—MISSION statement. https://oneworldonehealth.wcs.org/About-Us/Mission.aspx. Accessed 25.06.2020

Oyarzabal OA, Backert S (2012) Food safety resources. In: Microbial food safety, Springer, pp 235–239

Parkhill J, Wren B, Mungall K, Ketley J, Churcher C, Basham D, Chillingworth T, Davies R, Feltwell T, Holroyd S (2000) The genome sequence of the food-borne pathogen *Campylobacter jejuni* reveals hypervariable sequences. Nature 403(6770):665–668. https://doi.org/10.1038/35001088

Phongsisay V (2016) The immunobiology of *Campylobacter jejuni:* innate immunity and autoimmune diseases. Immunobiology 221(4):535–543. https://doi.org/10.1016/j.imbio.2015.12.005

Pickett CL, Whitehouse CA (1999) The cytolethal distending toxin family. Trends Microbiol 7(7):292–297. https://doi.org/10.1016/s0966-842x(99)01537-1

Pielsticker C, Glünder G, Rautenschlein S (2012) Colonization properties of *Campylobacter jejuni* in chickens. Eur J Microbiol Immunol (Bp) 2(1):61–65. https://doi.org/10.1556/EuJMI.2.2012.1.9

Pitter JG, Vokó Z, Jozwiak A, Berkics A, Engelhardt T (2018) *Campylobacter* control measures in indoor broiler chicken: conservative reassessment of cost-utility and analysis of putative barriers to implementation. In: IAFP European symposium on food safety, IAFP

Poly F, Guerry P (2008) Pathogenesis of *Campylobacter*. Curr Opin Gastroenterol 24(1):27–31. https://doi.org/10.1097/MOG.0b013e3282f1dcb1

Price A, Jewkes J, Sanderson P (1979) Acute diarrhoea: *Campylobacter* colitis and the role of rectal biopsy. J Clin Pathol 32(10):990–997. https://doi.org/10.1136/jcp.32.10.990

Public Health England (2015) The Global burden of disease – infographics. PHE publications gateway number: 2015301

Robertson SA, Care AS, Skinner RJ (2007) Interleukin 10 regulates inflammatory cytokine synthesis to protect against lipopolysaccharide-induced abortion and fetal growth restriction in mice. Biol Reprod 76(5):738–748. https://doi.org/10.1095/biolreprod.106.056143

Robertson SA, Skinner RJ, Care AS (2006) Essential role for IL-10 in resistance to lipopolysaccharide-induced preterm labor in mice. J Immunol 177(7):4888–4896. https://doi.org/10.4049/jimmunol.177.7.4888

Schmidt A-M, Escher U, Mousavi S, Tegtmeyer N, Boehm M, Backert S, Bereswill S, Heimesaat MM (2019) Immunopathological properties of the *Campylobacter jejuni* flagellins and the adhesin CadF as assessed in a clinical murine infection model. Gut Pathog 11:24–24. https://doi.org/10.1186/s13099-019-0306-9

Schwabe CW (1984) Veterinary medicine and human health. Williams & Wilkins, Baltimore

Skirrow M (1977) *Campylobacter* enteritis: a"new" disease. Br Med J 2(6078):9–11. https://doi.org/10.1136/bmj.2.6078.9

Smith OM, Snyder WE, Owen JP (2020) Are we overestimating risk of enteric pathogen spillover from wild birds to humans? Biol Rev 95(3):652–679. https://doi.org/10.1111/brv.12581

Snyder SL, Walker RI (1976) Inhibition of lethality in endotoxin-challenged mice treated with zinc chloride. Infect Immun 13(3):998–1000. https://doi.org/10.1128/IAI.13.3.998-1000.1976

Stahl M, Graef FA, Vallance BA (2017) Mouse models for *Campylobacter jejuni* colonization and infection. In: *Campylobacter jejuni*, Springer, pp 171–188. https://doi.org/10.1007/978-1-4939-6536-6_15

Stahl M, Ries J, Vermeulen J, Yang H, Sham HP, Crowley SM, Badayeva Y, Turvey SE, Gaynor EC, Li X, Vallance BA (2014) A novel mouse model of *Campylobacter jejuni* gastroenteritis reveals key pro-inflammatory and tissue protective roles for Toll-like receptor signaling during infection. PLoS Pathog 10(7):e1004264. https://doi.org/10.1371/journal.ppat.1004264
Stingl K, Knüver M-T, Vogt P, Buhler C, Krüger N-J, Alt K, Tenhagen B-A, Hartung M, Schroeter A, Ellerbroek L (2012) Quo vadis?—Monitoring *Campylobacter* in Germany. Eur J Microbiol Immunol 2(1):88–96. https://doi.org/10.1556/EuJMI.2.2012.1.12
Talukder R, Sutradhar S, Rahman K, Uddin M, Akhter H (2011) Guillian-Barre Syndrome. Mymensingh Med J MMJ 20(4):748–756
Tam C (2001) *Campylobacter* reporting at its peak year of 1998: don't count your chickens yet. Commun Dis Public Health 4(3):194
Taveira da Silva AM, Kaulbach HC, Chuidian FS, Lambert DR, Suffredini AF, Danner RL (1993) Brief report: shock and multiple-organ dysfunction after self-administration of *Salmonella* endotoxin. New Engl J Med 328(20):1457–1460. https://doi.org/10.1056/nejm199305203282005
Ternhag A, Asikainen T, Giesecke J, Ekdahl K (2007) A meta-analysis on the effects of antibiotic treatment on duration of symptoms caused by infection with *Campylobacter* species. Clin Infect Dis 44(5):696–700. https://doi.org/10.1086/509924
Thomas KM, de Glanville WA, Barker GC, Benschop J, Buza JJ, Cleaveland S, Davis MA, French NP, Mmbaga BT, Prinsen G (2020) Prevalence of *Campylobacter* and *Salmonella* in African food animals and meat: A systematic review and meta-analysis. Int J Food Microbiol 315:108382. https://doi.org/10.1016/j.ijfoodmicro.2019.108382
Van Lier E, Havelaar A (2007) Disease burden of infectious diseases in Europe: a pilot study. National Institute for Public Health and the Environment. RIVM; Bilthoven, Bilthoven
Vieira A, Seddon AM, Karlyshev AV (2015) *Campylobacter*-Acanthamoeba interactions. Microbiology 161(5):933–947. https://doi.org/10.2807/esm.12.12.00751-en
Wakerley BR, Uncini A, Yuki N, Attarian S, Barreira AA, Chan Y-C, Créange A, Hsieh S-T, Islam B, Kannan MA (2014) Guillain-Barré and Miller Fisher syndromes—new diagnostic classification. Nat Rev Neurol 10(9):537. https://doi.org/10.1038/nrneurol.2014.138
Walker RI, Caldwell MB, Lee EC, Guerry P, Trust TJ, Ruiz-Palacios GM (1986) Pathophysiology of *Campylobacter* enteritis. Microbiol Rev 50(1):81–94
Warren HS, Fitting C, Hoff E, Adib-Conquy M, Beasley-Topliffe L, Tesini B, Liang X, Valentine C, Hellman J, Hayden D, Cavaillon JM (2010) Resilience to bacterial infection: difference between species could be due to proteins in serum. J Infect Dis 201(2):223–232. https://doi.org/10.1086/649557
Wassenaar TM, Blaser MJ (1999) Pathophysiology of *Campylobacter jejuni* infections of humans. Microbes Infect 1(12):1023–1033. https://doi.org/10.1016/s1286-4579(99)80520-6
WHO (2020) World Health Organisation. *Campylobacter*.https://www.who.int/news-room/fact-sheets/detail/campylobacter. Accessed 04.06.2020
Young KT, Davis LM, Dirita VJ (2007) *Campylobacter jejuni:* molecular biology and pathogenesis. Nat Rev Microbiol 5(9):665–679. https://doi.org/10.1038/nrmicro1718
Zinsstag J, Schelling E, Waltner-Toews D, Tanner M (2011) From "one medicine" to "one health" and systemic approaches to health and well-being. Prev Vet Med 101(3–4):148–156. https://doi.org/10.1016/j.prevetmed.2010.07.003

The Data Behind Risk Analysis of *Campylobacter Jejuni* and *Campylobacter Coli* Infections

Racem Ben Romdhane and Roswitha Merle

Contents

Abstract *Campylobacter jejuni* and *Campylobacter coli* are major causes of food-borne enteritis in humans. Poultry meat is known to be responsible for a large proportion of cases of human campylobacteriosis. However, other food-borne, environmental and animal sources are frequently associated with the disease in humans as well. Human campylobacteriosis causes gastroenteritis that in most cases is self-limiting. Nevertheless, the burden of the disease is relatively large compared with other food-borne diseases, which is mostly due to rare but long-lasting symptoms related to immunological sequelae. In order to pave the way to improved surveillance and control of human campylobacteriosis, we review here the data that is typically used for risk analysis to quantify the risk and disease burden, identify specific surveillance strategies and assist in choosing the most effective control strategies. Such

The original version of this chapter was revised: The family name and given name of the author Racem Ben Romdhane have been corrected. The correction to this chapter is available at https://doi.org/10.1007/978-3-030-65481-8_12

R. Ben Romdhane · R. Merle (✉)
Faculty of Veterinary Medicine, Institute for Veterinary Epidemiology and Biostatistics, Freie Universität Berlin, Berlin, Germany
e-mail: roswitha.merle@fu-berlin.de

R. Ben Romdhane
e-mail: ben.romdhane.racem@fu-berlin.de

S. Backert (ed.), *Fighting Campylobacter Infections*, Current Topics in Microbiology and Immunology 431, https://doi.org/10.1007/978-3-030-65481-8_2

data are mostly collected from the literature, and their nature is discussed here, for each of the three processes that are essential for a complete risk analysis procedure: risk assessment, risk management and risk communication. Of these, the first, risk assessment, is most dependent on data, and this process is subdivided into the steps of hazard identification, hazard characterization, exposure assessment and risk characterization. For each of these steps of risk assessment, information from published material that is typically collected will be summarized here. In addition, surveillance data are highly valuable for risk assessments. Different surveillance systems are employed in different countries, which can make international comparison of data challenging. Risk analysis typically results in targeted control strategies, and these again differ between countries. The applied control strategies are as yet not sufficient to eradicate human campylobacteriosis. The surveillance tools of *Campylobacter* in humans and exposure sources in place in different countries are briefly reviewed to better understand the *Campylobacter* dynamics and guide control strategies. Finally, the available control measures on different risk factors and exposure sources are presented.

1 Introduction

Bacteria belonging to the genus *Campylobacter* are commensals in wild and domestic mammals and birds. The genus comprises 34 species, of which two species are mainly responsible for human campylobacteriosis (WHO 2012; OIE 2017): *Campylobacter jejuni* and *Campylobacter coli*. In combination, these two species are the main cause of human bacterial intestinal disease identified in many industrialized countries (Havelaar et al. 2013; Scallan et al. 2011). Over 80% of cases are caused by *C. jejuni*, and about 10% of cases are caused by *C. coli* (OIE 2017), though their relative frequency can vary between countries. These same two species (which in this chapter are collectively called '*Campylobacter*' unless stated otherwise) do not usually cause disease in their natural animal hosts, which are warm-blooded animals (birds and mammals) (OIE 2017). In 2018, an incidence of 82/100,000 inhabitants with a total of 67,872 cases was reported in Germany (RKI 2019). This number is an underestimate, as it is based on cases with laboratory confirmation only; the true incidence is assumed to be 7–100 times higher than reported cases (Kapperud 1994; Wheeler et al. 1999; Mead et al. 1999; Friedman et al. 2004; Stingl et al. 2012). In humans, *Campylobacter* infection typically presents as acute enteritis with severe abdominal pain that can last for up to two weeks (Blaser and Engberg 2008). The disease is mostly self-limiting, though occasionally more severe symptoms including bacteremia occur that require treatment. Post-infection sequelae such as Guillain-Barré syndrome, inflammatory bowel disease and reactive arthritis have all been described, and in rare cases the infection can be lethal (WHO 2012).

Risk analysis is an empirical process aiming to identify and estimate the risk of harmful events and to help control their occurrence or consequences. This field of research is implemented in various areas, from finances, commercial management and industrialized production processes to human and animal health issues. Risk analyses can be performed to investigate the impact of individual risk sources for human diseases and to guide control strategies. Here, we present a risk analysis approach for human campylobacteriosis, in order to provide a better understanding of the sources and pathways of human exposure and of the relative risks and consequences involved of these exposure routes. In addition, risk analysis can assist to identify effective ways to monitor *Campylobacter* in humans and the most common exposure sources, in order to improve existing control measures to limit human campylobacteriosis.

Several risk analysis studies of human campylobacteriosis are available from the literature, of which four concentrated on the assessment of the risk associated with consumption of poultry meat (Nauta et al. 2009; Chapman et al. 2016; Nauta and Christensen et al. 2001; ICRA 2011). These studies built on earlier publications on more general risk assessment of campylobacteriosis (Christensen et al. 2001; Hartnett et al. 2001) and new analyses are still being carried out (Dogan et al. 2019; Lee et al. 2019). Risk assessment approaches can assess either qualitative risks as by Horigan and colleagues (2014), or quantitative risks as by Chapman and co-workers (2016).

For risk analyses performed in the context of food production, in the Codex Alimentarius (FAO and WHO 2019) *risk* is defined as 'a function of the probability of an adverse health effect and the severity of that effect, consequential to a hazard(s) in food' and a *hazard* is defined as 'a biological, chemical or physical agent in, or condition of, food with the potential to cause an adverse health effect' (FAO and WHO 2019). Apart from the risk analysis process related to food safety presented in the Codex Alimentarius, an alternative approach is described in the OIE Terrestrial Animal Health Code (OIE 2019). The two methodologies followed in these guides are similar and equally valid (Vose et al. 2001). Whereas the OIE risk analysis methodology is designed to assess the risk related to (imported) animal diseases, the Codex Alimentarius methodology is specifically adapted for food-borne risks. Since human exposure to *Campylobacter* occurs through ingestion of bacteria and this is most commonly enabled by contaminated food, we consider the Codex Alimentarius approach as more appropriate for the risk assessment of human campylobacteriosis.

The methodology presented in the Codex Alimentarius recognizes three main components for a complete risk analysis process, which are risk assessment, risk management and risk communication. A complete risk analysis is thus a joined effort of risk assessors, risk managers and decision makers/public bodies, as the outcomes produced by risk assessors must be translated into appropriate risk management measures and implemented by regulatory bodies with proper communication strategies toward the general public. Risk assessment is again separated into four different steps, which are: (1) hazard identification, (2) hazard characterization, (3) exposure assessment and (4) risk characterization. Details and goal of each risk analysis steps are reviewed (Fabech et al. 2002). In the context of this chapter, the hazard identification can be described as particular pathogenic *Campylobacter* species (or strains therein) that are possibly present in food items. Hazard characterization then assesses the consequences that might occur upon exposure to these pathogenic bacteria, describing common symptoms and less common but more serious sequelae of human campylobacteriosis. Exposure assessment describes the probability that

Campylobacter is transmitted from food items to humans, and finally during risk characterization all available qualitative and quantitative data from these three steps are combined to estimate and quantify the risk. Risk management takes advantage of risk assessment findings and then proposes and assesses the efficacy of control strategies, aiming to reduce the estimated risk. Communication is a key point of risk analysis. This process should involve a variety of experts in the risk analysis work (risk assessors, decision makers, consumers, industry, etc.) in an iterative and continuous process among all the risk assessment steps, to provide guidance as well as to validate, interpret and diffuse its results. The individual steps for campylobacteriosis risk assessment are described in more detail in the next section.

2 Risk Assessment of Human Campylobacteriosis

2.1 Hazard Identification

The genus *Campylobacter* comprises 34 species, but not all species are of interest in human campylobacteriosis (WHO 2012; OIE 2017). In addition to *C. jejuni* and *C. coli*, other species that have been associated with human disease are, among others, *C. concisus, C. upsaliensis, C. ureolyticus* and *C. fetus*. The causal role and the clinical importance in humans of these collectively called 'emerging *Campylobacter* species' appear to be relatively low (Kaakoush et al. 2015). Within a single *Campylobacter* species, the pathogenicity can vary between isolates due to the production of specific toxins and other bacterial biological characteristics (e.g., flagella) (Dasti et al. 2010; Backert and Hofreuter 2013). More on the aspect of pathogenicity is described in Chapters "*Campylobacter* Virulence Factors and Molecular Host–Pathogen Interactions", "Diarrheal Mechanisms and the Role of Intestinal Barrier Dysfunction in *Campylobacter* Infections" and "Murine Models for the Investigation of Colonization Resistance and Innate Immune Responses in *Campylobacter Jejuni* Infections" of this book. The pathogenicity may further depend on the sensitivity and immunological status of the individual host, which is not further outlined here. *Campylobacter* can enter a viable but non-culturable (VBNC) state (Oliver 2005), in which case it is impossible to assess their pathogenicity. The bacteria may enter a VBNC state due to environment conditions, during their presence on food items, during food preparation or even during transport to the laboratory. The passage through the gastrointestinal tract may also in theory induce a VBNC state. It is unknown if VBNC bacteria can be revitalized and if so, under which conditions (Fakruddin et al. 2013; Li et al. 2014; Zhao et al. 2017). It cannot be excluded that these bacteria are pathogenic when ingested. Regarding the complexity in defining the pathogenicity of *Campylobacter* species, variation between strains and variation due to the viability state of the bacteria, risk assessment studies usually assume all *Campylobacter* bacteria present in a given situation are equivalently pathogenic (Cróinín and Backert 2012). However, it is disputable whether this reflects the reality.

In terms of transmission to humans, food of animal origin seems to play a major role, since a number of warm-blooded animals can carry *Campylobacter*, including food animals, although they usually do not show clinical symptoms. Some animal species common hosts are, in particular poultry, but also ruminants and pigs. Therefore, risk assessment studies mainly focus on food-borne campylobacteriosis, with emphasis on food of animal origin.

2.2 Hazard Characterization

In the context of campylobacteriosis, hazard characterization is defined by the disease that follows from exposure to the pathogen. The most common manifestation of human campylobacteriosis is gastroenteritis with watery or bloody diarrhea, abdominal pain, fever, headache, nausea and/or vomiting; these describe the qualitative primary hazard characteristics. For a quantitative assessment, it must be considered for how long symptoms last, which can be up to two weeks but is mostly in the range of one week (Blaser and Engberg 2008). Further, the full spectrum of mild symptoms that last only a few days and spontaneously resolve, to severe disease that requires hospitalization, needs to be taken into account, and if possible these various clinical outcomes must be stratified, so that their relative contributions can be assessed. Possibly, such stratifications depend on the population under study, and further aspects may have to been taken into consideration: Gender, age, professional exposure and seasonality have all been described to affect the frequency and/or severity of campylobacteriosis. Lastly, the quantitative effect of sequelae (reactive arthritis, irritable bowel syndrome, inflammatory bowel disease and Guillain-Barré syndrome) may be relatively minor, as these conditions only occur in a small minority of cases, but their effects may be long-lasting and severe, requiring intensive medical care and resulting in significant economic losses. Thus, the sequelae account for a large part to the overall burden of campylobacteriosis (Havelaar et al. 2000; Lackner et al. 2019).

2.3 Exposure Assessment

Assessment of human exposure to pathogenic campylobacters seems straightforward, but in practice this is muddled with difficulties. Qualitative aspects to be considered here include the state in which campylobacters reach the human gut, where the bacteria must colonize in order to cause disease. Not all bacteria present in a food item may be equally ‘fit’ to do so. A fraction of the bacteria may have been inactivated due to exposure to oxygen in the air, food preparation and handling practices including cooling/freezing, presence of preservatives and heating/cooking prior to consumption. Campylobacters are relatively fragile and due to their thermophilic and micro-aerobic growth requirements do not usually replicate in or on food. This means that over time the numbers present on a contaminated food item may stay constant or decrease, but rarely increase. This is in strong contrast to other food pathogens such as *Escherichia coli* or *Salmonella*, which can multiply rapidly in certain food items. Quantitative aspects of exposure assessment are further hampered by a lack of reliable dose–response curves, which are the basis on which risk assessors adjust their mathematical models.

The data available to risk modelers typically originate from human challenge studies, outbreak investigations and case control studies. Human challenge studies with *Campylobacter* are no longer considered ethical, so that only a few existing studies are available (Black et al. 1988; Kirkpatrick et al. 2013; Tribble et al. 2009, 2010), that were based on a limited number of species and strains therein. Outbreak investigations have illustrated that the outcome of exposure is individually highly variable, with some individuals falling ill while other equally exposed persons may not experience symptoms at all (reviewed in Teunis et al. 2018). Such differences must be incorporated in an exposure assessment. Most data are available from case control studies that were conducted for source attribution, as they can identify which food sources are responsible for clinically diagnosed campylobacteriosis cases. Such studies depend on molecular epidemiology that can identify presence of particular strains (genotypes) shared by human cases and by particular food sources, e.g., poultry meat. Sources of *Campylobacter* and their likely transmission routes to humans have been reviewed elsewhere (Pires et al. 2009; Mullner et al. 2009; WHO 2012; Newell et al. 2017; Mughini-Gras et al. 2018; Cody et al. 2019). Sources of human campylobacteriosis have also been described in former risk assessment studies, and these are excellently reviewed in Chapman and co-workers (2016). Obviously, national differences must be taken into account for some of the recognized food sources. For instance, the fraction of infections resulting from meat sources is smaller for populations with a large number of vegetarians, and cultures not eating pork will not experience swine-related cases. Unfortunately, due to a general sparsity of data, risk researchers tend to maximize their information sources, thereby sometimes ignoring cultural and national peculiarities. The consumption of chicken liver pate may be higher in the UK than in other countries; raw pork is consumed in Germany but not very often in the USA, and well water may be commonly consumed in remote regions of Canada but may be less common in France. Such differences should be taken into account when conducting an exposure assessment. From these examples, it is clear that the amount of data needed to take into consideration can be enormous for a well-balanced exposure assessment. These data are typically a combination of published literature and expert's opinions. As the vast majority of source attribution studies have associated handling and consumption of poultry meat with occurrence of campylobacteriosis, this will be considered in more detail next, before turning to other likely food sources.

2.3.1 Poultry Meat-Associated Campylobacteriosis

Campylobacter can be present in several domestic and wild mammals and birds (OIE 2017; EFSA 2007). Both *C. jejuni* and *C. coli* are frequently detected in live poultry flocks, both chickens (broilers) and turkeys. Despite the uncertainties about the relative contribution of the different *Campylobacter* sources to human infections, handling and consumption of poultry meat has been highlighted as a major source of human campylobacteriosis (EFSA 2010a). Studies in different countries estimated that 18.3–74% of cases are attributable to poultry and/or chicken (Havelaar et al. 2008; Wilson et al. 2008; Davidson et al. 2011; Batz et al. 2012; Ravel et al. 2017; Rosner et al. 2017). The wide range of these findings may be due to national (culinary) differences, or it may be due to differences in prevalence of *Campylobacter* in commercial poultry: A survey conducted in 28 European countries revealed flock prevalences ranging from 2 to 100% with a mean prevalence of 71.2% (EFSA 2010b). Whether *Campylobacter* present in living chickens will eventually be present on poultry meat at retail further depends on production processes, which may again introduce national differences.

The sources from which *Campylobacter* is introduced into poultry flocks are still not fully understood. Chicks are negative at hatch, as vertical transmission from mother hen to egg is uncommon. A flock typically becomes contaminated after a lag phase of several weeks, and once a single chick is positive, the complete flock will soon be colonized, too. Birds can be contaminated by bacteria that enter their housing with contaminated feed or water, fomites such as worker's clothes and shoes, or be transferred from the extern environment by insects and vermin (Bull et al. 2006; Newell and Fearnley 2003; Newell et al. 2011). Transfer between different flocks within a farm or from a previous flock to a next production round is also common. Katsma and co-workers (2007) estimated that a broiler flock has a probability of 68% to become infected when the flock present in the same house during the previous production cycle had also been infected.

When colonized, birds carry *Campylobacter* in their gut, crop, and on their feathers, often with high numbers occurring (EFSA 2010a). During slaughter and processing of carcasses, feces leakage may cross-contaminate flocks that entered the abattoir uncolonized, and contamination of carcass surfaces (skin) is common (Slader et al. 2002; Lee et al. 2017; Nauta et al. 2005). Despite a number of control measures implemented in the processing of poultry meat, contaminated meat products still end up at retail (Scherer et al. 2006). As a result, humans are exposed to *Campylobacter* by consumption or handling of contaminated poultry meat. The bacteria may survive in undercooked poultry meat, a risk that was shown to be increased when meat was prepared at barbecues (Allerberger et al. 2003; WHO 2009). Cross-contamination during handling of raw meat in the kitchen (or during a garden barbecue) may transfer bacteria via hands to kitchen tools and surfaces, and that way food items consumed raw may become contaminated as well.

Risk assessment of *Campylobacter* reaching humans via poultry meat usually aims to evaluate the risk of human contamination by its route. To achieve this goal, it is important to incorporate all available knowledge about the mechanisms leading to this particular human exposure. These mechanisms range from how the chickens are infected, how and to what degree the carcasses are contaminated, how the meat is packed, stored and prepared, and finally how individual humans react to an exposure event caused by contaminated meat. Information on all these components of the 'farm to fork' or 'stable to table' chain of events must be taken into account for a proper risk assessment. It is also important to mention here that the higher relative importance of poultry meat as a source of human campylobacteriosis has to be interpreted with a bit of caution. Since poultry meat has been indicated as a major source, several studies have focused on the link between poultry (especially chicken) and humans. These efforts have provided a large amount of data that confirmed this link, and feeding these data into risk assessment models can result in a self-fulfilling prophecy effect. In contrast, the paucity of data on other transmission routes necessarily leaves those other routes as less established and they may possibly be under-evaluated at present.

2.3.2 Campylobacteriosis Associated with Other Foods

Products from Animal Origin Other than Poultry Meat

Apart from meat, chickens also produce eggs, and since laying hens can be colonized by *Campylobacter*, it would not be unreasonable to assume eggs could in principle be contaminated, too. However, in contrast to *Salmonella*, *Campylobacter* is not typically found inside eggs. Although egg surfaces might become contaminated while passing through the cloaca, *Campylobacter* is rarely cultured from eggs or egg surfaces, so that consumption of eggs is not considered a risk (Wagenaar et al. 2008).

Campylobacter can also colonize ruminants (cattle and sheep) and pigs. In bovine production, the prevalence varies from 0 to 89.4% (Stanley et al. 1998; Wesley et al. 2000; Bywater et al. 2004; Englen et al. 2007; Pezzotti et al. 2003). Strains typically present in cattle are often genetically different from those in other hosts, as has been shown by, for instance, MLST (discussed in detail by Nachamkin et al. 2008). Beef and pork meat can carry *Campylobacter* due to fecal contamination that occurs during slaughtering, though this is not occurring at the same scale as with poultry production. Defeathering of poultry birds during slaughter may cause the bacteria to enter deeper into the skin, while removal of the hairs and skin of cattle and swine does not have this effect, and the resulting meat is drier, which supports lower survival rates of the bacteria on the surface (Shange et al. 2019). As a consequence, the prevalence of *Campylobacter* in red meat at retail is much lower than in poultry meat: It ranges from 1.3–4.7% in beef to 6.9–12.6% in lamb and mouton meat (Osano and Arimi 1999; Pezzotti et al. 2003; Wong et al. 2007; Little et al. 2008; Whyte et al. 2004). The prevalence of *Campylobacter* in pork meat at retail varies from 0.27 to 10.3% in European countries (Osano and Arimi 1999; Pezzotti et al. 2003; Wong et al. 2007; Little et al. 2008; Whyte et al. 2004; EFSA and ECDC 2015).

Campylobacteriosis cases directly related to beef or pork consumption are relatively rare. This is attributed mainly to a cleaner production process, with less cross-contamination between carcasses, and the drier surface of the meat compared to poultry meat. Nevertheless, the ruminant reservoir can be responsible for a relatively high proportion (up to 35%) of human campylobacteriosis. This relative importance varies depending on the used detection method, while only 0.4% is associated with pigs (Sheppard et al. 2009; Jonas et al. 2015; Mossong et al. 2016). Actually, the proportion of human cases attributed to non-poultry meat varies between countries and with regard to the source attribution method used: Sheppard and colleagues (2009) indicated that cattle and sheep contributed to less than 20% of human *C. jejuni* cases in Scotland in 2005–2006. Mossong and co-workers (2016) attributed 33.3% of the human *C. jejuni* and *C. coli* cases in Luxembourg between 2010 and 2013 to cattle using MLST typing data. Jonas and colleagues (2015) also used MLST typing to attribute sources of human campylobacteriosis cases in Switzerland in 2009 and estimated that 36% (*C. jejuni*) and 16% (*C. coli*) of the investigated human isolates were attributed to cattle. However, based on *flaB* typing data, these authors attributed only 18% and 0%, respectively, of *C. jejuni* and *C. coli* human isolates to cattle (Jonas et al. 2015), which demonstrates that the outcome of such studies can be heavily influenced by the typing method applied.

Human campylobacteriosis is also known to be associated with the consumption of contaminated raw milk. In developed countries, industrially produced milk is rarely contaminated by *Campylobacter* (Christidis et al. 2016), as the bacteria are effectively killed by milk pasteurization, but raw milk can contain viable bacteria (Fernandes et al. 2015; EFSA 2005). Not only fecal contamination during milking, but also clinical or subclinical mastitis in dairy cattle can be the reason of this presence (Orr et al. 1995). Campylobacteriosis has thus been associated with consumption of raw milk from cows and goats (Peterson 2003; Hutchinson et al. 1985; Davis et al. 2016).

Campylobacter lari has been isolated from clams and oysters (Endtz et al. 1997), and when consumed raw this can pose a significant risk. This contamination could be associated with the presence of sewage effluents or other sources of surface water contamination, such as wild bird droppings (Abeyta et al. 1993; Wilson and Moore 1996). Such droppings more often contain *C. lari* than *C. jejuni or C. coli* (Wilson and Moore 1996). *Campylobacter* has also been identified in fresh meat of crabs in the USA, but the number of detected bacteria was very low (Reinhard et al. 1996). Fish and seafood other than clams and oysters are not generally recognized as significant risks for food-borne campylobacteriosis.

Fruit and Vegetables

The presence of *Campylobacter* in raw vegetables has been described in several studies. In the Netherlands, 30 out of 5640 vegetable and fruit samples were positive for *Campylobacter*, giving a prevalence of 0.23% (Verhoeff-Bakkenes et al. 2011). The prevalence was higher in packaged products (0.36%) than in unpackaged products (0.07%). Possibly, the bacteria survive longer in the protective atmosphere that is used to limit growth of Enterobacteriaceae. The same study estimated that the detected prevalence was responsible for 5.3×10^5 infections per year in the Netherlands (3% of a total of 1.6×10^7 estimated cases) (Verhoeff-Bakkenes et al. 2011). A study in Canada was conducted in the Ottawa area to investigate contamination of vegetables from farmers' outdoor markets (Park and Sanders 1992). The researchers reported a *C. jejuni* prevalence of around 3% in spinach and lettuce, 2.7–2.4% in radish, green onions and parsley (all of which are likely to be consumed raw), and 1.6% in potatoes. In India, 3.57% of 56 fruit and vegetable items investigated were *C. jejuni* positive (Kumar et al. 2001). *Campylobacter* spp. have also been detected in mushrooms (Doyle and Shoeni 1986). A recent review (Mohammadpour et al. 2018) highlighted that the average prevalence of *Campylobacter* in vegetables, fruits and fresh produce is generally low, estimated to be 0.53% only. The same study also highlighted important differences between countries, with the highest prevalence (33.4%) being reported in Asia.

Since *Campylobacter* does not seem able to infect plants, such contamination is most likely the result of using contaminated water in agricultural production of produce washing and rinsing. Exposure to *Campylobacter* present in the environment was indeed considered the most likely explanation for detection of these bacteria on produce (EFSA 2005). Other risk factors that may be responsible for the contamination of fruits and vegetables by *Campylobacter* include improper hygiene of workers handling the items, contaminated harvesting equipment, contact with wild or domestic animals, contamination introduced from processing equipment or transport containers; improper storage and packaging may also introduce cross-contamination at retail level (Beuchat 1996). In conclusion, although not commonly contaminated with *Campylobacter*, the consumption of vegetables and fruits may still provide a risk as viable bacteria can survive on these products and the products are mainly consumed raw (Castillo and Escartin 1994; Kärenlampi and Hänninen 2004).

2.3.3 Campylobacteriosis from Non-food Sources

Pets

Very few *C. jejuni* or *C. coli* populations will multiply outside a living animal (except for laboratory conditions), so they are classical zoonoses. Not only food animals, but also wild animals and pets can be colonized, often without symptoms, at least for adult animals, although young animals can get diarrhea. *Campylobacter* species have been isolated from dogs and cats in a large number of studies, but these hosts more frequently carry *C. helveticus* or *C. upsaliensis*, with *C. jejuni* and *C. coli* being less frequently identified. A review summarizing various enteropathogens summarizes literature data on prevalence of cats and dogs, which can be as high as 87% in the latter (Marks et al. 2011). However, that fraction represents the sum of all detected *Campylobacter* species in dogs, a host that frequency carries *C. upsaliensis* or *C. helveticus*. The prevalence of *C. jejuni* or *C. coli* is often lower, though some studies have reported up to 45% of dogs being positive for *C. jejuni* (Marks et al. 2011; Pintar et al. 2015; Acke 2018). When a household has both an infected pet and an infected individual, it is not always clear in which way the transfer took place, but it has been shown that pets have infected humans, and having a pet in the household is considered a risk factor, in particular a puppy or kitten (Mughini-Gras et al. 2013; Thépault et al. 2020). Pets can become infected through ingestion of undercooked or raw food, drinking unpasteurized milk, contact with feces, vectors and other environmental exposure (Acke 2018).

Environment

Campylobacter jejuni and *C. coli* are not typical inhabitants of soil or surface, but their presence can be detected, mostly as a result of (recent) fecal contamination of animal or human origin. Surface water and groundwater have been shown to contain *Campylobacter*. In Germany, up to 103 *Campylobacter* per 100 ml have been detected in water collected from streams in mountains (Stelzer and Jacob 1992). In Canada, 413 water samples collected in summer were positive for *C. jejuni* (33.9%) (Guy et al. 2018). Contaminated surface water can lead to human exposure during recreational activities such as swimming (Schönberg-Norio et al. 2004; Doorduyn et al. 2010; Viau et al. 2011). Contaminated well water that is consumed untreated can also lead to infections, and this has led to a number of outbreaks (Kramer et al. 1996; Bruce-Grey-Owen Sound Health Unit 2000; Smith et al. 2006). Sunlight (more specifically, UV light) inactivates the bacteria over time, as was shown in various studies (Obiri-Danso et al. 2001; Mattioli et al. 2017; Boehm et al. 2018). That surface water nevertheless frequently contains live bacteria indicates that there are many reservoirs (anthropogenic or not) that regularly recontaminate water bodies.

Campylobacter can also be present in the soil, from which it can reach the rhizosphere of plants. Protected from sunlight, the bacteria can survive longer: Brandl and co-workers (2004) demonstrated that a 4-log reduction in *Campylobacter* numbers in the spinach rhizosphere is reached after 21 days, compared to just 2 days for bacteria present on the leaves. This indicates a potentially long survival of *Campylobacter* in soil around plant roots. It has long been recognized that poultry flocks not only 'breed' *Campylobacter*, but unwillingly also spread this into the environment. For instance, in a recent study *C. coli* was isolated from the direct environment of broiler farms, with a 34.4% prevalence (31 of 90 samples) from the flock environment, 66.7% (10/15) from the manure storage area and 53.3% (8/15) from used litter (Mohammed and Abdel Aziz 2019). These results illustrate the potential role of broiler farms in releasing *Campylobacter* in the environment (soil, surface water, groundwater, etc.). Even when the temperature is increased to above 50 °C, *C. jejuni* could survive for 2–5 days in stockpiles of solid farmyard manure (Nicholson et al. 2005).

The spread by flies was already mentioned, and indeed, insects can play a role in *Campylobacter* transmission to both humans and animals. A potential role of *Alphitobius diaperinus* and *Musca domestica* has been recognized (Jonsson et al. 2012; Strother et al. 2005; Förster et al. 2009; Rosef and Kapperud 1983). The insect transmission route is assumed to play an important role in *Campylobacter* transmission, especially as a means for the bacteria to reach poultry flocks from an environmental source. Fly screens have been applied to reduce the risk of flocks becoming positive for *Campylobacter*, with limited success, as poultry houses require intense ventilation that is difficult to combine with fly-proof barriers (Hald et al. 2007). What the contribution of flies is in direct transmissions to humans (most likely via food) has not yet been fully investigated.

One striking characteristic of human campylobacteriosis is that it follows a seasonal pattern, with a peak of cases in summer/late summer. This can be observed in temperate zones of both hemispheres (Geissler et al. 2017; Altekruse et al. 1999; Lake et al. 2019; Jore et al. 2010; McCarthy et al. 2012). A slight variation regarding the start and the duration of the high incidence season exists between countries (Lake et al. 2019). The reason for this seasonality is not completely understood, with several explanations being put forward: The warm season could be associated with increased exposure to surface water during recreational activities; the seasonality in incidence of human cases may be caused by seasonal peaks in bacterial loads in poultry (Smith et al. 2019; Jorgensen et al. 2011; Williams et al. 2015; Jore et al. 2010), barbecues are more popular in summer, and flies are not around in winter. However, summer is also associated with more UV exposure, so that environmental sources may be more rapidly depleted.

Human-To-Human Spread

Lastly, the likelihood of human-to-human transmission must be mentioned. Large outbreaks of *Campylobacter* are rare, and human-to-human transmission is mostly confined to household members, but it has been recognized as a potential risk factor of human campylobacteriosis in several studies (Rao 2001; Havelaar et al. 2008). Various studies conducted in the UK, the Netherlands, New Zealand and Australia have estimated that 3–5% of human campylobacteriosis cases can be attributed to human-to-human transmission (Little et al. 2010; Gilpin et al. 2013; Mughini-Gras et al. 2014). One study even described human-to-human transmission of *C. jejuni* via the unusual route of sexual contact (Gaudreau et al. 2015). However, overall, human-to-human transmission route is relatively rare, as was concluded in the early literature (Altekruse et al. 1999; Allos 2001) and that conclusion still holds.

2.4 Risk Characterization

Characterization of the risk of campylobacteriosis is typically done by estimating the total disease burden, expressed in disability-adjusted life years, or DALYs, which is a commonly used indicator for the overall health of a population (Devleesschauwer et al. 2017). This parameter is the sum of two disease burdens: the years of life lost (YLL), indicating how many years an individual has lost when dying prematurely from a hazard (here, campylobacteriosis) and the years lost due to disability (YLD). The latter expresses how many days, weeks, months or years a patient is hampered to live his or her usual live, due to the disease. Obviously, as relatively few patients die of campylobacteriosis, the YLL is relatively low. And with the majority of cases being self-limiting, a disease duration of one to two weeks does not add up to a large

Table 1 Estimated median rates of illness caused by *Campylobacter jejuni* or *C. coli*, together with the incidence of deaths and disability-adjusted life years (DALYs) by geographic region

Geographic region	Illness per 100,000 (95% UI)	Deaths per 100,000 (95% UI)	DALYs per 100,000 (95% UI)
Africa	2221 (335–8482)	0.8 (0.4–1)	70 (41–112)
Americas	1389 (490–3207)	0.07 (0.04–0.1)	13 (8–18)
Eastern Mediterranean	1873 (488–5608)	1 (0.6–1)	90 (56–130)
Europe	522 (363–687)	0.05 (0.03–0.09)	9 (6–13)
South East Asia	1152 (200–3372)	0.4 (0.1–0.9)	33 (9–83)
Western Pacific	876 (359–3855)	0.04 (0.02–0.1)	10 (4–17)
Global	1390 (752–2576)	0.3 (0.2–0.5)	31 (22–46)

Source WHO 2015

YLD either. However, long-term consequences of a campylobacteriosis result in a high YLD, even when such consequences are relatively uncommon.

A comprehensive study estimating of the burden of human campylobacteriosis was performed for the World Health Organization in 2015 (Haagsma et al. 2015). Based on data from 2010, those authors estimated that globally there were on average 166,175,078 cases of campylobacteriosis per year (a surprisingly high accuracy for an estimate), of which 37,604 were estimated to result in deaths yearly. Together with estimates of YLLs, this resulted in an estimated total of 3,733,822 DALYs related to yearly global cases. Of the estimated cases, 58% (with an uncertainty interval, UI, of 44–69%) were considered likely to have been caused by food-borne sources, which leaves a considerable fraction to be caused by non-food-related sources. The same study highlighted the variation in the burden of the disease between different regions around the world, which are summarized in Table 1. The burden of the disease was highest in Africa, which resulted in the highest estimates for illness frequency and deaths, whereas the DALYs were highest for Eastern Mediterranean regions. A more recent study estimated an overall burden of 8,811 DALYs (10.85 DALYs per 100,000) by *Campylobacter*-related diseases in Germany based on German data from 2014 (Lackner et al. 2019). That study estimated that, of all possible clinical outcomes, gastroenteritis caused the lowest disease burden, with 0.001 DALY per case. The highest disease burden by case was estimated for inflammatory bowel disease (8.817 DALY per case) followed by Guillain-Barré syndrome (7.747 DALY per case), which is related to their high YLL consequences. Another study assessed the burden of disease of seven pathogens that are all commonly transmitted through food in Denmark, and this resulted in *Campylobacter* to be estimated to have the highest burden (1709 DALYs) (Pires et al. 2019).

3 Current Surveillance Strategies of *Campylobacter*

As will be clear from the previous sections, surveillance studies provide highly valuable data that are required for risk analysis studies. Surveillance has been defined as 'the systematic collection, analysis and interpretation of data on specific diseases within a determinate population in order to guide the actions and decisions in the field of Public Health' (Thacker 2010). *Campylobacter* surveillance is an instrument to locally and timely observe the dynamics of a disease and apply effective control strategies to reduce disease occurrence and impact (Taylor and Batz 2008). In this meaning, surveillance mainly focuses on humans and the sources of their exposure that leads to disease. Legislation is in place to enable the required surveillance of *Campylobacter* enteritis in Europe, as the condition is listed in 'Communicable diseases and related special health issues to be covered by the epidemiological surveillance network' (EC 2018). Surveillance is mostly performed on a national basis. For a national surveillance system to be complete, ideally it should include data from all regions, collecting notifications of human cases and their associated microbiological data, together with animal cases and sources of exposure, if appropriate (Facciolà et al. 2017). Such a surveillance system can provide a broad view on the national and sometimes local situation on campylobacteriosis and assists to identify sources responsible for human outbreaks. However, not one country currently has a complete surveillance system in place.

A first step in disease surveillance is to provide a precise case definition, which seems easy enough, but can be fraught with difficulties. When different definitions are in use between countries, data become less comparable, which is exactly the problem we face. In the USA and in Canada, for instance, a confirmed case is defined as a case with bacterial isolation of *Campylobacter* spp. from a clinical specimen (CDC 2015; Public Health Agency of Canada 2009). Thus, this definition only includes cases for which bacteriological identification was performed, severely limiting the practicality of the definition. In contrast, in the European Union (EU) and in Australia, a confirmed case is defined as a case with isolation of *Campylobacter* spp. or identification of its nucleic acid (as by PCR amplification) from a clinical specimen (Department of Health - AUS 2004; EC 2018). This definition allows identification of the bacteria by PCR as well as from bacterial culture, which allows inclusion of cases that in the USA or Canada would not be included. Apart from cases, surveillance also captures 'probable cases.' A probable case as defined in the EU (and defined similarly elsewhere) is recognized when a person suffering from diarrhea, pain or fever can be epidemiologically linked to a likely source of *Campylobacter*. Moreover, both confirmed and probable cases are notifiable in Europe. Again, there are national differences in this practice, as in the USA, Canada and Australia only confirmed cases are notifiable. The national surveillance systems in those countries are based on passive surveillance of cases (resulting from a compilation of already existing data). In the EU, the surveillance systems are also passive and in addition case-based (tracing of individual cases). A further difference exists in that notification is not always mandatory. Within the EU, 21 member states have mandatory notification systems for human campylobacteriosis, but for instance in Belgium, France, Greece, Italy, Luxembourg and the Netherlands, this notification is based on a voluntary system (ECDC 2018). In other regions of the world, surveillance systems have not been implemented or are not nationwide rolled out, resulting in highly incomplete data. As a result of these current local situations, data can only be compared and interpreted between countries with such differences in surveillance practices in mind. This is not always recognized or as clearly stated as would be desirable.

Within the EU, European Directive (EC 2003) dictates that member states are obliged to collect data on the occurrence of zoonoses, zoonotic agents, animal populations as well as food-borne outbreaks. Therefore, the reporting of food-borne outbreaks of campylobacteriosis is mandatory within the EU. The European Food Safety Agency (EFSA) analyzes this data and publishes annual reports in cooperation with the European Centre for Disease Prevention and Control (ECDC). The European Commission is authorized to organize harmonized EU-wide baseline surveys. In such a case, EFSA is responsible for the definition and data analysis of the survey (e.g., EFSA 2010b). Similar baseline surveys exist in other countries and regions (e.g., SARDI 2010).

4 Risk Management

Risk analysis efforts are conducted to eventually reduce the risk of a given hazard, which means findings must somehow be translated into actions. This is what the last procedure in the process concentrates on. Risk management translates the findings into appropriate measures and implementations by regulatory bodies. Regarding the relatively high incidence of human campylobacteriosis and the considerate burden of the disease, control measures have been implemented in different countries. These control measures essentially target the different sources and pathways of human exposure. The measures to control food-borne campylobacteriosis include interventions to: (i) reduce the prevalence and the level of contamination in relevant food animals, (ii) reduce or suppress the contamination of foods from animal origins and (iii) reduce the effect of exposure to contaminated food. As poultry meat is the main source of food-borne campylobacteriosis, a large part of existing control measures focuses on this source.

4.1 Control in Poultry Meat

A recent paper reviewed the different prevention and mitigation strategies for *Campylobacter* in poultry products (Alter 2017). Measures to control *Campylobacter* in chicken meat are summarized in a guideline document (Codex Alimentarius

Commission 2011). Proposed measures concern the entire chicken meat production chain, from grandparent flocks to the consumer.

4.1.1 At the Farm Level

A highly effective control strategy would be to prevent *Campylobacter* from reaching farmed poultry in the first place. However, since the bacteria do not result in any overt effects in birds and are well adapted to colonize the chicken intestinal tract (where they can be considered commensals), this is not easily achieved. Nevertheless, regarding the role of flock colonization in contamination of poultry meat products, poultry farms are the first step in the meat production chain that are targeted by *Campylobacter* control measures. In 2016, Meunier et al. (2016) reviewed existing control measures to reduce poultry flock colonization by *Campylobacter*. Biosecurity measures were identified as the most efficient measures to prevent poultry flock colonization (Wagenaar et al. 2013; De Giessen et al. 1998; Gibbens et al. 2001). These measures aim to avoid, or decrease, the introduction of *Campylobacter* in poultry flocks, with variable success. Proposed measures include the implementation of hygienic measures by workers (e.g., hand washing facilities, cloth changes between flocks and decontamination of boots) (Hansson et al. 2007). Further, inclusion of physical barriers and treatments can avoid entrance of insects, wild animals and rodents that may all act as vectors to introduce the bacteria (Hald et al. 2007; Newell et al. 2011; Nesbit et al. 2001). Additionally, these barriers contribute to limit the numbers of persons, including workers and potential unauthorized persons, who by visiting the flocks may unintentionally introduce *Campylobacter*, especially when they have previously been in contact with positive flocks (EFSA 2011). Thinning (also termed depopulation) involves personnel entering the chicken house to remove a number of birds to adjust their density. This practice has been recognized as a high risk associated with *Campylobacter* introduction into a flock (Allen et al. 2008), but it is still being practiced. Drinking water and feed can also contribute to the flock colonization and should be controlled (Hansson et al. 2007). The chlorination of drinking water was early recognized as an option to eliminate *Campylobacter* (Pearson et al. 1993). As *Campylobacter* can persist in the rearing building and material and has been shown to be transmitted between consecutive flocks, proper disinfection between rearing periods is recommended to avoid such transmission (Newell et al. 2011; Berrang et al. 2003). In free-range chickens, biosecurity measure is very difficult to implement due to their exposure to the environment (Klein et al. 2015).

The immunization of birds would also be a favorable control option. The effectiveness of vaccination to immunize animals and potentially enhance *Campylobacter* control has been shown in several studies (Widders et al. 1996; Rice 1997; de Zoete et al. 2007; Meunier et al. 2016). However, due to the limited cross-strain protection and its duration, effective vaccines are not yet commercially available (Alter 2017). Possibly, young chicks are passively immunized through parental antibodies, but the effectiveness does not last, and whether this can be actively enhanced to be effective still needs to be further studied (Hermans et al. 2014). As an alternative option to immunization, it was proposed to select for animals or breeds that are resistant to *Campylobacter* colonization (Laisney et al. 2004). However, given the lack of symptoms and the commensal relationship between the bacteria and this host, it is questionable whether this is feasible.

Other control measures have concentrated to decrease, rather than prevent, the animals' colonization, and this has shown some promising results. This can be achieved by introducing changes in the intestinal microflora of the birds that result in conditions that are less favorable for *Campylobacter* colonization. It has been modeled that a reduction of 3 log-10 units of *Campylobacter* in the ceca of living chickens would be sufficient to reduce the burden to human health by 58% (Koutsoumanis et al. 2020). A decrease in bird colonization was achieved by competitive exclusion through probiotic bacteria or by adding prebiotics to the feed to adjust the intestinal flora (Mead et al. 1996; Schoeni and Wong 1994; Morishita et al. 1997; Smialek et al. 2018; Ghareeb et al. 2012; Messaoudi et al. 2013; Gaggìa et al. 2010). This work is ongoing and may lead to further applications, provided they are practically applicable at low costs. An alternative to reduce numbers of *Campylobacter* is by feeding the birds with bacteriophages that prey on the bacteria. Studies have provided evidence of the benefits of phages to decrease *Campylobacter* numbers both in the intestines and in feces. For instance, recent in vivo experiments showed a decrease of up to 2.4 $\log_{10}$ CFU/g of cecal content in phage-treated animals compared to control animals (Richards et al. 2019). In a field study, Kittler and coworkers (2013) showed a decrease of up to 3.2 $\log_{10}$ CFU/g of feces by phage cocktail administration to broilers. However, many phages are strain-specific, and the practicality of this measure at an industrial scale still has to be demonstrated. The cost factor must also be taken into account. Chapter Phage Biocontrol of *Campylobacter*: A One Health Approach in this book deals with phage therapy in detail.

Bacteriocins are antimicrobial peptides produced by bacteria (Saint-Cyr et al. 2016), and several bacteriocins have been identified that can reduce *Campylobacter* colonization in poultry (Meunier et al. 2016). Purified bacteriocins are in general more effective in reducing intestinal load of *Campylobacter* in poultry than feeding the bacteria that produce them (Alter 2017). Other feed and water supplements that have been investigated for decreasing effects of *Campylobacter* colonization of poultry include organic short-chain fatty acids, plant-derived substances and essential oils with variable effectiveness (reviewed in Meunier et al. 2016). Some interventions have been shown to be particularly effective when administered prior to harvesting the birds. For instance, feed withdrawal of broilers prior to harvesting induces an increase of the bacteria present in crops and carcasses, so that food withdrawal is best avoided (Northcutt et al. 2003; Byrd et al. 1998), and adding lactic acid to drinking water for a short period pre-slaughter was also shown to decrease the load of *Campylobacter* in crops (Byrd et al. 2001).

Transport of the birds to the slaughterhouse has been proposed to contribute to contamination of the birds, including their feathers. Crates used to transport animals can be heavily contaminated despite their systematic cleaning (Slader et al. 2002).

Catching and transporting the birds induce stress, which may increase fecal dropping and thus the shedding of *Campylobacter* (Whyte et al. 2001b). Stern and co-workers (1995) investigated the effect of transport on external contamination of animals from 10 broiler farms. That study highlighted that the prevalence of contaminated animals increased (from 12.1 to 56%) as well as the number of bacteria that were present on the animal carcasses (from $10^{2.71}$ to $10^{5.15}$ CFU per carcass) after slaughter.

4.1.2 Control Measures at Meat Processing and Storage Level

Slaughter and processing of poultry on a commercial scale are not as clean as slaughter practices of large mammals. Flocks that enter an abattoir in a non-colonized state may result in meat leaving the premise with *Campylobacter* present, as cross-contamination is very common. In particular, the defeathering and evisceration steps often result in feces leakage that contaminates the carcass, and when the equipment involved in these steps induces skin lesions, the bacteria cannot effectively be removed by the subsequent rinsing steps. Skin lesions are frequently introduced when birds are incorrectly positioned or when their size is suboptimal for the equipment. Ideally, the defeathering and evisceration equipment should be adapted to animal sizes and avoid harming the skin, and flocks should consist of equally sized birds, but in practice these requirements are not always met.

Various strategies have been tested to minimize meat contamination during slaughter and to avoid cross-contamination between flocks. It has been tested if feces leakage during slaughtering can be prevented by plugging the cloaca of the birds (Musgrove et al. 1997), but the feasibility of such measures in large commercial processing systems can be questioned. Several risk assessment studies investigated the effect of logistic slaughter on reducing the campylobacteriosis risk in humans in Europe (reviewed by Nauta and Havelaar 2008). A risk assessment study in Japan estimated that logistic slaughter could decrease the risk of human infection by 44% (Sasaki et al. 2013).

If contamination during slaughter cannot be avoided, decontamination of carcasses is the next option. Decontamination can be performed in various ways, including chemical and physical treatment of carcasses (WHO 2012). Chemical treatments include use of chlorinated and electrolyzed water in carcass washing (Park et al. 2002; Yang et al. 2001; Bashor et al. 2004; Berrang and Bailey 2009), dipping pre-washed carcasses in an acidified solution containing sodium chlorite (Kemp et al. 2000), immersion in acid or triphosphate solution (Stern et al. 1985; Whyte et al. 2001b), and modified atmosphere packaging of poultry products (Meredith et al. 2014; Phebus et al. 1991; Phillips 1998; Rajkovic et al. 2010). Physical treatments include increased water temperature during scalding (Lehner et al. 2014; Purnell et al. 2004; Yang et al. 2002), freezing or heat treatment of contaminated carcasses (Reiersen et al. 2002; Hofshagen 2003; Corry et al. 2003), hot water rinsing of carcasses (Li et al. 2002; Purnell et al. 2004), hydrostatic high-pressure processing (Martínez-Rodriguez and Mackey 2005; Solomon and Hoover 2004; Bièche et al. 2012), and irradiation of carcasses and meat (Farkas 1998; Lewis et al. 2002;

Haughton et al. 2012). Some of these chemical and physical treatments may be not feasible due to specific legislations that vary between countries. There is also variation in acceptance by consumers: The 'chlorine chickens' mentioned in the popular press and some political debates are related to decontamination practices allowed in the USA that are not used in the EU.

Depending how poultry meat is processed (marinating and fermentation) or stored (freezing and protective atmosphere packaging), *Campylobacter* numbers may decrease over time to different degrees (Alter et al. 2006; Lee et al. 1998; Borkelsson et al. 2003; Björkroth 2005). More research is needed to establish packaging strategies that would reduce survival of *Campylobacter* without allowing other pathogens to increase or resulting in loss of quality of the meat during storage.

4.2 Control of **Campylobacter** *Sources Other Than Poultry Meat*

Controlling *Campylobacter* in pork and red meat starts at the rearing and slaughtering levels through applying biosecurity and hygienic measures. Since these animals represent higher values than chickens or turkeys, measures are more often economically feasible compared to poultry farms. A higher attention is needed in pork rearing due to their more intensive production compared to cattle or sheep, but so far pork has not been identified as a common source of *Campylobacter*, most likely due to the slaughter process and meat properties, as discussed above. Good hygiene practices during milk collection and an appropriate heat treatment are sufficient to avoid human exposure to campylobacter through milk. Drinking water that is produced and treated by conventional methods (i.e., chlorination) effectively eliminates *Campylobacter* (Lund 1996).

Fruit and vegetables can become contaminated, but *Campylobacter* cannot multiply on these matrices. Their contamination is imputable to transfer of bacteria from environment, wild animals, irrigation water and organic fertilizers. Sewage sludge treatment was shown to be effective in eliminating *Campylobacter* (Stampi et al. 1999). Good hygienic and production practice in vegetables and fruits can limit the contamination of fruit and vegetable products. Chemical treatment such as rinsing with chlorinated water was also effective to reduce the number of *Campylobacter* on fruits and vegetables (Nguyen-The and Carlin 2000; Park and Sanders 1992). It seems that, although the alternative routes by which *Campylobacter* can reach humans amount to between 26 and 72%, not one single source can be identified that is responsible for a large fraction of these cases. As a consequence, current control measures remain relatively sparse and their effect is only marginable. Clearly, more work is needed to fill in this knowledge gap so that better control measures can be implemented that are targeted to other contamination routes than poultry meat.

5 Risk Communication

The last step of a complete risk analysis is risk communication, and possibly this is the activity at which most effect can currently still be gained. Consumers are key players in avoiding campylobacteriosis (Nauta and Havelaar 2008; EFSA 2011). In many countries, *Campylobacter* is now better known than, say, twenty years ago, but in comparison with the level of knowledge of the general population regarding *Salmonella* and food-poisoning *E. coli*, the risk of campylobacteriosis is still not generally recognized. In part, this may be due to the name of the organism (*Salmonella* is so much easier to remember), and to the lack of large outbreaks with many patients involved and considerable numbers of fatalities, as has been experienced with pathogenic *E. coli*. The burden of campylobacteriosis is mostly due to the serious sequelae that can be long-lasting but are also relatively rare. In combination, these conditions result in less awareness by the general public than would be desired for proper management of the risk.

Risk communication can be performed in many ways. An example is the warning on poultry products that meat must be thoroughly cooked prior to consumption. The informed reader knows this warning mostly relates to *Campylobacter*, but the species is not mentioned anywhere. Several countries invested in consumer information, education and training to reduce exposure to *Campylobacter* from poultry meat preparation and consumption (Reiersen et al. 2002; MacRitchie et al. 2014; Altekruse et al. 1996). Such information campaigns presenting the risk factors and providing recommendation on good practices could avoid human infection when exposed to *Campylobacter* (Lammerding 1997; Schlundt 1999). Still, more can be done. A television cook may be shown cutting poultry meat in one scene and tomatoes in the next, without the scene that showed she changed knife and cutting board in between—such programs could be much more educative if a remark about kitchen hygiene were added. The means of risk communication are far more diverse than leaflets and information campaigns. Alternative media, including blogs, social media, etc., could be implemented to raise public awareness that may result in strong benefits to public health (Fig. 1).

6 Concluding Remarks

The literature on which risk analysis of *Campylobacter* is based has matured and is now extensive. Nevertheless, some data gaps remain regarding sources and pathways leading to human exposure to *Campylobacter* spp.: For example, transmission from the environment, the role of insects in human and animal exposure, or the effect of meat packaging strategies could be further investigated. The efficacy of national or regional surveillance systems would gain from designed centralized and harmonized surveillance of both human campylobacteriosis and sources of human exposure (Batz and Morris 2010). Enhancing *Campylobacter* surveillance in humans and exposure

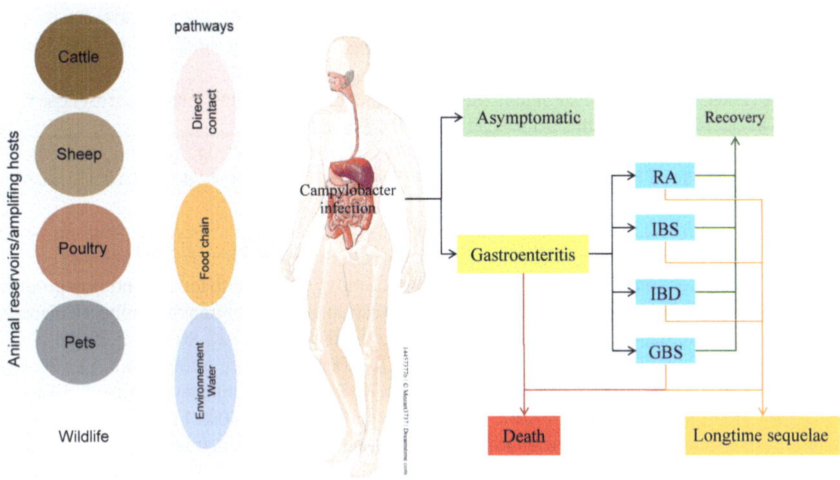

Fig. 1 Reservoirs, routes of transmission and clinical manifestations associated with *Campylobacter* species. This figure was adapted from Pires (2014) and WHO (2012). Abbreviations used: IBD, inflammatory bowel disease; IBS, irritable bowel syndrome; RA, reactive arthritis; GBS, Guillain-Barré syndrome

source would allow filling the gaps in knowledge regarding *Campylobacter* dynamics in humans, within and between animal hosts, and the environment. Several control measures have been individually studied. However, the effectiveness of combined control measures was rarely investigated and is anticipated to be larger than their sum, as one measure can enforce the effect of another. Control measures have been investigated mostly in vitro and need to be backed up in vivo, under conditions that are feasible and economical. The field effectiveness in commercial food production as well as the cost effectiveness of these measures needs to be further studied.

Acknowledgements This work is supported by the German Federal Ministry of Education and Research (BMBF) in frame of the zoonoses research consortium PAC-*Campylobacter* to R.M. (project IP10/01KI1725A).

References

Abeyta C, Deeter FG, Kaysner CA, Stott RF, Wekell MM (1993) *Campylobacter jejuni* in a Washington state shellfish growing bed associated with illness. J Food Prot 56:323–325. https://doi.org/10.4315/0362-028X-56.4.323

Acke E (2018) Campylobacteriosis in dogs and cats: a review. NZ Vet J 66:221–228. https://doi.org/10.1080/00480169.2018.1475268

Allen VM, Weaver H, Ridley AM, Harris JA, Sharma M, Emery J, Sparks N, Lewis M, Edge S (2008) Sources and spread of thermophilic *Campylobacter* spp. during partial depopulation of broiler chicken flocks. J Food Prot 71:264–270. https://doi.org/10.4315/0362-028x-71.2.264

Allerberger F, Al-Jazrawi N, Kreidl P, Dierich MP, Feierl G, Hein I, Wagner M (2003) Barbecued chicken causing a multi-state outbreak of *Campylobacter jejuni* enteritis. Infection 31:19–23. https://doi.org/10.1007/s15010-002-3088-8

Allos BM (2001) *Campylobacter jejuni* infections: update on emerging issues and trends. Clin Infect Dis 32:1201–1206. https://doi.org/10.1086/319760

Altekruse SF, Street DA, Fein SB, Levy AS (1996) Consumer knowledge of foodborne microbial hazards and food-handling practices. J Food Prot 59:287–294. https://doi.org/10.4315/0362-028X-59.3.287

Altekruse SF, Stern NJ, Fields PI, Swerdlow DL (1999) *Campylobacter jejuni*—an emerging foodborne pathogen. Emerg Infect Dis 5:28–35. https://doi.org/10.3201/eid0501.990104

Alter T (2017) Prevention and mitigation strategies for *Campylobacter* with focus on poultry production. In: G. Klein (Ed.) *Campylobacter*: features, detection, and prevention of foodborne disease. Academic, Amsterdam, pp 111–129. https://doi.org/10.1016/B978-0-12-803623-5.00006-X

Alter T, Bori A, Hamedi A, Ellerbroek L, Fehlhaber K (2006) Influence of inoculation levels and processing parameters on the survival of *Campylobacter jejuni* in German style fermented Turkey sausages. Food Microbiol 23:701–707. https://doi.org/10.1016/j.fm.2005.11.001

Backert S, Hofreuter D (2013) Molecular methods to investigate adhesion, transmigration, invasion and intracellular survival of the foodborne pathogen *Campylobacter jejuni*. J Microbiol Methods 95(1):8–23. https://doi.org/10.1016/j.mimet.2013.06.031

Bashor MP, Curtis PA, Keener KM, Sheldon BW, Kathariou S, Osborne JA (2004) Effects of carcass washers on *Campylobacter* contamination in large broiler processing plants. Poultry Sci 83:1232–1239. https://doi.org/10.1093/ps/83.7.1232

Batz MB, Hoffmann S, Morris JG (2012) Ranking the disease burden of 14 pathogens in food sources in the United States using attribution data from outbreak investigations and expert elicitation. J Food Prot 75:1278–1291. https://doi.org/10.4315/0362-028X.JFP-11-418

Batz M, Morris JG Jr. (2010) Building the science foundation of a modern food safety system—lessons from Denmark, the Netherlands, and the United Kingdom on creating a more coordinated and integrated approach to food safety information—a report for the produce safety project, Washington DC

Berrang ME, Bailey JS (2009) On-line brush and spray washers to lower numbers of *Campylobacter* and *Escherichia coli* and presence of *Salmonella* on broiler carcasses during processing. J Appl Poultry Res 18:74–78. https://doi.org/10.3382/japr.2008-00067

Berrang ME, Northcutt JK, Fletcher DL, Cox NA (2003) Role of dump cage fecal contamination in the transfer of *Campylobacter* to carcasses of previously negative broilers. J Appl Poultry Res 12:190–195. https://doi.org/10.1093/japr/12.2.190

Beuchat LR (1996) Pathogenic microorganisms associated with fresh produce. J Food Prot 59:204–216. https://doi.org/10.4315/0362-028X-59.2.204

Bièche C, de Lamballerie M, Chevret D, Federighi M, Tresse O (2012) Dynamic proteome changes in *Campylobacter jejuni* 81–176 after high pressure shock and subsequent recovery. J Proteomics 75:1144–1156. https://doi.org/10.1016/j.jprot.2011.10.028

Björkroth J (2005) Microbiological ecology of marinated meat products. Meat Sci 70:477–480. https://doi.org/10.1016/j.meatsci.2004.07.018

Black RE, Levine MM, Clements ML, Hughes TP, Blaser MJ (1988) Experimental *Campylobacter jejuni* infection in humans. J Inf Dis 157:472–479. https://doi.org/10.1093/infdis/157.3.472

Blaser MJ, Engberg J (2008) Clinical aspects of *Campylobacter jejuni* and *Campylobacter coli* infections. In: Nachamkin I, Szymanski CM, Blaser MJ (eds) Campylobacter. American Society of Microbiology, Third Edition, pp 97–121

Boehm AB, Graham KE, Jennings WC (2018) Can we swim yet? Systematic review, meta-analysis, and risk assessment of aging sewage in surface waters. Environ Sci Technol 52:9634–9645. https://doi.org/10.1021/acs.est.8b01948

Borkelsson Á, Georgsson F, Guðmundsdóttir E, Harðardóttir H, Reiersen J (2003) Freezing of poultry contaminated with *Campylobacter* as intervention against disease in humans. Int J Med Microbiol 293:147

Brandl MT, Haxo AF, Bates AH, Mandrell RE (2004) Comparison of survival of *Campylobacter jejuni* in the phyllosphere with that in the rhizosphere of spinach and radish plants. Appl Environ Microbiol 70:1182–1189. https://doi.org/10.1128/AEM.70.2.1182-1189.2004
Bruce-Grey-Owen Sound Health Unit (2000) The investigative report of thewalkerton outbreak of waterborne gastroenteritis: May–June, 2000. Available from https://www.publichealthgreybruce.on.ca/Portals/0/Topics/WaterSafety/Public_Drinking/Walkerton/WalkertonReport/REPORT_Oct00.PDF
Bull SA, Allen VM, Domingue G, Jørgensen F, Frost JA, Ure R, Whyte R, Tinker D, Corry JEL, Gillard-King J, Humphrey TJ (2006) Sources of *Campylobacter* spp. colonizing housed broiler flocks during rearing. Appl Environ Microbiol 72:645–652. https://doi.org/10.1128/AEM.72.1.645-652.2006
Byrd JA, Corrier DE, Hume ME, Bailey RH, Stanker LH, Hargis BM (1998) Effect of feed withdrawal on *Campylobacter* in the crops of market-age broiler chickens. Avian Dis 42:802–806
Byrd JA, Hargis BM, Caldwell DJ, Bailey RH, Herron KL, McReynolds JL, Brewer RL, Anderson RC, Bischoff KM, Callaway TR, Kubena LF (2001) Effect of lactic acid administration in the drinking water during preslaughter feed withdrawal on *Salmonella* and *Campylobacter* contamination of broilers. Poultry Sci 80:278–283. https://doi.org/10.1093/ps/80.3.278
Bywater R, Deluyker H, Deroover E, de Jong A, Marion H, McConville M, Rowan T, Shryock T, Shuster D, Thomas V, Vallé M, Walters J (2004) A European survey of antimicrobial susceptibility among zoonotic and commensal bacteria isolated from food-producing animals. J Antimicrob Chemother 54:744–754. https://doi.org/10.1093/jac/dkh422
Castillo A, Escartin EF (1994) Survival of *Campylobacter jejuni* on sliced watermelon and papaya. J Food Prot 57:166–168. https://doi.org/10.4315/0362-028X-57.2.166
CDC (2015) Campylobacteriosis—2015 case definition. Available from: https://wwwn.cdc.gov/nndss/conditions/campylobacteriosis/case-definition/2015. Accessed 14 April 2020
Chapman B, Otten A, Fazil A, Ernst N, Smith BA (2016) A review of quantitative microbial risk assessment and consumer process models for *Campylobacter* in broiler chickens. Microb Risk Anal 2–3:3–15. https://doi.org/10.1016/j.mran.2016.07.001
Christensen B, Sommer HM, Rosenquist H, Nielsen N (2001) Risk assessment on *Campylobacter jejuni* in chicken products, 1st edn. Division of Microbial Safety, Institute of Food Safety and Toxicology, DVFA
Christidis T, Pintar KDM, Butler AJ, Nesbitt A, Thomas MK, Marshall B, Pollari F (2016) *Campylobacter* spp. Prevalence and levels in raw milk: a systematic review and meta-analysis. J Food Prot 79:1775–1783. https://doi.org/10.4315/0362-028X.JFP-15-480
Codex Alimentarius Commission (2011): Guidelines for the control of *Campylobacter* and *Salmonella* in chicken meat—CAC/GL 78–2011. Edited by Food and Agriculture Organization (FAO), Rome
Cody AJ, Maiden MC, Strachan NJ, McCarthy ND (2019) A systematic review of source attribution of human campylobacteriosis using multilocus sequence typing. Euro Surveill 24. https://doi.org/10.2807/1560-7917.ES.2019.24.43.1800696
Corry JEL, James C, O'Neill D, Yaman H, Kendall A, Howell M (2003) Physical methods, readily adapted to existing commercial processing plants, for reducing numbers of campylobacters, on raw poultry. Int J Med Microbiol 293:32
Dasti JI, Tareen AM, Lugert R, Zautner AE, Gross U (2010) *Campylobacter jejuni*: a brief overview on pathogenicity-associated factors and disease-mediating mechanisms. Int J Med Microbiol 300:205–211. https://doi.org/10.1016/j.ijmm.2009.07.002
Davidson VJ, Ravel A, Nguyen TN, Fazil A, Ruzante JM (2011) Food-specific attribution of selected gastrointestinal illnesses: estimates from a Canadian expert elicitation survey. Foodborne Pathog Dis 8:983–995. https://doi.org/10.1089/fpd.2010.0786
Davis KR, Dunn AC, Burnett C, McCullough L, Dimond M, Wagner J, Smith L, Carter A, Willardson S, Nakashima AK (2016) *Campylobacter jejuni* infections associated with raw milk consumption—Utah, 2014. MMWR Morb Mortal Wkly Rep 65:301–305. https://doi.org/10.15585/mmwr.mm6512a1

de Zoete MR, van Putten JPM, Wagenaar JA (2007) Vaccination of chickens against *Campylobacter*. Vaccine 25:5548–5557. https://doi.org/10.1016/j.vaccine.2006.12.002

Department of Health—AUS (2004) Australian gouvernment—Department of Health—Campylobacteriosis case definition. https://www1.health.gov.au/internet/main/publishing.nsf/Content/cda-surveil-nndss-casedefs-cd_campy.htm. Accessed 14 April 2020

Devleesschauwer B, Bouwknegt M, Mangen M-JJ, Havelaar AH (2017) Chapter 2—Health and economic burden of *Campylobacter*. In: Klein G (ed) *Campylobacter*: features, detection, and prevention of foodborne disease. Academic, Amsterdam, pp 27–40

Dogan OB, Clarke J, Mattos F, Wang B (2019) A quantitative microbial risk assessment model of *Campylobacter* in broiler chickens: Evaluating processing interventions. Food Control 100:97–110. https://doi.org/10.1016/j.foodcont.2019.01.003

Doorduyn Y, Van Den Brandhof WE, Van Duynhoven YTHP, Breukink BJ, Wagenaar JA, van Pelt W (2010) Risk factors for indigenous *Campylobacter jejuni* and *Campylobacter coli* infections in The Netherlands: a case-control study. Epidemiol Infect 138:1391–1404. https://doi.org/10.1017/S095026881000052X

Doyle MP, Schoeni JL (1986) Isolation of *Campylobacter jejuni* from retail mushrooms. Appl Environ Microbiol 151:449–450. https://doi.org/10.1128/AEM.51.2.449-450.1986

EC (2003) Directive 2003/99/EC of the European Parliament and of the Council of 17 November 2003 on the monitoring of zoonoses and zoonotic agents, amending Council Decision 90/424/EEC and repealing Council Directive 92/117/EEC—OJ L 325, 12.12.2003, pp 31–40

EC (2018) Commission implementing decision (EU) 2018/945 of 22 June 2018 on the communicable diseases and related special health issues to be covered by epidemiological surveillance as well as relevant case definitions (Text with EEA relevance). C/2018/3868. OJ L 170, 6.7.2018, pp 1–74

ECDC (2018) European centre for disease prevention and control annual epidemiological report for 2017. Surveillance systems overview for 2017. Stockholm

EFSA (2005) Opinion of the scientific panel on biological hazards (BIOHAZ) related to *Campylobacter* in animals and foodstuffs. EFSA J 3:173. https://doi.org/10.2903/j.efsa.2005.173

EFSA (2007) The community summary report on trends and sources of zoonoses, zoonotic agents, antimicrobial resistance and foodborne outbreaks in the European Union in 2005. EFSA J 4:94R. https://doi.org/10.2903/j.efsa.2006.94r

EFSA (2010a) Scientific opinion on quantification of the risk posed by broiler meat to human campylobacteriosis in the EU. EFSA J 8:1437. https://doi.org/10.2903/j.efsa.2010.1437

EFSA (2010b) Analysis of the baseline survey on the prevalence of *Campylobacter* in broiler batches and of *Campylobacter* and *Salmonella* on broiler carcasses, in the EU, 2008—Part B: analysis of factors associated with *Campylobacter* colonisation of broiler batches and. EFSA J 8:1522. https://doi.org/10.2903/j.efsa.2010.1522

EFSA (2011) Scientific opinion on *Campylobacter* in broiler meat production: control options and performance objectives and/or targets at different stages of the food chain. EFSA J 9:2105. https://doi.org/10.2903/j.efsa.2011.2105

EFSA and ECDC (2015) The European Union summary report on trends and sources of zoonoses, zoonotic agents and food-borne outbreaks in 2014. EFSA J 13:4329. https://doi.org/10.2903/j.efsa.2015.4329

Endtz HP, Vliegenthart JS, Vandamme P, Weverink HW, van den Braak NP, Verbrugh HA, van Belkum A (1997) Genotypic diversity of *Campylobacter lari* isolated from mussels and oysters in The Netherlands. Int J Food Microbiol 34:79–88. https://doi.org/10.1016/s0168-1605(96)01174-9

Englen MD, Hill AE, Dargatz DA, Ladely SR, Fedorka-CRAY PJ (2007) Prevalence and antimicrobial resistance of *Campylobacter* in US dairy cattle. J Appl Microbiol 102:1570–1577. https://doi.org/10.1111/j.1365-2672.2006.03189.x

Facciolà A, Riso R, Avventuroso E, Visalli G, Delia SA, Laganà P (2017) *Campylobacter*: from microbiology to prevention. J Prev Med Hyg 58:E79–E92

FAO and WHO (2019) Codex Alimentarius Commission—procedural manual twenty-seventh edition, Rome

Fabech B, Bryhni K, Forshell LP, Georgsson F, Gry J, Hansen BT, Hallström H, Hatakka M, Holene E, Kapperud G, Kristinsson J, Maijala R, Nielsen NL, Nordström U, Schultz AC, Solheim C, Thorkelsson ÀE (2002) A practical approach to the application of the risk analysis process: illustrated with two models, Caffeine and *Campylobacter*. Food, 2002:510. Nordic Council of Ministers, København

Fakruddin M, Mannan KSB, Andrews S (2013) Viable but nonculturable bacteria: food safety and public health perspective. ISRN Microbiol 2013:703813. https://doi.org/10.1155/2013/703813

Farkas J (1998) Irradiation as a method for decontaminating food. Int J Food Microbiol 44:189–204. https://doi.org/10.1016/S0168-1605(98)00132-9

Fernandes AM, Balasegaram S, Willis C, Wimalarathna HML, Maiden MC, McCarthy ND (2015) Partial failure of milk pasteurization as a risk for the transmission of *Campylobacter* from cattle to humans. Clin Infect Dis 61:903–909. https://doi.org/10.1093/cid/civ431

Förster M, Sievert K, Messler S, Klimpel S, Pfeffer K (2009) Comprehensive study on the occurrence and distribution of pathogenic microorganisms carried by synanthropic flies caught at different rural locations in Germany. J Med Entomol 46:1164–1166. https://doi.org/10.1603/033.046.0526

Friedman CR, Hoekstra RM, Samuel M, Marcus R, Bender J, Shiferaw B, Reddy S, Ahuja SD, Helfrick DL, Hardnett F, Carter M, Anderson B, Tauxe RV (2004) Risk factors for sporadic *Campylobacter* infection in the United States: a case-control study in FoodNet sites. Clin Infect Dis 38(Suppl 3):S285–S296. https://doi.org/10.1086/381598

Gaggìa F, Mattarelli P, Biavati B (2010) Probiotics and prebiotics in animal feeding for safe food production. Int J Food Microbiol 141(Suppl 1):S15-28. https://doi.org/10.1016/j.ijfoodmicro.2010.02.031

Gaudreau C, Rodrigues-Coutlée S, Pilon PA, Coutlée F, Bekal S (2015) Long-lasting outbreak of erythromycin- and ciprofloxacin-resistant *Campylobacter jejuni* subspecies *jejuni* from 2003 to 2013 in men who have sex with men, Quebec, Canada. Clin Inf Dis 61:1549–1552. https://doi.org/10.1093/cid/civ570

Geissler AL, Bustos Carrillo F, Swanson K, Patrick ME, Fullerton KE, Bennett C, Barrett K, Mahon BE (2017) Increasing *Campylobacter* infections, outbreaks, and antimicrobial resistance in the United States, 2004–2012. Clin Inf Dis 65:1624–1631. https://doi.org/10.1093/cid/cix624

Ghareeb K, Awad WA, Mohnl M, Porta R, Biarnés M, Böhm J, Schatzmayr G (2012) Evaluating the efficacy of an avian-specific probiotic to reduce the colonization of *Campylobacter jejuni* in broiler chickens. Poultry Sci 91:1825–1832. https://doi.org/10.3382/ps.2012-02168

Gibbens JC, Pascoe SJ, Evans SJ, Davies RH, Sayers AR (2001) A trial of biosecurity as a means to control *Campylobacter* infection of broiler chickens. Prev Vet Med 48:85–99. https://doi.org/10.1016/s0167-5877(00)00189-6

Gilpin BJ, Walshe G, Walsh G, On SL, Smith D, Marshall JC, French NP (2013) Application of molecular epidemiology to understanding campylobacteriosis in the Canterbury region of New Zealand. Epidemiol Infect 141:1253–1266. https://doi.org/10.1017/S0950268812001719

Guy RA, Arsenault J, Kotchi SO, Gosselin-Théberge M, Champagne M-J, Berthiaume P (2018) *Campylobacter* in recreational lake water in southern Quebec, Canada: presence, concentration, and association with precipitation and ruminant farm proximity. J Water Health 16:516–529. https://doi.org/10.2166/wh.2018.222

Haagsma JA, Maertens de Noordhout C, Polinder S, Vos T, Havelaar AH, Cassini A, Devleesschauwer B, Kretzschmar ME, Speybroeck N, Salomon JA (2015) Assessing disability weights based on the responses of 30,660 people from four European countries. Pop Health Metrics 13:10. https://doi.org/10.1186/s12963-015-0042-4

Hald B, Sommer HM, Skovgård H (2007) Use of fly screens to reduce *Campylobacter* spp. introduction in broiler houses. Emerg Inf Dis 13:1951–1953. https://doi.org/10.3201/eid1312.070488

Hansson I, Vågsholm I, Svensson L, Olsson Engvall E (2007) Correlations between *Campylobacter* spp. prevalence in the environment and broiler flocks. J Appl Microbiol 103:640–649. https://doi.org/10.1111/j.1365-2672.2007.03291.x

Hartnett E, Kelly L, Newell D, Wooldridge M, Gettinby G (2001) A quantitative risk assessment for the occurrence of *Campylobacter* in chickens at the point of slaughter. Epidem Infect 127:195–206. https://doi.org/10.1017/S0950268801005866

Haughton PN, Grau EG, Lyng J, Cronin D, Fanning S, Whyte P (2012) Susceptibility of *Campylobacter* to high intensity near ultraviolet/visible 395±5 nm light and its effectiveness for the decontamination of raw chicken and contact surfaces. Int J Food Microbiol 159:267–273. https://doi.org/10.1016/j.ijfoodmicro.2012.09.006

Havelaar AH, Galindo AV, Kurowicka D, Cooke RM (2008) Attribution of foodborne pathogens using structured expert elicitation. Foodborne Path Dis 5:649–659. https://doi.org/10.1089/fpd.2008.0115

Havelaar AH, Ivarsson S, Löfdahl M, Nauta MJ (2013) Estimating the true incidence of campylobacteriosis and salmonellosis in the European Union, 2009. Epidemiol Infect 141:293–302. https://doi.org/10.1017/S0950268812000568

Havelaar AH, de Wit MA, van Koningsveld R, van Kempen E (2000) Health burden in the Netherlands due to infection with thermophilic *Campylobacter* spp. Epidemiol Infect 125:505–522. https://doi.org/10.1017/S0950268800004933

Hermans D, van Steendam K, Verbrugghe E, Verlinden M, Martel A, Seliwiorstow T, Heyndrickx M, Haesebrouck F, de Zutter L, Deforce D, Pasmans F (2014) Passive immunization to reduce *Campylobacter jejuni* colonization and transmission in broiler chickens. Vet Res. 45:27. https://doi.org/10.1186/1297-9716-45-27

Hofshagen (2003) Two years with the Norwegian action plan against *Campylobacter* spp. in broilers. Int J Med Microbiol 293:28

Horigan V, Davies RH, Kelly LA, Mead GC, Irvine RM, Simons RRL (2014) A qualitative risk assessment of the microbiological risks to consumers from the production and consumption of uneviscerated and eviscerated small game birds in the UK. Food Control 45:127–137. https://doi.org/10.1016/j.foodcont.2014.04.040

Hutchinson DN, Bolton FJ, Jelley WCN, Mathews WG, Telford DR, Counter DE, Jessop EG, Horsley SD (1985) *Campylobacter* enteritis associated with consumption of raw goat's milk. Lancet 325:1037–1038. https://doi.org/10.1016/S0140-6736(85)91632-0

ICRA (2011) Interactive online catalogue on risk assessment (ICRA). https://icra.foodrisk.org/. Accessed 4 May 2020

Jonas R, Kittl S, Overesch G, Kuhnert P (2015) Genotypes and antibiotic resistance of bovine *Campylobacter* and their contribution to human campylobacteriosis. Epidem Infect 143:2373–2380. https://doi.org/10.1017/S0950268814003410

Jonsson ME, Chriél M, Norström M, Hofshagen M (2012) Effect of climate and farm environment on *Campylobacter* spp. colonisation in Norwegian broiler flocks. Prev Vet Med 107:95–104. https://doi.org/10.1016/j.prevetmed.2012.05.002

Jore S, Viljugrein H, Brun E, Heier BT, Borck B, Ethelberg S, Hakkinen M, Kuusi M, Reiersen J, Hansson I, Olsson Engvall E, Løfdahl M, Wagenaar JA, van Pelt W, Hofshagen M (2010) Trends in *Campylobacter* incidence in broilers and humans in six European countries, 1997–2007. Prev Vet Med 93:33–41. https://doi.org/10.1016/j.prevetmed.2009.09.015

Jorgensen F, Ellis-Iversen J, Rushton S, Bull SA, Harris SA, Bryan SJ, Gonzalez A, Humphrey TJ (2011) Influence of season and geography on *Campylobacter jejuni* and *C. coli* subtypes in housed broiler flocks reared in Great Britain. Appl Environ Microbiol 77:3741–3748. https://doi.org/10.1128/AEM.02444-10

Kaakoush NO, Castaño-Rodríguez N, Mitchell HM, Man SM (2015) Global epidemiology of *Campylobacter* infection. Clin Microbiol Rev 28:687–720. https://doi.org/10.1128/CMR.00006-15

Kapperud G (1994) *Campylobacter* infections. Epidemiology, risk factors and preventive measures (in Norwegian). Tidsskr Nor Laegeforen 114:795

Kärenlampi R, Hänninen M-L (2004) Survival of *Campylobacter jejuni* on various fresh produce. Int J Food Microbiol 97:187–195. https://doi.org/10.1016/j.ijfoodmicro.2004.04.019
Katsma WEA, de Koeijer AA, Jacobs-Reitsma WF, Mangen M-JJ, Wagenaar JA (2007) Assessing interventions to reduce the risk of *Campylobacter* prevalence in broilers. Risk Anal 27:863–876. https://doi.org/10.1111/j.1539-6924.2007.00928.x
Kemp GK, Aldrich ML, Waldroup AL (2000) Acidified sodium chlorite antimicrobial treatment of broiler carcasses. J Food Prot 63:1087–1092. https://doi.org/10.4315/0362-028X-63.8.1087
Kirkpatrick BD, Lyon CE, Porter CK, Maue AC, Guerry P, Pierce KK, Carmolli MP, Riddle MS, Larsson CJ, Hawk D, Dill EA, Fingar A, Poly F, Fimlaid KA, Hoq F, Tribble DR (2013) Lack of homologous protection against *Campylobacter jejuni* CG8421 in a human challenge model. Clin Inf Dis 57:1106–1113. https://doi.org/10.1093/cid/cit454
Kittler S, Fischer S, Abdulmawjood A, Glünder G, Klein G (2013) Effect of bacteriophage application on *Campylobacter jejuni* loads in commercial broiler flocks. Appl Environ Microbiol 79:7525–7533. https://doi.org/10.1128/AEM.02703-13
Klein G, Jansen W, Kittler S, Reich F (2015) Mitigation strategies for *Campylobacter* spp. in broiler at pre-harvest and harvest level. Berl Munch Tierarztl Wochenschr 128:132–140
Koutsoumanis K, Allende A, Alvarez-Ordóñez A, Bolton D, Bover-Cid S, Davies R, De Cesare A, Herman L, Hilbert F, Lindqvist R, Nauta M, Peixe L, Ru G, Simmons M, Skandamis P, Suffredini E, Alter T, Crotta M, Ellis-Iversen J, Hempen M, Messens W, Chemaly M (2020) EFSA panel on biological hazards (BIOHAZ). Update and review of control options for Campylobacter in broilers at primary production. EFSA J 18:e06090. https://doi.org/10.2903/j.efsa.2020.6090
Kramer MH, Herwaldt BL, Craun GF, Calderon RL, Juranek DD (1996) Surveillance for waterborne-disease outbreaks–United States, 1993–1994. MMWR CDC Surveill Summ 45:1–33
Kumar A, Agarwal RK, Bhilegaonkar KN, Shome BR, Bachhil VN (2001) Occurrence of *Campylobacter jejuni* in vegetables. Int J Food Microbiol 67:153–155. https://doi.org/10.1016/S0168-1605(01)00433-0
Lackner J, Weiss M, Müller-Graf C, Greiner M (2019) The disease burden associated with *Campylobacter* spp in Germany 2014. PLoS ONE 14:e0216867. https://doi.org/10.1371/journal.pone.0216867
Laisney MJ, Gillard MO, Salvat G (2004) Influence of bird strain on competitive exclusion of *Campylobacter jejuni* in young chicks. Brit Poultry Sci 45:49–54. https://doi.org/10.1080/00071660410001668851
Lake IR, Colón-González FJ, Takkinen J, Rossi M, Sudre B, Dias JG, Tavoschi L, Joshi A, Semenza JC, Nichols G (2019) Exploring *Campylobacter* seasonality across Europe using The European Surveillance System (TESSy), 2008–2016. Eurosurveillance 24. https://doi.org/10.2807/1560-7917.ES.2019.24.13.180028
Lammerding AM (1997) An overview of microbial food safety risk assessment. J Food Prot 60:1420–1425. https://doi.org/10.4315/0362-028X-60.11.1420
Lee A, Smith SC, Coloe PJ (1998) Survival and growth of *Campylobacter jejuni* after artificial inoculation onto chicken skin as a function of temperature and packaging conditions. J Food Prot 61:1609–1614. https://doi.org/10.4315/0362-028x-61.12.1609
Lee SK, Park HJ, Lee JH, Lim JS, Seo KH, Heo EJ, Kim YJ, Wee SH, Moon JS (2017) Distribution and molecular characterization of *Campylobacter* species at different processing stages in two poultry processing plants. Foodborne Pathog Dis 14:141–147. https://doi.org/10.1089/fpd.2016.2218
Lee J, Lee H, Lee S, Kim S, Ha J, Choi Y, Oh H, Kim Y, Lee Y, Yoon K-S, Seo K, Yoon Y (2019) Quantitative microbial risk assessment for *Campylobacter jejuni* in ground meat products in Korea. Food Sci Anim Resour 39:565–575. https://doi.org/10.5851/kosfa.2019.e39
Lehner Y, Reich F, Klein G (2014) Influence of process parameter on *Campylobacter* spp. counts on poultry meat in a slaughterhouse environment. Curr Microbiol 69:240–244. https://doi.org/10.1007/s00284-014-0575-y
Lewis SJ, Velásquez A, Cuppett SL, McKee SR (2002) Effect of electron beam irradiation on poultry meat safety and quality. Poultry Sci 81:896–903. https://doi.org/10.1093/ps/81.6.896

Li L, Mendis N, Trigui H, Oliver JD, Faucher SP (2014) The importance of the viable but non-culturable state in human bacterial pathogens. Front Microbiol 5:258. https://doi.org/10.3389/fmicb.2014.00258

Li Y, Yang H, Swem BL (2002) Effect of high-temperature inside-outside spray on survival of *Campylobacter jejuni* attached to prechill chicken carcasses. Poultry Sci 81:1371–1377. https://doi.org/10.1093/ps/81.9.1371

Little CL, Richardson JF, Owen RJ, de Pinna E, Threlfall EJ (2008) *Campylobacter* and *Salmonella* in raw red meats in the United Kingdom: prevalence, characterization and antimicrobial resistance pattern, 2003–2005. Food Microbiol 25:538–543. https://doi.org/10.1016/j.fm.2008.01.001

Little CL, Gormley FJ, Rawal N, Richardson JF (2010) A recipe for disaster: outbreaks of campylobacteriosis associated with poultry liver pâté in England and Wales. Epidemiol Infect 138:1691–1694. https://doi.org/10.1017/s0950268810001974

Lund V (1996) Evaluation of *E. coli* as an indicator for the presence of *Campylobacter jejuni* and *Yersinia enterocolitica* in chlorinated and untreated oligotrophic lake water. Water Res 30:1528–1534. https://doi.org/10.1016/0043-1354(96)00034-6

MacRitchie LA, Hunter CJ, Strachan NJC (2014) Consumer acceptability of interventions to reduce *Campylobacter* in the poultry food chain. Food Control 35:260–266. https://doi.org/10.1016/j.foodcont.2013.06.005

Marks SL, Rankin SC, Byrne BA, Weese JS (2011) Enteropathogenic bacteria in dogs and cats: diagnosis, epidemiology, treatment, and control. J Vet Intern Med 25:1195–1208. https://doi.org/10.1111/j.1939-1676.2011.00821.x

Martínez-Rodriguez A, Mackey BM (2005) Factors affecting the pressure resistance of some *Campylobacter* species. L Appl Microbiol 41:321–326. https://doi.org/10.1111/j.1472-765X.2005.01768.x

Mattioli MC, Sassoubre LM, Russell TL, Boehm AB (2017) Decay of sewage-sourced microbial source tracking markers and fecal indicator bacteria in marine waters. Water Res 108:106–114. https://doi.org/10.1016/j.watres.2016.10.066

McCarthy ND, Gillespie IA, Lawson AJ, Richardson J, Neal KR, Hawtin PR, Maiden MCJ, O'brien SJ (2012) Molecular epidemiology of human *Campylobacter jejuni* shows association between seasonal and international patterns of disease. Epidemiol Infect 140:2247–2255. https://doi.org/10.1017/S0950268812000192

Mead GC, Scott MJ, Humphrey TJ, McAlpine K (1996) Observations on the control of *Campylobacter jejuni* infection of poultry by 'competitive exclusion' . Avian Pathol 25:69–79. https://doi.org/10.1080/03079459608419121

Mead PS, Slutsker L, Dietz V, McCaig LF, Bresee JS, Shapiro C, Griffin PM, Tauxe RV (1999) Food-related illness and death in the United States. Emerg Infect Dis 5:607–625. https://doi.org/10.3201/eid0505.990502

Meredith H, Valdramidis V, Rotabakk BT, Sivertsvik M, McDowell D, Bolton DJ (2014) Effect of different modified atmospheric packaging (MAP) gaseous combinations on *Campylobacter* and the shelf-life of chilled poultry fillets. Food Microbiol 44:196–203. https://doi.org/10.1016/j.fm.2014.06.005

Messaoudi S, Manai M, Kergourlay G, Prévost H, Connil N, Chobert J-M, Dousset X (2013) *Lactobacillus salivarius*: bacteriocin and probiotic activity. Food Microbiol 36:296–304. https://doi.org/10.1016/j.fm.2013.05.010

Meunier M, Guyard-Nicodème M, Dory D, Chemaly M (2016) Control strategies against *Campylobacter* at the poultry production level: biosecurity measures, feed additives and vaccination. J Appl Microbiol 120:1139–1173. https://doi.org/10.1111/jam.12986

Mohammadpour H, Berizi E, Hosseinzadeh S, Majlesi M, Zare M (2018) The prevalence of *Campylobacter* spp. in vegetables, fruits, and fresh produce: a systematic review and meta-analysis. Gut Pathog 10:41. https://doi.org/10.1186/s13099-018-0269-2

Mohammed AN, Abdel Aziz SAA (2019) The prevalence of *Campylobacter* species in broiler flocks and their environment: assessing the efficiency of chitosan/zinc oxide nanocomposite for adopting control strategy. Environ Sci Pollut Res 26:30177–30187. https://doi.org/10.1007/s11356-019-06030-z

Morishita TY, Aye PP, Harr BS, Cobb CW, Clifford JR (1997) Evaluation of an avian-specific probiotic to reduce the colonization and shedding of *Campylobacter jejuni* in broilers. Avian Dis 41:850–855

Mossong J, Mughini-Gras L, Penny C, Devaux A, Olinger C, Losch S, Cauchie H-M, van Pelt W, Ragimbeau C (2016) Human campylobacteriosis in Luxembourg, 2010–2013: A case-control study combined with Multilocus Sequence Typing for source attribution and risk factor analysis. Sci Rep 6:20939. https://doi.org/10.1038/srep20939

Mughini Gras L, Smid JH, Wagenaar JA, Koene MGJ, Havelaar AH, Friesema IHM, French NP, Flemming C, Galson JD, Graziani C, Busani L, van Pelt W (2013) Increased risk for *Campylobacter jejuni* and *C. coli* infection of pet origin in dog owners and evidence for genetic association between strains causing infection in humans and their pets. Epidemiol Infect 141:2526–2535. https://doi.org/10.1017/S0950268813000356

Mughini-Gras L, Smid JH, Wagenaar JA, de Boer A, HavelaarAH FIHM, French NP, Graziani C, Busani L, van Pelt W (2014) Campylobacteriosis in returning travellers and potential secondary transmission of exotic strains. Epidemiol Infect 142:1277–1288. https://doi.org/10.1017/S0950268813002069

Mughini-Gras L, Kooh P, Augustin J-C, David J, Fravalo P, Guillier L, Jourdan-Da-Silva N, Thébault A, Sanaa M, Watier L (2018) Source attribution of foodborne diseases: potentialities, hurdles, and future expectations. Front Microbiol 9:1983. https://doi.org/10.3389/fmicb.2018.01983

Mullner P, Jones G, Noble A, Spencer SEF, Hathaway S, French NP (2009) Source attribution of food-borne zoonoses in New Zealand: a modified Hald model. Risk Anal 29:970–984. https://doi.org/10.1111/j.1539-6924.2009.01224.x

Musgrove MT, Cason JA, Fletcher DL, Stern NJ, Cox NA, Bailey JS (1997) Effect of cloacal plugging on microbial recovery from partially processed broilers. Poultry Sci 76:530–533. https://doi.org/10.1093/ps/76.3.530

Nachamkin I, Szymanski CM, Blaser MJ (Eds) (2008) *Campylobacter*, 3rd edition. American Society of Microbiology

Nauta M, Christensen B (2011) The impact of consumer phase models in microbial risk analysis. Risk Anal 31:255–265. https://doi.org/10.1111/j.1539-6924.2010.01481.x

Nauta M, Hill A, Rosenquist H, Brynestad S, Fetsch A, van der Logt P, Fazil A, Christensen B, Katsma E, Borck B, Havelaar A (2009) A comparison of risk assessments on *Campylobacter* in broiler meat. Int J Food Microbiol 129:107–123. https://doi.org/10.1016/j.ijfoodmicro.2008.12.001

Nauta MJ, Havelaar AH (2008) Risk-based standards for *Campylobacter* in the broiler meat chain. Food Control 19:372–381. https://doi.org/10.1016/j.foodcont.2007.04.016

Nauta MJ, Jacobs-Reitsma WF, Evers EG, van Pelt W, Havelaar AH (2005) Risk assessment of *Campylobacter* in the Netherlands via broiler meat and other routes—RIVM report 250911006/2005

Nesbit EG, Gibbs P, Dreesen DW, Lee MD (2001) Epidemiologic features of *Campylobacter jejuni* isolated from poultry broiler houses and surrounding environments as determined by use of molecular strain typing. Am J Vet Res 62:190–194. https://doi.org/10.2460/ajvr.2001.62.190

Newell DG, Fearnley C (2003) Sources of *Campylobacter* colonization in broiler chickens. Appl Environ Microbiol 69:4343–4351. https://doi.org/10.1128/AEM.69.8.4343-4351.2003

Newell DG, Elvers KT, Dopfer D, Hansson I, Jones P, James S, Gittins J, Stern NJ, Davies R, Connerton I, Pearson D, Salvat G, Allen VM (2011) Biosecurity-based interventions and strategies to reduce *Campylobacter* spp. on poultry farms. Appl Environ Microbiol 77:8605–8614. https://doi.org/10.1128/AEM.01090-10

Newell DG, Mughini-Gras L, Kalupahana RS, Wagenaar JA (2017) *Campylobacter* epidemiology—sources and routes of transmission for human infection. In: Klein G (ed) *Campylobacter*: features, detection, and prevention of foodborne disease. Academic, Amsterdam, pp 85–110

Nguyen-The (2000) The microbiological quality and safety of food: fresh and processed vegetables. 620

Nicholson FA, Groves SJ, Chambers BJ (2005) Pathogen survival during livestock manure storage and following land application. Bioresour Technol 96:135–143. https://doi.org/10.1016/j.biortech.2004.02.030

Northcutt JK, Berrang ME, Dickens JA, Fletcher DL, Cox NA (2003) Effect of broiler age, feed withdrawal, and transportation on levels of coliforms, *Campylobacter*, *Escherichia coli* and *Salmonella* on carcasses before and after immersion chilling. Poultry Sci 82:169–173. https://doi.org/10.1093/ps/82.1.169

Cróinín TÓ, Backert S (2012) Host epithelial cell invasion by *Campylobacter jejuni*: trigger or zipper mechanism? Front Cell Infect Microbiol 2:25. https://doi.org/10.3389/fcimb.2012.00025

Obiri-Danso K, Paul N, Jones K (2001) The effects of UVB and temperature on the survival of natural populations and pure cultures of *Campylobacter jejuni*, *C. coli*, *C. lari* and urease-positive thermophilic *campylobacters* (UPTC) in surface waters. J Appl Microbiol 90:256–267. https://doi.org/10.1046/j.1365-2672.2001.01239.x

OIE (2017) OIE terrestrial manual

OIE (2019) Terrestrial animal health code

Oliver (2005) The viable but nonculturable state in bacteria. J Microbiol 43:93

Orr KE, Lightfoot NF, Sisson PR, Harkis BA, Tweddle JL, Boyd P, Carroll A, Jackson CJ, Wareing DR, Freeman R (1995) Direct milk excretion of *Campylobacter jejuni* in a dairy cow causing cases of human enteritis. Epidemiol Infect 114:15–24. https://doi.org/10.1017/s0950268800051876

Osano O, Arimi SM (1999) Retail poultry and beef as sources of *Campylobacter jejuni*. East Afr Med J 76:141–143

Park H, Hung Y-C, Brackett RE (2002) Antimicrobial effect of electrolyzed water for inactivating *Campylobacter jejuni* during poultry washing. Int J Food Microbiol 72:77–83. https://doi.org/10.1016/S0168-1605(01)00622-5

Park CE, Sanders GW (1992) Occurrence of thermotolerant campylobacters in fresh vegetables sold at farmers' outdoor markets and supermarkets. Can J Microbiol 38:313–316. https://doi.org/10.1139/m92-052

Pearson AD, Greenwood M, Healing TD, Rollins D, Shahamat M, Donaldson J, Colwell RR (1993) Colonization of broiler chickens by waterborne *Campylobacter jejuni*. Appl Environ Microbiol 59:987–996. https://doi.org/10.1128/AEM.59.4.987-996.1993

Peterson MC (2003) *Campylobacter jejuni* enteritis associated with consumption of raw milk. J Environ Health 65:20–26

Pezzotti G, Serafin A, Luzzi I, Mioni R, Milan M, Perin R (2003) Occurrence and resistance to antibiotics of *Campylobacter jejuni* and *Campylobacter coli* in animals and meat in northeastern Italy. Int J Food Microbiol 82:281–287. https://doi.org/10.1016/S0168-1605(02)00314-8

Phebus RK, Draughon FA, Mount JR (1991) Survival of *Campylobacter jejuni* in modified atmosphere packaged turkey roll. J Food Prot 54:194–199. https://doi.org/10.4315/0362-028X-54.3.194

Phillips CA (1998) The isolation of *Campylobacter* spp. from modified atmosphere packaged foods. Int J Environ Health Res 8:215–221. https://doi.org/10.1080/09603129873499

Pintar KD, Christidis T, Thomas MK, Anderson M, Nesbitt A, Keithlin J, Marshall B, Pollari F (2015) A systematic review and meta-analysis of the *Campylobacter* spp. prevalence and concentration in household pets and petting zoo animals for use in exposure assessments. PLoS ONE 10(12):e0144976. https://doi.org/10.1371/journal.pone.0144976

Pires SM, Evers EG, van Pelt W, Ayers T, Scallan E, Angulo FJ, Havelaar A, Hald T (2009) Attributing the human disease burden of foodborne infections to specific sources. Foodborne Pathog Dis 6:417–424. https://doi.org/10.1089/fpd.2008.0208

Pires SM (2014) Burden of disease of foodborne pathogens in Denmark. Technical Report. Technical University of Denmark - National Food Institute

Pires SM, Jakobsen LS, Ellis-Iversen J, Pessoa J, Ethelberg S (2019) Burden of disease estimates of seven pathogens commonly transmitted through foods in Denmark, 2017. Foodborne Pathog Dis 17:322–339. https://doi.org/10.1089/fpd.2019.2705

Public Health Agency of Canada (2009) Case definitions for communicable diseases under National surveillance. CCDR

Purnell G, Mattick K, Humphrey T (2004) The use of 'hot wash' treatments to reduce the number of pathogenic and spoilage bacteria on raw retail poultry. J Food Eng 62:29–36. https://doi.org/10.1016/S0260-8774(03)00168-7

Rajkovic A, Tomic N, Smigic N, Uyttendaele M, Ragaert P, Devlieghere F (2010) Survival of *Campylobacter jejuni* on raw chicken legs packed in high-oxygen or high-carbon dioxide atmosphere after the decontamination with lactic acid/sodium lactate buffer. Int J Food Microbiol 140:201–206. https://doi.org/10.1016/j.ijfoodmicro.2010.03.034

Rao MR (2001) Pathogenicity and convalescent excretion of *Campylobacter* in rural Egyptian children. Am J Epidemiol 154:166–173. https://doi.org/10.1093/aje/154.2.166

Ravel A, Hurst M, Petrica N, David J, Mutschall SK, Pintar K, Taboada EN, Pollari F (2017) Source attribution of human campylobacteriosis at the point of exposure by combining comparative exposure assessment and subtype comparison based on comparative genomic fingerprinting. PLoS ONE 12:e0183790. https://doi.org/10.1371/journal.pone.0183790

Reiersen J, Briem H, Hardardottir H, Gunnarsson E, Georgsson F, Gudmundsdottir E, Kristinsson KG (2002) Human campylobacteriosis epidemic in Iceland 1998–2000 and effect of interventions aimed at poultry and humans. In: FAO/WHO global forum of food safety regulators

Reinhard RG, Mcadam TJ, Flick GJ, Croonenberghs RE, Wittman RF, Diallo AA, Fernandes C (1996) Analysis of *Campylobacter jejuni, Campylobacter coli, Salmonella, Klebsiella pneumoniae*, and *Escherichia coli* O157:H7 in fresh hand-picked blue crab (*Callinectes sapidus*) meat. J Food Prot 59:803–807. https://doi.org/10.4315/0362-028X-59.8.803

Rice B (1997) *Campylobacter jejuni* in broiler chickens: colonization and humoral immunity following oral vaccination and experimental infection. Vaccine 15:1922–1932. https://doi.org/10.1016/S0264-410X(97)00126-6

Richards PJ, Connerton PL, Connerton IF (2019) Phage biocontrol of *Campylobacter jejuni* in chickens does not produce collateral effects on the gut microbiota. Front Microbiol 10:476. https://doi.org/10.3389/fmicb.2019.00476

RKI (2019): Infektionsepidemiologisches Jahrbuch meldepflichtiger Krankheiten für 2018. Robert Koch Institute

Rosef O, Kapperud G (1983) House flies (*Musca domestica*) as possible vectors of *Campylobacter fetus* subsp. *jejuni*. Appl Environ Microbiol 45:381–383. https://doi.org/10.1128/aem.45.2.381-383.1983

Rosner BM, Schielke A, Didelot X, Kops F, Breidenbach J, Willrich N, Gölz G, Alter T, Stingl K, Josenhans C, Suerbaum S, Stark K (2017) A combined case-control and molecular source attribution study of human Campylobacter infections in Germany, 2011–2014. Sci Rep 7:5139. https://doi.org/10.1038/s41598-017-05227-x

SARDI (2010) Baseline survey on the prevalence and concnetration of *Salmonella* and *Campylobacter* in chicken meat on-farm and at primary processing. Available from www.foodstandards.gov.au.

Sasaki Y, Maruyama N, Zou B, Haruna M, Kusukawa M, Murakami M, Asai T, Tsujiyama Y, Yamada Y (2013) *Campylobacter* cross-contamination of chicken products at an abattoir. Zoon Pub Health 60:134–140. https://doi.org/10.1111/j.1863-2378.2012.01509.x

Saint-Cyr MJ, Guyard-Nicodème M, Messaoudi S, Chemaly M, Cappelier J-M, Dousset X, Haddad N (2016) Recent advances in screening of anti-*Campylobacter* activity in probiotics for use in poultry. Front Microbiol 7:553. https://doi.org/10.3389/fmicb.2016.00553

Scallan E, Hoekstra RM, Angulo FJ, Tauxe RV, Widdowson MA, Roy SL, Jones JL, Griffin PM (2011) Foodborne illness acquired in the United States–major pathogens. Emerg Infect Dis 17(1):7–15. https://doi.org/10.3201/eid1701.p11101

Scherer K, Bartelt E, Sommerfeld C, Hildebrandt G (2006) Quantification of *Campylobacter* on the surface and in the muscle of chicken legs at retail. J Food Prot 69:757–761. https://doi.org/10.4315/0362-028x-69.4.757

Schlundt J (1999) Principles of food safety risk management. Food Control 10:299–302. https://doi.org/10.1016/S0956-7135(99)00050-X

Schoeni JL, Wong AC (1994) Inhibition of *Campylobacter jejuni* colonization in chicks by defined competitive exclusion bacteria. Appl Environ Microbiol 60:1191–1197

Schönberg-Norio D, Takkinen J, Hänninen M-L, Katila M-L, Kaukoranta S-S, Mattila L, Rautelin H (2004) Swimming and *Campylobacter* infections. Emerg Infect Dis 10:1474–1477. https://doi.org/10.3201/eid1008.030924

Shange N, Gouws P, Hoffman LC (2019) *Campylobacter* and *Arcobacter* species in food-producing animals: prevalence at primary production and during slaughter. World J Microbiol Biotechnol 35:45. https://doi.org/10.1007/s11274-019-2722-x

Sheppard SK, Dallas JF, Strachan NJC, MacRae M, McCarthy ND, Wilson DJ, Gormley FJ, Falush D, Ogden ID, Maiden MCJ, Forbes KJ (2009) *Campylobacter* genotyping to determine the source of human infection. Clin Infect Dis 48:1072–1078. https://doi.org/10.1086/597402

Slader J, Domingue G, Jørgensen F, McAlpine K, Owen RJ, Bolton FJ, Humphrey TJ (2002) Impact of transport crate reuse and of catching and processing on *Campylobacter* and *Salmonella* contamination of broiler chickens. Appl Environ Microbiol 68:713–719. https://doi.org/10.1128/AEM.68.2.713-719.2002

Smialek M, Burchardt S, Koncicki A (2018) The influence of probiotic supplementation in broiler chickens on population and carcass contamination with *Campylobacter* spp.—field study. Res Vet Sci 118:312–316. https://doi.org/10.1016/j.rvsc.2018.03.009

Smith A, Reacher M, Smerdon W, Adak GK, Nichols G, Chalmers RM (2006) Outbreaks of waterborne infectious intestinal disease in England and Wales, 1992–2003. Epidemiol Infect 134:1141–1149. https://doi.org/10.1017/S0950268806006406

Smith BA, Meadows S, Meyers R, Parmley EJ, Fazil A (2019) Seasonality and zoonotic foodborne pathogens in Canada: relationships between climate and *Campylobacter*, *E. coli* and *Salmonella* in meat products. Epidemiol Infect 147:e190. https://doi.org/10.1017/S0950268819000797

Solomon EB, Hoover DG (2004) Inactivation of *Campylobacter jejuni* by high hydrostatic pressure. Lett Appl Microbiol 38:505–509. https://doi.org/10.1111/j.1472-765X.2004.01527.x

Stampi S, de Luca G, Varoli O, Zanetti F (1999) Occurrence, removal and seasonal variation of thermophilic *campylobacters* and *Arcobacter* in sewage sludge. Zentralbl Hyg Umweltmed 202:19–27. https://doi.org/10.1016/S0934-8859(99)80048-0

Stanley KN, Wallace JS, Currie JE, Diggle PJ, Jones K (1998) The seasonal variation of thermophilic *campylobacters* in beef cattle, dairy cattle and calves. J Appl Microbiol 85:472–480. https://doi.org/10.1046/j.1365-2672.1998.853511.x

Stelzer W, Jacob J (1992) Das Vorkommen von *Campylobacter* in einem Mittelgebirgsbach. Zentralbl Für Mikrobiol 147:45–50. https://doi.org/10.1016/S0232-4393(11)80362-8

Stern NJ, Clavero MR, Bailey JS, Cox NA, Robach MC (1995) *Campylobacter* spp. in broilers on the farm and after transport. Poultry Sci 74:937–941. https://doi.org/10.3382/ps.0740937

Stern NJ, Rothenberg PJ, Stone JM (1985) Enumeration and Reduction of *Campylobacter jejuni* in poultry and red meats. J Food Prot 48:606–610. https://doi.org/10.4315/0362-028X-48.7.606

Stingl K, Knüver M-T, Vogt P, Buhler C, Krüger N-J, Alt K, Tenhagen B-A, Hartung M, Schroeter A, Ellerbroek L, Appel B, Käsbohrer A (2012) Quo vadis?—monitoring *Campylobacter* in Germany. Eur J Microbiol Immunol 2:88–96. https://doi.org/10.1556/eujmi.2.2012.1.12

Strother KO, Steelman CD, Gbur EE (2005) Reservoir competence of lesser mealworm (*Coleoptera: Tenebrionidae*) for *Campylobacter jejuni* (Campylobacterales: Campylobacteraceae). J Med Entomol 42:42–47. https://doi.org/10.1093/jmedent/42.1.42

Taylor MR, Batz MB (2008). Harnessing knowledge to ensure food safety: opportunities to improve the nation's food safety information infrastructure (FSRC report 08–01). Gainesville, Florida: Food Safety Research Consortium; Washington, D.C.: School of Public Health and Health Services, The George Washington University

Thacker SB (2010) Historical development. In: Principles and practice of public health surveillance. Oxford University Press, pp 1–17. https://doi.org/10.1093/acprof:oso/9780195372922.001.0001

Thépault A, Rose V, Queguiner M, Chemaly M, Rivoal K (2020) Dogs and Cats: reservoirs for highly diverse *Campylobacter jejuni* and a potential source of human exposure. Animals (Basel) 10:838. https://doi.org/10.3390/ani10050838

Teunis PFM, Bonačić Marinović A, Tribble DR, Porter CK, Swart A (2018) Acute illness from *Campylobacter jejuni* may require high doses while infection occurs at low doses. Epidemics 24:1–20. https://doi.org/10.1016/j.epidem.2018.02.001

Tribble DR, Baqar S, Carmolli MP, Porter C, Pierce KK, Sadigh K, Guerry P, Larsson CJ, Rockabrand D, Ventone CH, Poly F, Lyon CE, Dakdouk S, Fingar A, Gilliland T, Daunais P, Jones E, Rymarchyk S, Huston C, Darsley M, Kirkpatrick BD (2009) *Campylobacter jejuni* strain CG8421: a refined model for the study of campylobacteriosis and evaluation of *Campylobacter* vaccines in human subjects. Clin Inf Dis 49:1512–1519. https://doi.org/10.1086/644622

Tribble DR, Baqar S, Scott DA, Oplinger ML, Trespalacios F, Rollins D, Walker RI, Clements JD, Walz S, Gibbs P, Burg EF, Moran AP, Applebee L, Bourgeois AL (2010) Assessment of the duration of protection in *Campylobacter jejuni* experimental infection in humans. Infect Immun 78:1750–1759. https://doi.org/10.1128/IAI.01021-09

Van de Giessen A, Tilburg J, Ritmeester W, van der Plas J (1998) Reduction of *Campylobacter* infections in broiler flocks by application of hygiene measures. Epidemiol Infect 121:57–66. https://doi.org/10.1017/S0950268898008899

Verhoeff-Bakkenes L, Jansen HAPM, in't Veld PH, Beumer RR, Zwietering MH, van Leusden FM (2011) Consumption of raw vegetables and fruits: a risk factor for *Campylobacter* infections. Int J Food Microbiol 144:406–412. https://doi.org/10.1016/j.ijfoodmicro.2010.10.027

Viau EJ, Goodwin KD, Yamahara KM, Layton BA, Sassoubre LM, Burns SL, Tong H-I, Wong SHC, Lu Y, Boehm AB (2011) Bacterial pathogens in Hawaiian coastal streams–associations with fecal indicators, land cover, and water quality. Water Res 45:3279–3290. https://doi.org/10.1016/j.watres.2011.03.033

Vose D, Acar J, Anthony F, Franklin A, Gupta R, Nicholls T, Tamura Y, Thompson S, Threlfall EJ, van Vuuren M, White DG, Wegener HC, Costarrica ML (2001) Antimicrobial resistance risk analysis methodology for the potential impact on public health of antimicrobial resistant bacteria of animal origin. Rev Sci Tech 20:811–827. https://doi.org/10.20506/rst.20.3.1319

Wagenaar JA, French NP, Havelaar AH (2013) Preventing *Campylobacter* at the source: why is it so difficult? Clin Inf Dis 57:1600–1606. https://doi.org/10.1093/cid/cit555

Wagenaar JA, Jacobs-Reitsma W, Hofshagen M, Newell D (2008) Poultry colonization with *Campylobacter* and its control at the primary production level. In: I Nachamkin, CM Szymanski, MJ Blaser (eds.) *Campylobacter*, 3rd edition. American Society of Microbiology, pp 667–678. https://doi.org/10.1128/9781555815554.ch37

Wesley IV, Wells SJ, Harmon KM, Green A, Schroeder-Tucker L, Glover M, Siddique I (2000) Fecal shedding of *Campylobacter* and *Arcobacter* spp. in dairy cattle. Appl Environ Microbiol 66:1994–2000. https://doi.org/10.1128/aem.66.5.1994-2000.2000

Wheeler JG, Sethi D, Cowden JM, Wall PG, Rodrigues LC, Tompkins DS, Hudson MJ, Roderick PJ (1999) Study of infectious intestinal disease in England: rates in the community, presenting to general practice, and reported to national surveillance. The infectious intestinal disease study executive. BMJ 318:1046–1050. https://doi.org/10.1136/bmj.318.7190.1046

WHO (2012) The global view of campylobacteriosis: report of an expert consultation, Utrecht, Netherlands, 9–11 July 2012. World Health Organization, Geneva, Switzerland

WHO (2015) WHO estimates of the global burden of foodborne diseases. World Health Organization, Geneva, Switzerland

WHO (World Health Organization), FAO (Food And Agriculture Organization) (2009) Risk Assessment of *Campylobacter* spp. in broiler chickens: technical report. In: Microbiological risk assessment Series No. 12. World Health Organization, Geneva

Whyte P, Collins JD, McGill K, Monahan C, O'Mahony H (2001) The effect of transportation stress on excretion rates of campylobacters in market-age broilers. Poultry Sci 80:817–820. https://doi.org/10.1093/ps/80.6.817

Whyte P, Collins JD, McGill K, Monahan C, O'Mahony H (2001) Quantitative investigation of the effects of chemical decontamination procedures on the microbiological status of broiler carcasses during processing. J Food Prot 64:179–183. https://doi.org/10.4315/0362-028X-64.2.179

Whyte P, McGill K, Cowley D, Madden RH, Moran L, Scates P, Carroll C, O'Leary A, Fanning S, Collins JD, McNamara E, Moore JE, Cormican M (2004) Occurrence of *Campylobacter* in retail foods in Ireland. Int J Food Microbiol 95:111–118. https://doi.org/10.1016/j.ijfoodmicro.2003.10.018

Widders PR, Perry R, Muir WI, Husband AJ, Long KA (1996) Immunisation of chickens to reduce intestinal colonisation with *Campylobacter jejuni*. Brit Poultry Sci 37:765–778. https://doi.org/10.1080/00071669608417906

Williams MS, Golden NJ, Ebel ED, Crarey ET, Tate HP (2015) Temporal patterns of *Campylobacter* contamination on chicken and their relationship to campylobacteriosis cases in the United States. Int J Food Microbiol 208:114–121. https://doi.org/10.1016/j.ijfoodmicro.2015.05.018

Wilson IG, Moore JE (1996) Presence of *Salmonella* spp. and *Campylobacter* spp. in shellfish. Epidemiol Infect 116:147–153. https://doi.org/10.1017/s0950268800052377

Wilson DJ, Gabriel E, Leatherbarrow AJH, Cheesbrough J, Gee S, Bolton E, Fox A, Fearnhead P, Hart CA, Diggle PJ (2008) Tracing the source of campylobacteriosis. PLoS Genet 4:e1000203. https://doi.org/10.1371/journal.pgen.1000203

Wong TL, Hollis L, Cornelius A, Nicol C, Cook R, Hudson JA (2007) Prevalence, numbers, and subtypes of *Campylobacter jejuni* and *Campylobacter coli* in uncooked retail meat samples. J Food Prot 70:566–573. https://doi.org/10.4315/0362-028x-70.3.566

Yang H, Li Y, Johnson MG (2001) Survival and death of *Salmonella typhimurium* and *Campylobacter jejuni* in processing water and on chicken skin during poultry scalding and chilling. J Food Prot 64:770–776. https://doi.org/10.4315/0362-028X-64.6.770

Yang H, Wang S, Li Y, Johnson MG (2002) Predictive models for the survival/death of *Campylobacter jejuni* and *Salmonella Typhimurium* in poultry scalding and chilling. J Food Sci 67:1836–1843. https://doi.org/10.1111/j.1365-2621.2002.tb08731.x

Zhao X, Zhong J, Wei C, Lin C-W, Ding T (2017) Current perspectives on viable but non-culturable state in foodborne pathogens. Front Microbiol 8:580. https://doi.org/10.3389/fmicb.2017.00580

Population Biology and Comparative Genomics of *Campylobacter* Species

Lennard Epping, Esther-Maria Antão, and Torsten Semmler

Contents

Abstract The zoonotic pathogen *Campylobacter* is the leading cause for bacterial foodborne infections in humans. *Campylobacters* are most commonly transmitted via the consumption of undercooked poultry meat or raw milk products. The decreasing costs of whole genome sequencing enabled large genome-based analyses of the evolution and population structure of this pathogen, as well as the development of novel high-throughput molecular typing methods. Here, we review the evolutionary development and the population diversity of the two most clinically relevant *Campylobacter* species; *C. jejuni* and *C. coli.* The state-of-the-art phylogenetic studies showed clustering of *C. jejuni* lineages into host specialists and generalists with

The original version of this chapter was revised: This chapter has been changed to open access under the terms of the Creative Commons Attribution 4.0 International License. The correction to this chapter is available at https://doi.org/10.1007/978-3-030-65481-8_12

L. Epping · T. Semmler (✉)
Microbial Genomics, Robert Koch Institute, Nordufer 20, 13353 Berlin, Germany
e-mail: SemmlerT@rki.de

L. Epping
e-mail: EppingL@rki.de

E.-M. Antão
Robert Koch Institute, Nordufer 20, 13353 Berlin, Germany
e-mail: AntaoE@rki.de

S. Backert (ed.), *Fighting Campylobacter Infections*, Current Topics in Microbiology and Immunology 431, https://doi.org/10.1007/978-3-030-65481-8_3

coexisting lifestyles in chicken and livestock-associated hosts, as well as the separation of *C. coli* isolates of riparian origin (waterfowl, water) from *C. coli* isolated from clinical and farm-related samples. We will give an overview of recombination between both species and the potential impact of horizontal gene transfer on host adaptation in *Campylobacter*. Additionally, this review briefly places the current knowledge of the population structure of other *Campylobacter* species such as *C. lari*, *C. concisus* and *C. upsaliensis* into perspective. We also provide an overview of how molecular typing methods such as multilocus sequence typing (MLST) and whole genome MLST have been used to detect and trace *Campylobacter* outbreaks along the food chain.

1 Introduction

Campylobacter is one of the most common causes of foodborne infections worldwide (Kaakoush et al. 2015). To date, the genus *Campylobacter* includes 32 formally described species and 9 subspecies (Costa and Iraola 2019) and is part of the natural microbiota in the intestines of farm and wild animals (Altekruse et al. 1999). The most commonly known species are *Campylobacter jejuni* and *Campylobacter coli* that are mainly associated with campylobacteriosis in humans (Møller Nielsen 1997; Gillespie et al. 2002). *Campylobacter lari*, *Campylobacter concisus* and *Campylobacter upsaliensis* are less important for human gastrointestinal infections, but still can be frequently isolated from clinically relevant samples (Man 2011). Most notably, their multi-host lifestyles and ability for adaptation make *C. jejuni* and *C. coli* dangerous pathogens that are typically transmitted through the food chain (Oyarzabal and Backert 2012). Mainly spread through undercooked chicken meet or raw milk, these bacteria infect around 550 million people annually as reported by the World Health Organization (WHO), resulting in worldwide healthcare costs and economy loss of billions of dollars (Kaakoush et al. 2015).

Since the first complete genome sequence of the *Campylobacter* species *C. jejuni* published in 2000 (Parkhill et al. 2000), the functionality of whole genome sequencing (WGS) such as next-generation sequencing (NGS) or long read sequencing technology, namely Oxford Nanopore Technology (ONT) and Pacific Bioscience (PacBio), has massively improved. Time-consuming and low-resolution methods like pulsed-field gel electrophoresis (PFGE) (Yan et al. 1991; Potturi-Venkata et al. 2007) and *flaA* typing (Nachamkin et al. 1993) have been replaced by multilocus sequence typing (MLST) or whole/core genome MLST (wgMLST, cgMLST) that have since been frequently used for epidemiological studies (Tagini and Greub 2017). Instead of only analyzing a small part of the genome, e.g., a single gene (*flaA* typing) or MLST, which accounts for only 0.2% of the genome (Sheppard and Maiden 2015), wgMLST differentiates isolates by using all coding regions of the genomes incorporating hundreds of genes. This high discriminatory power even allows to link transmission events in epidemiological studies. Thus, high-throughput sequencing has become a time- and cost-effective method for typing, transmission-tracing, evolutionary analyses and surveillance of *Campylobacter*.

Besides comprehensive typing methods, NGS provides a broad range of possibilities to study genetic variations with respect to phenotypic difference. Powerful tools such as pan-genomic studies (Medini et al. 2005) or genome-wide association studies (GWAS), which were recently applied to microbial genomics (Falush 2016), allow very detailed correlation of the presence/absence and the allelic variants of all genes within a bacterial species population with specific phenotypes (see Sect. 4.1 below). These WGS-driven approaches enable researchers to effectively study the important aspects of host-specificity and adaptation of *Campylobacter* and help to understand the transmission and emergence of *Campylobacter* infections.

In this review, we give a broad overview of the historical evolution of *Campylobacter* and how the current population structure has been formed by niche adaptation together with inter- and intra-*Campylobacter* species recombination. Furthermore, we describe the huge potential of high-throughput and computational methods used to study relationships of *Campylobacter* strains in an agricultural and clinical environment that have provided new evidence regarding host and niche segregation.

2 Evolution Theory and Concepts for the Genus *Campylobacter*

In order to understand evolutionary and ecological processes within bacterial evolution, it is important to measure the molecular rate of mutations per replication event, also known as a molecular clock (Duchêne et al. 2016). The mutation rate of bacteria can be influenced by several different evolutionary processes such as selection pressure, genetic drift or the bottleneck effect that might play an important role in a host-adapted species like *C. jejuni* (Toft and Andersson 2010). The general approach of Ochman and Wilson (1987) to analyze the molecular clock is based on ancestral diversification calculated by 1% divergence in 16S rRNA nucleotides per 50 million years. Using this method, the divergence time of the genus *Campylobacter* was estimated to have started around 10 million years ago and clade formation of *C. coli* around 2.5 million years ago (Sheppard and Maiden 2015). However, *Campylobacter* was identified to evolve more rapidly than *Escherichia coli* and *Salmonella Typhimurium*, which have been used by Ochman and Wilson. *Campylobacter* has an unusually high rate of recombination, as horizontal gene transfer was estimated to generate two times more genetic diversity than de novo mutations (Wilson et al. 2009). Furthermore, bacterial lineages accumulate genetic substitutions more rapidly while they undergo adaptive evolution (Eyre-Walker and Keightley 2007). For all these reasons, Wilson et al. (2009) proposed a novel approach to estimate divergence in *Campylobacter* population by applying a more rapid rate of the molecular clock. They estimated the divergence of *C. coli* and *C. jejuni* to 6,580 years ago, with 95% confidence intervals (CI) of 3,580–12,400. This estimate fits within the time frame of the first domestication of wild animals during the agricultural revolution (Neolithic Revolution). The Neolithic Revolution started around 10,000–12,000 BC

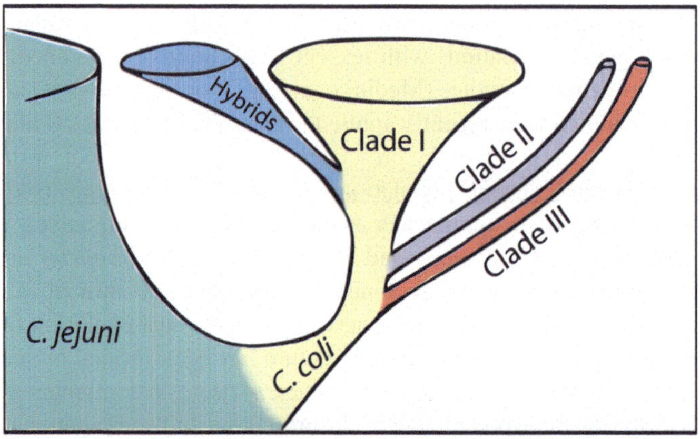

Fig. 1 Schematic representation of an evolutionary scenario of *C. coli* and *C. jejuni* (adapted from Sheppard et al. 2013a). *C. coli* and *C. jejuni* separated into two species. Due to different ecological niches *C. coli* differentiated into three clades (I–III) (Sheppard et al. 2008, 2013a). Recent recombination between strains from *C. coli* clade I and *C. jejuni* lead to the development of *C. coli* hybrid strains with substantial genomic introgression from *C. jejuni* (Sheppard et al. 2008; Golz et al. 2020)

in the Middle East and spread to central Europe 3,000–5,000 BC, providing novel niches and possibilities to emerge for commensal and pathogenic bacteria (Mira et al. 2006). The divergence of *C. coli* into three distinct clades was estimated to 1,000–1,700 years ago, and clonal complexes of *C. jejuni* started to evolve 400 years ago (Fig. 1). This timeline indicates that the emergence of *C. jejuni* and *C. coli* as individual species is a very recent event compared to *E. coli* where the main population without members of related genera has been formed around five million years ago (Wirth et al. 2006).

Independent of the model used, it is clear that the clonal complexes and clade forming lineages separated after the ancestral split of the genus into these major species that currently play a significant role in clinical and foodborne diseases. However, the development of two distinct species did not force a strict recombination barrier between them (Sheppard et al. 2013a). While the speciation within the genus *Campylobacter* was probably triggered by the agricultural revolution thousands of years ago, methods of the modern food industry, globalization or environmental changes form novel evolutionary niches and selection pressure for bacteria in general (de Mazancourt et al. 2008; Van Alfen 2015; Caniça et al. 2019). In case of *Campylobacter*, there is evidence that *C. coli* started to converge toward *C. jejuni* due to a change in their ecology, e.g., by colonizing the same niche or host (Sheppard et al. 2008), which has been facilitating recombination between these species, as will be discussed below.

3 Population Structure

Since *Campylobacter* spp. have become more and more relevant for public health, high-throughput molecular typing plays an important role in surveillance programs and outbreak control. Most importantly, MLST and NGS provide a generic approach and, additionally, have a massive impact on understanding the population structure of *Campylobacter*. MLST is a generic scheme based on allelic variants from seven housekeeping genes used to classify bacteria into related or distant lineages (Maiden et al. 1998). *C. jejuni* and *C. coli* are characterized by the same MLST scheme which analyzes allelic variants of the same orthologous loci in both species, enabling the possibility of directly comparing the species with each other (Dingle et al. 2001; Miller et al. 2005). With the advent of high-throughput NGS, epidemiological studies made use of more detailed and complex schemes and methods developed for comparative genomics, which generated in-depth knowledge about the population structure of microbes. In this section, we will describe the population structure of both *C. jejuni* and *C. coli* that have an average nucleotide identity (ANI) of 85% (Fig. 2) (Dingle et al. 2005). Furthermore, we will give an overview of recombination events between these species, which resulted in the emergence of "hybrid" strains.

3.1 Diversity and Population Structure of C. Jejuni *and* C. Coli

C. jejuni is a natural part of the gut microbiota in a wide range of hosts such as chicken, cattle, pigs or wild birds and can also be found in environmental reservoirs such as water (Altekruse et al. 1999). This multi-host lifestyle is reflected by its broad diversity, which can even be detected by a low-resolution method like MLST, representing less than 1% of the genomic DNA in *Campylobacter*. Based on phylogenetic analyses (Fig. 3), resulting from a concatenated alignment of the genes used for cgMLST, *C. jejuni* forms a weak clonal complex structure (Dingle et al. 2001; Suerbaum et al. 2001). The clonal complexes CC-45 and CC-21 harbor the most relevant clinical and outbreak strains and are among the most prevalent isolates at PubMLST database (https://pubmlst.org/), with 24% and 9% of the entries, respectively, emphasizing their importance. Isolates belonging to these complexes are known to be "host-generalist" that can colonize cattle, chicken or human hosts (Manning et al. 2003; Dearlove et al. 2016). Their ability to switch rapidly between hosts makes them a dangerous threat for human health through consumption of contaminated milk and of undercooked chicken products. Geographical signatures in *Campylobacter* are relatively weak as "identical" host associated lineages emerge all over the world (Pascoe et al. 2017). However, the frequency of specific STs can vary between countries. For example, ST-22 has been identified in Finland (Revez et al. 2011), ST-4526 in Japan (Asakura et al. 2012), and ST-190 and ST-474 were observed to emerge rapidly in New Zealand (McTAVISH et al. 2008; Mohan et al.

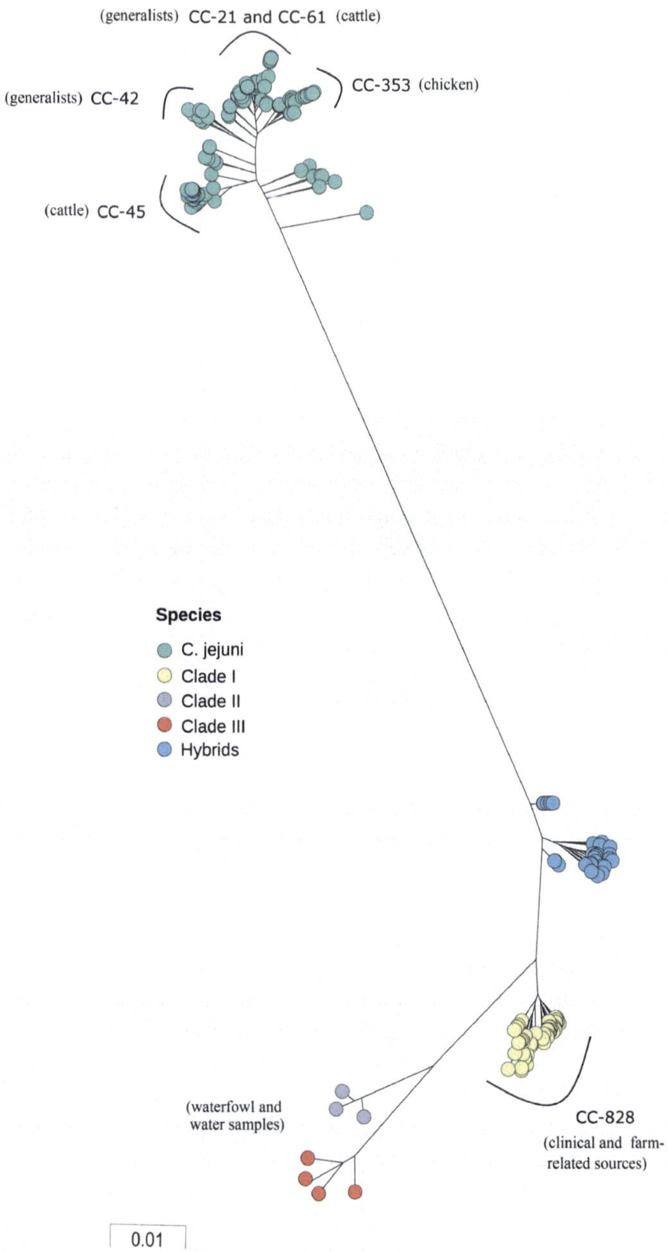

Fig. 2 Graphical visualization of pairwise ANI values of *C. coli* and *C. jejuni* genomes. *C. coli* Clade I (yellow), Clade II (red) and Clade III (purple) are clearly separated based on ANI. *C. jejuni* (turquoise) and *C. coli* are distinct species, with approximately 85% ANI. Hybrid strains formed a separate cluster but were classified as *C. coli* based on 97% ANI, in contrast to 88% ANI between the hybrid strains and *C. jejuni*. Data were taken from (Sheppard et al. 2013a, b; Golz et al. 2020). ANI was calculated using FastANI (Jain et al. 2018) and visualized with pheatmap (Kolde 2015)

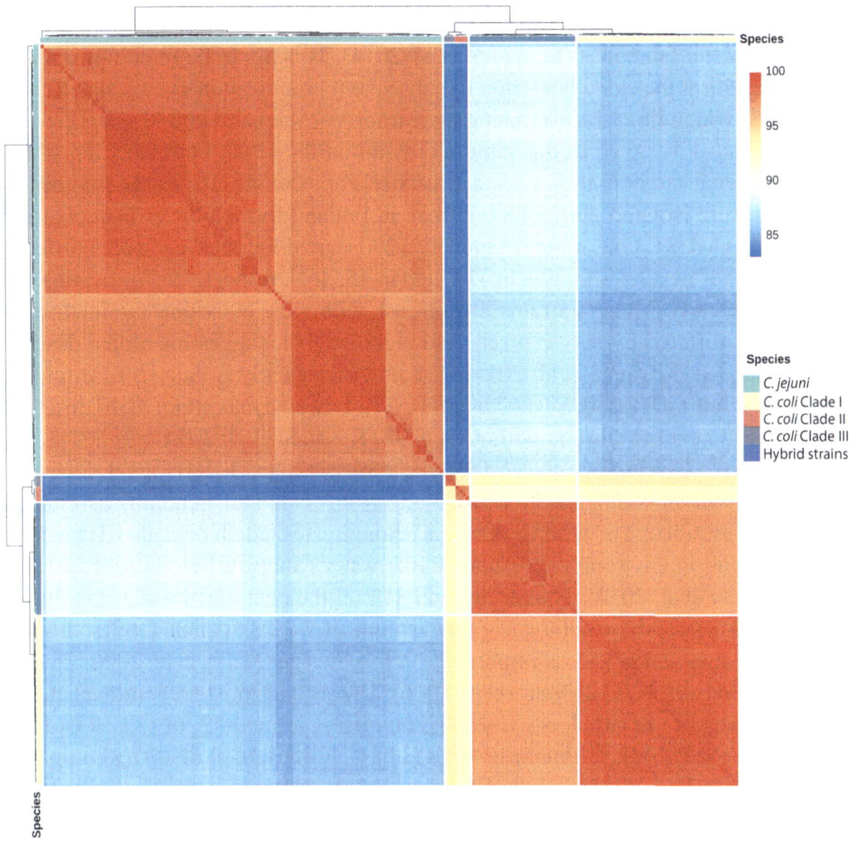

Fig. 3 Core genome-based phylogeny of *C. coli, C. jejuni* and hybrid strains. *C. jejuni* (turquoise) shows a diverse lineage-specific population structure with CC-21 and CC-45 (both host generalists), CC-42 and CC-61 (predominantly isolated from cattle), and CC-353 (from chicken). *C. coli* shows a three-clade structure with Clade I (yellow: from clinical- and farm-related sources), Clade II (purple) and Clade III (red: both from waterfowl and water samples). Clade I mainly consists of CC-828. Hybrid genomes with high DNA introgression from *C. jejuni* are colored in blue. Data were taken from (Sheppard et al. 2013a, b; Golz et al. 2020), and the phylogenetic tree was created with FastTree v2.1 (Price et al. 2010) based on 874 core genes including 123,227 variable sites

2013). Besides host generalists, repetivitve some lineages of host specialists can also cause human infections through food products. Those include CC-42 and CC-61 that are associated with cattle and sheep (Colles et al. 2003), and several different STs and CCs associated with chicken, including CC-257, CC-353 or CC-443 (Sheppard et al. 2011a, 2014). Other lineages such as CC-177 and CC-682 can be isolated from wild birds and water, causing the so-called water-born *Campylobacter* infections (Colles et al. 2009; Mohan et al. 2013). *C. jejuni* also shows a high level of diversity within the same barn or herd—e.g., isolates belonging to more than 10 distinct CCs have been found within a single chicken flock (Colles et al. 2008; Vidal et al. 2016).

However, *C. jejuni* CCs may be subject to a strong recombination barrier even if they colonize the same host (Sheppard et al. 2014). This might be forced by a niche separation within the same host, due to subsequent colonization events at different time points, which limit the horizontal gene transfer (Sheppard and Maiden 2015).

Even isolates assigned to the same ST based on the seven housekeeping genes can vary to great extent in their genetic diversity. For example, 16 strains assigned to ST-45 that were isolated during an outbreak in Finland formed three distinct strain clusters in wgMLST. Out of approximately 1200 shared loci, these clusters differed from each other by alleles in 293, in 414, and in 453 loci, respectively, indicating the presence of clearly different strains. In contrast, within the individual strain clusters the genomes differed by between zero and eighteen loci, suggesting clonal descent of those isolates (Kovanen et al. 2014). The other frequently isolated STs from this outbreak, including ST-230, ST-267 and ST-677, showed a maximum of 40 different alleles among genome clusters within each ST (Kovanen et al. 2014).

In contrast to *C. jejuni*, *C. coli* forms three distinct clades (I-III) (Figs. 1, 2 and 3), colonizing different ecological niches. Isolates from clade I are generally associated with an agricultural origin, whereas isolates belonging to clade II or clade III can most likely be found in environmental sources like water (Sheppard et al. 2008, 2013a; Skarp-de Haan et al. 2014). To date, around 81% of the genotyped isolates included in the PubMLST database belong to clonal complex CC-828 of clade I, reflecting the clinical relevance and industrial importance of this lineage (Miller et al. 2006; Thakur et al. 2006; Cody et al. 2012; Nohra et al. 2016). The second-most predominant clonal complex, also part of clade I, is CC-1150, comprising around 5% of *C. coli* isolates submitted to the PubMLST database. Clade I has a lower rate of diversity compared to *C. coli* clade II, to *C. coli* clade III, or to the general population structure of *C. jejuni* (Duim et al. 1999; Dingle et al. 2005; Sheppard et al. 2010b). The relatively low variation within the housekeeping genes as well as the lack of a proper lineage separation, especially in clade I, indicate the effect of a recent bottleneck and thus an early phase of lineage separation in the *C. coli* population (Sheppard et al. 2010b). Due to the distinct ecological niches, an ecological recombination barrier might have led to the development of three clades in *C. coli* (Sheppard et al. 2010b). However, recombination between *C. coli* clade I and *C. jejuni* resulted in hybrid strains (Figs. 1 and 3), as has been shown in several studies (Sheppard et al. 2008, 2013a; Sheppard and Maiden 2015; Golz et al. 2020).

3.2 Inter Species Recombination and Hybrid Species

Bacterial evolution is highly influenced by horizontal or lateral gene transfer (HGT or LGT) through transformation, transduction or conjugation. For recombination events, one has to distinguish between DNA introgression of complete genes or gene loci and intragenic recombination between loci leading to new mosaic allelic variants. Mosaic alleles consist of sequence content derived from different evolutionary and ancestral backgrounds (Smith 1992). As previously mentioned, early inter-species recombination, especially between *C. jejuni* and *C. coli*, plays a major role in the evolution of the genus *Campylobacter*, which might compensate for the small genome size of this genus (Suerbaum et al. 2001). Indeed, about 18.6% of the allelic variants of the seven MLST genes in *C. coli* exhibit *C. jejuni* ancestry, whereas just 2.3% of *C. jejuni* alleles were acquired from *C. coli*, indicating asymmetric gene flow between the two species (Sheppard et al. 2008). A more detailed analysis of the mosaic ancestry patterns among the seven housekeeping genes revealed an average inter-species gene flow of around 8.3% from *C. jejuni* to *C. coli* clade I, but less than 0.5% from *C. coli* clade I to *C. jejuni* (Sheppard et al. 2011b). Even in *C. coli* clade I, the genome-wide DNA introgression rate differs substantially among the predominant clonal complexes. CC-828 showed an overall introgression of approximately 10% whereas CC-1150 was found to contain up to 23% of its genome acquired from *C. jejuni* in agriculture-associated samples. Recombination mainly happened in agriculturally relevant isolates rather than in non-agricultural *C. coli* isolates and thus might be an important adaptation and niche aggregation factor. In *C. coli* clade II and clade III, genome-wide recombination with *C. jejuni* played a minor role as those isolated had only 0.2–1.2% inferred *C. jejuni* ancestry (Sheppard et al. 2013a).

Apart from the single allele exchanges, it is possible that multiple loci in the genomes have been exchanged between *C. jejuni* and *C. coli*. This would lead to the appearance of several hybrid strains (Fig. 1) that cannot clearly be identified by routine polymerase chain reaction (PCR) typing with single species differentiation marker genes and need to be investigated further by WGS. Several of such untypeable *Campylobacter* strains were isolated from egg shells of chickens in Germany (Golz et al. 2020). These isolates showed a DNA introgression of up to 15% from *C. jejuni*. However, they were still identified as *C. coli* as they exhibited 97% average nucleotide identity with *C. coli* clade I, but only 88% ANI with *C. jejuni* (Fig. 2). Furthermore, detailed genome analysis provided evidence that these recombination events are not distributed randomly across the chromosome. Instead, they particularly affect genes that are involved in general stress response, in DNA repair and in cell wall synthesis mechanisms and thus might enhance the fitness of *C. coli* for survival under harsh environmental conditions.

3.3 Additional Species

C. jejuni and *C. coli* are the most prevalent species concerning food contamination and clinical *Campylobacter* infections. Besides these, 13 additional *Campylobacter* species, sporadically causing clinically relevant symptoms, have been summarized (Costa and Iraola 2019). In the following subsection, we exemplarily describe the population structure of *C. lari*, *C. upsaliensis* and *C. concisus* that are frequently found in gastroenteritis patients (Man 2011).

C. lari is usually found in coastal regions and marine environments. It is mainly associated with shorebirds, like gulls, albatrosses, redshanks, to name a few, but also with marine mammals and shellfish, and occasionally causes gastroenteritis infections (Costa and Iraola 2019). However, the species definition of *C. lari* is an ongoing process, and several *C. lari*-like species have been described, including *Campylobacter insulaenigrae*, *Campylobacter peloridis*, *Campylobacter subantarcticus* and *Campylobacter volucris*. In 2009, *C. lari* was divided into two subspecies, namely *C. lari* subsp. *lari* and *C. lari* subsp. *concheus* (Debruyne et al. 2009). All *C. lari* and *C. lari*-like species are summarized as *Campylobacter lari* group (Miller et al. 2014).

C. concisus colonizes the human oral cavity and consists of two genetically distinct genomospecies (GS1 and GS2) that cannot be distinguished on the phenotypic level despite DNA binding values of only 42–50% in DNA-DNA hybridization experiments (Vandamme et al. 1989; Aabenhus et al. 2005). However, both genomospecies include multiple strains that have been isolated from healthy as well as diarrheic patients, which makes it difficult to make a general assumption on its pathogenicity (Chung et al. 2016). In particular, *C. concisus* GS2 seems to be more pathogenic as it is more often isolated from clinical patients with bloody diarrhea (Kalischuk and Inglis 2011). In addition, a recent study discovered novel genomic markers and a specific plasmid which are associated with *C. concisus* GS2 from patients suffering from Crohn's Disease (Liu et al. 2018).

C. upsaliensis is commonly found in domestic animals like cats and dogs (Goossens et al. 1990), but has also been isolated all over the world from clinical cases of bloody diarrhea (Bourke et al. 1998). This *Campylobacter* species is closely related to *C. coli* and *C. jejuni* based on 16S rRNA comparison (Vandamme et al. 1991). In contrast to *C. concisus*, *C. upsaliensis* shows a homogenous population structure with 80–96% DNA-DNA hybridization between strains (Sandstedt et al. 1983), even though it possesses a high degree of diversity on a genotypic level (Lentzsch et al. 2004). Besides this, little is known about the emergence of *C. upsaliensis*, which needs to be investigated in further studies.

4 Host Association of *Campylobacter*

Comparative genomic methods not only had a major influence on our understanding of population structures, but also advanced our knowledge and understanding of host adaptive mechanisms of *Campylobacter*. Besides the MLST and cgMLST schemes, (pan-genome) approaches and genome-wide association studies have opened the door for large-scale genome analyses of these traits.

4.1 Impact of Genomic High-Throughput Methods

Pan-genomic analyses have become powerful tools to study a variety of bacterial species (Rouli et al. 2015). The term "pan-genome" describes the entire set of genes composed of core and accessory genes within a bacterial population. Genes that occur in at least 99% of the population are marked as core genes whereas accessory genes only have to occur at least once in the population. Core genes mostly encode proteins that are involved in housekeeping functions of the organisms. Accessory genes on the other hand can have an adaptive function toward a specific environment or selection pressure and are usually acquired by HGT. Therefore, it is highly probable that these parts of the genome are involved in niche or host adaptation of *Campylobacter*. CgMLST and wgMLST make use of the concept of pan-genomes and establish a novel typing scheme for bacterial strains that, in contrast to MLST, includes all core genes of a species and thereby provides a high resolution by comprising the whole genetic diversity (Sheppard et al. 2013b). Similar to the MLST scheme for *C. jejuni* and *C. coli,* the cgMLST scheme combines *C. jejuni* and *C. coli* and utilizes 1343 gene loci to describe the genetic variation among the strains (Cody et al. 2017).

Due to decreasing costs in WGS and a subsequent increase in bacterial genome sequencing, the concept of GWAS has emerged in the field of microbial genomics (Chen and Shapiro 2015; Lees and Bentley 2016). GWAS is a statistical concept to compare two different phenotypes in order to identify trait-associated genomic compounds. This can be generally used to analyze epidemiology-, resistance- or, in case of *C. jejuni* host-related determinants based on WGS data. Different methods have been developed to apply this method either on entire genes, *k-mer* (word of length *k*), or single nucleotide polymorphism (SNP) level to bacterial populations. In comparison to GWAS tools that are made for human genetic research, these take into account the clonal and lineage-related phylogenetic structure of bacterial populations (Brynildsrud et al. 2016; Power et al. 2017). In order to investigate the host association of *C. jejuni*, a couple of GWAS have been applied in this field of research, mainly for the clinically relevant lineages CC-21 and CC-45 (Sheppard et al. 2013b; Yahara et al. 2017; Thépault et al. 2017; Buchanan et al. 2017). These complexes contain isolates from different hosts of predominantly avian and ruminant origin. Thus, these strains need to adapt frequently to varying environments. For example, chicken and cattle hosts substantially differ in their body temperature, pH level or in the microbiome of their digestive tract. In addition, bacterial cells are exposed to oxidative stress outside the host gut (Kim et al. 2015) during transmission to a new host. Intentionally, many of these studies used a gene-by-gene approach (Yahara et al. 2017; Buchanan et al. 2017), whereas others also keep in mind that core genome adaptation might play a role in host adaptation, especially in host-adapted lineages. Therefore, a *k-mer* approach (Sheppard et al. 2013b; Lees et al. 2018) can not only be applied in order to detect the presence of entire genes but also to identify specific alleles of core genes that may be involved in host adaptation.

4.2 Source Attribution in Clinical and Agricultural Setting

Host-adapted clonal lineages can be observed in several different bacterial pathogens, such as *C. jejuni*, *Staphylococcus aureus* or *Salmonella enterica* on different genetic levels (2010a, 2011a; Weinert et al. 2012; Hayward et al. 2016; Sheppard et al. 2018). Gene sets of these lineages are affected by several factors, including the host, the composition of food, and by antibiotics and interactions with the host microbiome that can either lead to a temporary or to a permanent adaptation. Genetic mechanisms like DNA replication errors that lead to point mutations, insertions, deletions or recombination events may result in rapid adaptation and the formation of host-specific lineages. In general, *Campylobacter* species are distributed differentially among livestock animals; *C. coli* is dominant in pig-associated samples (Thakur et al. 2006) whereas *C. jejuni* is more abundant in cattle and chicken hosts. Additionally, there might also exist a geographic factor. For example, *Campylobacter* cases in France are more likely to be caused by isolates from ruminant hosts than in other countries (Thépault et al. 2017).

Several studies investigated host adaptation, especially from *C. jejuni*, as this species shows a well-defined lineage separation based on MLST data that distinguish the population into host-specialist and host-generalist clonal complexes (Sheppard et al. 2014). Several colonization studies revealed that modification and differential transcription of motility genes in *C. jejuni* play a key role in adaptation and transmission (Hermans et al. 2011; de Vries et al. 2017; Ren et al. 2018). These data were supported by in vitro experiments as well as by genomic data and by RNA sequencing. Apart from traditional WGS analysis, the novel concepts of GWAS provided great in-depth knowledge about host adaptation, colonization and clinically relevant factors of *C. jejuni*. The group of Sheppard and co-workers discovered multiple genes involved in vitamin B5 biosynthesis and iron uptake within cattle-related strains of the CC-45 complex by applying a *k-mer*-based GWAS (Sheppard et al. 2013b). These genes might be related to different nutrition of cattle host in comparison to poultry. Independently, the same genes have also been detected within a set of 25 diagnostic marker genes by a pan-genome approach leading to the identification of clinically relevant *C. jejuni* isolates with up to 90% accuracy (Buchanan et al. 2017). However, even strains of the clinically relevant complexes CC-21 and CC-45, isolated from poultry processing chains, show substantially different genotypes and carry different genes involved in lipooligosaccharide synthesis (*kpsC, kpsD)*, metabolic processes (*glmS*), oxidative stress response (*nuoK* and *fumC)* as well as genes involved in nucleotide salvage (cj1377c*)* and antimicrobial resistance like efflux proteins (cj1375*)* (Yahara et al. 2017). A pan-genomic approach by Thèpault and co-workers identified 15 additional host-segregation markers in *C. jejuni* isolates from France that might aid to determine the source of clinical cases. Those genes are mainly involved in metabolic processes and nucleotide metabolism. These markers had been utilized to trace back the source of *C. jejuni* infections with an average accuracy of 80.7% for chicken-induced cases and of 68.2% for ruminant-caused cases (Thépault et al. 2017). While numerous studies have focused on the source of *Campylobacter* infections, less work has been dedicated to understanding the genetic mechanisms behind livestock- and environment-specific STs in chickens, cattle or water sources. However, this might generate valuable insights into the evolution and relevant host-specific factors of *Campylobacter* in order to deal with the spread and contamination in livestock environment and further understand the process of adaptation toward clinically relevant pathogens.

4.3 Relevance for Public Health (Applications)

WGS has not only improved our general understanding and knowledge of bacterial populations, adaptations and recombination to date, but is also an important part of routine high-throughput diagnostics for hospitals, for animal husbandry and for surveillance programs of foodborne diseases (Gerner-Smidt et al. 2019). Due to these programs, it is possible to detect a sudden increase in case numbers within a specific time interval. When applying WGS-driven approaches to outbreak detection and source tracking, it is important to distinguish between geographically restricted point-source outbreaks and clusters of cases that are not necessarily related to each other geographically (Llarena et al. 2017). Most outbreaks are diffuse and show a spatial and time-dependent clustering of *Campylobacter* genotypes or subtypes within livestock and clinical cases (Llarena et al. 2017). These outbreaks can spread across several countries, but can be linked to contaminated food products with a low level of contamination. The difficulty in detecting these outbreaks is to be able to distinguish them from sporadic *Campylobacter* cases and to handle the high rate of genetic exchange and recombination within the species (Llarena et al. 2017). This might be achieved by WGS-based molecular characterization in combination with wgMLST or cgMLST that provide the necessary resolution for the genomic comparisons of closely related strains (Deurenberg et al. 2017). For example, a recent wgMLST-based study on the genomic diversity of *C. jejuni* isolates from Israel detected 29 diffuse clusters of genetically related strains that have shown a low variance in allelic differences (Rokney et al. 2018). Importantly, this study further identified adapted clones that kept causing infections over the span of several years. Another study from Finland showed that *C. jejuni* infections, which increased during the summer, were mainly related to three STs with 16 to 37 allelic differences between the cluster, and thus, due to the short period of time, probably belonged to the same source (Kovanen et al. 2014).

In addition to such diffuse outbreaks, point-source outbreaks can also occur; however, those are less frequent and are usually locally restricted. They are mostly related to restaurant meals (Glashower et al. 2017), canteen food (Moffatt et al. 2016) or farming communities (Forbes et al. 2009) and are associated with a high level of contamination within the food products. An appropriate methodology to identify these outbreaks is based on single nucleotide variants (SNVs), because diversity in general should be low and resulting in only a small amount of allelic variants. This approach has been successfully applied in several studies. Moffatt et al. showed a high level of identity in a chicken-related outbreak in Australia with two different genotypes with a SNV difference of only 3–8 SNPs and 30 SNPs, respectively. Additionally, several studies conducted by Revez and colleagues (Revez et al. 2014) demonstrated how wgMLST can be applied in outbreak investigations and source tracing. Patient isolates from milk-born outbreaks shared 1432 loci with isolates from a milk source and only showed three SNPs difference between the strains. Just like for many other bacterial pathogens, WGS-based methods provide a great benefit for *Campylobacter*-related public health applications. However, in contrast to other bacteria, the high species diversity of *Campylobacter* within the same host often requires an adapted approach.

5 Concluding Remarks

The species of genus *Campylobacter* show a very individual population structure ranging from less clonal diversity to strictly separated clonal lineages. Horizontal gene transfer and recombination events may occur at various levels within the individual population but also between the *Campylobacter* species. Even "hybrid" strains exist that contain large proportion of genomic elements from two species. Modern next-generation sequencing-based methods paved the way for high-resolution molecular typing of outbreak and disease-related strains by applying a standardized typing scheme based on the whole core genome and, additionally, the pangenome. Further, they also allowed the identification of genomic factors that contribute to host adaptation of individual lineages on the gene and allele level and to trace the source of several *Campylobacter* lineages. This contribution to tracing and unraveling transmission and infection chains results in important public health applications to contain this important zoonotic pathogen.

Acknowledgements This work is supported by the German Federal Ministry of Education and Research (BMBF) in frame of the zoonoses research consortium PAC-Campylobacter to T. S. (project IP10B/01KI1725F)

References

Aabenhus R, On SLW, Siemer BL, Permin H, Andersen LP (2005) Delineation of *Campylobacter concisus* genomospecies by amplified fragment length polymorphism analysis and correlation of results with clinical data. J Clin Microbiol 43:5091–5096. https://doi.org/10.1128/JCM.43.10.5091-5096.2005

Altekruse SF, Stern NJ, Fields PI, Swerdlow DL (1999) *Campylobacter jejuni*—an emerging foodborne pathogen. Emerg Infect Dis 5:28–35. https://doi.org/10.3201/eid0501.990104

Asakura H, Brüggemann H, Sheppard SK, Ekawa T, Meyer TF, Yamamoto S, Igimi S (2012) Molecular evidence for the thriving of *Campylobacter jejuni* ST-4526 in Japan. PLoS ONE 7:e48394. https://doi.org/10.1371/journal.pone.0048394

Bourke B, Chan VL, Sherman P (1998) *Campylobacter upsaliensis*: waiting in the wings. Clin Microbiol Rev 11:440–449. https://doi.org/10.1128/CMR.11.3.440

Brynildsrud O, Bohlin J, Scheffer L, Eldholm V (2016) Erratum to: rapid scoring of genes in microbial pan-genome-wide association studies with Scoary. Genome Biol 17:262. https://doi.org/10.1186/s13059-016-1132-8

Buchanan CJ, Webb AL, Mutschall SK, Kruczkiewicz P, Barker DORR, Hetman BM, Gannon VPJJ, Abbott DW, Thomas JE, Inglis GD, Taboada EN (2017) A genome-wide association study to identify diagnostic markers for human pathogenic *Campylobacter jejuni* strains. Front Microbiol 8:1224. https://doi.org/10.3389/fmicb.2017.01224

Caniça M, Manageiro V, Abriouel H, Moran-Gilad J, Franz CMAP (2019) Antibiotic resistance in foodborne bacteria. Trends Food Sci Technol 84:41–44. https://doi.org/10.1016/j.tifs.2018.08.001

Chen PE, Shapiro BJ (2015) The advent of genome-wide association studies for bacteria. Curr Opin Microbiol 25:17–24. https://doi.org/10.1016/j.mib.2015.03.002

Chung HKL, Tay A, Octavia S, Chen J, Liu F, Ma R, Lan R, Riordan SM, Grimm MC, Zhang L (2016) Genome analysis of *Campylobacter concisus* strains from patients with inflammatory bowel disease and gastroenteritis provides new insights into pathogenicity. Sci Rep 6:38442. https://doi.org/10.1038/srep38442

Cody AJ, Bray JE, Jolley KA, McCarthy ND, Maiden MCJ (2017) Core genome multilocus sequence typing scheme for stable, comparative analyses of *Campylobacter jejuni* and *C. coli* human disease isolates. J Clin Microbiol 55:2086–2097. https://doi.org/10.1128/JCM.00080-17

Cody AJ, McCarthy NM, Wimalarathna HL, Colles FM, Clark L, Bowler ICJW, Maiden MCJ, Dingle KE (2012) A Longitudinal 6-year study of the molecular epidemiology of clinical *Campylobacter* isolates in Oxfordshire, United Kingdom. J Clin Microbiol 50:3193–3201. https://doi.org/10.1128/JCM.01086-12

Colles FM, Jones K, Harding RM, Maiden MCJ (2003) Genetic diversity of *Campylobacter jejuni* isolates from farm animals and the farm environment. Appl Environ Microbiol 69:7409–7413. https://doi.org/10.1128/AEM.69.12.7409-7413.2003

Colles FM, Jones TA, McCarthy ND, Sheppard SK, Cody AJ, Dingle KE, Dawkins MS, Maiden MCJJ (2008) *Campylobacter* infection of broiler chickens in a free-range environment. Environ Microbiol 10:2042–2050. https://doi.org/10.1111/j.1462-2920.2008.01623.x

Colles FM, McCarthy ND, Howe JC, Devereux CL, Gosler AG, Maiden MCJ (2009) Dynamics of *Campylobacter* colonization of a natural host, *Sturnus vulgaris* (European Starling). Environ Microbiol 11:258–267. https://doi.org/10.1111/j.1462-2920.2008.01773.x

Costa D, Iraola G (2019) Pathogenomics of emerging *Campylobacter* species. Clin Microbiol Rev 32:0001. https://doi.org/10.1128/CMR.00072-18

de Mazancourt C, Johnson E, Barraclough TG (2008) Biodiversity inhibits species' evolutionary responses to changing environments. Ecol Lett 11:380–388. https://doi.org/10.1111/j.1461-0248.2008.01152.x

de Vries SPW, Gupta S, Baig A, Wright E, Wedley A, Jensen AN, Lora LL, Humphrey S, Skovgard H, MacLeod K, Pont E, Wolanska DP, L'Heureux J, Mobegi FM, Smith DGEE, Everest P, Zomer A, Williams N, Wigley P, Humphrey T, Maskel DJ, Grant AJ, Skovgård H, MacLeod K, Pont E, Wolanska DP, L'Heureux J, Mobegi FM, Smith DGEE, Everest P, Zomer A, Williams N, Wigley P, Humphrey T, Maskell DJ, Grant AJ, Skovgard H, MacLeod K, Pont E, Wolanska DP, L'Heureux J, Mobegi FM, Smith DGEE, Everest P, Zomer A, Williams N, Wigley P, Humphrey T, Maskel DJ, Grant AJ, Skovgård H, MacLeod K, Pont E, Wolanska DP, L'Heureux J, Mobegi FM, Smith DGEE, Everest P, Zomer A, Williams N, Wigley P, Humphrey T, Maskell DJ, Grant AJ (2017) Genome-wide fitness analyses of the foodborne pathogen *Campylobacter jejuni* in in vitro and in vivo models. Sci Rep 7:1251. https://doi.org/10.1038/s41598-017-01133-4

Dearlove BL, Cody AJ, Pascoe B, Méric G, Wilson DJ, Sheppard SK (2016) Rapid host switching in generalist *Campylobacter* strains erodes the signal for tracing human infections. ISME J 10:721–729. https://doi.org/10.1038/ismej.2015.149

Debruyne L, On SLW, De Brandt E, Vandamme P (2009) Novel *Campylobacter lari*-like bacteria from humans and molluscs: description of *Campylobacter peloridis* sp. nov., *Campylobacter lari* subsp. *concheus* subsp. nov. and *Campylobacter lari* subsp. *lari* subsp. nov. Int J Syst Evol Microbiol 59:1126–1132. https://doi.org/10.1099/ijs.0.000851-0

Deurenberg RH, Bathoorn E, Chlebowicz MA, Couto N, Ferdous M, García-Cobos S, Kooistra-Smid AMD, Raangs EC, Rosema S, Veloo ACM, Zhou K, Friedrich AW, Rossen JWA (2017) Application of next generation sequencing in clinical microbiology and infection prevention. J Biotechnol 243:16–24. https://doi.org/10.1016/j.jbiotec.2016.12.022

Dingle KE, Colles FM, Falush D, Maiden MCJ (2005) Sequence typing and comparison of population biology of *Campylobacter coli* and *Campylobacter jejuni*. J Clin Microbiol 43:340–347. https://doi.org/10.1128/JCM.43.1.340-347.2005

Dingle KE, Colles FM, Wareing DRAA, Ure R, Fox AJ, Bolton FE, Bootsma HJ, Willems RJL, Urwin R, Maiden MCJ (2001) Multilocus sequence typing system for *Campylobacter jejuni*. J Clin Microbiol 39:14–23. https://doi.org/10.1128/JCM.39.1.14-23.2001

Duchêne S, Holt KE, Weill F-X, Le Hello S, Hawkey J, Edwards DJ, Fourment M, Holmes EC (2016) Genome-scale rates of evolutionary change in bacteria. Microbial Genomics 2:e000094. https://doi.org/10.1099/mgen.0.000094

Duim B, Wassenaar TM, Rigter A, Wagenaar J (1999) High-resolution genotyping of *Campylobacter* strains isolated from poultry and humans with amplified fragment length polymorphism fingerprinting. Appl Environ Microbiol 65:2369–2375. https://doi.org/10.1128/AEM.65.6.2369-2375.1999

Eyre-Walker A, Keightley PD (2007) The distribution of fitness effects of new mutations. Nat Rev Genet 8:610–618. https://doi.org/10.1038/nrg2146

Falush D (2016) Bacterial genomics: microbial GWAS coming of age. Nat Microbiol 1:16059. https://doi.org/10.1038/nmicrobiol.2016.59

Forbes KJ, Gormley FJ, Dallas JF, Labovitiadi O, MacRae M, Owen RJ, Richardson J, Strachan NJC, Cowden JM, Ogden ID, McGuigan CC (2009) *Campylobacter* immunity and coinfection following a large outbreak in a farming community. J Clin Microbiol 47:111–116. https://doi.org/10.1128/JCM.01731-08

Gerner-Smidt P, Besser J, Concepción-Acevedo J, Folster JP, Huffman J, Joseph LA, Kucerova Z, Nichols MC, Schwensohn CA, Tolar B (2019) Whole genome sequencing: bridging one-health surveillance of foodborne diseases. Front Public Health 7:0001. https://doi.org/10.3389/fpubh.2019.00172

Gillespie IA, O'Brien SJ, Frost JA, Adak GK, Horby P, Swan AV, Painter MJ, Neal KR (2002) A case-case Comparison of *Campylobacter coli* and *Campylobacter jejuni* infection: a tool for generating hypotheses. Emerg Infect Dis 8:937–942. https://doi.org/10.3201/eid0809.010817

Glashower D, Snyder J, Welch D, McCarthy S (2017) Notes from the field: outbreak of *Campylobacter jejuni* associated with consuming undercooked chicken liver Mousse—Clark County, Washington, 2016. MMWR Morbidity and Mortality Weekly Report 66:1027. https://doi.org/10.15585/mmwr.mm6638a4

Golz JC, Epping L, Knüver M-T, Borowiak M, Hartkopf F, Deneke C, Malorny B, Semmler T, Stingl K (2020) Whole genome sequencing reveals extended natural transformation in *Campylobacter* impacting diagnostics and the pathogens adaptive potential. Sci Rep 10:3686. https://doi.org/10.1038/s41598-020-60320-y

Goossens H, Vlaes L, De Boeck M, Levy J, De Mol P, Butzler J-P, Kersters K, Pot B, Vandamme P (1990) Is "*Campylobacter upsaliensis*" an unrecognised cause of human diarrhoea? Lancet 335:584–586. https://doi.org/10.1016/0140-6736(90)90359-D

Hayward MR, Petrovska L, Jansen VAA, Woodward MJ (2016) Population structure and associated phenotypes of *Salmonella enterica* serovars Derby and Mbandaka overlap with host range. BMC Microbiol 16:15. https://doi.org/10.1186/s12866-016-0628-4

Hermans D, Van Deun K, Martel A, Van Immerseel F, Messens W, Heyndrickx M, Haesebrouck F, Pasmans F (2011) Colonization factors of *Campylobacter jejuni* in the chicken gut. Vet Res 42:82. https://doi.org/10.1186/1297-9716-42-82

Jain C, Rodriguez-R LM, Phillippy AM, Konstantinidis KT, Aluru S (2018) High throughput ANI analysis of 90 K prokaryotic genomes reveals clear species boundaries. Nat Commun 9:5114. https://doi.org/10.1038/s41467-018-07641-9

Kaakoush NO, Castaño-Rodríguez N, Mitchell HM, Man SM (2015) Global epidemiology of *Campylobacter* infection. Clin Microbiol Rev 28:687–720. https://doi.org/10.1128/CMR.00006-15

Kalischuk LD, Inglis GD (2011) Comparative genotypic and pathogenic examination of *Campylobacter concisus* isolates from diarrheic and non-diarrheic humans. BMC Microbiol 11:53. https://doi.org/10.1186/1471-2180-11-53

Kim J-C, Oh E, Kim J, Jeon B (2015) Regulation of oxidative stress resistance in *Campylobacter jejuni*, a microaerophilic foodborne pathogen. Front Microbiol 6:0001. https://doi.org/10.3389/fmicb.2015.00751

Kolde R (2015) Pheatmap : pretty heatmaps. R package version 1.0.8

Kovanen SM, Kivisto RI, Rossi M, Schott T, Karkkainen U-M, Tuuminen T, Uksila J, Rautelin H, Hanninen M-L (2014) Multilocus sequence typing (MLST) and whole-genome MLST of *Campylobacter jejuni* isolates from human infections in three districts during a seasonal peak in Finland. J Clin Microbiol 52:4147–4154. https://doi.org/10.1128/JCM.01959-14

Lees JA, Bentley SD (2016) Bacterial GWAS: not just gilding the lily. Nat Rev Microbiol 14:406. https://doi.org/10.1038/nrmicro.2016.82

Lees JA, Galardini M, Bentley SD, Weiser JN, Corander J (2018) pyseer: a comprehensive tool for microbial pangenome-wide association studies. Bioinformatics 34:4310–4312. https://doi.org/10.1093/bioinformatics/bty539

Lentzsch P, Rieksneuwohner B, Wieler LH, Hotzel H, Moser I (2004) High-resolution genotyping of *Campylobacter upsaliensis* strains originating from three continents. J Clin Microbiol 42:3441–3448. https://doi.org/10.1128/JCM.42.8.3441-3448.2004

Liu F, Ma R, Tay CYA, Octavia S, Lan R, Chung HKL, Riordan SM, Grimm MC, Leong RW, Tanaka MM, Connor S, Zhang L (2018) Genomic analysis of oral *Campylobacter concisus* strains identified a potential bacterial molecular marker associated with active Crohn's disease. Emer Micro Inf 7:1–14. https://doi.org/10.1038/s41426-018-0065-6

Llarena A-K, Taboada E, Rossi M (2017) Whole-genome sequencing in epidemiology of *Campylobacter jejuni* infections. J Clin Microbiol 55:1269–1275. https://doi.org/10.1128/JCM.00017-17

Maiden MCJ, Bygraves JA, Feil E, Morelli G, Russell JE, Urwin R, Zhang Q, Zhou J, Zurth K, Caugant DA, Feavers IM, Achtman M, Spratt BG (1998) Multilocus sequence typing: a portable approach to the identification of clones within populations of pathogenic microorganisms. Proc Natl Acad Sci 95:3140–3145. https://doi.org/10.1073/pnas.95.6.3140

Man SM (2011) The clinical importance of emerging *Campylobacter* species. Nat Rev Gastroenterol Hepatol 8:669–685. https://doi.org/10.1038/nrgastro.2011.191

Manning G, Dowson CG, Bagnall MC, Ahmed IH, West M, Newell DG (2003) Multilocus sequence typing for comparison of veterinary and human isolates of *Campylobacter jejuni*. Appl Environ Microbiol 69:6370–6379. https://doi.org/10.1128/AEM.69.11.6370-6379.2003

McTavish SM, Pope CE, Nicol C, Sexton K, French N, Carter PE (2008) Wide geographical distribution of internationally rare *Campylobacter* clones within New Zealand. Epidemiol Infect 136:1244–1252. https://doi.org/10.1017/S0950268807009892

Medini D, Donati C, Tettelin H, Masignani V, Rappuoli R (2005) The microbial pan-genome. Curr Opin Genet Dev 15:589–594. https://doi.org/10.1016/j.gde.2005.09.006

Miller WG, Englen MD, Kathariou S, Wesley IV, Wang G, Pittenger-Alley L, Siletz RM, Muraoka W, Fedorka-Cray PJ, Mandrell RE (2006) Identification of host-associated alleles by multilocus sequence typing of *Campylobacter coli* strains from food animals. Microbiology 152:245–255. https://doi.org/10.1099/mic.0.28348-0

Miller WG, On SLW, Wang G, Fontanoz S, Lastovica AJ, Mandrell RE (2005) Extended multilocus sequence typing system for *Campylobacter coli*, *C. lari*, *C. upsaliensis*, and *C. helveticus*. J Clin Microbiol 43:2315–2329. https://doi.org/10.1128/JCM.43.5.2315-2329.2005

Miller WG, Yee E, Chapman MH, Smith TPL, Bono JL, Huynh S, Parker CT, Vandamme P, Luong K, Korlach J (2014) Comparative genomics of the *Campylobacter lari* Group. Genome Biol Evol 6:3252–3266. https://doi.org/10.1093/gbe/evu249

Mira A, Pushker R, Rodríguez-Valera F (2006) The Neolithic revolution of bacterial genomes. Trends Microbiol 14:200–206. https://doi.org/10.1016/j.tim.2006.03.001

Moffatt CRM, Greig A, Valcanis M, Gao W, Seemann T, Howden BP, Kirk MD (2016) A large outbreak of *Campylobacter jejuni* infection in a university college caused by chicken liver pâté, Australia, 2013. Epidemiol Infect 144:2971–2978. https://doi.org/10.1017/S0950268816001187

Mohan V, Stevenson M, Marshall J, Fearnhead P, Holland BR, Hotter G, French NP (2013) *Campylobacter jejuni* colonization and population structure in urban populations of ducks and starlings in New Zealand. MicrobiologyOpen 2:659–673. https://doi.org/10.1002/mbo3.102

Møller Nielsen E (1997) Distribution of serotypes of *Campylobacter jejuni* and *C. coli* from Danish patients, poultry, cattle and swine. FEMS Immunol Med Microbiol 19:47–56. https://doi.org/10.1016/S0928-8244(97)00049-7

Nachamkin I, Bohachick K, Patton CM (1993) Flagellin gene typing of *Campylobacter jejuni* by restriction fragment length polymorphism analysis. J Clin Microbiol 31:1531–1536. https://doi.org/10.1128/JCM.31.6.1531-1536.1993

Nohra A, Grinberg A, Midwinter AC, Marshall JC, Collins-Emerson JM, French NP (2016) Molecular epidemiology of *Campylobacter coli* strains isolated from different sources in New Zealand between 2005 and 2014. Appl Environ Microbiol 82:4363–4370. https://doi.org/10.1128/AEM.00934-16

Ochman H, Wilson AC (1987) Evolution in bacteria: evidence for a universal substitution rate in cellular genomes. J Mol Evol 26:74–86. https://doi.org/10.1007/BF02111283

Oyarzabal OA, Backert S (2012) Microbial food safety: an introduction (food science text series). Springer, Heidelberg

Parkhill J, Wren BW, Mungall K, Ketley JM, Churcher C, Basham D, Chillingworth T, Davies RM, Feltwell T, Holroyd S, Jagels K, Karlyshev AV, Moule S, Pallen MJ, Penn CW, Quail MA, Rajandream M-A, Rutherford KM, van Vliet AHM, Whitehead S, Barrell BG (2000) The genome sequence of the food-borne pathogen *Campylobacter jejuni* reveals hypervariable sequences. Nature 403:665–668. https://doi.org/10.1038/35001088

Pascoe B, Méric G, Yahara K, Wimalarathna H, Murray S, Hitchings MD, Sproston EL, Carrillo CD, Taboada EN, Cooper KK, Huynh S, Cody AJ, Jolley KA, Maiden MCJ, McCarthy ND, Didelot X, Parker CT, Sheppard SK (2017) Local genes for local bacteria: evidence of allopatry in the genomes of transatlantic *Campylobacter* populations. Mol Ecol. https://doi.org/10.1111/mec.14176

Potturi-Venkata LP, Backert S, Lastovica AJ, Vieira SL, Norton RA, Miller RS, Pierce S, Oyarzabal OA (2007) Evaluation of different plate media for direct cultivation of *Campylobacter* species from live broilers. Poult Sci 86:1304–1311. https://doi.org/10.1093/ps/86.7.1304

Power RA, Parkhill J, de Oliveira T (2017) Microbial genome-wide association studies: lessons from human GWAS. Nat Rev Genet 18:41–50. https://doi.org/10.1038/nrg.2016.132

Price MN, Dehal PS, Arkin AP (2010) FastTree 2—approximately maximum-likelihood trees for large alignments. PLoS ONE 5:e9490. https://doi.org/10.1371/journal.pone.0009490

Ren F, Li X, Tang H, Jiang Q, Yun X, Fang L, Huang P, Tang Y, Li Q, Huang J, Jiao X (2018) Insights into the impact of flhF inactivation on *Campylobacter jejuni* colonization of chick and mice gut. BMC Microbiol 18:149. https://doi.org/10.1186/s12866-018-1318-1

Revez J, Rossi M, Ellström P, de Haan C, Rautelin H, Hänninen M-L (2011) Finnish *Campylobacter jejuni* strains of multilocus sequence type ST-22 complex have two lineages with different characteristics. PLoS ONE 6:e26880. https://doi.org/10.1371/journal.pone.0026880

Revez J, Zhang J, Schott T, Kivistö R, Rossi M, Hänninen ML (2014) Genomic variation between *Campylobacter jejuni* isolates associated with milk-borne-disease outbreaks. J Clin Microbiol 52(8):2782–6. https://doi.org/10.1128/JCM.00931-14.

Rokney A, Valinsky L, Moran-Gilad J, Vranckx K, Agmon V, Weinberger M (2018) Genomic epidemiology of *Campylobacter jejuni* transmission in Israel. Front Microbiol 9:0001. https://doi.org/10.3389/fmicb.2018.02432

Rouli L, Merhej V, Fournier P-E, Raoult D (2015) The bacterial pangenome as a new tool for analysing pathogenic bacteria. New Microbes New Infections 7:72–85. https://doi.org/10.1016/j.nmni.2015.06.005

Sandstedt K, Ursing J, Walder M (1983) Thermotolerant *Campylobacter* with no or weak catalase activity isolated from dogs. Curr Microbiol 8:209–213. https://doi.org/10.1007/BF01579548

Sheppard SK, Cheng L, Méric G, De Haan CPA, Llarena AK, Marttinen P, Vidal A, Ridley A, Clifton-Hadley F, Connor TR, Strachan NJC, Forbes K, Colles FM, Jolley KA, Bentley SD, Maiden MCJ, Hänninen ML, Parkhill J, Hanage WP, Corander J (2014) Cryptic ecology among host generalist *Campylobacter jejuni* in domestic animals. Mol Ecol 23:2442–2451. https://doi.org/10.1111/mec.12742

Sheppard SK, Colles F, Richardson J, Cody AJ, Elson R, Lawson A, Brick G, Meldrum R, Little CL, Owen RJ, Maiden MCJ, McCarthy ND (2010a) Host association of *Campylobacter* genotypes transcends geographic variation. Appl Environ Microbiol 76:5269–5277. https://doi.org/10.1128/AEM.00124-10

Sheppard SK, Colles FM, McCarthy ND, Strachan NJC, Ogden ID, Forbes KJ, Dallas JF, Maiden MCJ (2011a) Niche segregation and genetic structure of *Campylobacter jejuni* populations from wild and agricultural host species. Mol Ecol 20:3484–3490. https://doi.org/10.1111/j.1365-294X.2011.05179.x

Sheppard SK, Dallas JF, Wilson DJ, Strachan NJCC, McCarthy ND, Jolley KA, Colles FM, Rotariu O, Ogden ID, Forbes KJ, Maiden MCJ (2010b) Evolution of an agriculture-associated disease causing *Campylobacter coli* clade: evidence from national surveillance data in Scotland. PLoS ONE 5:e15708. https://doi.org/10.1371/journal.pone.0015708

Sheppard SK, Didelot X, Jolley KA, Darling AE, Pascoe B, Meric G, Kelly DJ, Cody A, Colles FM, Strachan NJC, Ogden ID, Forbes K, French NP, Carter P, Miller WG, McCarthy ND, Owen R, Litrup E, Egholm M, Affourtit JP, Bentley SD, Parkhill J, Maiden MCJ, Falush D (2013a) Progressive genome-wide introgression in agricultural *Campylobacter coli*. Mol Ecol 22:1051–1064. https://doi.org/10.1111/mec.12162

Sheppard SK, Didelot X, Meric G, Torralbo A, Jolley KA, Kelly DJ, Bentley SD, Maiden MCJ, Parkhill J, Falush D (2013b) Genome-wide association study identifies vitamin B5 biosynthesis as a host specificity factor in *Campylobacter*. Proc Natl Acad Sci USA 110:11923–11927. https://doi.org/10.1073/pnas.1305559110

Sheppard SK, Guttman DS, Fitzgerald JR (2018) Population genomics of bacterial host adaptation. Nat Rev Genet 19:549–565. https://doi.org/10.1038/s41576-018-0032-z

Sheppard SK, Maiden MCJ (2015) The Evolution of *Campylobacter jejuni* and *Campylobacter coli*. Cold Spring Harb Perspect Biol 7:a018119. https://doi.org/10.1101/cshperspect.a018119

Sheppard SK, McCarthy ND, Falush D, Maiden MCJ (2008) Convergence of *Campylobacter* species: implications for bacterial evolution. Science 320:237–239. https://doi.org/10.1126/science.1155532

Sheppard SK, McCarthy ND, Jolley KA, Maiden MCJ (2011b) Introgression in the genus *Campylobacter*: generation and spread of mosaic alleles. Microbiology 157:1066–1074. https://doi.org/10.1099/mic.0.045153-0

Skarp-de Haan CPA, Culebro A, Schott T, Revez J, Schweda EKH, Hänninen M-L, Rossi M (2014) Comparative genomics of unintrogressed *Campylobacter coli* clades 2 and 3. BMC Genom 15:129. https://doi.org/10.1186/1471-2164-15-129

Smith J (1992) Analyzing the mosaic structure of genes. J Mol Evol 34:0001. https://doi.org/10.1007/BF00182389

Suerbaum S, Lohrengel M, Sonnevend A, Ruberg F, Kist M (2001) Allelic Diversity and recombination in *Campylobacter jejuni*. J Bacteriol 183:2553–2559. https://doi.org/10.1128/JB.183.8.2553-2559.2001

Tagini F, Greub G (2017) Bacterial genome sequencing in clinical microbiology: a pathogen-oriented review. Eur J Clin Microbiol Infect Dis 36:2007–2020. https://doi.org/10.1007/s10096-017-3024-6

Thakur S, Morrow WEM, Funk JA, Bahnson PB, Gebreyes WA (2006) Molecular Epidemiologic investigation of *Campylobacter coli* in swine production systems, using multilocus sequence typing. Appl Environ Microbiol 72:5666–5669. https://doi.org/10.1128/AEM.00658-06

Thépault A, Méric G, Rivoal K, Pascoe B, Mageiros L, Touzain F, Rose V, Béven V, Chemaly M, Sheppard SK (2017) Genome-wide identification of host-segregating epidemiological markers for source attribution in *Campylobacter jejuni*. Appl Environ Microbiol 83:e03085–e03116. https://doi.org/10.1128/AEM.03085-16

Toft C, Andersson SGE (2010) Evolutionary microbial genomics: insights into bacterial host adaptation. Nat Rev Genet 11:465–475. https://doi.org/10.1038/nrg2798

Van Alfen NK (2015) Encyclopedia of agriculture and food systems. Choice Rev Online 52:52-2875–52-2875. https://doi.org/10.5860/CHOICE.188216

Vandamme P, Falsen E, Pot B, Hoste B, Kersters K, De Ley J (1989) Identification of EF group 22 campylobacters from gastroenteritis cases as *Campylobacter concisus*. J Clin Microbiol 27:1775–1781. https://doi.org/10.1128/JCM.27.8.1775-1781.1989

Vandamme P, Falsen E, Rossau R, Hoste B, Segers P, Tytgat R, De Ley J (1991) Revision of *Campylobacter*, *Helicobacter*, and *Wolinella* taxonomy: emendation of generic descriptions and proposal of Arcobacter gen. nov. Int J Syst Bacteriol 41:88–103. https://doi.org/10.1099/00207713-41-1-88

Vidal AB, Colles FM, Rodgers JD, McCarthy ND, Davies RH, Maiden MCJ, Clifton-Hadley FA (2016) Genetic diversity of *Campylobacter jejuni* and *Campylobacter coli* isolates from conventional broiler flocks and the impacts of sampling strategy and laboratory method. Appl Environ Microbiol 82:2347–2355. https://doi.org/10.1128/AEM.03693-15

Weinert LA, Welch JJ, Suchard MA, Lemey P, Rambaut A, Fitzgerald JR (2012) Molecular dating of human-to-bovid host jumps by *Staphylococcus aureus* reveals an association with the spread of domestication. Biol Let 8:829–832. https://doi.org/10.1098/rsbl.2012.0290

Wilson DJ, Gabriel E, Leatherbarrow AJH, Cheesbrough J, Gee S, Bolton E, Fox A, Hart CA, Diggle PJ, Fearnhead P (2009) Rapid evolution and the importance of recombination to the gastroenteric pathogen *Campylobacter jejuni*. Mol Biol Evol 26:385–397. https://doi.org/10.1093/molbev/msn264

Wirth T, Falush D, Lan R, Colles F, Mensa P, Wieler LH, Karch H, Reeves PR, Maiden MCJ, Ochman H, Achtman M (2006) Sex and virulence in *Escherichia coli*: an evolutionary perspective. Mol Microbiol 60:1136–1151. https://doi.org/10.1111/j.1365-2958.2006.05172.x

Yahara K, Méric G, Taylor AJ, de Vries SPW, Murray S, Pascoe B, Mageiros L, Torralbo A, Vidal A, Ridley A, Komukai S, Wimalarathna H, Cody AJ, Colles FM, McCarthy N, Harris D, Bray JE, Jolley KA, Maiden MCJ, Bentley SD, Parkhill J, Bayliss CD, Grant A, Maskell D, Didelot X, Kelly DJ, Sheppard SK (2017) Genome-wide association of functional traits linked with *Campylobacter jejuni* survival from farm to fork. Environ Microbiol 19:361–380. https://doi.org/10.1111/1462-2920.13628

Yan W, Chang N, Taylor DE (1991) Pulsed-field gel electrophoresis of *Campylobacter jejuni* and *Campylobacter coli* genomic DNA and its epidemiologic application. J Infect Dis 163:1068–1072. https://doi.org/10.1093/infdis/163.5.1068

Management Strategies for Prevention of *Campylobacter* Infections Through the Poultry Food Chain: A European Perspective

Thomas Alter and Felix Reich

Contents

Abstract Numerous studies point out that at present, a complete elimination of *Campylobacter* species in the poultry food chain is not feasible. Thus, the current aim should be to establish control measures and intervention strategies to minimize the occurrence of *Campylobacter* spp. in livestock (esp. poultry flocks) and to reduce the quantitative *Campylobacter* burden along the food chain in animals and subsequently in foods. The most effective measures to mitigate *Campylobacter* focus on the primary production stage. Nevertheless, measures applied during slaughter and processing complement the general meat hygiene approaches by reducing fecal contamination during slaughtering and processing and as a consequence help to reduce *Campylobacter* in poultry meat. Such intervention measures at slaughter and processing level would include general hygienic improvements, technological innovations and/or decontamination measures that are applied at single slaughter or processing steps. In particular, approaches that do not focus on a single intervention measure would need to be based on a thorough process of evaluation, and potential combinatory effects have to be modeled and tested. Finally, the education of all stakeholders (including retailers, food handlers and consumers) is required and

T. Alter (✉)
Center for Veterinary Public Health, Institute of Food Safety and Food Hygiene, Free University Berlin, Koenigsweg 69, Berlin 14163, Germany
e-mail: thomas.alter@fu-berlin.de

F. Reich
German Federal Institute for Risk Assessment, Max-Dohrn-Strasse 8-10, Berlin 10589, Germany
e-mail: felix.reich@bfr.bund.de

S. Backert (ed.), *Fighting Campylobacter Infections*, Current Topics in Microbiology and Immunology 431, https://doi.org/10.1007/978-3-030-65481-8_4

will help to increase awareness for the presence of foodborne pathogens in raw meat and meat products and can thus aid in the development of the required good kitchen hygiene.

1 Introduction

Human campylobacteriosis is a foodborne disease strongly associated with poultry and poultry meat (Humphrey et al. 2007). The options to reduce the burden of *Campylobacter* will require evaluating the food chain as a whole (Fig. 1). This includes management options on the farm, which would firstly result in a reduced proportion of slaughter batches carrying *Campylobacter* and could in turn produce meat free of *Campylobacter*. Secondly, if keeping a batch free from *Campylobacter* is not possible, managing should be aimed toward reducing the concentration in the intestines of the birds (Nauta et al. 2009a; Rosenquist et al. 2003). This would in turn reduce the contamination of carcasses during slaughter. At the slaughterhouse, preventive operations should focus on minimizing fecal contamination of carcasses and, when it occurs, to clean off that new contamination efficiently. This will help to reduce the proportion of meat contaminated with high levels of bacteria. Finally, the education of consumers should be adequate to enable consumers to evaluate the potential risk associated with fresh or minimally processed food and to handle it with care in the kitchen to avoid cross-contamination and lower the risk of becoming infected. It is generally agreed that application of control options on farm level is most cost efficient. However, EFSAs updated model (EFSA 2020) resulted in lower estimates of impact than the model used in the previous EFSA opinion (EFSA 2011). In the 2020 EFSA opinion, a 3 $\log_{10}$ CFU *Campylobacter* reduction in broiler cecal concentrations was estimated to reduce the relative risk of human campylobacteriosis

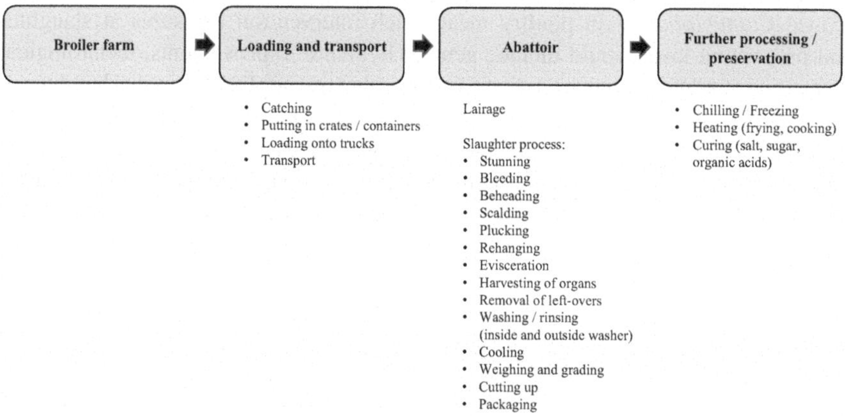

Fig. 1 Steps of the broiler meat production chain

in the EU attributable to broiler meat by 58% compared to an estimate larger than 90% in the previous EFSA opinion (2011).

In recent years, several comprehensive reviews in this area were published, largely focusing on farm level intervention and management practices (EFSA 2020; Nastasijevic et al. 2020; Wales et al. 2019). However, in our opinion, intervention measures on abattoir level and post-harvest approaches should be included when developing management strategies to combat *Campylobacter* infections. This review article is therefore designed to especially focus on harvest and post-harvest intervention strategies.

2 Management Practices and Control Options

2.1 Farm Level

As described above, a number of excellent and comprehensive reviews published within the last years largely focused on farm level approaches describing the impact of different mitigation strategies on *Campylobacter* colonization of poultry on farm. Risk factor analysis for poultry flock colonization demonstrated that increased animal age, the number of houses on the farm, production type, stocking density, flock size, the presence of other animals on the farm, partial depopulation (thinning) or the type of nipple drinkers are associated with the degree of *Campylobacter* colonization (EFSA 2011; Näther et al. 2009). The common denominator explaining most of these risk factors is weak biosecurity or the lack of such measures (Wagenaar et al. 2013). In general, pre-harvest intervention strategies can be divided into three main groups: (i) reduction or elimination of environmental exposure (by biosecurity and hygienic measures), (ii) combating *Campylobacter* colonization and minimizing the bacterial load (by, e.g., application of bacteriocins or bacteriophages) and iii) improving host resistance (by vaccination, probiotic application, competitive exclusion, stimulating the immune system, genetic selection). High biosecurity and hygiene levels on poultry farms can prevent or at least reduce the introduction of *Campylobacter* into a poultry flock. However, this does not guarantee a *Campylobacter*-free flock at slaughter. If consequently applied, hygiene barriers can contribute massively to reduce the risk of poultry colonization by *Campylobacter* (Gibbens et al. 2001; Newell et al. 2011). Single biosecurity measures are well investigated, e.g., fly screens/insect controls, rodent controls, hygienic anteroom designs, effective cleaning and disinfection, clean litter. Nonetheless, it has yet to be demonstrated if such data, usually generated under specific geographical (climatic) and structural (farming) conditions, can be generalized or transferred as such to other countries or regions, since prevalence of *Campylobacter*, climatic conditions and poultry production systems may differ. For example, data on the application of fly screens were intensively generated in the Nordic countries (Hald et al. 2007). Additionally, data on the combined effect of different biosecurity measures are still lacking. Furthermore, some biosecurity measures can be in conflict with other goals of sustainable farming, e.g., outdoor farming or free-range farming (Klein et al. 2015).

Complementary to enhancing biosecurity and hygiene, non-biosecurity-based approaches are also required to maximize the reduction of *Campylobacter*-positive flocks at farm level. A growing number of studies is available on the efficacy of specific intervention measures, such as vaccination (Meunier et al. 2017), passive immunization (Cawthraw and Newell 2010; Hermans et al. 2014), bacteriophage application (Connerton et al. 2011; Hammerl et al. 2014; Kittler et al. 2013), the use of probiotic bacteria (e.g., application of *Bacillus* spp., *Lactobacillus* spp., *Enterococcus faecium, Bifidobacterium longum)* (Manes-Lazaro et al. 2017; Sikic Pogacar et al. 2020), competitive exclusion (Schneitz and Hakkinen 2016), bacteriocins (Hansson et al. 2018; Lin 2009; Saint-Cyr et al. 2016) or feed and water additives (e.g., secondary bile acid, short-chain organic acids and medium-chain fatty acids, ferric tyrosine, essential oils, plant extracts, carvacrol, ß-resorcylic acid) (Alrubaye et al. 2019; Guyard-Nicodeme et al. 2016; Kelly et al. 2017; Khattak et al. 2018; Wagle et al. 2017). Unfortunately, most of these studies were performed under experimental and/or laboratory conditions (animal trials with a limited number of birds under non-industry conditions). Consequently, data obtained from large studies performed under industry conditions for most of these measures are still missing. Differences between successful application of single agents under experimental conditions and results in field trials were highlighted for instance by Huneau-Salaun et al. (2018) evaluating the effect of a patented feed additive (ion-exchanged compound) on *Campylobacter* contamination in broilers reared under commercial conditions. Even though successful in experimental studies, the application of feed additives under commercial conditions to poultry did not have a significant impact on *Campylobacter* load in the chicken ceca.

2.2 *Abattoir Level*

Currently, the slaughter process still contributes to intense cross-contamination occurring at different stages of the slaughter line. Heavily contaminated broiler carcasses originate mostly from *Campylobacter*-positive flocks. Such contaminated flocks might additionally act as source of cross-contamination during the slaughter process. Based on the data from the EU *Campylobacter* baseline survey, it was concluded that a *Campylobacter*-colonized broiler batch was about 30 times more likely to have the sampled carcass contaminated with *Campylobacter*, compared to a non-colonized batch, and a higher *Campylobacter* count on carcasses was strongly associated with *Campylobacter* colonization of the batch (EFSA 2010b). EFSA (2020) recently reviewed the literature for the association of *Campylobacter* concentrations in the ceca and on broiler skin or meat after processing: 15 studies were analyzed with six of these studies reporting no significant correlation and nine that found values of the slope of the regression line between 0.21 and 1.15. Based on these studies, the uncertainty of the slope was expressed as a BetaPert distribution with minimum of 0,0 to maximum of 0.7 and a most likely value of 0.27. Even though, slaughter and processing measures appear to be less effective than farm actions in reducing *Campylobacter* in broilers (Skarp et al. 2016), EFSA's (2020) calculations highlight the need to include intervention measures at slaughter and processing level in overall management strategies in order to prevent pathogen colonization and contamination. Table 1 summarizes the processing steps of poultry slaughter where potential intervention approaches can be applied. At the abattoir, the meat is harvested from life poultry, which originate from various farms. Life poultry are transported to the slaughterhouse and in general are slaughtered consecutively in batches consisting of birds of the same age and from the same farm. The microbiological status of a batch is influenced by its origin; thus, batches colonized with *Campylobacter* will introduce the pathogen into the slaughter line. At this stage, the

Table 1 Potential intervention options during poultry slaughter

Processing steps	Approach	References
General slaughter process evaluation	Factor analysis/explanatory variables	Pacholewicz et al. (2016b), Seliwiorstow et al. (2016)
Scalding	Scalding water temperature	Wempe et al. (1983), Yang et al. (2001), Lehner et al. (2014)
	Scald tank design (multiple tanks, counterflow)	Berrang and Dickens (2000)
Plucking	Cloacal plugging	Buhr et al. (2003), Musgrove et al. (1997)
Evisceration	Adjustment of eviszeration machine to bird size	Malher et al. (2011)
(After evisceration)	Sonosteam, hot steam	James et al. (2007), Boysen and Rosenquist (2009), Musavian et al. (2014)
Washing/rinsing (inside and outside washer)	Design	Wang et al. (2018)
	Hot water	Purnell et al. (2004)
	Chlorinated water	Bashor et al. (2004), Berrang and Bailey (2009), Northcutt et al. (2005)
	Electrolyzed water	Northcutt et al. (2007), Wang et al. (2018)
	Sodium hypochlorite solutions	Northcutt et al. (2007)
	High-pressure spray	Giombelli et al. (2015)
Cooling	Dry air versus immersion chilling	Berrang et al. (2008), Huezo et al. (2007)
	Post-chilling water treatment (peracetic acid)	Nagel et al. (2013)

management needs to focus on the processing of batches and the technology applied during slaughtering and meat harvesting to reduce the counts of *Campylobacter* on the meat and as a consequence to reduce the exposure of the consumer from meat contaminated with high levels of *Campylobacter*. This will include considerations on slaughter logistics as well as process technology-related options, from mere modification of the existing equipment up to the implementation of devices with specific antibacterial effect.

2.2.1 Logistic Slaughtering and Scheduling

Logistic slaughtering and scheduling are aimed toward managing poultry batches prior to processing. Logistic slaughtering means separate slaughter of non-colonized batches early, before the slaughter of colonized batches. This should reduce cross-contamination between batches. As a result, the meat harvested from non-colonized batches would be free from *Campylobacter*. It was suggested being an effective additional effect together with implementing increased hygiene in the processing environment in some studies (Sasaki et al. 2013, 2014). The effect on quantitative reduction, which should be the most important achievement in *Campylobacter* mitigation, is very low though. When assessed for its quantitative impact to avoid relevant *Campylobacter* contamination levels on the meat in different studies, only very limited effects were seen. From a practical point of view, the additional effect of logistic slaughtering was considered to be negligible, as in a controlled slaughter process the relevant batch-to-batch contamination would be very limited (Johannessen et al. 2007; Nauta et al. 2009a; Pless et al. 2012). Therefore, it cannot be considered as an effective intervention. Scheduling on the other hand is targeted toward batches with high levels of *Campylobacter*, as the meat of those batches would most likely carry higher contamination levels after slaughtering and processing. As such, batches with a certain *Campylobacter* carriage level would be deviated to treatments, reducing the bacterial load on the meat. This was considered as an efficient management approach when the potential health benefit was put into relation with assumed cost reduction for not having to treat all batches (Havelaar et al. 2007; Nauta and Havelaar 2008). A later evaluation of testing and scheduling resulted in the conclusion that this approach was not as efficient as expected, because there was no reliable correlation found between the *Campylobacter* concentrations in feces or ceca content and the resulting contamination levels on skinned breast meat (Nauta et al. 2009b). The authors further pointed toward different aspects that might have influenced this discrepancy to the earlier assumptions on the efficacy of this approach, which included limitations by model complexity or the understanding of variables like contamination dynamics during slaughter, measurement errors or limitations of the sampling protocol to underestimate within batch variability, etc. Still, they stressed the important need for tests to predict highly contaminated meat (Nauta et al. 2009b). Consequently, the early identification of batches close to slaughter carrying high concentrations of *Campylobacter* is a prerequisite for proper decision making concerning preventive measures, but this would require the availability of quick and reliable tests (Havelaar et al. 2007). As of now, the quantification of *Campylobacter* via direct plating in accordance with the international standard ISO 10272-2:2017 (ISO 2017) is time consuming and requires at least two days for a presumptive estimate of the concentration. A promising approach for faster action is the application of qPCR assays that could even be combined with a live/dead discrimination by applying propidium monoazide (PMA) to identify living bacteria and would allow the quantification of live *Campylobacter* within one working day (Josefsen et al. 2010; Pacholewicz et al. 2019; Stingl et al. 2015).

2.2.2 Influence of Poultry Meat Processing on *Campylobacter* Contamination of Meat

Broiler meat processing is a mostly automated operation in nowadays industry in most parts of the world, although the scale and throughput of abattoirs might be different. Independent of the level of automation, the slaughtering of poultry and processing of carcasses follows a general approach (Löhren 2012). In short, live birds are delivered to the slaughterhouse where animals are stunned and killed by bleeding. Subsequently, carcasses need to be scalded in a water bath before defeathering and cutting of the feet. Birds are then introduced into the evisceration line, where the body cavity is opened, and the viscera are drawn from the birds. After this stage, official meat inspection is in place. When meat is fit for consumption, carcasses are chilled and optional cutting into parts may follow. Line speeds of about 13,000 broilers per hour can be reached (Löhren 2012); modern large throughput plants can even exceed this line speed.

Poultry carry *Campylobacter* in their intestines, and contamination of the feathers and skin already occurs during fattening at the farm. Thus, birds enter the slaughter operation with contamination on the outside (Kotula and Pandya 1995), and high levels of *Campylobacter* are present in the feces and intestinal content of colonized birds (Hansson et al. 2010; Reich et al. 2018). During slaughtering, the contamination of the meat is influenced by the meat processing technology; each step along this line can affect the contamination levels with *Campylobacter*. The scalding step is the application of a hot water bath early in the slaughter line where a reduction in bacterial contamination levels occurs by inactivating *Campylobacter* through heat (Pacholewicz et al. 2015). The reduction effect is larger if the temperature is higher (Yang et al. 2001). An increase in *Campylobacter* counts with varying intensity can be seen after defeathering related to leakage of feces from the cloacae or after evisceration as a consequence of ruptured intestines (Huang et al. 2017; Pacholewicz et al. 2015; Zweifel et al. 2015). For the management options in the slaughterhouse, a consensus derived from quantitative microbiological risk assessments is to focus on the quantitative reduction of *Campylobacter* present on meat. It was identified that lowering the counts of *Campylobacter* on meat would have a much higher effect on reducing human cases of campylobacteriosis than to just lower the prevalence of *Campylobacter* (Nauta et al. 2009a; Rosenquist et al. 2003).

Food business operators processing broiler meat are responsible for their product quality, and they are obliged to test broiler meat for *Campylobacter* at the end of processing, for instance, in Australia, New Zealand, the USA and in the European Union (EU) as part of process hygiene control (EC 2005; FSIS 2015; MPI 2018). Another aspect worth considering is the sampling approach, which is mostly linked with the existing protocols. For example, testing for *Campylobacter* has been introduced in Europe as a hygiene criterion with weekly testing of five samples of neck skin, it is also the only quantitative microbiological criterion implemented for broiler processing until now. So, with the knowledge of *Campylobacter* seasonality, a weekly sampling of one random batch might be of limited value to identify out of control conditions related to the slaughtering process in areas or at times when *Campylobacter* prevalence is rather low, and thus, it might be worth considering to adapt sampling protocols (Reich et al. 2018). Another approach would be to use suitable indicator bacteria like *Escherichia coli,* which was shown to be an adequate candidate to measure the slaughter and processing-related fecal contamination of broiler meat as a substitute for *Campylobacter* (Boysen et al. 2016). *E. coli* is easy to measure and regularly present in batches of broilers. It would therefore be appropriate as an indicator, if the level of contamination changes along processing in the same way as *Campylobacter*, which has been found in several studies. However, if deviations are observed in certain processing steps where the indicator behaves differently, this must be adequately taken into account (Boysen et al. 2016; Pacholewicz et al. 2015; Roccato et al. 2018). Novel technologies such as whole genome sequencing can aid in the identification of suitable indicator bacteria, for instance, by analysis of microbiome changes after processing steps or intervention treatments (Kim et al. 2017). The use of indicator bacteria could be of particular interest in situations where *Campylobacter* prevalence is low and feasible sampling plans are of limited effect, so that process control based on a more widespread bacterial species would be better suited to determine proper process control as mentioned above.

In any case, data from quantitative microbiological testing need to be evaluated with statistical tools for proper decision making. This could include recognizing the limitations of certain sampling plans (Reich et al. 2018) or the determination of the best type of sample to be used to estimate the probability of high levels of contamination to be expected under the current processing practices (Duque et al. 2018). When common processing practices or sampling plans are changed, this has an influence on the interpretation of results. A recent study from the United Kingdom (UK) showed the difference in *Campylobacter* counts when sampling is based on neck skin, breast skin or combinations thereof, with higher counts found in neck skin samples than in breast skin (Hutchison et al. 2019). It was also found that processors in the UK started to remove the neck skin, and it was questionable whether the reported reduction in *Campylobacter* counts was not partly due to the inclusion of less contaminated breast skin in the samples (Hutchison et al. 2019). While results from statutory sampling plans give an indication for when action should be taken, the effort for process control is on the side of the food processor and requires more in-depth analysis of samples in connection with process-related data to evaluate their influence on the contamination of the meat with *Campylobacter*. When evaluating quantitative microbiological data, means are usually calculated. For the evaluation of the data, in addition to the pure consideration of the means, it is important to consider the distribution of the data and to identify and take measures for the higher contaminated meat (Habib et al. 2012). Reducing the counts of *Campylobacter* on the meat at the slaughterhouse level would be possible by several approaches: i) hygienic improvement by optimizing the existing processing technologies, which can lead to a reduction in additional contamination and ii) intervention, which is specifically focused toward pathogen reduction by means of specialized treatments to lower the counts on the meat. The decision on the appropriate solution would also have to include the legal framework as to whether a particular treatment is allowed.

2.2.3 Hygienic Improvement

Microbiological testing is part of the program to verify process control. If limits are consecutively exceeded, quality managers at poultry processing plants are forced to take action in general (EC 2005; FSIS 2013). However, EC does not specify which corrective action to take, and it is up to the food business operator to choose effective corrective actions and implement them. The initial requirement is to improve hygiene during processing. Identifying processing steps at the slaughterhouse that have an influence on the level of *Campylobacter* contamination of the meat is a prerequisite to achieve this goal. Scalding in particular is known for its cross-contamination potential between carcasses, but at the same time the levels of *Campylobacter* on the meat are significantly reduced during this processing step (Duffy et al. 2014). On the other hand, it is important to identify processing steps which lead to increasing contamination levels of the meat. Such critical points at the slaughterhouse resulting in increased levels of *Campylobacter* on carcasses were the plucking or evisceration steps, although this was not observed in all investigated premises in the same way (Pacholewicz et al. 2015; Seliwiorstow et al. 2015). In general, it is needed to identify plant-specific explanatory variables and combine these with microbiological testing to identify the critical processing steps for contamination that are relevant to the particular plant. A plant-specific approach is necessary as differences between plants were observed for instance during the European baseline study (EFSA 2010a). A few examples of risk factor analysis studies on *Campylobacter* contamination levels during poultry processing were published recently (Pacholewicz et al. 2016b; Seliwiorstow et al. 2016). The approach was based on explanatory variables of the plant but also included batch-related components. The results of Seliwiorstow et al. (2016) pointed toward different technical aspects influencing the contamination of the meat, like the type of unloading systems used, electrical stunning and scalding temperature but also identified different steps along the evisceration line. The slaughter batch-related factors included uniformity of bird size or weight, and internal or external load of *Campylobacter* on arrival at the plant. In particular, Seliwiorstow et al. (2016) recommended the use of drawer unloading systems and the use of gas stunning and to further evaluate to highest possible scalding temperature. Further, they pointed toward the importance to have a good evisceration control with adjusting settings best suited to the batch slaughtered as effective to lower *Campylobacter* levels in broilers after evaluation of six Belgian broiler slaughterhouses. In a study from the Netherlands, Pacholewicz et al. (2016b) pointed toward different relevant factors for each of the two slaughterhouses visited. After relevant factors at a slaughterhouse are identified, observational studies are necessary to establish appropriate changes to processing steps for lowering the contamination of the meat. An example was given in a recent study for evaluation of the evisceration operation in a broiler slaughterhouse, with a specific focus on compliance of slaughterhouse staff (Pacholewicz et al. 2016a). Indeed, the presence of visual fecal contamination was more often recognized in one slaughterhouse together with higher bacterial contamination of the meat where the employees were less complying with required evisceration process control. It was thus suggested that improving compliance at this stage would lead to a reduction in the bacterial contamination of the meat (Pacholewicz et al. 2016a). Another specific option for hygienic improvement is to clean carcasses from contamination by washing. Besides the standard inside outside washer that is located at the end of processing before chilling, additional washing cabinets can be placed along the processing line where contamination could occur. Studies on their efficiency for lowering bacterial contamination levels on carcasses came to different results (Giombelli et al. 2015; Lehner et al. 2014; Stopforth et al. 2007). Therefore, the implementation of additional devices necessitates inclusion of the whole slaughter process, but also the legal aspect of whether antibacterial chemicals such as chlorine are approved for use in, e.g., water-based sprays.

2.2.4 Intervention Measures During Slaughter and Processing

After evaluation and implementation of possible managerial approaches, specific interventions would represent an additional step to improve food safety (Oyarzabal and Backert 2012). Such measures would be considered one-point measures comparable to a critical control point (CCP) as part of a HACCP based system, where relevant reductions in bacterial counts can be achieved. Such steps would most likely be positioned just before the end of processing, close to portioning and packaging to avoid any recontamination after their application. There are two main types of interventions currently recognized, which include physical or chemical treatment of the meat. Physical intervention is usually the application of temperature change, either by cold or heat treatment of carcasses. In addition, UV light and gamma or x-ray irradiation have been evaluated. For chemical decontamination, application of chlorine or acidic compounds has been tested. Legal requirements might limit the domestic use of some interventions. While application of cold treatment, for instance, freezing or crust freezing or heat treatment, like hot water sprays and washes of chicken carcasses is usually possible, the general treatment of slaughter carcasses or meat with irradiation or chlorine is, for example, forbidden in the EU. Consumer acceptance is also an important factor, with the use of chemical washing or irradiation being the least acceptable (MacRitchie et al. 2014). Therefore, additional measures, in particular irradiation or chemical washes/sprays, should always be considered only after the potential for hygienic improvement has been assessed and should be weighed against the effectiveness of simple water rinses. The latest knowledge on established and innovative intervention approaches and technologies applicable to meat processing and retail, not only in relation to *Campylobacter*, was compiled in several recent publications (Lu et al. 2019; Projahn et al. 2018; Umaraw et al. 2017).

To date, several hygienically relevant processing steps were reported in different studies usually conducted during routine processing operations, as discussed above for the risk factor analysis. This contrasts with studies focusing on new technological interventions, which were often made at pilot plants or on a limited scale or with only small sample numbers. There are usually good reasons for the small sample numbers included in the studies, such as the need to have a standardized trial setup to allow experiments to be repeated and the workload that limits the extent of such a study because of limited resources and funding. On the other hand, it is not necessarily possible to draw direct conclusions about the effectiveness of a tested intervention on the operation in every slaughter plant. The effectiveness of interventions can be influenced by differences between plants, but it is also likely that variability between and within the contamination levels of batches of broilers will also influence the potential impact of intervention measures. It therefore makes sense to evaluate the effectiveness of such interventions in routine operations. The variability of bacterial counts is an important aspect to consider when evaluating the effectiveness of quantitative microbiological interventions. This is of particular importance for methods that cause only small quantitative reductions, as has been shown in an example of *Salmonella* reduction in pig meat production (Duarte et al. 2016). Studies on quantitative mitigation that only showed aforementioned small reductions in bacterial counts are interesting from an academic viewpoint. Their effectiveness alone may not be sufficient to meet the expectations of risk managers and should therefore be evaluated as part of a hurdle concept. In our opinion, such measures should be considered as possible additions in the context of changes in process technology and hygienic optimization based on risk factor analysis. Validation of the efficiency of findings from pilot studies or small-scale studies transferred to commercial slaughterhouses would require large-scale studies during routine processing together with routine testing.

2.3 Post-harvest Level

Poultry meat is processed at the slaughterhouse to kitchen-ready state. This includes the oven-ready broiler or a diverse range of poultry cuts, such as legs, wings or breast meat. Poultry meat is usually packed for sale and is either marketed as fresh, frozen or processed meat. In relation to *Campylobacter*, most focus is on fresh and frozen meat. During processing of fresh poultry meat, there is no treatment or processing step in place that will kill *Campylobacter* reliably, even if the process is well controlled. Storage at refrigeration temperatures for approx. 3–7 days results in a reduction in *Campylobacter* cell count of up to 2.94 CFU/g chicken meat (Bhaduri and Cottrell 2004; Gruntar et al. 2015). In recent years, research focused on the role of viable but not culturable (VBNC) state *Campylobacter*. Non-lethal changes of intrinsic or extrinsic conditions (e.g., refrigeration temperatures, salting, changes in pH) might enable *Campylobacter* to enter a VBNC state (Chaisowwong et al. 2012). For frozen meat on the other hand, it was shown that freezing is an option for reducing *Campylobacter* counts (Rosenquist et al. 2006; Sampers et al. 2010; Zhao et al. 2003). Freeze-thawing cycles significantly reduce the survival, but *C. jejuni* remains viable for at least three freeze-thaw cycles regardless if frozen at −70 or −20 °C (Lee et al. 1998). Different factors, like ice crystal formation, ice nucleation and dehydration, have been implicated in the freeze-induced injury of bacterial cells. Oxidative stress has been shown to contribute to the freeze–thawing induced killing of *Campylobacter* as well (Stead and Park 2000). A majority of studies investigating the susceptibility of *Campylobacter* to meat freezing showed a decrease in live pathogens by 1 to 3 $\log_{10}$ units within the first days of storage (El-Shibiny et al. 2009; Moorhead and Dykes 2002). Application of freezing is rather limited in the EU, as fresh poultry meat must not be frozen at any stage of the processing or sales without losing its marketability as fresh meat (EC 2013).

Options to control *Campylobacter* contamination in fresh poultry meat after processing currently are prone to packaging technologies that were reported to be able to lower counts of *Campylobacter* during storage in some instances and are widely used as a well-established option. Other post-harvest treatments are available but may not be yet widely established because of legal limitations. An in-depth review on post-harvest measures such as modified atmosphere packaging (MAP), irradiation, high hydrostatic pressure, bio-preservation and antimicrobial food packaging with effect on poultry meat microbiota and sensorial properties can be found in Silva et al. (2018).

2.4 Retail and Consumer Phase

An additional approach was chosen in the UK, by putting pressure explicitly on the retail level. The Food Standards Agency (FSA) encouraged major retailers to publish *Campylobacter* contamination data at retail level. According to Wales et al. (2019) this seems to have changed the willingness to adopt incentive schemes and innovation in producer practices in the UK. In our opinion, these data and that approach are very promising and could be adopted by other countries. Even though multiplication of *Campylobacter* at proper storage conditions at retail (storing meat frozen or chilled) can be excluded, handling of fresh meat at butchers shops and external contamination of chicken packaging should be considered as other potential transmission pathways. These aspects were highlighted by Harrison et al. (2001) and Burgess et al. (2005).

At the consumer level, meat is prepared in a kitchen environment to prepare whole meals usually containing different ingredients. Kitchen hygiene has a major influence on the level of exposure to *Campylobacter*. When live *Campylobacter* bacteria are present on the meat, contamination of the kitchen environment is possible during handling and can subsequently result in cross-contamination of other foods which has been considered an important source for infection with *Campylobacter*, with even higher effects on the rate of human campylobacteriosis than meat that was not thoroughly cooked (Luber 2009). Appropriate handling of raw meat is thus important and requires knowledge along all steps of the consumer phase, which includes proper transport of raw meat after shopping to the actual handling in the kitchen to avoid bacterial cross-contamination by washing hands and equipment in the right way. Thus, consumer education is a tool to raise awareness of the fact that raw food may contain pathogenic bacteria, and this should encourage people to handle such food with care to help reduce foodborne diseases that arise from the home environment. Various studies have shown that there are gaps in consumer knowledge of how to handle raw chicken meat, which could lead to an increased risk of illness (Bearth et al. 2014; Katiyo et al. 2019; Koppel et al. 2015). Considerable knowledge gaps were, for example, identified in a recent study from Germany, where 68.3% of the respondents had never heard of *Campylobacter*. Although 20.2% had heard of *Campylobacter,* they did not know how to protect themselves, and only 11.5% said they knew how to protect themselves from *Campylobacter*. Slightly more than half (52.2%) of the respondents who at least had heard of *Campylobacter* knew that *Campylobacter* was transmissible via meat (Henke et al. 2020). Additionally, a study from South Africa showed that a considerable part of consumers is unaware of the correct storage temperature of chicken meat, and, also knowledge and practices in kitchen hygiene were lacking in almost two-thirds of study participants (Katiyo et al. 2019). In a study from Switzerland, an expert consultation resulted in a questionnaire addressed toward consumers to test their awareness and knowledge about *Campylobacter* and foodborne hazards related to poultry meat (Bearth et al. 2014). According to experts, relevant aspects to avoid *Campylobacter* infection by handling raw poultry meat in the kitchen were avoiding cross-contamination, high standards for handwashing and washing of kitchen utensils, omitting of rinsing poultry meat under tap water, proper storage and thorough heating to 70 °C. Consumer knowledge evaluation, based on questionnaires, showed that the risk of pathogenic bacteria on meat was ranked as the lowest in this Swiss study. Further on, eating food prepared at restaurants was perceived a higher risk compared to homemade meals. Consequently, the study classified three major groups of consumers “unsafe cooks,” “intermediate cooks” and “safe cooks.” Interestingly, both the unsafe and intermediate cooks showed the same risk perception related to their own cooking. “Unsafe cooks” on the other hand showed considerable deviations from expert-based good habits especially in regard to avoiding cross-contamination. These cooks were not aware of their lack of knowledge and would benefit most from risk communication. The main goals to be achieved would have to address knowledge gaps and increase the personal risk perception. Due to the low interest to search for information, it might be difficult to reach that group of cooks, “intermediate cooks” showed deviations in some cases, would need education to overcome certain behavior and might be easier to target, as it was assumed that this group would have more interest in learning from new information. Lastly, the “safe cooks” showed overall good risk perception and behavior but shared the general misperception that organic or locally produced poultry meat would be generally safer and less contaminated compared to other sources (Bearth et al. 2014). The general need is to increase awareness of the presence of foodborne risks, as the presence of *Campylobacter* on meat is not well known. Furthermore, there is a need to educate about deficiencies in kitchen hygiene and bad habits, as they were common in the two studies mentioned above. Similar shortcomings were also found in several other studies from Canada (Murray et al. 2017), Germany (Henke et al. 2020), Slovenia (Sternisa et al. 2018) or the USA (Bruhn 2014). Effective education would need to be specifically tailored to vulnerable groups or groups by level of knowledge, and research needs to be done on how best to address different target groups (Kosa et al. 2019). Most studies agree that higher education levels (Carbas et al. 2013; Henke et al. 2020; Samapundo et al. 2016), higher household incomes (Lin et al. 2005) and older people (Henke et al. 2020; Zorba and Kaptan 2011) are significantly better informed about *Campylobacter* and kitchen hygiene measures. Effective education could be combined with food safety labeling of chicken meat to raise consumer awareness directly at the point of sale and was considered a valuable but underused tool to inform consumers about the food safety risk posed by *Campylobacter* in relation with fresh chicken meat (Allan et al. 2018). When transferring food safety messages to consumers, it has to be kept in mind that consumer practices are habitual and motivated by other needs than safety as well (Langsrud et al. 2020). Besides the proper handling of fresh meat to avoid cross-contamination, the second most important factor for food safety is thorough cooking at consumer level. The correct judgment of meat doneness is not always easy when relying on color alone and can be influenced by the light source in the kitchen (Maughan et al. 2019). The use of thermometers is recommended for making sure the meat is thoroughly cooked (Chambers et al. 2018; Maughan et al. 2019). Although their use is well known, application of meat thermometers may be limited by habitual practices or personal attitude and should also be addressed in consumer education (Feng and Bruhn 2019).

2.5 Management and Responsibilities

The food business operator is responsible for supplying safe food to the consumer. When a food is not safe, by rule, the food business operator must take measures based on corresponding legislation. To control *Campylobacter* at the production level, legislation requires testing at the poultry slaughterhouse as mentioned above. Several studies investigated the willingness of stakeholders of the food production chain to take preventive measures. In an online survey in Belgium by Lupo et al. (2016) among stakeholders of the food production chain, 60% of the respondents identified microbial pathogens as the main hazard. When asked about responsibility for risk prevention, respondents mentioned competent authorities (75%), food business operators (62%) and sector representatives (54%). The majority of participants stated that preventive measures should be made compulsory (75%) (Lupo et al. 2016). Wijnen et al. (2019) studied the awareness of abattoir workers and their willingness to take measure to mitigate bacterial foodborne hazards in the meat production chain in Belgium and the Netherlands. There was awareness of *Campylobacter* as an important foodborne hazard in the majority of poultry slaughterhouses (91%, $n = 23$), but only nine out of 23 abattoirs thought that additional preventive measures could reduce the pathogen load. Although they felt that additional measures were necessary (19 out of 23), poultry slaughterhouses were less willing to implement such additional measures compared to slaughterhouses for ruminants, for example. When asked about the responsibility for implementing preventive measures, 12 out of 23 pointed toward government bodies, while nine addressed the slaughterhouse or the sector as a whole and two preferred both. Other factors influencing the willingness of industry to implement measures were a scientifically proven effectiveness of such measures or evidence that they are applicable and feasible (Wijnen et al. 2019). On the basis of the studies cited, it can be concluded that there should be sufficient awareness of the relevant foodborne hazards among the consulted stakeholders, but it became clear that cooperation with official or governmental bodies is important for the implementation of preventive measures and that appropriate incentives need to be identified and some compulsory requirements will be necessary. Any intervention to reduce consumer exposure to *Campylobacter* will have costs (this includes risk factor analyses at the farm or at the slaughterhouse leading to applying targeted hygiene measures, process verification through microbiological testing or applying specific interventions to reduce the bacterial load on meat). These financial costs will largely arise at the level of meat production. Measures leading to fewer cases of campylobacteriosis will result in cost savings at the public health and economic sectors. The question is how such costs will be shared between the sectors in the future and if consumers will accept the intervention measures applied in the poultry production chain. Regarding the acceptable costs, only limited data are available from consumer surveys. In an older study by Gilbert and Cressey (2008), a quarter of the respondents were willing to pay a 10–20% premium on safe chicken (by improved on-farm biosecurity).

Consumer's perception of different intervention strategies was currently summarized by Nastasijevic et al. (2020). The authors identified four factors that can influence consumers' perception toward intervention strategies: (i) the level of concern associated with individual intervention strategies (e.g., controversial methods such as irradiation), (ii) public awareness, (iii) willingness to voluntarily accept the method, (iv) extend of the consequences for the consumers, if a specific intervention method was not applied. In general, the least favorable measure for the interviewed consumers was decontamination of carcasses, and the most favorable was stricter farm management.

3 Concluding Remarks

Combinations of management and intervention strategies on the pre-harvest and harvest level of poultry are important to limit the *Campylobacter* load finally present on the meat. Although this review is partly based on European requirements, the approaches to be adopted are similar in different countries and production types. Based on the approach chosen by the EU, a baseline study is first needed to assess the prevalence of *Campylobacter* in production and then to identify risk factors associated with *Campylobacter* transmission in meat production from farm to fork. The application of best practice, together with the optimization of biosecurity on the farm and hygienic procedures in the slaughterhouse, must form the basis. Specific interventions and treatments can be implemented in line with local legislation to mitigate or reduce the residual risk. Finally, the education of retailers, food handlers and the consumers will help to increase awareness for the presence of foodborne pathogens in raw meat and can thus aid in the development of good habits in the kitchen and in the preparation of meals at home.

Acknowledgements This work is supported by the German Federal Ministry of Education and Research (BMBF) in frame of the zoonoses research consortium PAC-Campylobacter to TA (01KI1725A) and FR (project 01KI1725B).

References

Allan PD, Palmer C, Chan F, Lyons R, Nicholson O, Rose M, Hales S, Baker MG (2018) Food safety labelling of chicken to prevent *campylobacteriosis*: consumer expectations and current practices. BMC Public Health 18:414. https://doi.org/10.1186/s12889-018-5322-z

Alrubaye B, Abraha M, Almansour A, Bansal M, Wang H, Kwon YM, Huang Y, Hargis B, Sun X (2019) Microbial metabolite deoxycholic acid shapes microbiota against *Campylobacter jejuni* chicken colonization. PLoS ONE 14:e0214705. https://doi.org/10.1371/journal.pone.0214705

Bashor MP, Curtis PA, Keener KM, Sheldon BW, Kathariou S, Osborne JA (2004) Effects of carcass washers on *Campylobacter* contamination in large broiler processing plants. Poult Sci 83:1232–1239. https://doi.org/10.1093/ps/83.7.1232

Bearth A, Cousin ME, Siegrist M (2014) Poultry consumers' behaviour, risk perception and knowledge related to campylobacteriosis and domestic food safety. Food Control 44:166–176. https://doi.org/10.1016/j.foodcont.2014.03.055

Berrang ME, Bailey JS (2009) On-line brush and spray washers to lower numbers of *Campylobacter* and *Escherichia coli* and presence of *Salmonella* on broiler carcasses during processing. J Appl Poultry Res 18:74–78. https://doi.org/10.3382/japr.2008-00067

Berrang ME, Dickens JA (2000) Presence and level of *Campylobacter* spp. on broiler carcasses throughout the processing plant. J Appl Poultry Res 9:43–47. https://doi.org/10.1093/japr/9.1.43

Berrang ME, Meinersmann RJ, Smith DP, Zhuang H (2008) The effect of chilling in cold air or ice water on the microbiological quality of broiler carcasses and the population of *Campylobacter*. Poult Sci 87:992–998. https://doi.org/10.3382/ps.2007-00406

Bhaduri S, Cottrell B (2004) Survival of cold-stressed *Campylobacter jejuni* on ground chicken and chicken skin during frozen storage. Appl Environ Microbiol 70:7103–7109. https://dx.doi.org/10.1128%2FAEM.70.12.7103-7109.2004

Boysen L, Rosenquist H (2009) Reduction of thermotolerant *Campylobacter* species on broiler carcasses following physical decontamination at slaughter. J Food Prot 72:497–502. https://doi.org/10.4315/0362-028X-72.3.497

Boysen L, Nauta M, Rosenquist H (2016) *Campylobacter* spp. and *Escherichia coli* contamination of broiler carcasses across the slaughter line in Danish slaughterhouses. Microbial Risk Analysis 2–3:63–67. https://doi.org/10.1016/j.mran.2016.05.005

Bruhn CM (2014) Chicken preparation in the home: an observational study. Food Prot Trends 34:318–330

Buhr RJ, Berrang ME, Cason JA (2003) Bacterial recovery from breast skin of genetically feathered and featherless broiler carcasses immediately following scalding and picking. Poult Sci 82:1641–1647. https://doi.org/10.1093/ps/82.10.1641

Burgess F, Little CL, Allen G, Williamson K, Mitchelli RT (2005) Prevalence of *Campylobacter*, *Salmonella*, and *Escherichia coli* on the external packaging of raw meat. J Food Prot 68:469–475. https://doi.org/10.4315/0362-028x-68.3.469

Carbas B, Cardoso L, Coelho AC (2013) Investigation on the knowledge associated with foodborne diseases in consumers of northeastern Portugal. Food Control 30:54–57. https://doi.org/10.1016/j.foodcont.2012.06.028

Cawthraw SA, Newell DG (2010) Investigation of the presence and protective effects of maternal antibodies against *Campylobacter jejuni* in chickens. Avian Dis 54:86–93. https://doi.org/10.1637/9004-072709-Reg.1

Chaisowwong W, Kusumoto A, Hashimoto M, Harada T, Maklon K, Kawamoto K (2012) Physiological characterization of *Campylobacter jejuni* under cold stresses conditions: its potential for public threat. J Vet Med Sci 74:43–50. https://doi.org/10.1292/jvms.11-0305

Chambers E, Godwin S, Terry T (2018) Recipes for determining doneness in poultry do not provide appropriate information based on US government guidelines. Foods 7:126. https://doi.org/10.3390/foods7080126

Connerton PL, Timms AR, Connerton IF (2011) *Campylobacter* bacteriophages and bacteriophage therapy. J Appl Microbiol 111:255–265. https://doi.org/10.1111/j.1365-2672.2011.05012.x

Duarte ASR, Nauta MJ, Aabo S (2016) Variation in the effect of carcass decontamination impacts the risk for consumers. Food Control 59:12–19. https://doi.org/10.1016/j.foodcont.2015.05.015

Duffy LL, Blackall PJ, Cobbold RN, Fegan N (2014) Quantitative effects of in-line operations on *Campylobacter* and *Escherichia coli* through two Australian broiler processing plants. Int J Food Microbiol 188:128–134. https://doi.org/10.1016/j.ijfoodmicro.2014.07.024

Duque B, Daviaud S, Guillou S, Haddad N, Membre JM (2018) Quantification of *Campylobacter jejuni* contamination on chicken carcasses in France. Food Res Int 106:1077–1085. https://doi.org/10.1016/j.foodres.2017.12.017

EC (2005) Commission Regulation (EC) No 2073/2005 on microbiological criteria for foodstuffs. Official Journal of the European Union L 338/1–26

EC (2013) Commission Regulation (EU) No 1308/2013 establishing a common organisation of the markets in agricultural products and repealing Council Regulations (EEC) No 922/72, (EEC) No 234/79, (EC) No 1037/2001 and (EC) No 1234/2007. Official Journal of the European Union L 347/671–854

EFSA (2010a) Analysis of the baseline survey on the prevalence of *Campylobacter* in broiler batches and of *Campylobacter* and *Salmonella* on broiler carcasses in the EU, 2008, Part B: analysis of factors associated with *Campylobacter* colonisation of broiler batches and with *Campylobacter* contamination of broiler carcasses; and investigation of the culture method diagnostic characteristics used to analyse broiler carcass samples. EFSA J 8:8. https://doi.org/10.2903/j.efsa.2010.1522

EFSA (2010b) Scientific opionion on quantification of the risk posed by broiler meat to human campylobacteriosis in the EU. EFSA J 8:1. https://doi.org/10.2903/j.efsa.2010.1437

EFSA (2011) Scientific Opinion on *Campylobacter* in broiler meat production: control options and performance obejctives and/or targets at different stages of the food chain. EFSA J 9:4. https://doi.org/10.2903/j.efsa.2011.2105

EFSA (2020) Update and review of control options for *Campylobacter* in broilers at primary production. EFSA J 18:4. https://doi.org/10.2903/j.efsa.2020.6090

El-Shibiny A, Connerton P, Connerton I (2009) Survival at refrigeration and freezing temperatures of *Campylobacter coli* and *Campylobacter jejuni* on chicken skin applied as axenic and mixed inoculums. Int J Food Microbiol 131:197–202. https://doi.org/10.1016/j.ijfoodmicro.2009.02.024

Feng YH, Bruhn CM (2019) Motivators and barriers to cooking and refrigerator thermometer use among consumers and food workers: a review. J Food Prot 82:128–150. https://doi.org/10.4315/0362-028X.JFP-18-245

FSIS (2013) *Salmonella* and *Campylobacter* verification program for raw meat and poultry products. FSIS Directive 10,250.1. United States Department of Agriculture, Food Safety and Inspection Service

FSIS (2015) Changes to the *Salmonella* and *Campylobacter* verification testing program: Proposed performance standards for *Salmonella* and *Campylobacter* in not-ready-to-eat comminuted chicken and turkey products and raw chicken parts and related agency verification procedures and other changes to agency sampling. United States Department of Agriculture, Food Safety and Inspection Service

Gibbens JC, Pascoe SJS, Evans SJ, Davies RH, Sayers AR (2001) A trial of biosecurity as a means to control *Campylobacter* infection of broiler chickens. Prevent Veteri Med 48:85–99. https://doi.org/10.1016/s0167-5877(00)00189-6

Gilbert S, Cressey P (2008) Consumer knowledge, attitudes and beliefs with respect to *Campylobacter* in poultry and food-borne campylobacteriosis in humans. https://www.mpi.govt.nz/dmsdocument/25763-sis-consumer-knowledge-attitudes-and-beliefs-with-respect-to-campylobacter-in-poultry-and-food-borne-campylobacteriosis-in-humans

Giombelli A, Hammerschmitt D, Cerutti MF, Chiarini E, Landgraf M, Franco B, Destro MT (2015) High pressure spray with water shows similar efficiency to trimming in controlling microorganisms on poultry carcasses. Poult Sci 94:2589–2595. https://doi.org/10.3382/ps/pev235

Gruntar I, Biasizzo M, Kusar D, Pate M, Ocepek M (2015) *Campylobacter jejuni* contamination of broiler carcasses: population dynamics and genetic profiles at slaughterhouse level. Food Microbiol 50:97–101. https://doi.org/10.1016/j.fm.2015.03.007

Guyard-Nicodeme M, Keita A, Quesne S, Amelot M, Poezevara T, Le Berre B, Sanchez J, Vesseur P, Martin A, Medel P, Chemaly M (2016) Efficacy of feed additives against *Campylobacter* in live broilers during the entire rearing period. Poult Sci 95:298–305. https://doi.org/10.3382/ps/pev303

Habib I, Berkvens D, De Zutter L, Dierick K, Van Huffel X, Speybroeck N, Geeraerd AH, Uyttendaele M (2012) *Campylobacter* contamination in broiler carcasses and correlation with slaughterhouses operational hygiene inspection. Food Microbiol 29:105–112. https://doi.org/10.1016/j.fm.2011.09.004

Hald B, Sommer HM, Skovgard H (2007) Use of fly screens to reduce *Campylobacter* spp. introduction in broiler houses. Emerg Infect Dis 13:1951–1953. https://doi.org/10.3201/eid1312.070488

Hammerl JA, Jackel C, Alter T, Janzcyk P, Stingl K, Knuver MT, Hertwig S (2014) Reduction of *Campylobacter jejuni* in broiler chicken by successive application of Group II and Group III phages. PLoS ONE 9:e114785. https://doi.org/10.1371/journal.pone.0114785

Hansson I, Pudas N, Harbom B, Engvall EO (2010) Within-flock variations of *Campylobacter* loads in ceca and on carcasses from broilers. Int J Food Microbiol 141:51–55. https://doi.org/10.1016/j.ijfoodmicro.2010.04.019

Hansson I, Sandberg M, Habib I, Lowman R, Engvall EO (2018) Knowledge gaps in control of *Campylobacter* for prevention of campylobacteriosis. Transbound Emerg Dis 65:30–48. https://doi.org/10.1111/tbed.12870

Harrison WA, Griffith CJ, Tennant D, Peters AC (2001) Incidence of *Campylobacter* and *Salmonella* isolated from retail chicken and associated packaging in South Wales. Lett Appl Microbiol 33:450–454. https://doi.org/10.1046/j.1472-765x.2001.01031.x

Havelaar AH, Mange MJ, de Koeijer AA, Bogaardt MJ, Evers EG, Jacobs-Reitsma WF, van Pelt W, Wagenaar JA, de Wit GA, van der Zee H, Nauta MJ (2007) Effectiveness and efficiency of controlling *Campylobacter* on broiler chicken meat. Risk Anal 27:831–844. https://doi.org/10.1111/j.1539-6924.2007.00926.x

Henke KA, Alter T, Doherr MG, Merle R (2020) Comparison of consumer knowledge about *Campylobacter*, *Salmonella* and *Toxoplasma* and their transmissibility via meat: results of a consumer study in Germany. BMC Public Health 20:336. https://doi.org/10.1186/s12889-020-08476-0

Hermans D, Van Steendam K, Verbrugghe E, Verlinden M, Martel A, Seliwiorstow T, Heyndrickx M, Haesebrouck F, De Zutter L, Deforce D, Pasmans F (2014) Passive immunization to reduce *Campylobacter jejuni* colonization and transmission in broiler chickens. Vet Res 45:27. https://doi.org/10.1186/1297-9716-45-27

Huang JL, Zang XQ, Zhai WH, Guan CA, Lei TY, Jiao XN (2017) Quantitative analysis of *Campylobacter* spp. contamination in chicken slaughtering lines by "label tracking method" in eastern China. Food Control 80:67–73. https://doi.org/10.1016/j.foodcont.2017.03.052

Huezo R, Northcutt JK, Smith DP, Fletcher DL, Ingram KD (2007) Effect of dry air or immersion chilling on recovery of bacteria from broiler carcasses. J Food Prot 70:1829–1834. https://doi.org/10.4315/0362-028x-70.8.1829

Humphrey T, O'Brien S, Madsen M (2007) Campylobacters as zoonotic pathogens: a food production perspective. Int J Food Microbiol 117:237–257. https://doi.org/10.1016/j.ijfoodmicro.2007.01.006

Huneau-Salaun A, Guyard-Nicodeme M, Benzoni G, Gautier X, Quesne S, Poezevara T, Chemaly M (2018) Randomized control trial to test the effect of a feed additive on *Campylobacter* contamination in commercial broiler flocks up to slaughter. Zoonoses Public Health 65:404–411. https://doi.org/10.1111/zph.12447

Hutchison ML, Tchorzewska MA, Harrison D, Madden RH, Corry JEL (2019) Consequences of using two types of skin samples from chilled chicken broiler carcasses to measure the degree of contamination by *Campylobacter* spp. J Food Prot 82:1124–1129. https://doi.org/10.4315/0362-028X.JFP-18-559

ISO (2017) ISO 10272-2:2017 Microbiology of the food chain—Horizontal method for detection and enumeration of *Campylobacter* spp.—Part 2: Colony-count technique

James C, James SJ, Hannay N, Purnell G, Barbedo-Pinto C, Yaman H, Araujo M, Gonzalez ML, Calvo J, Howell M, Corry JE (2007) Decontamination of poultry carcasses using steam or hot water in combination with rapid cooling, chilling or freezing of carcass surfaces. Int J Food Microbiol 114:195–203. https://doi.org/10.1016/j.ijfoodmicro.2006.09.019

Johannessen GS, Johnsen G, Okland M, Cudjoe KS, Hofshagen M (2007) Enumeration of thermotolerant *Campylobacter* spp. from poultry carcasses at the end of the slaughter-line. Lett Appl Microbiol 44:92–97. https://doi.org/10.1111/j.1472-765X.2006.02026.x

Josefsen MH, Lofstrom C, Hansen TB, Christensen LS, Olsen JE, Hoorfar J (2010) Rapid quantification of viable *Campylobacter* bacteria on chicken carcasses, using real-time PCR and propidium monoazide treatment, as a tool for quantitative risk assessment. Appl Environ Microbiol 76:5097–5104. https://doi.org/10.1128/AEM.00411-10

Katiyo W, de Kock HL, Coorey R, Buys EM (2019) Assessment of safety risks associated with handling chicken as based on practices and knowledge of a group of South African consumers. Food Control 101:104–111. https://doi.org/10.1016/j.foodcont.2019.02.027

Kelly C, Gundogdu O, Pircalabioru G, Cean A, Scates P, Linton M, Pinkerton L, Magowan E, Stef L, Simiz E, Pet I, Stewart S, Stabler R, Wren B, Dorrell N, Corcionivoschi N (2017) The *in vitro* and *in vivo* effect of carvacrol in preventing *Campylobacter* infection, colonization and in improving productivity of chicken broilers. Foodborne Pathog Dis 14:341–349. https://doi.org/10.1089/fpd.2016.2265

Khattak F, Paschalis V, Green M, Houdijk JGM, Soultanas P, Mahdavi J (2018) TYPLEX® Chelate, a novel feed additive, inhibits *Campylobacter jejuni* biofilm formation and cecal colonization in broiler chickens. Poult Sci 97:1391–1399. https://doi.org/10.3382/ps/pex413

Kim SA, Park SH, Lee SI, Owens CM, Ricke SC (2017) Assessment of chicken carcass microbiome responses during processing in the presence of commercial antimicrobials using a next generation sequencing approach. Sci Rep 7:43354. https://doi.org/10.1038/srep43354

Kittler S, Fischer S, Abdulmawjood A, Glunder G, Klein G (2013) Effect of bacteriophage application on *Campylobacter jejuni* loads in commercial broiler flocks. Appl Environ Microbiol 79:7525–7533. https://doi.org/10.1128/AEM.02703-13

Klein G, Jansen W, Kittler S, Reich F (2015) Mitigation strategies for *Campylobacter* spp. in broiler at pre-harvest and harvest level. Berl Munch Tierarztl Wochenschr 128:132–140 PMID: 25876273

Koppel K, Timberg L, Shalimov R, Vazquez-Araujo L, Carbonell-Barracchina AA, Di Donfrancesco B, Chambers E (2015) Purchase, storage, and preparation of eggs and poultry in selected European countries. British Food Journal 117:749–765. https://doi.org/10.1108/BFJ-01-2014-0021

Kosa KM, Cates SC, Brophy J, Godwin S, Chambers D, Chambers E (2019) Older adults and parents of young children have different handling practices for raw poultry. J Food Prot 82:200–206. https://doi.org/10.4315/0362-028X.JFP-18-323

Kotula KL, Pandya Y (1995) Bacterial contamination of broiler chickens before scalding. J Food Prot 58:1326–1329. https://doi.org/10.4315/0362-028X-58.12.1326

Langsrud S, Sørheim O, Skuland SE, Almli VL, Jensen MR, Grøvlen MS, Ueland Ø, Møretrø T (2020) Cooking chicken at home: common or recommended approaches to judge doneness may not assure sufficient inactivation of pathogens. PLoS ONE 15:e0230928. https://doi.org/10.1371/journal.pone.0230928

Lee A, Smith SC, Coloe PJ (1998) Survival and growth of *Campylobacter jejuni* after artificial inoculation onto chicken skin as a function of temperature and packaging conditions. J Food Prot 61:1609–1614. https://doi.org/10.4315/0362-028X-61.12.1609

Lehner Y, Reich F, Klein G (2014) Influence of process parameter on *Campylobacter* spp. counts on poultry meat in a slaughterhouse environment. Curr Microbiol 69:240–244. https://doi.org/10.1007/s00284-014-0575-y

Lin CTJ, Jensen KL, Yen ST (2005) Awareness of foodborne pathogens among US consumers. Food Qual Prefer 16:401–412. https://doi.org/10.1016/j.foodqual.2004.07.001

Lin J (2009) Novel approaches for *Campylobacter* control in poultry. Foodborne Pathog Dis 6:755–765. https://doi.org/10.1089/fpd.2008.0247

Löhren U (2012) Overview on current practices of poultry slaughtering and poultry meat inspection. Supporting Publications EN-298. https://doi.org/10.2903/sp.efsa.2012.EN-298

Lu T, Marmion M, Ferone M, Wall P, Scannell AGM (2019) Processing and retail strategies to minimize *Campylobacter* contamination in retail chicken. J Food Process Preserv 43:0001. https://doi.org/10.1111/jfpp.14251

Luber P (2009) Cross-contamination versus undercooking of poultry meat or eggs—which risks need to be managed first? Int J Food Microbiol 134:21–28. https://doi.org/10.1016/j.ijfoodmicro.2009.02.012

Lupo C, Wilmart O, Van Huffel X, Dal Pozzo F, Saegerman C (2016) Stakeholders' perceptions, attitudes and practices towards risk prevention in the food chain. Food Control 66:158–165. https://doi.org/10.1016/j.foodcont.2016.02.003

Malher X, Simon M, Charnay V, des Deserts RD, Lehebel A, Belloc C (2011) Factors associated with carcass contamination by *Campylobacter* at slaughterhouse in cecal-carrier broilers. Int J Food Microbiol 150:8–13. https://doi.org/10.1016/j.ijfoodmicro.2011.07.007

Manes-Lazaro R, Van Diemen PM, Pin C, Mayer MJ, Stevens MP, Narbad A (2017) Administration of *Lactobacillus johnsonii* FI9785 to chickens affects colonisation by *Campylobacter jejuni* and the intestinal microbiota. Brit Poult Sci 58:373–381. https://doi.org/10.1080/00071668.2017.1307322

MacRitchie LA, Hunter CJ, Strachan NJC (2014) Consumer acceptability of interventions to reduce *Campylobacter* in the poultry food chain. Food Control 35:260–266. https://doi.org/10.1016/j.foodcont.2013.06.005

Maughan C, Chambers E, Godwin S, Chambers D (2019) Changes in lighting source can produce inaccurate assessment of visual poultry doneness and induce consumers to eat undercooked ground turkey patties. J Food Prot 82:528–534. https://doi.org/10.4315/0362-028X.JFP-18-392

Meunier M, Guyard-Nicodeme M, Vigouroux E, Poezevara T, Beven V, Quesne S, Bigault L, Amelot M, Dory D, Chemaly M (2017) Promising new vaccine candidates against *Campylobacter* in broilers. PLoS ONE 12:e0188472. https://doi.org/10.1371/journal.pone.0188472

Moorhead SM, Dykes GA (2002) Survival of *Campylobacter jejuni* on beef trimmings during freezing and frozen storage. Lett Appl Microbiol 34:72–76. https://doi.org/10.1046/j.1472-765x.2002.01043.x

MPI (2018) Specifications for National Microbiological Database Programme. Ministry of Primary Industries, New Zealand

Murray R, Glass-Kaastra S, Gardhouse C, Marshall B, Ciampa N, Franklin K, Hurst M, Thomas MK, Nesbitt A (2017) Canadian Consumer Food Safety Practices and Knowledge: Foodbook Study. J Food Prot 80:1711–1718. https://doi.org/10.4315/0362-028X.JFP-17-108

Musavian HS, Krebs NH, Nonboe U, Corry JE, Purnell G (2014) Combined steam and ultrasound treatment of broilers at slaughter: a promising intervention to significantly reduce numbers of naturally occurring campylobacters on carcasses. Int J Food Microbiol 176:23–28. https://doi.org/10.1016/j.ijfoodmicro.2014.02.001

Musgrove MT, Cason JA, Fletcher DL, Stern NJ, Cox NA, Bailey JS (1997) Effect of cloacal plugging on microbial recovery from partially processed broilers. PoultSci 76:530–533. https://doi.org/10.1093/ps/76.3.530

Nagel GM, Bauermeister LJ, Bratcher CL, Singh M, McKee SR (2013) *Salmonella* and *Campylobacter* reduction and quality characteristics of poultry carcasses treated with various antimicrobials in a post-chill immersion tank. Int J Food Microb 165:281–286. https://doi.org/10.1016/j.ijfoodmicro.2013.05.016

Nastasijevic I, Proscia F, Boskovic M, Glisic M, Blagojevic B, Sorgentone S, Kirbis A, Ferri M (2020) The European Union control strategy for *Campylobacter* spp. in the broiler meat chain. J Food Safety 22. https://doi.org/10.1111/jfs.12819

Näther G, Alter T, Martin A, Ellerbroek L (2009) Analysis of risk factors for *Campylobacter* species infection in broiler flocks. Poult Sci 88:1299–1305. https://doi.org/10.3382/ps.2008-00389

Nauta M, Havelaar A (2008) Risk-based standards for *Campylobacter* in the broiler meat chain. Food Control 19:372–381. https://doi.org/10.1016/j.foodcont.2007.04.016

Nauta M, Hill A, Rosenquist H, Brynestad S, Fetsch A, van der Logt P, Fazil A, Christensen B, Katsma E, Borck B, Havelaar A (2009a) A comparison of risk assessments on *Campylobacter* in broiler meat. Int J Food Microbiol 129:107–123. https://doi.org/10.1016/j.ijfoodmicro.2008.12.001

Nauta MJ, van der Wal FJ, Putirulan FF, Post J, van de Kassteele J, Bolder NM (2009b) Evaluation of the "testing and scheduling" strategy for control of *Campylobacter* in broiler meat in The Netherlands. Int J Food Microbiol 134:216–222. https://doi.org/10.1016/j.ijfoodmicro.2009.06.014

Newell DG, Elvers KT, Dopfer D, Hansson I, Jones P, James S, Gittins J, Stern NJ, Davies R, Connerton I, Pearson D, Salvat G, Allen VM (2011) Biosecurity-based interventions and strategies to reduce *Campylobacter* spp. on poultry farms. Appl Environ Microbiol 77:8605–8614. https://doi.org/10.1128/AEM.01090-10

Northcutt J, Smith D, Ingram KD, Hinton A, Musgrove M (2007) Recovery of bacteria from broiler carcasses after spray washing with acidified electrolyzed water or sodium hypochlorite solutions. Poult Sci 86:2239–2244. https://doi.org/10.1093/ps/86.10.2239

Northcutt JK, Smith DP, Musgrove MT, Ingram KD, Hinton A Jr (2005) Microbiological impact of spray washing broiler carcasses using different chlorine concentrations and water temperatures. Poult Sci 84:1648–1652. https://doi.org/10.1093/ps/84.10.1648

Oyarzabal OA, Backert S (2012) Microbial food safety: an introduction (food science text series). Springer, Heidelberg

Pacholewicz E, Barus AS, Swart A, Havelaar AH, Lipman LJA, Luning PA (2016a) Influence of food handlers' compliance with procedures of poultry carcasses contamination: a case study concerning evisceration in broiler slaughterhouses. Food Control 68:367–378. https://doi.org/10.1016/j.foodcont.2016.04.009

Pacholewicz E, Swart A, Schipper M, Gortemaker BG, Wagenaar JA, Havelaar AH, Lipman LJ (2015) A comparison of fluctuations of *Campylobacter* and *Escherichia coli* concentrations on broiler chicken carcasses during processing in two slaughterhouses. Int J Food Microbiol 205:119–127. https://doi.org/10.1016/j.ijfoodmicro.2015.04.006

Pacholewicz EWA, Swart A, Wagenaar JA, Lipman LJA, Havelaar AH (2016b) Explanatory variables associated with *Campylobacter* and *Escherichia coli* concentrations on broiler chicken carcasses during processing in two slaughterhouses. J Food Prot 79:2038–2047. https://doi.org/10.4315/0362-028X.JFP-16-064

Pacholewicz E, Buhler C, Wulsten IF, Kraushaar B, Luu HQ, Iwobi AN, Huber I, Stingl K (2019) Internal sample process control improves cultivation-independent quantification of thermotolerant *Campylobacter*. Food Microbiol 78:53–61. https://doi.org/10.1016/j.fm.2018.09.017

Pless P, Matt M, Wagner P (2012) *Campylobacter* contamination of poultry meat: influence of slaughter logistics and process hygiene on the *Campylobacter* spp. impact of chicken carcasses. Fleischwirtschaft 92:88–92

Projahn M, Pacholewicz E, Becker E, Correia-Carreira G, Bandick N, Kaesbohrer A (2018) Reviewing interventions against *Enterobacteriaceae* in broiler processing: Using old techniques for meeting the new challenges of ESBL *E. coli*? Biomed Res Int. 2018:7309346. https://doi.org/10.1155/2018/7309346

Purnell G, Mattick K, Humphrey T (2004) The use of 'hot wash' treatments to reduce the number of pathogenic and spoilage bacteria on raw retail poultry. J Food Eng 62:29–36. https://doi.org/10.1016/S0260-8774(03)00168-7

Reich F, Valero A, Schill F, Bungenstock L, Klein G (2018) Characterisation of *Campylobacter* contamination in broilers and assessment of microbiological criteria for the pathogen in broiler slaughterhouses. Food Control 87:60–69. https://doi.org/10.1016/j.foodcont.2017.12.013

Roccato A, Mancin M, Barco L, Cibin V, Antonello K, Cocola F, Ricci A (2018) Usefulness of indicator bacteria as potential marker of *Campylobacter* contamination in broiler carcasses. Int J Food Microbiol 276:63–70. https://doi.org/10.1016/j.ijfoodmicro.2018.04.003

Rosenquist H, Nielsen NL, Sommer HM, Norrung B, Christensen BB (2003) Quantitative risk assessment of human campylobacteriosis associated with thermophilic *Campylobacter* species in chickens. Int J Food Microbiol 83:87–103. https://doi.org/10.1016/s0168-1605(02)00317-3

Rosenquist H, Sommer HM, Nielsen NL, Christensen BB (2006) The effect of slaughter operations on the contamination of chicken carcasses with thermotolerant *Campylobacter*. Int J Food Microbiol 108:226–232. https://doi.org/10.1016/j.ijfoodmicro.2005.12.007

Saint-Cyr MJ, Guyard-Nicodeme M, Messaoudi S, Chemaly M, Cappelier JM Dousset, X Haddad N (2016) Recent Advances in Screening of Anti-*Campylobacter* Activity in Probiotics for Use in Poultry. Front Microbiol 7:553. https://doi.org/10.3389/fmicb.2016.00553

Samapundo S, Thanh TNC, Xhaferi R, Devlieghere F (2016) Food safety knowledge, attitudes and practices of street food vendors and consumers in Ho Chi Minh city. Vietnam. Food Control 70:79–89. https://doi.org/10.1016/j.foodcont.2016.05.037

Sampers I, Habib I, De Zutter L, Dumoulin A, Uyttendaele M (2010) Survival of *Campylobacter* spp. in poultry meat preparations subjected to freezing, refrigeration, minor salt concentration, and heat treatment. Int J Food Microbiol 137:147–153. https://doi.org/10.1016/j.ijfoodmicro.2009.11.013

Sasaki Y, Haruna M, Mori T, Kusukawa M, Murakami M, Tsujiyama Y, Ito K, Toyofuku H, Yamada Y (2014) Quantitative estimation of *Campylobacter* cross-contamination in carcasses and chicken products at an abattoir. Food Control 43:10–17. https://doi.org/10.1016/j.foodcont.2014.02.015

Sasaki Y, Maruyama N, Zou B, Haruna M, Kusukawa M, Murakami M, Asai T, Tsujiyama Y, Yamada Y (2013) *Campylobacter* cross-contamination of chicken products at an abattoir. Zoonoses Public Health 60:134–140. https://doi.org/10.1111/j.1863-2378.2012.01509.x

Schneitz C, Hakkinen M (2016) The efficacy of a commercial competitive exclusion product on *Campylobacter* colonization in broiler chickens in a 5-week pilot-scale study. Poult Sci 95:1125–1128. https://doi.org/10.3382/ps/pew020

Seliwiorstow T, Bare J, Berkvens D, Van Damme I, Uyttendaele M, De Zutter L (2016) Identification of risk factors for *Campylobacter* contamination levels on broiler carcasses during the slaughter process. Int J Food Microbiol 226:26–32. https://doi.org/10.1016/j.ijfoodmicro.2016.03.010

Seliwiorstow T, Bare J, Van Damme I, Uyttendaele M, De Zutter L (2015) *Campylobacter* carcass contamination throughout the slaughter process of *Campylobacter*-positive broiler batches. Int J Food Microbiol 94:25–31. https://doi.org/10.1016/j.ijfoodmicro.2014.11.004

Sikic Pogacar M, Langerholc T, Micetic-Turk D, Mozina SS, Klancnik A (2020) Effect of *Lactobacillus* spp. on adhesion, invasion, and translocation of *Campylobacter jejuni* in chicken and pig small-intestinal epithelial cell lines. BMC Vet Res 16:34. https://doi.org/10.1186/s12917-020-2238-5

Silva F, Domingues FC, Nerin C (2018) Trends in microbial control techniques for poultry products. Crit Rev Food Sci Nutr 58:591–609. https://doi.org/10.1080/10408398.2016.1206845

Skarp CPA, Hanninen ML, Rautelin HIK (2016) *Campylobacteriosis*: the role of poultry meat. Clin Microbiol Infect 22:103–109. https://doi.org/10.1016/j.cmi.2015.11.019

Stead D, Park SF (2000) Roles of Fe superoxide dismutase and catalase in resistance of *Campylobacter coli* to freeze-thaw stress. Appl Environ Microbiol 66:3110–3112. https://doi.org/10.1128/aem.66.7.3110-3112.2000

Sternisa M, Mozina SS, Levstek S, Kukec A, Raspor P, Jevsnik M (2018) Food safety knowledge, self-reported practices and attitude of poultry meat handling among Slovenian consumers. British Food Journal 120:1344–1357. https://doi.org/10.1108/BFJ-06-2017-0360

Stingl K, Buhler C, Kruger NJ (2015) *Campylobacter* detection along the food chain - towards improved quantitative risk analysis by live/dead discriminatory culture-independent methods. Berl Munch Tierarztl Wochenschr 128:122–129 PMID: 25876272

Stopforth JD, O'Connor R, Lopes M, Kottapalli B, Hill WE, Samadpour M (2007) Validation of individual and multiple-sequential interventions for reduction of microbial populations during processing of poultry carcasses and parts. J Food Prot 70:1393–1401. https://doi.org/10.4315/0362-028x-70.6.1393

Umaraw P, Prajapati A, Verma AK, Pathak V, Singh VP (2017) Control of *Campylobacter* in poultry industry from farm to poultry processing unit: a review. Crit Rev Food Sci Nutr 57:659–665. https://doi.org/10.1080/10408398.2014.935847

Wagenaar JA, French NP, Havelaar AH (2013) Preventing *Campylobacter* at the source: why is it so difficult? Clin Infect Dis 57:1600–1606. https://doi.org/10.1093/cid/cit555

Wagle BR, Upadhyay A, Arsi K, Shrestha S, Venkitanarayanan K, Donoghue AM, Donoghue DJ (2017) Application of beta-resorcylic acid as potential antimicrobial feed additive to reduce *Campylobacter* colonization in broiler chickens. Front Microbiol 8:599. https://doi.org/10.3389/fmicb.2017.00599

Wales AD, Vidal AB, Davies RH, Rodgers JD (2019) Field interventions against colonization of broilers by *Campylobacter*. Comprehens Rev Food Sci Food Safety 18:167–188. https://doi.org/10.1111/1541-4337.12397

Wang HH, Qi J, Duan DB, Dong Y, Xu XL, Zhou GZ (2018) Combination of a novel designed spray cabinet and electrolyzed water to reduce microorganisms on chicken carcasses. Food Control 86:200–206. https://doi.org/10.1016/j.foodcont.2017.11.027

Wempe JM, Genigeorgis CA, Farver TB, Yusufu HI (1983) Prevalence of *Campylobacter jejuni* in two California chicken processing plants. Appl Environ Microbiol 45:355–359. https://doi.org/10.1128/aem.45.2.355-359.1983

Wijnen LI, Biasino W, Verbeke W, De Zutter L, Seliwiorstow T, Van Damme I (2019) Attitudes and opinions of the abattoir sector toward the control and prevention of microbiological foodborne pathogens. J Food Safety 39:5. https://doi.org/10.1111/jfs.12667

Yang H, Li Y, Johnson MG (2001) Survival and death of *Salmonella* Typhimurium and *Campylobacter jejuni* in processing water and on chicken skin during poultry scalding and chilling. J Food Prot 64:770–776. https://doi.org/10.4315/0362-028x-64.6.770

Zhao T, Ezeike GO, Doyle MP, Hung YC, Howell RS (2003) Reduction of *Campylobacter jejuni* on poultry by low-temperature treatment. J Food Prot 66:652–655. https://doi.org/10.4315/0362-028x-66.4.652

Zorba NND, Kaptan M (2011) Consumer food safety perceptions and practices in a Turkish community. J Food Prot 74:1922–1929. https://doi.org/10.4315/0362-028x.jfp-11-126

Zweifel C, Althaus D, Stephan R (2015) Effects of slaughter operations on the microbiological contamination of broiler carcasses in three abattoirs. Food Control 51:37–42. https://doi.org/10.1016/j.foodcont.2014.11.002

Emission Sources of *Campylobacter* from Agricultural Farms, Impact on Environmental Contamination and Intervention Strategies

Vanessa Szott and Anika Friese

Contents

Abstract Although extensive research has been carried out to describe the transmission pathways of *Campylobacter* entering livestock farms, the role of livestock farms as source of *Campylobacter* contamination of the environment is still poorly investigated. It is assumed that *Campylobacter*-positive livestock farms contribute to an environmental contamination, depending on the animal species on the farm, their *Campylobacter* status, the housing system, manure management as well as their general farm hygienic and biosecurity management. Different emission sources, like manure, air, water, insects and rodents as well as personnel, including equipment and vehicles, contribute to *Campylobacter* emission into the environment. Even though *Campylobacter* are rather fastidious bacteria, they are able to survive in the environment for even a longer period of time, when environmental conditions enable survival in specific niches. We conclude that a significant reduction of *Campylobacter* emission in the environment can be successfully achieved if various intervention strategies, depending on the farm type, are applied simultaneously, including proper

V. Szott · A. Friese (✉)
Institute for Animal Hygiene and Environmental Health, Free University Berlin, Centre for Infection Medicine, Robert-von-Ostertag-Str. 7-13, 14163 Berlin, Germany
e-mail: anika.friese@fu-berlin.de

V. Szott
e-mail: vanessa.szott@fu-berlin.de

S. Backert (ed.), *Fighting Campylobacter Infections*, Current Topics in Microbiology and Immunology 431, https://doi.org/10.1007/978-3-030-65481-8_5

general and personal hygiene, establishing of hygienic barriers, insect controls, manure management and hygienization of stables, barns and exhaust air.

1 Introduction

Campylobacter is the most prevalent bacterial foodborne zoonotic pathogen in humans in the industrialized world. *Campylobacter* (*C.*) *jejuni* and *C. coli* are the main causative agents of human campylobacteriosis. Clinical symptoms manifest from watery to hemorrhagic diarrhea with high fever and abdominal pain. Infections are usually self-limiting. However, Guillain-Barré syndrome and reactive arthritis are serious sequelae that have been reported in a small proportion of patients (Dasti et al. 2010; Crushell et al. 2004; Zilbauer et al. 2008; Oyarzabal and Backert 2012). *C. jejuni* and *C. coli* naturally occur in clinically healthy domestic animals and are assumed to be commensals of the intestinal microbiome of birds (with highest numbers found in the ceca) and various mammals (Osimani et al. 2017). Poultry meat is considered the most important source of human infection (Vetchapitak and Misawa 2019; Stafford et al. 2008; Meldrum et al. 2005; Facciola et al. 2017). To address this situation, suitable mitigation strategies, especially focusing on the poultry meat production chain, are required in order to diminish *Campylobacter* colonization in poultry. Chapter 4 of this book summarizes these approaches for the poultry meat sector.

Previous studies examined risk factors and entry routes for *Campylobacter* into livestock herds, mainly for broiler chicken (Babacan et al. 2020; Hasan et al. 2020; Adkin et al. 2006). Recent findings indicate a significant role of the environment for the persistence of *Campylobacter* in the close proximity of the farms and for the introduction or re-introduction into the livestock farms (Agunos et al. 2014). Various measures to lower the risk of a flock or herd to become *Campylobacter* positive are already discussed in detail by different authors (Hald et al. 2007; Ridley et al. 2011). However, it is worth changing the perspective and focusing on the description of the complex interaction between *Campylobacter* and the environment. A specific focus on farm emission of *Campylobacter* might be useful since *Campylobacter* emission is related to intensive livestock farming (Agunos et al. 2014; de Rooij et al. 2019). Therefore, intensive husbandry systems might contribute to a potentially high environmental *Campylobacter* contamination. This chapter reviews (i) current information on the on-farm sources of *Campylobacter* emission, (ii) their impact on the environment and (iii) preventive measures to reduce *Campylobacter* emissions from livestock farms.

2 Colonization of *Campylobacter* in Livestock

Campylobacter spp., in particular *C. jejuni* and *C. coli*, are frequently isolated in clinically healthy livestock. These agents are considered as commensal bacteria in various animals (Golz et al. 2014; Rukambile et al. 2019). In some cases, *C. jejuni* can cause clinical symptoms like abortions in sheep, cattle and goats (Sahin et al. 2017) or diarrhea in some chicken breeds due to a prolonged inflammatory response (Humphrey et al. 2014). In general, *Campylobacter* species and prevalence differ between animal species, animal age and their physiological status (Rukambile et al. 2019).

Most cases of human campylobacteriosis are poultry associated (Mullner et al. 2010; Wilson et al. 2008). EFSA reported for 2018 a *Campylobacter* prevalence of 26% in broiler chicken (from $n = 13{,}636$ sample units) and 71.6% in turkeys (from $n = 1{,}174$ sample units) in Europe. In 2,452 investigated samples from broilers *C. jejuni* and *C. coli* were dominating (EFSA and ECDC 2019). German data from 2016 determined a single animal *Campylobacter* prevalence for broiler chickens of 43.5% ($n = 446$) and a prevalence of 73.7% for turkeys ($n = 502$). The flock prevalence was 47.5% for broiler chickens ($n = 61$) and 57.6% for turkey flocks ($n = 66$) (BfR 2019). In broilers, a distinct seasonality with prevalence peaks for *Campylobacter* in summer and autumn is observed (Sahin et al. 2015; Hartnack et al. 2009; Meldrum et al. 2005). For instance, Denmark showed the highest prevalence with up to 78% positive chicken flocks during June to November and a very low prevalence of 7% during December to May (Rosenquist et al. 2009). The number of *Campylobacter*-positive chicken within each flock also depends on the animals' age. It is well established that day-old chicks are usually *Campylobacter* negative and most authors agree that vertical transmission is rather negligible (Sahin et al. 2003). *Campylobacter* colonization naturally occurs in two- or three-week-old chickens (Humphrey et al. 2014). This is in line with recent data, where increasing broilers' age was determined to correlate with *Campylobacter* colonization. Subsequently, the level of colonization and fecal shedding increases toward the end of the rearing period (Agunos et al. 2014; Hald et al. 2000). The concentration of *Campylobacter* in cecal contents of chickens ranges from 10^6 to 10^8 colony-forming units per gram (CFU/g) (Hermans et al. 2012; Rosenquist et al. 2006). During the slaughtering process, 60–80% of the carcasses are contaminated with campylobacters, largely due to fecal contamination of the carcasses (Hermans et al. 2012). In Germany, 76.9% of 130 analyzed neck skin samples were tested positive for *Campylobacter* spp. (BfR 2019). These data highlight the relevance of poultry as the major source of *Campylobacter* within the food chain.

In cattle, the prevalence of *Campylobacter* in 4,220 investigated samples from 10 European countries was rather low with a prevalence of 3.2%. Similarly, beef showed a low prevalence of 0.5% ($n = 589$ sample units) (EFSA and ECDC 2019). In contrast to the European data, 86.6% of dairy cattle herds ($n = 82$) were tested positive for *C. jejuni/coli* in 2019 (Ocejo et al. 2019), and 72.1% of fecal samples ($n = 2298/3184$) from beef cattle were *Campylobacter* positive in the USA (Tang et al.

2017). Also, Abley and colleagues found 77% of fecal samples from steers of one farm in the USA to be positive for *Campylobacter* with a concentration of 3.7×10^4 CFU/g (Abley et al. 2012b). When focusing on young stock, also calves were shown to have a rather high *Campylobacter* prevalence (Stanley and Jones 2003). Klein and colleagues found 14.9% out of 382 calves positive for *Campylobacter* spp., whereas a study by Johnson and coworkers revealed a higher prevalence of 46% ($n = 74$ calves) (Johnsen et al. 2006; Klein et al. 2013). In contrast to poultry, the relevance of beef as transmission route to humans is quite low. The most common species in cattle is considered *C. jejuni*; however, other species like *C. coli*, *C. fetus* or *C. hyointestinales* also occur (Milnes et al. 2008; Ocejo et al. 2019; Sahin et al. 2017; Sproston et al. 2011).

Pigs are considered a natural reservoir of *Campylobacter* with prevalences ranging between 38.1 and 69.3% positive animals. Here, *C. coli* is the predominating species (Alter et al. 2005; Milnes et al. 2008; Nathues et al. 2013; Mdegela et al. 2011). A study investigating the prevalence on herd level reported nine out of 17 pig herds to be *Campylobacter* positive (Oporto et al. 2007). Pigs excrete *Campylobacter* levels of up to 10^7 CFU/g feces (Abley et al. 2012a). In contrast, the European Union (EU) One Health Zoonoses Report 2018 stated 2,481 sample units from pigs, of those 2% were found to be *Campylobacter* positive (EFSA and ECDC 2019).

3 Emission Sources

Livestock farms might have a significant impact on *Campylobacter* emissions into the environment. Schets and colleagues (2017) observed *Campylobacter*-positive environmental samples (soil, air and wastewater) acquired at or close to poultry farms with *Campylobacter*-positive flocks. Feces of *Campylobacter*-positive livestock (cattle, swine or poultry) usually contain high bacterial loads (Abley et al. 2012a, b; Hermans et al. 2012; Rosenquist et al. 2006). Since fecal material can contaminate various matrices in the barn or the farm, *Campylobacter* (originating from animal feces) might contaminate the environment via different emission sources (Fig. 1), described in detail in the following section.

3.1 Manure

The fecal waste generated by *Campylobacter*-positive livestock represents an important emission source of these bacteria. After broiler removal from the barns, manure is utilized for crop growth or in general as fertilizer (right away or rather after an initial storage time). Another purpose for manure is its anaerobic fermentation in biogas plants. Hutchinson and coworkers, for example, studied the prevalence of *Campylobacter* in stored manure from pigs, cattle, poultry and sheep and found between 7.7 and 11.1% of the samples to be *Campylobacter* positive, with a *Campylobacter*

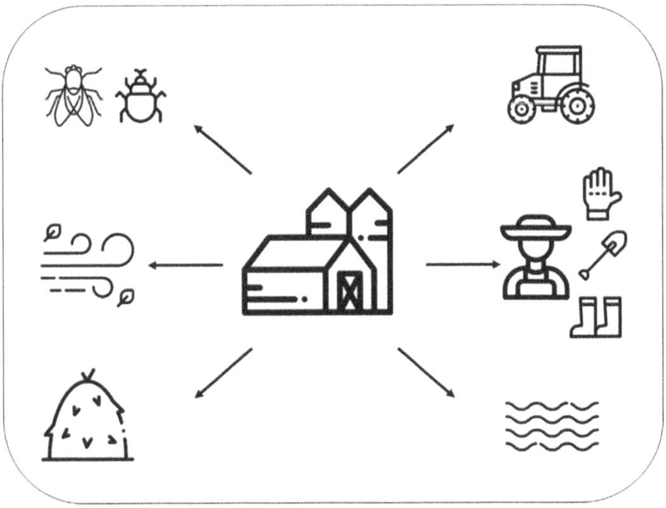

Fig. 1 Various emission sources of *Campylobacter* from livestock farms into the environment: insects, air, manure, vehicles, personnel and equipment, process water

concentration ranging from 2.6×10^2 to 1.6×10^3 CFU/g manure (Hutchison et al. 2005b). In general, performed studies demonstrated a rather low persistence of *Campylobacter* in solid manure (Gilpin et al. 2009; Nicholson et al. 2005; Sinton et al. 2007b). In this context, it is important to note that *Campylobacter* detection in manure is dependent on the manure's consistency. In conventional animal husbandries, broilers and turkeys are typically kept in floor houses with litter. The litter is removed as solid manure from the animal house after fattening. Interestingly, the soil close to manure storage sheds was found to be positive for *Campylobacter* on poultry farms in the Netherlands (Schets et al. 2017). Also Mohammed and coworkers (2019) found ten out of 15 sampled manure storage areas and eight out of 15 litter samples from broiler farms positive for *C. coli* (Mohammed and Abdel Aziz 2019). Cattle generate about 50 L of liquid feces per day, which cumulate to slurry or litter, depending on the farming system used. Most often slurry is stored for longer periods in tanks. Laboratory studies showed that *Campylobacter* can remain viable in stored slurry over a longer period of time with different *D*-values between seven and even more than 112 days (Jones 2001; Hutchison et al. 2005c; Kearney et al. 1993b). A study conducted with *Campylobacter* naturally present in fresh cow feces showed a *D*-value of 2.2 days (Gilpin et al. 2009). The season seems to influence the inactivation time. In winter, culturable *Campylobacter* were longer detectable in feces in comparison to summer months (Sinton et al. 2007b). In dairy slurry, stored in tanks, campylobacters were found right before it was applied on the field. In winter, *Campylobacter* were culturable in samples taken five days after application on farmland, whereas *Campylobacter* was not culturable after 24 h in the summer (Gerritsen et al. 1991).

Storing manure in slurry tanks is also common in conventional pig husbandries. *Campylobacter* was detected in slurry with a *D*-value of about 11 days in winter and seven days in summer (Hutchison et al. 2005c). In a recent study, *Campylobacter* was found in pig manure (10–10^2 CFU/g) before field fertilization but also in the soil afterward, showing amounts of 0.1–10^2 CFU/g soil (Van den Meersche et al. 2020). Since feces and manure attract insects and wild birds, such matrices can act as vectors, carrying *Campylobacter* and promoting a further distribution (Hald et al. 2008). In addition, the occurrence of *Campylobacter* in rivers was shown to be correlated with agricultural locations or events like emptying slurry tanks or spraying slurry on farmland adjacent to the rivers (Jones 2001). This underlines the importance of manure as a source of environmental contamination.

3.2 Air

Air as an emission source of infectious agents from agricultural farms is of high interest. A recent study conducted in the Netherlands detected *C. jejuni* DNA in fresh air samples at greater distances of 250 m to approximately 1,000 m from livestock farms in residential sites, especially in poultry dense areas (de Rooij et al. 2019). Barn exhaust air is usually emitted from agricultural farms without any treatment, so airborne bacteria originating from livestock can enter the environment. With regard to *Campylobacter*, however, the role of an airborne emission and transmission is currently not well understood. There are several studies examining the occurrence of *Campylobacter* in air samples, although the focus of these studies was primarily on poultry farms, using different sampling and detection methods. In contrast, only one study exemplarily investigated air samples in a pig stable and found *Campylobacter* DNA in the tested air samples (Julich et al. 2016). Detailed results of *Campylobacter* detection in air samples are summarized in Table 1. Ahmed and colleagues applied two different air sampling systems simultaneously, in order to detect culturable *Campylobacter* in barn air of *C. jejuni*-positive laying hen flocks by using an AGI-30 impinger and a Coriolis®μ Air Sampler (cyclone). Bacteria were collected immediately in a fluid which represents a rather gentle sampling procedure. Interestingly, *Campylobacter* could not be cultivated from all examined samples independent of the air sampler used. To detect *C. jejuni*-specific DNA, they filtered air and found in 15 out of 18 (83%) samples specific *C. jejuni* DNA (Ahmed et al. 2013a). Two studies using exclusively PCR assays showed that *Campylobacter* is found in air samples of chicken houses before a detection of *Campylobacter* DNA in boot swab samples was possible (Olsen et al. 2009; Sondergaard et al. 2014). Similar results were demonstrated in another study conducted in *Campylobacter*-positive broiler houses (Olsen et al. 2009). By applying electrostatic capture, *Campylobacter* was detected in very small air volumes (1.8 L) by using real-time PCR. However, a positive PCR result does not necessarily demonstrate the presence of live campylobacters; that limitation of the used real-time PCR approaches was highlighted by the authors: A cultivation of *Campylobacter* failed after having carried out sedimentation by using agar plates without a lid on the same farm. Similar observations were made using gelatin filters for the detection of *Campylobacter* in air samples. There, *Campylobacter* detection was successful using PCR in barn air samples of two-week-old broiler chickens (Sondergaard et al. 2014). Applying equal sampling procedures inside and outside of chicken houses in Australia, only one air sample from the inside was tested positive for *Campylobacter* by cultivation, although numerous samplings were performed (Chinivasagam et al. 2009). A recent study was carried out in Europe investigated 44 broiler flocks using boot swabs and air samples (gelatin filter method). *Campylobacter* was culturable from air samples from three flocks, whereas PCR detected *Campylobacter* from 14 flocks. The latter was similar to the results of the simultaneous cultivation of boot swabs (Johannessen et al. 2020). Bull and coresearchers used three different air sampling methods (described in Table 1) and found 6% from 248 air samples from the inside of the barns to be *Campylobacter* positive dependending on the sampling method used. Additionally, *Campylobacter* was detectable in four out of 18 air samples downwind of broiler houses (up to 30 m). Based on current knowledge available to us, published data lead to the conclusion that air might not play an important role as an emission source. However, *Campylobacter* isolates found in the air outside the broiler house were similar to isolates found inside the housing (Bull et al. 2006), which might indicate an airborne emission. Moreover, O'Mahony and coauthors (2011) found identical *C. jejuni* strains in broiler barn air and feces as well as in a puddle and in soil in the surrounded farm environment. The relevance of contaminated air is also highlighted in an experimental study by Zhao and coworkers (2011). Experimentally infected broiler chickens were allocated in a cage in the middle of four different rooms. In each of the four rooms, susceptible *Campylobacter*-negative animals were placed in single cages close to the infected animals in a distance of approximately 0.75 m. In each room, at least one susceptible animal was colonized with *Campylobacter*. The authors speculated that an airborne transmission occurred, although none of the air samples taken by three different methods showed culturable *Campylobacter* (Zhao et al. 2011).

3.3 Insects and Rodents

Especially in summer, a significant number of insects can be found on livestock farms. Insects can carry campylobacters and subsequently represent a potential source of emission. Hald and coresearchers (2004) captured flies in the environment of broiler houses and found culturable *C. jejuni* in 8.2% (4/49) of flies, and in 70.2% when using PCR assays. In another study investigating flying insects outside the broiler houses, *Campylobacter* was cultivated from three out of 291 samples. Here, the authors found only one isolate corresponding to isolates from the inside of the broiler houses (Hansson et al. 2007). Another approach investigated flies on eight poultry farms multiple times but found no culturable *Campylobacter* within the flies (Schets et al. 2017). Similar observations were made in Switzerland, where the investigation of 15 broiler flocks revealed not a single fly to be *Campylobacter* positive

(Zweifel et al. 2008). A study of three broiler flocks in Ireland revealed comparable results (O'Mahony et al. 2011). In contrast, 66.6% of the investigated flies from a turkey farm carried culturable *Campylobacter* bearing the same sequence type as isolates from the animals (Piccirillo et al. 2018). This high variability of prevalences was also demonstrated by Agunos and coworkers (2014), who summarized a collection of studies and reported an overall prevalence of 28.9% (95% confidence interval: 12.1–58.8%) *Campylobacter*-positive beetles and 7.1% (95% confidence interval: 1.6–26.0%) positive flies. Concerning rodents, the prevalence was significantly higher with about 49.6% (95% confidence interval: 24.0–75.5%) in this study. Molecular epidemiological approaches using meta-analysis demonstrated a link between broiler flock strains and strains found in insects and rodents (Agunos et al. 2014). Unfortunately, most research so far has focused on poultry farms and only few studies included other animal species than poultry. In Scotland, flies nearby a cattle and sheep farm were investigated for *Campylobacter* carriage. They carried *Campylobacter* with a prevalence of 5.8%, and the detected sequence types corresponded to the sequence types found in feces (Sproston et al. 2010). Also pig herds in Germany showed a comparable *Campylobacter* prevalence in flies of 6.7% (3/45), whereas rodent droppings ($n = 20$) were all negative (Nathues et al. 2013). To sum up, the emission of *Campylobacter* from the animal houses especially by flying insects varies but seems relevant for the overall environmental contamination by farm emission.

3.4 Personnel, Equipment, Vehicles

Various studies investigated contaminated vehicles, personnel and equipment for their potential role in the introduction and spread of *Campylobacter* on farms. A number of studies demonstrated that vehicles, personnel and equipment can be frequently contaminated with *Campylobacter*, highlighting the role of these matrices for *Campylobacter* transmission (Ridley et al. 2011; Mohammed and Abdel Aziz 2019; Allen et al. 2008). When testing broiler farms, Ramabu and colleagues found that more than 50% of the sampled catchers' and lorry drivers' boots were positive for *C. jejuni*. Fomites, which were *Campylobacter* positive, appeared wet and still visibly contaminated with feces even after cleaning. The authors concluded that *Campylobacter* survived due to favorable conditions in the fecal matter in combination with a humid environment. Therefore, equipment and personnel is assumed to be a relevant emission source for *Campylobacter* (Ramabu et al. 2004). Moreover, five out of 30 farm workers' hands were also tested *Campylobacter*-positive (Mohammed and Abdel Aziz 2019). In particular, the thinning process in broiler flocks is supposed to contribute to *Campylobacter* emission by contaminated personnel, equipment or vehicles leaving the farm. Thinning is a popular practice, where a proportion of the broiler flock is removed for slaughter, allowing the remaining broilers to occupy the full space and grow until slaughter age. After thinning, the personnel, their equipment used for thinning and the vehicles were found to be *Campylobacter* positive. Furthermore, the same *Campylobacter* strains were found in tire marks of vehicles leaving the farm (Allen et al. 2008). Therefore, it is reasonable to conclude that emission by personnel or related equipment might take place continuously on poultry farms. Even though these studies have largely focused on broiler farms, it can be speculated that similar scenarios take place on other animal species farms, when personnel or vehicles enter and leave the farms. The relevance of that emission route probably depends on the *Campylobacter* prevalence within the specific animal species: A study including cattle showed only 2.2% positive animals, and 25 swabs of the transport vehicle were negative after transportation (Beach et al. 2002). However, further research concerning other animals than broiler chicken is needed.

3.5 Waterborne Emission

Campylobacter can survive in the aquatic environment, for example, in fresh water like ponds, rivers and drinking water or in waste water, sewage, sludge and slurry (Jones 2001). Poultry house cleaning processes generate large amounts of wastewater. In general, they are stored in tanks or discharged into the sewage system. However, some small amounts may run out of the barn and lead to a contamination of the environment. A study by Schets and coworkers (2017) highlighted the role of waterborne emission by detecting *Campylobacter* isolates of the same sequence type (ST) in cecal material within poultry houses as well as in surface water around poultry houses, in wastewater and soil. Another study showed that waterborne emission might also occur indirectly through microbial run-off from farmland fertilized with manure. This phenomenon usually takes place after heavy rainfalls (Brooks et al. 2009). In addition, *Campylobacter* were found sporadically in water used to wash broiler transport boxes on the farm or at the slaughterhouse (Slader et al. 2002; Northcutt and Berrang 2006).

4 Tenacity of *Campylobacter* in the Environment

Campylobacter are rather fastidious bacteria; however, they are able to survive outside a suitable host for longer period of time (Murphy et al. 2006; Nicholson et al. 2005). Due to the above-described emission processes, *Campylobacter* might spread in the farms' environment (Jones 2001; Agunos et al. 2014). If artificially inoculated poultry feces are spread directly on grass pasture, *Campylobacter* can survive up to 42 days (Hutchison et al. 2005d). In contrast, in artificially inoculated cow pats, an average *C. jejuni* inactivation of 90% was observed after 6.2 days after deposition (Sinton et al. 2007b). In stored pig slurry, *Campylobacter* was culturable up to three months (Nicholson et al. 2005). In contrast, studies showed survival of *C. jejuni* in laying hen feces for up to six days (Ahmed et al. 2013b) while it survived considerably less (only two days) in broiler feces (Berrang et al. 2004). However, persistence in litter was shown to be significantly shorter (only 4 h) (Smith et al. 2016). This was quite different within another study, where dairy cattle cow pads were artificially inoculated with *Campylobacter* and stored with litter in heaps. *Campylobacter* was detectable until 62 days after storage (Hutchison et al. 2005a). The different study outcomes are difficult to compare or interpret, since manure is a highly variable material that can be influenced by factors like bacterial flora, bedding material, animal diet, relative humidity, temperature of storage or housing type. For this reason, results vary widely (Hutchison et al. 2005b; Smith et al. 2016).

There is a seasonal impact, showing higher *Campylobacter* detection rates in the environment during the winter months (Hansson et al. 2007). Possible explanations might be lower temperatures and decreased ultraviolet radiation levels (Obiri-Danso et al. 2001; Sinton et al. 2007a). Experimental studies have already shown reduced *C. jejuni* and *C. coli* counts if media were stored in the presence of light and air (Bolton et al. 1984). The formation of reactive oxygen intermediates might damage nucleic acids, proteins and membranes (Park 2002). Thermophilic *Campylobacter* prefer temperatures between 37 and 42 °C. *Campylobacter* cease proliferation at temperatures below 30 °C. However, they can hibernate, e.g., in the aquatic environment or at colder temperatures such as 4 and 10 °C in comparison to 22 and 37 °C (Buswell et al. 1998). Furthermore, the tenacity and the spread of *Campylobacter* in the environment correlates with humidity (damp and rainy climate), leading to a higher contamination of farm surroundings during a humid climate (Hansson et al. 2007). Indeed, there is an observable association of heavy rainfalls with an increased risk of *Campylobacter*-positive broiler flocks (Jonsson et al. 2012). Also, horizontal transmission of *Campylobacter* is elevated if chickens are kept in housing conditions with high relative humidity (Line 2006).

5 Viable but Non-culturable Form of *Campylobacter* in the Environment

When exposed to stress factors like low temperature, desiccation, toxic oxygen >5%, starvation, acid or salt treatment, *Campylobacter* may devolve into a viable but non-culturable state (VBNC) (Silva et al. 2011; Pinto et al. 2015). In the VBNC state, *Campylobacter* are not able to proliferate in culture media; however, they still exhibit metabolic activity and membrane integrity (Ramamurthy et al. 2014). Although *Campylobacter* have shown to persist as VBNC forms for several months, they can recover their cultivability if conditions become favorable and can regain their full infective phenotypes (Baffone et al. 2006; Cappelier et al. 1999; Lazaro et al. 1999).

Various conditions in the farm environment, e.g., temperature, oxygen stress and starvation, may probably initiate the VBNC state conversion. This is of high importance for the persistence in the environment. Few studies forced transformation of *C.*

jejuni into the VBNC state. Talibart and colleagues (2000) showed that an incubation in water at 4 °C for 46–48 days (Patrone et al. 2013) and even shorter (14–21 days) was sufficient to induce a VBNC state in *Campylobacter* (Talibart et al. 2000).

The detection of VBNC *Campylobacter* is possible using real-time PCR. For preparation, samples are treated with propidium monoazide (PMA) or ethidium monoazide (EMA) prior to DNA extraction. PMA and EMA can enter dead cells with compromised cell membranes, then intercalate with double-helical DNA and cross-link the DNA when exposed to light (Josefsen et al. 2010; Pacholewicz et al. 2019; Kruger et al. 2014).

The detection rates of culturable *Campylobacter* in environmental samples of broiler farms' surroundings differ between studies (Agunos et al. 2014). The contribution of VBNC *Campylobacter* sustaining in the environment to animal colonization and human outbreaks remains unknown. There are some studies investigating VBNC *Campylobacter* in artificially inoculated water samples (Bae and Wuertz 2012; Seinige et al. 2014). However, the assessment of environmental samples using PMA-PCR is challenging due to the interference of high particle concentrations and complex biological/chemical matrices (Bae and Wuertz 2009). A study investigating EMA-treated compost samples subsequently analyzed with qPCR showed *Campylobacter* cells with intact and with compromised cell membranes (Inglis et al. 2010). Another study analyzed feces with qPCR after PMA treatment. However, results revealed lower mean counts in comparison to qPCR without PMA treatment, but higher counts in comparison to direct culture (Seliwiorstow et al. 2015). To our knowledge, there are no systematic studies about the occurrence and prevalence of VBNC state *Campylobacter* in the environment of livestock farms. Further research is necessary to evaluate the relevance of VBNC *Campylobacter* and the complexity of *Campylobacter* transmission between environment, animals and humans.

6 Intervention Against Emission

As described above, there are different sources of *Campylobacter* emissions from livestock farms. To reduce emission of *Campylobacter* into the environment, first of all, colonization of livestock should be prevented. Much research has been carried out to describe the different approaches to achieve that goal (Agunos et al. 2014; Gibbens et al. 2001; van de Giessen et al. 1998). However, in this review, we focus on prevention strategies to reduce *Campylobacter* emissions from agricultural farms. We suggest to implement different intervention measures simultaneously to decrease the risk of emission. All intervention strategies should be tailored to the respective farm conditions.

The storage of manure and their field application might be relevant as an emission source. Fortunately, there are miscellaneous effective manure treatments, which inactivate bacteria successfully in general (Martens and Bohm 2009), for example, creating dung heaps out of solid manure. This process leads to a rapid decline of fecal bacteria concentration due to high temperatures of 50–60° C in the center of the heap

(Siller et al. 2020; Erickson et al. 2010). A study using artificially contaminated broiler waste, stored in heaps, demonstrated that *C. jejuni* concentration fell below the limit of detection after eight days of storage (Hutchison et al. 2005c). If contaminated livestock waste was spread on grass pasture, *Campylobacter* was detected up to 42 days (Hutchison et al. 2005d). Another study, investigating composting of poultry manure, found culturable *Campylobacter* for up to two weeks (Esperon et al. 2020).

With regard to liquid manure from cattle or pigs, a heap building is not possible. Thus, the mechanical separation of the fluid was shown to be effective to treat *Campylobacter*-positive pig slurry. After separation, *Campylobacter* was not detectable in the solid manure; however, detection was still possible at prevalences of 38.5% and 28.6% in the fluid and in unseparated manure, respectively (Watabe et al. 2003). Another method is a mesophilic anaerobic digestion of manure using a biogas plant. Manyi-Loh et al. showed a reduction of *Campylobacter* in cattle manure under the detection limit within 18 days, starting with a concentration of 10.1×10^3 CFU/g to concentrations below the detection limit of 10^2 CFU/g (Manyi-Loh et al. 2014). In contrast, in another study the time necessary for a *Campylobacter* reduction of one log CFU/g was even 438 days (Kearney et al. 1993a). Using thermophilic anaerobic digestion, operating with temperatures above 50 °C, *C. jejuni* did not survive longer than 24 h (Wagner et al. 2008). Chemical and microbiological procedures are also considered to reduce *Campylobacter* concentration in manure; however, for economic reasons, most of all composting and anaerobic treatment as described are relevant for conventional farms (Martens and Bohm 2009).

Hygienization of stables or exhaust air is usually not applied on livestock farms. Furthermore, this is rather impossible for open stable systems. However, an application in barns with specific ventilation systems is feasible. Unfortunately, there are no studies focusing specifically on airborne *Campylobacter.* One approach is to treat the air inside the barn, another to clean exhaust air before emission in the environment. Tenzin et al. investigated the efficacy of air decontamination in a pig barn by nebulizing electrochemically activated water. By fogging the solution in the animal house, they achieved a total bacterial load reduction of more than two log units per m^3 air within three hours (Tenzin et al. 2019). This is in the line with a Chinese study where acidic electrolyzed water was sprayed slightly in laying hen houses. Here, a reduced airborne microbial level was demonstrated successfully (Hao et al. 2014). With regard to exhaust air decontamination, a possible method is biofiltration. Several studies investigated that topic in pig facilities. However, results revealed a rather low quantitative reduction of total bacteria in air samples ranging between 11 and 75% (Seedorf and Hartung 1999; Tymczyna et al. 2011) or up to 90–95% (Martens et al. 2001; Clauss et al. 2013). Similar efficiencies were obtained using an air cleaner consisting of a washer in combination with an UV-irradiation unit. Using this method in a fattening pig stable, a quantitative reduction up to 96% of mesophilic airborne bacteria was achieved (Schulz et al. 2013). Since the bedding material in animal houses contributes to the airborne dust in animal houses with 55–68% (Seedorf 2004), hygienic treatment like litter acidification in poultry houses might help to reduce airborne *Campylobacter* (Line 2002; Line and Bailey 2006).

Another feasible intervention strategy to prevent *Campylobacter* emission is insect control. Hygienic barriers around a broiler barn present a highly protective practice (Agunos et al. 2014). For instance, installation of fly screens resulted in a decrease of *Campylobacter* colonization from 51.4 to 15.4% positive broiler houses in Denmark (Hald et al. 2007). This is in the line with another study where the *Campylobacter* prevalence in 99 broiler chicken flocks decreased from 41.4% positive flocks to 10.3% after installation of fly screens (Bahrndorff et al. 2013). However, the ventilation system influences the efficiency of fly screens (Sahin et al. 2015). Other chemical or non-chemical control strategies to decrease the number of insects inside the barn could also be applied. However, there are no data available concerning their effect on *Campylobacter* transmission.

The general and personal hygiene is a key tool to reduce emission processes on livestock farms. There are several studies, which dealt with biosecurity and hygiene measures to prevent *Campylobacter* from entering an animal house (Gibbens et al. 2001; van de Giessen et al. 1998). However, special attention to hygiene measures is required after visiting a *Campylobacter*-positive flock in order to lower an environmental contamination. The process of thinning of a broiler flock is a recurrent and important topic. During this procedure, the catching crew, their equipment and their vehicles might transfer *Campylobacter* out of the farm (Ramabu et al. 2004; Allen et al. 2008). Ridley and colleagues implemented new strategies to reduce such contaminations: cleaning and disinfection of the vehicles, providing facilities for hand hygiene as well as fresh clothing and shoes, should be particularly emphasized. Those implemented biosecurity measures significantly lowered the incidence of *Campylobacter* detection on equipment, vehicles, hands and footwear of the catching crew. Since footwear is quite difficult to clean, an implementation of a good hygiene practice is recommended. Moreover, transport boxes should be cleaned and disinfected more efficiently. The necessity is highlighted by the fact that boxes are frequently *Campylobacter* positive, even after washing in a factory. For this reason, boxes should be cleaned thoroughly including the removal of all fecal residues in order to ensure an effective disinfection afterward (Slader et al. 2002). A recent study examined an optimized cleaning system for poultry transport crates. The combination of a high-performance washer fitted with high volume, high-pressure nozzles, heating the water and using chemicals showed a significant reduction of *Campylobacter* on the crates base in comparison to conventional crate washing systems (Atterbury et al. 2020).

7 Concluding Remarks

Campylobacter are present in different livestock species with varying prevalence. *Campylobacter* contamination of the farms' environment is usually higher if positive flocks are present in the corresponding farm (Schets et al. 2017). Furthermore, farm-emission processes as described here are regarded to be contributing factors to a general *Campylobacter* contamination of the environment. Since emission is considered a complex process, Fig. 1 summarizes different emission sources and routes. The relevance of these sources depends on the animal species, the housing system, the manure management as well as the general hygienic and biosecurity management. We conclude that a significant reduction of *Campylobacter* emission in the environment can be successful if various intervention strategies are applied simultaneously. However, efficacy is also depending on the farm type. We highly recommend an effective manure treatment prior to a field application as well as a strict hygiene policy for personnel or equipment leaving a *Campylobacter*-positive barn. If such a strict hygiene protocol is implemented and followed, we assume that an emission prevention from infected flocks might be achievable. However, further studies focusing on the importance of VBNC state *Campylobacter* for an environmental contamination and studying the implementation of specific measures concerning an emission reduction on farms are necessary.

Table 1 Summary of studies investigating air samples for culturable *Campylobacter* and *Campylobacter* DNA by PCR

Air sample location	Air sampling method and volume	Detection method	Results[a]	References
Poultry dense area, outside air	Filtration: Havard Impactor, 50.4 m^3	PCR	Positive (26 from 61 sites)	de Rooij et al. (2019)
Laying hens inside air	Impingement: AGI-30, 300 l	Microbial culture	Negative (0/18)	Ahmed et al. (2013a)
	Cyclone: Coriolis®µ, 900 l		Negative (0/18)	
	Filtration: IOM, 300 l	PCR	Positive (15/18)	
Broiler, inside air	Sedimentation	Microbial culture	Negative (0/6)	Olsen et al. (2009)
	Electrostatic capture, 1.8 l	PCR	Positive (6/6)	
	Filtration: MD8 AirPort, 750 l	Microbial culture and PCR	Negative by culture (0/29) Positive by PCR (29/29)	Sondergaard et al. (2014)

(continued)

Table 1 (continued)

Air sample location	Air sampling method and volume	Detection method	Results[a]	References
	MD8 Airport, gelatin filter	Microbial culture and PCR	Positive by culture (3/144) Positive by PCR (14/44)	Johannessen et al. (2020)
Broiler, inside and outside air	Sedimentation Cyclone, 7.5–11.25 m^3 Impaction: 100 l	Microbial culture	Positive inside (15/248) Positive outside (4/18)	Bull et al. (2006)
	Filtration: MD8 AirPort, 4.6–6.0 m^3	Microbial culture	Positive inside (1/48) Negative outside (0/48)	Chinivasagam et al. (2009)
	Impaction, 500 l	Microbial culture	Positive inside (1/70) Negative outside	O'Mahony et al. (2011)
Broiler, inside air Pig, inside air	Cyclone: Coriolis®μ, 900 l	PCR with chip-assisted DNA purification	Positive (23/32) Positive (25/32)	Julich et al. (2016)

[a]Positive or negative for detection of *Campylobacter* by the mentioned detection method and from air (inside and/or outside) as mentioned within the same table row

Acknowledgements This work is supported by the German Federal Ministry of Education and Research (BMBF) in frame of the zoonoses research consortium PAC-Campylobacter to A.F. (project IP1/01KI1725A). Icons in Fig. 1 were made by Pixel perfect from www.flaticon.com.

References

Abley MJ, Wittum TE, Moeller SJ, Zerby HN, Funk JA (2012) Quantification of campylobacter in swine before, during, and after the slaughter process. J Food Prot 75(1):139–143. https://doi.org/10.4315/0362-028X.JFP-11-334

Abley MJ, Wittum TE, Zerby HN, Funk JA (2012) Quantification of Campylobacter and Salmonella in cattle before, during, and after the slaughter process. Foodborne Pathog Dis 9(2):113–119. https://doi.org/10.1089/fpd.2011.0931

Adkin A, Hartnett E, Jordan L, Newell D, Davison H (2006) Use of a systematic review to assist the development of Campylobacter control strategies in broilers. J Appl Microbiol 100(2):306–315. https://doi.org/10.1111/j.1365-2672.2005.02781.x

Agunos A, Waddell L, Leger D, Taboada E (2014) A systematic review characterizing on-farm sources of Campylobacter spp. for broiler chickens. PLoS One 9 (8):e104905. https://doi.org/10.1371/journal.pone.0104905

Ahmed MF, Schulz J, Hartung J (2013) Air samplings in a Campylobacter jejuni positive laying hen flock. Ann Agric Environ Med 20(1):16–20

Ahmed MF, Schulz J, Hartung J (2013) Survival of Campylobacter jejuni in naturally and artificially contaminated laying hen feces. Poult Sci 92(2):364–369. https://doi.org/10.3382/ps.2012-02496

Allen VM, Weaver H, Ridley AM, Harris JA, Sharma M, Emery J, Sparks N, Lewis M, Edge S (2008) Sources and spread of thermophilic Campylobacter spp. during partial depopulation of broiler chicken flocks. J Food Prot 71(2):264–270. https://doi.org/10.4315/0362-028x-71.2.264

Alter T, Gaull F, Kasimir S, Gurtler M, Mielke H, Linnebur M, Fehlhaber K (2005) Prevalences and transmission routes of Campylobacter spp. strains within multiple pig farms. Vet Microbiol 108(3–4):251–261. https://doi.org/10.1016/j.vetmic.2005.03.004

Atterbury RJ, Gigante AM, Tinker D, Howell M, Allen VM (2020) An improved cleaning system to reduce microbial contamination of poultry transport crates in the United Kingdom. J Appl Microbiol 128(6):1776–1784. https://doi.org/10.1111/jam.14576

Babacan O, Harris SA, Pinho RM, Hedges A, Jorgensen F, Corry JEL (2020) Factors affecting the species of Campylobacter colonizing chickens reared for meat. J Appl Microbiol. https://doi.org/10.1111/jam.14651

Bae S, Wuertz S (2009) Discrimination of viable and dead fecal Bacteroidales bacteria by quantitative PCR with propidium monoazide. Appl Environ Microbiol 75(9):2940–2944. https://doi.org/10.1128/AEM.01333-08

Bae S, Wuertz S (2012) Survival of host-associated bacteroidales cells and their relationship with Enterococcus spp., Campylobacter jejuni, Salmonella enterica serovar Typhimurium, and adenovirus in freshwater microcosms as measured by propidium monoazide-quantitative PCR. Appl Environ Microbiol 78(4):922–932. https://doi.org/10.1128/AEM.05157-11

Baffone W, Casaroli A, Citterio B, Pierfelici L, Campana R, Vittoria E, Guaglianone E, Donelli G (2006) Campylobacter jejuni loss of culturability in aqueous microcosms and ability to resuscitate in a mouse model. Int J Food Microbiol 107(1):83–91. https://doi.org/10.1016/j.ijfoodmicro.2005.08.015

Bahrndorff S, Rangstrup-Christensen L, Nordentoft S, Hald B (2013) Foodborne disease prevention and broiler chickens with reduced Campylobacter infection. Emerg Infect Dis 19(3):425–430. https://doi.org/10.3201/eid1903.111593

Beach JC, Murano EA, Acuff GR (2002) Prevalence of Salmonella and Campylobacter in beef cattle from transport to slaughter. J Food Prot 65(11):1687–1693. https://doi.org/10.4315/0362-028x-65.11.1687

Berrang ME, Northcutt JK, Cason JA (2004) Recovery of Campylobacter from broiler feces during extended storage of transport cages. Poult Sci 83(7):1213–1217. https://doi.org/10.1093/ps/83.7.1213

BfR, (2019) Erreger von Zoonosen in Deutschland im Jahr 2016. Bundesinstitut Für Risikobewertung. https://doi.org/10.17590/20190326-112654

Bolton FJ, Coates D, Hutchinson DN (1984) The ability of campylobacter media supplements to neutralize photochemically induced toxicity and hydrogen peroxide. J Appl Bacteriol 56(1):151–157. https://doi.org/10.1111/j.1365-2672.1984.tb04707.x

Brooks JP, Adeli A, Read JJ, McLaughlin MR (2009) Rainfall simulation in greenhouse microcosms to assess bacterial-associated runoff from land-applied poultry litter. J Environ Qual 38(1):218–229. https://doi.org/10.2134/jeq2008.0029

Bull SA, Allen VM, Domingue G, Jorgensen F, Frost JA, Ure R, Whyte R, Tinker D, Corry JE, Gillard-King J, Humphrey TJ (2006) Sources of Campylobacter spp. colonizing housed broiler flocks during rearing. Appl Environ Microbiol 72(1):645–652. https://doi.org/10.1128/AEM.72.1.645-652.2006

Buswell CM, Herlihy YM, Lawrence LM, McGuiggan JT, Marsh PD, Keevil CW, Leach SA (1998) Extended survival and persistence of Campylobacter spp. in water and aquatic biofilms and their detection by immunofluorescent-antibody and -rRNA staining. Appl Environ Microbiol 64(2):733–741. https://doi.org/10.1128/AEM.64.2.733-741.1998

Cappelier JM, Minet J, Magras C, Colwell RR, Federighi M (1999) Recovery in embryonated eggs of viable but nonculturable Campylobacter jejuni cells and maintenance of ability to adhere to HeLa cells after resuscitation. Appl Environ Microbiol 65(11):5154–5157. https://doi.org/10.1128/AEM.65.11.5154-5157.1999

Chinivasagam HN, Tran T, Maddock L, Gale A, Blackall PJ (2009) Mechanically ventilated broiler sheds: a possible source of aerosolized Salmonella, Campylobacter, and Escherichia coli. Appl Environ Microbiol 75(23):7417–7425. https://doi.org/10.1128/AEM.01380-09

Clauss M, Schulz J, Stratmann-Selke J, Decius M, Hartung J (2013) Reduction of livestock-associated methicillin-resistant staphylococcus aureus (LA-MRSA) in the exhaust air of two piggeries by a bio-trickling filter and a biological three-step air cleaning system. Berl Munch Tierarztl Wochenschr 126(3–4):137–142

Crushell E, Harty S, Sharif F, Bourke B (2004) Enteric campylobacter: purging its secrets? Pediatr Res 55(1):3–12. https://doi.org/10.1203/01.PDR.0000099794.06260.71

Dasti JI, Tareen AM, Lugert R, Zautner AE, Gross U (2010) Campylobacter jejuni: a brief overview on pathogenicity-associated factors and disease-mediating mechanisms. Int J Med Microbiol 300(4):205–211. https://doi.org/10.1016/j.ijmm.2009.07.002

de Rooij MMT, Hoek G, Schmitt H, Janse I, Swart A, Maassen CBM, Schalk M, Heederik DJJ, Wouters IM (2019) Insights into livestock-related microbial concentrations in air at residential level in a livestock dense area. Environ Sci Technol 53(13):7746–7758. https://doi.org/10.1021/acs.est.8b07029

EFSA, ECDC (2019) The European Union one health 2018 zoonoses report. EFSA J 17. https://doi.org/10.2903/j.efsa.2019.5926

Erickson MC, Liao J, Boyhan G, Smith C, Ma L, Jiang X, Doyle MP (2010) Fate of manure-borne pathogen surrogates in static composting piles of chicken litter and peanut hulls. Bioresour Technol 101(3):1014–1020. https://doi.org/10.1016/j.biortech.2009.08.105

Esperon F, Albero B, Ugarte-Ruiz M, Dominguez L, Carballo M, Tadeo JL, Del Mar DM, Moreno MA, de la Torre A (2020) Assessing the benefits of composting poultry manure in reducing antimicrobial residues, pathogenic bacteria, and antimicrobial resistance genes: a field-scale study. Environ Sci Pollut Res Int 27(22):27738–27749. https://doi.org/10.1007/s11356-020-09097-1

Facciola A, Riso R, Avventuroso E, Visalli G, Delia SA, Lagana P (2017) Campylobacter: from microbiology to prevention. J Prev Med Hyg 58(2):E79–E92

Gerritsen ME, Schwarz SM, Medow MS (1991) Glucocorticoid-mediated alterations in fluidity of rabbit cardiac muscle microvessel endothelial cell membranes: influences on eicosanoid release. Biochim Biophys Acta 1065(1):63–68. https://doi.org/10.1016/0005-2736(91)90011-v

Gibbens JC, Pascoe SJ, Evans SJ, Davies RH, Sayers AR (2001) A trial of biosecurity as a means to control Campylobacter infection of broiler chickens. Prev Vet Med 48(2):85–99. https://doi.org/10.1016/s0167-5877(00)00189-6

Gilpin BJ, Robson B, Scholes P, Nourozi F, Sinton LW (2009) Survival of Campylobacter spp. in bovine faeces on pasture. Lett Appl Microbiol 48(2):162–166. https://doi.org/10.1111/j.1472-765X.2008.02496.x

Golz G, Rosner B, Hofreuter D, Josenhans C, Kreienbrock L, Lowenstein A, Schielke A, Stark K, Suerbaum S, Wieler LH, Alter T (2014) Relevance of Campylobacter to public health—the need for a one health approach. Int J Med Microbiol 304(7):817–823. https://doi.org/10.1016/j.ijmm.2014.08.015

Hald B, Skovgard H, Bang DD, Pedersen K, Dybdahl J, Jespersen JB, Madsen M (2004) Flies and Campylobacter infection of broiler flocks. Emerg Infect Dis 10(8):1490–1492. https://doi.org/10.3201/eid1008.040129

Hald B, Skovgard H, Pedersen K, Bunkenborg H (2008) Influxed insects as vectors for Campylobacter jejuni and Campylobacter coli in Danish broiler houses. Poult Sci 87(7):1428–1434. https://doi.org/10.3382/ps.2007-00301

Hald B, Sommer HM, Skovgard H (2007) Use of fly screens to reduce Campylobacter spp. introduction in broiler houses. Emerg Infect Dis 13(12):1951–1953. https://doi.org/10.3201/eid1312.070488

Hald B, Wedderkopp A, Madsen M (2000) Thermophilic Campylobacter spp. in Danish broiler production: a cross-sectional survey and a retrospective analysis of risk factors for occurrence in broiler flocks. Avian Pathol 29(2):123–131. https://doi.org/10.1080/03079450094153

Hansson I, Vagsholm I, Svensson L, Olsson Engvall E (2007) Correlations between Campylobacter spp. prevalence in the environment and broiler flocks. J Appl Microbiol 103(3):640–649. https://doi.org/10.1111/j.1365-2672.2007.03291.x

Hao X, Cao W, Li B, Zhang Q, Wang C, Ge L (2014) Slightly acidic electrolyzed water for reducing airborne microorganisms in a layer breeding house. J Air Waste Manag Assoc 64(4):494–500. https://doi.org/10.1080/10962247.2013.870940

Hartnack S, Doherr MG, Alter T, Toutounian-Mashad K, Greiner M (2009) Campylobacter monitoring in German broiler flocks: an explorative time series analysis. Zoonoses Public Health 56(3):117–128. https://doi.org/10.1111/j.1863-2378.2008.01184.x

Hasan MM, Talukder S, Mandal AK, Tasmim ST, Parvin MS, Ali MY, Sikder MH, Islam MT (2020) Prevalence and risk factors of Campylobacter infection in broiler and cockerel flocks in Mymensingh and Gazipur districts of Bangladesh. Prev Vet Med 180:105034. https://doi.org/10.1016/j.prevetmed.2020.105034

Hermans D, Pasmans F, Messens W, Martel A, Van Immerseel F, Rasschaert G, Heyndrickx M, Van Deun K, Haesebrouck F (2012) Poultry as a host for the zoonotic pathogen Campylobacter jejuni. Vector Borne Zoonotic Dis 12(2):89–98. https://doi.org/10.1089/vbz.2011.0676

Humphrey S, Chaloner G, Kemmett K, Davidson N, Williams N, Kipar A, Humphrey T, Wigley P (2014) Campylobacter jejuni is not merely a commensal in commercial broiler chickens and affects bird welfare. mBio 5(4):e01364–01314. https://doi.org/10.1128/mBio.01364-14

Hutchison ML, Walters LD, Avery SM, Moore A (2005) Decline of zoonotic agents in livestock waste and bedding heaps. J Appl Microbiol 99(2):354–362. https://doi.org/10.1111/j.1365-2672.2005.02591.x

Hutchison ML, Walters LD, Avery SM, Munro F, Moore A (2005) Analyses of livestock production, waste storage, and pathogen levels and prevalences in farm manures. Appl Environ Microbiol 71(3):1231–1236. https://doi.org/10.1128/AEM.71.3.1231-1236.2005

Hutchison ML, Walters LD, Moore A, Avery SM (2005) Declines of zoonotic agents in liquid livestock wastes stored in batches on-farm. J Appl Microbiol 99(1):58–65. https://doi.org/10.1111/j.1365-2672.2005.02585.x

Hutchison ML, Walters LD, Moore T, Thomas DJ, Avery SM (2005) Fate of pathogens present in livestock wastes spread onto fescue plots. Appl Environ Microbiol 71(2):691–696. https://doi.org/10.1128/AEM.71.2.691-696.2005

Inglis GD, McAllister TA, Larney FJ, Topp E (2010) Prolonged survival of Campylobacter species in bovine manure compost. Appl Environ Microbiol 76(4):1110–1119. https://doi.org/10.1128/AEM.01902-09

Johannessen GS, Garofolo G, Di Serafino G, Kolackova I, Karpiskova R, Wieczorek K, Osek J, Christensen J, Torp M, Hoorfar J (2020) Campylobacter in chicken—critical parameters for international, multicentre evaluation of air sampling and detection methods. Food Microbiol 90:103455. https://doi.org/10.1016/j.fm.2020.103455

Johnsen G, Zimmerman K, Lindstedt BA, Vardund T, Herikstad H, Kapperud G (2006) Intestinal carriage of Campylobacter jejuni and Campylobacter coli among cattle from south-western Norway and comparative genotyping of bovine and human isolates by amplified-fragment length polymorphism. Acta Vet Scand 48:4. https://doi.org/10.1186/1751-0147-48-4

Jones K (2001) Campylobacters in water, sewage and the environment. Symp Ser Soc Appl Microbiol 30:68S-79S. https://doi.org/10.1046/j.1365-2672.2001.01355.x

Jonsson ME, Chriel M, Norstrom M, Hofshagen M (2012) Effect of climate and farm environment on Campylobacter spp. colonisation in Norwegian broiler flocks. Prev Vet Med 107(1–2):95–104. https://doi.org/10.1016/j.prevetmed.2012.05.002

Josefsen MH, Lofstrom C, Hansen TB, Christensen LS, Olsen JE, Hoorfar J (2010) Rapid quantification of viable Campylobacter bacteria on chicken carcasses, using real-time PCR

and propidium monoazide treatment, as a tool for quantitative risk assessment. Appl Environ Microbiol 76(15):5097–5104. https://doi.org/10.1128/AEM.00411-10
Julich S, Hotzel H, Gartner C, Trouchet D, El Metwaly F, Ahmed M, Kemper N, Tomaso H (2016) Evaluation of a microfluidic chip system for preparation of bacterial DNA from swabs, air, and surface water samples. Biologicals 44(6):574–580. https://doi.org/10.1016/j.biologicals.2016.06.013
Kearney TE, Larkin MJ, Frost JP, Levett PN (1993) Survival of pathogenic bacteria during mesophilic anaerobic digestion of animal waste. J Appl Bacteriol 75(3):215–219. https://doi.org/10.1111/j.1365-2672.1993.tb02768.x
Kearney TE, Larkin MJ, Levett PN (1993) The effect of slurry storage and anaerobic digestion on survival of pathogenic bacteria. J Appl Bacteriol 74(1):86–93. https://doi.org/10.1111/j.1365-2672.1993.tb03000.x
Klein D, Alispahic M, Sofka D, Iwersen M, Drillich M, Hilbert F (2013) Prevalence and risk factors for shedding of thermophilic Campylobacter in calves with and without diarrhea in Austrian dairy herds. J Dairy Sci 96(2):1203–1210. https://doi.org/10.3168/jds.2012-5987
Kruger NJ, Buhler C, Iwobi AN, Huber I, Ellerbroek L, Appel B, Stingl K (2014) Limits of control—crucial parameters for a reliable quantification of viable campylobacter by real-time PCR. PLoS ONE 9(2):e88108. https://doi.org/10.1371/journal.pone.0088108
Lazaro B, Carcamo J, Audicana A, Perales I, Fernandez-Astorga A (1999) Viability and DNA maintenance in nonculturable spiral Campylobacter jejuni cells after long-term exposure to low temperatures. Appl Environ Microbiol 65(10):4677–4681. https://doi.org/10.1128/AEM.65.10.4677-4681.1999
Line JE (2002) Campylobacter and Salmonella populations associated with chickens raised on acidified litter. Poult Sci 81(10):1473–1477. https://doi.org/10.1093/ps/81.10.1473
Line JE (2006) Influence of relative humidity on transmission of Campylobacter jejuni in broiler chickens. Poult Sci 85(7):1145–1150. https://doi.org/10.1093/ps/85.7.1145
Line JE, Bailey JS (2006) Effect of on-farm litter acidification treatments on Campylobacter and Salmonella populations in commercial broiler houses in northeast Georgia. Poult Sci 85(9):1529–1534. https://doi.org/10.1093/ps/85.9.1529
Manyi-Loh CE, Mamphweli SN, Meyer EL, Okoh AI, Makaka G, Simon M (2014) Inactivation of selected bacterial pathogens in dairy cattle manure by mesophilic anaerobic digestion (balloon type digester). Int J Environ Res Public Health 11(7):7184–7194. https://doi.org/10.3390/ijerph110707184
Martens W, Bohm R (2009) Overview of the ability of different treatment methods for liquid and solid manure to inactivate pathogens. Bioresour Technol 100(22):5374–5378. https://doi.org/10.1016/j.biortech.2009.01.014
Martens W, Martinec M, Zapirain R, Stark M, Hartung E, Palmgren U (2001) Reduction potential of microbial, odour and ammonia emissions from a pig facility by biofilters. Int J Hyg Environ Health 203(4):335–345. https://doi.org/10.1078/1438-4639-00035
Mdegela RH, Laurence K, Jacob P, Nonga HE (2011) Occurrences of thermophilic Campylobacter in pigs slaughtered at Morogoro slaughter slabs Tanzania. Trop Anim Health Prod 43(1):83–87. https://doi.org/10.1007/s11250-010-9657-4
Meldrum RJ, Griffiths JK, Smith RM, Evans MR (2005) The seasonality of human campylobacter infection and Campylobacter isolates from fresh, retail chicken in Wales. Epidemiol Infect 133(1):49–52. https://doi.org/10.1017/s0950268804003188
Milnes AS, Stewart I, Clifton-Hadley FA, Davies RH, Newell DG, Sayers AR, Cheasty T, Cassar C, Ridley A, Cook AJ, Evans SJ, Teale CJ, Smith RP, McNally A, Toszeghy M, Futter R, Kay A, Paiba GA (2008) Intestinal carriage of verocytotoxigenic Escherichia coli O157, Salmonella, thermophilic Campylobacter and Yersinia enterocolitica, in cattle, sheep and pigs at slaughter in Great Britain during 2003. Epidemiol Infect 136(6):739–751. https://doi.org/10.1017/S0950268807009223
Mohammed AN, Abdel Aziz SAA (2019) The prevalence of Campylobacter species in broiler flocks and their environment: assessing the efficiency of chitosan/zinc oxide nanocomposite for

adopting control strategy. Environ Sci Pollut Res Int 26(29):30177–30187. https://doi.org/10.1007/s11356-019-06030-z

Mullner P, Collins-Emerson JM, Midwinter AC, Carter P, Spencer SE, van der Logt P, Hathaway S, French NP (2010) Molecular epidemiology of Campylobacter jejuni in a geographically isolated country with a uniquely structured poultry industry. Appl Environ Microbiol 76(7):2145–2154. https://doi.org/10.1128/AEM.00862-09

Murphy C, Carroll C, Jordan KN (2006) Environmental survival mechanisms of the foodborne pathogen Campylobacter jejuni. J Appl Microbiol 100(4):623–632. https://doi.org/10.1111/j.1365-2672.2006.02903.x

Nathues C, Gruning P, Fruth A, Verspohl J, Blaha T, Kreienbrock L, Merle R (2013) Campylobacter spp., Yersinia enterocolitica, and Salmonella enterica and their simultaneous occurrence in German fattening pig herds and their environment. J Food Prot 76(10):1704–1711. https://doi.org/10.4315/0362-028X.JFP-13-076

Nicholson FA, Groves SJ, Chambers BJ (2005) Pathogen survival during livestock manure storage and following land application. Bioresour Technol 96(2):135–143. https://doi.org/10.1016/j.biortech.2004.02.030

Northcutt JK, Berrang ME (2006) Influence of a chicken transport cage-washing system on wastewater characteristics and bacteria recovery from cage flooring. J Appl Poult Res 15(3):457–463. https://doi.org/10.1093/japr/15.3.457

O'Mahony E, Buckley JF, Bolton D, Whyte P, Fanning S (2011) Molecular epidemiology of Campylobacter isolates from poultry production units in southern Ireland. PLoS ONE 6(12):e28490. https://doi.org/10.1371/journal.pone.0028490

Obiri-Danso K, Paul N, Jones K (2001) The effects of UVB and temperature on the survival of natural populations and pure cultures of Campylobacter jejuni, Camp. coli, Camp. lari and urease-positive thermophilic campylobacters (UPTC) in surface waters. J Appl Microbiol 90(2):256–267. https://doi.org/10.1046/j.1365-2672.2001.01239.x

Ocejo M, Oporto B, Hurtado A (2019) Occurrence of Campylobacter jejuni and Campylobacter coli in cattle and sheep in Northern Spain and changes in antimicrobial resistance in two studies 10-years apart. Pathogens 8(3). https://doi.org/10.3390/pathogens8030098

Olsen KN, Lund M, Skov J, Christensen LS, Hoorfar J (2009) Detection of Campylobacter bacteria in air samples for continuous real-time monitoring of Campylobacter colonization in broiler flocks. Appl Environ Microbiol 75(7):2074–2078. https://doi.org/10.1128/AEM.02182-08

Oporto B, Esteban JI, Aduriz G, Juste RA, Hurtado A (2007) Prevalence and strain diversity of thermophilic campylobacters in cattle, sheep and swine farms. J Appl Microbiol 103(4):977–984. https://doi.org/10.1111/j.1365-2672.2007.03328.x

Osimani A, Aquilanti L, Pasquini M, Clementi F (2017) Prevalence and risk factors for thermotolerant species of Campylobacter in poultry meat at retail in Europe. Poult Sci 96(9):3382–3391. https://doi.org/10.3382/ps/pex143

Oyarzabal OA, Backert S (2012) Microbial Food Safety: An Introduction (Food Science Text Series). Springer, Heidelberg

Pacholewicz E, Buhler C, Wulsten IF, Kraushaar B, Luu HQ, Iwobi AN, Huber I, Stingl K (2019) Internal sample process control improves cultivation-independent quantification of thermotolerant Campylobacter. Food Microbiol 78:53–61. https://doi.org/10.1016/j.fm.2018.09.017

Park SF (2002) The physiology of Campylobacter species and its relevance to their role as foodborne pathogens. Int J Food Microbiol 74(3):177–188. https://doi.org/10.1016/s0168-1605(01)00678-x

Patrone V, Campana R, Vallorani L, Dominici S, Federici S, Casadei L, Gioacchini AM, Stocchi V, Baffone W (2013) CadF expression in Campylobacter jejuni strains incubated under low-temperature water microcosm conditions which induce the viable but non-culturable (VBNC) state. Antonie Van Leeuwenhoek 103(5):979–988. https://doi.org/10.1007/s10482-013-9877-5

Piccirillo A, Giacomelli M, Niero G, De Luca C, Carraro L, Ortali G, Mughini-Gras L (2018) Multilocus sequence typing of Campylobacter jejuni and Campylobacter coli to identify potential sources of colonization in commercial turkey farms. Avian Pathol 47(5):455–466. https://doi.org/10.1080/03079457.2018.1487529

Pinto D, Santos MA, Chambel L (2015) Thirty years of viable but nonculturable state research: unsolved molecular mechanisms. Crit Rev Microbiol 41(1):61–76. https://doi.org/10.3109/1040841X.2013.794127

Ramabu SS, Boxall NS, Madie P, Fenwick SG (2004) Some potential sources for transmission of Campylobacter jejuni to broiler chickens. Lett Appl Microbiol 39(3):252–256. https://doi.org/10.1111/j.1472-765X.2004.01573.x

Ramamurthy T, Ghosh A, Pazhani GP, Shinoda S (2014) Current Perspectives on viable but non-culturable (VBNC) pathogenic bacteria. Front Public Health 2:103. https://doi.org/10.3389/fpubh.2014.00103

Ridley A, Morris V, Gittins J, Cawthraw S, Harris J, Edge S, Allen V (2011) Potential sources of Campylobacter infection on chicken farms: contamination and control of broiler-harvesting equipment, vehicles and personnel. J Appl Microbiol 111(1):233–244. https://doi.org/10.1111/j.1365-2672.2011.05038.x

Rosenquist H, Boysen L, Galliano C, Nordentoft S, Ethelberg S, Borck B (2009) Danish strategies to control Campylobacter in broilers and broiler meat: facts and effects. Epidemiol Infect 137(12):1742–1750. https://doi.org/10.1017/S0950268809002684

Rosenquist H, Sommer HM, Nielsen NL, Christensen BB (2006) The effect of slaughter operations on the contamination of chicken carcasses with thermotolerant Campylobacter. Int J Food Microbiol 108(2):226–232. https://doi.org/10.1016/j.ijfoodmicro.2005.12.007

Rukambile E, Sintchenko V, Muscatello G, Kock R, Alders R (2019) Infection, colonization and shedding of Campylobacter and Salmonella in animals and their contribution to human disease: a review. Zoonoses Public Health 66(6):562–578. https://doi.org/10.1111/zph.12611

Sahin O, Kassem II, Shen Z, Lin J, Rajashekara G, Zhang Q (2015) Campylobacter in poultry: ecology and potential interventions. Avian Dis 59(2):185–200. https://doi.org/10.1637/11072-032315-Review

Sahin O, Kobalka P, Zhang Q (2003) Detection and survival of Campylobacter in chicken eggs. J Appl Microbiol 95(5):1070–1079. https://doi.org/10.1046/j.1365-2672.2003.02083.x

Sahin O, Yaeger M, Wu Z, Zhang Q (2017) Campylobacter-associated diseases in animals. Annu Rev Anim Biosci 5:21–42. https://doi.org/10.1146/annurev-animal-022516-022826

Schets FM, Jacobs-Reitsma WF, van der Plaats RQJ, Heer LK, van Hoek A, Hamidjaja RA, de Roda Husman AM, Blaak H (2017) Prevalence and types of Campylobacter on poultry farms and in their direct environment. J Water Health 15(6):849–862. https://doi.org/10.2166/wh.2017.119

Schulz J, Bao E, Clauss M, Hartung J (2013) The potential of a new air cleaner to reduce airborne microorganisms in pig house air: preliminary results. Berl Munch Tierarztl Wochenschr 126(3–4):143–148

Seedorf J (2004) An emission inventory of livestock-related bioaerosols for Lower Saxony Germany. Atmos Environ 38(38):6565–6581. https://doi.org/10.1016/j.atmosenv.2004.08.023

Seedorf J, Hartung J (1999) Reduction efficiencies of a biofilter and a bio-scrubber as bio-aerosols in two piggeries. Berl Munch Tierarztl Wochenschr 112(12):444–447

Seinige D, von Kockritz-Blickwede M, Krischek C, Klein G, Kehrenberg C (2014) Influencing factors and applicability of the viability EMA-qPCR for a detection and quantification of Campylobacter cells from water samples. PLoS ONE 9(11):e113812. https://doi.org/10.1371/journal.pone.0113812

Seliwiorstow T, Duarte A, Bare J, Botteldoorn N, Dierick K, Uyttendaele M, De Zutter L (2015) Comparison of sample types and analytical methods for the detection of highly campylobacter-colonized broiler flocks at different stages in the poultry meat production chain. Foodborne Pathog Dis 12(5):399–405. https://doi.org/10.1089/fpd.2014.1894

Siller P, Daehre K, Thiel N, Nubel U, Roesler U (2020) Impact of short-term storage on the quantity of extended-spectrum beta-lactamase-producing Escherichia coli in broiler litter under practical conditions. Poult Sci 99(4):2125–2135. https://doi.org/10.1016/j.psj.2019.11.043

Silva J, Leite D, Fernandes M, Mena C, Gibbs PA, Teixeira P (2011) Campylobacter spp. as a foodborne pathogen: a review. Front Microbiol 2:200. https://doi.org/10.3389/fmicb.2011.00200

Sinton L, Hall C, Braithwaite R (2007) Sunlight inactivation of Campylobacter jejuni and Salmonella enterica, compared with Escherichia coli, in seawater and river water. J Water Health 5(3):357–365. https://doi.org/10.2166/wh.2007.031

Sinton LW, Braithwaite RR, Hall CH, Mackenzie ML (2007) Survival of indicator and pathogenic bacteria in bovine feces on pasture. Appl Environ Microbiol 73(24):7917–7925. https://doi.org/10.1128/AEM.01620-07

Slader J, Domingue G, Jorgensen F, McAlpine K, Owen RJ, Bolton FJ, Humphrey TJ (2002) Impact of transport crate reuse and of catching and processing on Campylobacter and Salmonella contamination of broiler chickens. Appl Environ Microbiol 68(2):713–719. https://doi.org/10.1128/aem.68.2.713-719.2002

Smith S, Meade J, Gibbons J, McGill K, Bolton D, Whyte P (2016) The impact of environmental conditions on Campylobacter jejuni survival in broiler faeces and litter. Infect Ecol Epidemiol 6:31685. https://doi.org/10.3402/iee.v6.31685

Sondergaard MS, Josefsen MH, Lofstrom C, Christensen LS, Wieczorek K, Osek J, Hoorfar J (2014) Low-cost monitoring of Campylobacter in poultry houses by air sampling and quantitative PCR. J Food Prot 77(2):325–330. https://doi.org/10.4315/0362-028X.JFP-13-268

Sproston EL, Ogden ID, MacRae M, Dallas JF, Sheppard SK, Cody AJ, Colles FM, Wilson MJ, Forbes KJ, Strachan NJ (2011) Temporal variation and host association in the Campylobacter population in a longitudinal ruminant farm study. Appl Environ Microbiol 77(18):6579–6586. https://doi.org/10.1128/AEM.00428-11

Sproston EL, Ogden ID, MacRae M, Forbes KJ, Dallas JF, Sheppard SK, Cody A, Colles F, Wilson MJ, Strachan NJ (2010) Multi-locus sequence types of Campylobacter carried by flies and slugs acquired from local ruminant faeces. J Appl Microbiol 109(3):829–838. https://doi.org/10.1111/j.1365-2672.2010.04711.x

Stafford RJ, Schluter PJ, Wilson AJ, Kirk MD, Hall G, Unicomb L, OzFoodNet Working G (2008) Population-attributable risk estimates for risk factors associated with Campylobacter infection, Australia. Emerg Infect Dis 14(6):895–901. https://doi.org/10.3201/eid1406.071008

Stanley K, Jones K (2003) Cattle and sheep farms as reservoirs of Campylobacter. J Appl Microbiol 94(Suppl):104S-113S. https://doi.org/10.1046/j.1365-2672.94.s1.12.x

Talibart R, Denis M, Castillo A, Cappelier JM, Ermel G (2000) Survival and recovery of viable but noncultivable forms of Campylobacter in aqueous microcosm. Int J Food Microbiol 55(1–3):263–267. https://doi.org/10.1016/s0168-1605(00)00201-4

Tang Y, Meinersmann RJ, Sahin O, Wu Z, Dai L, Carlson J, Plumblee Lawrence J, Genzlinger L, LeJeune JT, Zhang Q (2017) Wide but variable distribution of a hypervirulent campylobacter jejuni clone in beef and dairy cattle in the United States. Appl Environ Microbiol 83(24). https://doi.org/10.1128/AEM.01425-17

Tenzin S, Ogunniyi AD, Khazandi M, Ferro S, Bartsch J, Crabb S, Abraham S, Deo P, Trott DJ (2019) Decontamination of aerosolised bacteria from a pig farm environment using a pH neutral electrochemically activated solution (Ecas4 anolyte). PLoS ONE 14(9):e0222765. https://doi.org/10.1371/journal.pone.0222765

Tymczyna L, Chmielowiec-Korzeniowska A, Raczynska J, Drabik A (2011) Removal of microbial contaminants from pig house air using biofilter organic media. Ann Anim Sci 11(3):453–464

van de Giessen AW, Tilburg JJ, Ritmeester WS, van der Plas J (1998) Reduction of campylobacter infections in broiler flocks by application of hygiene measures. Epidemiol Infect 121(1):57–66. https://doi.org/10.1017/s0950268898008899

Van den Meersche T, Rasschaert G, Vanden Nest T, Haesebrouck F, Herman L, Van Coillie E, Van Weyenberg S, Daeseleire E, Heyndrickx M (2020) Longitudinal screening of antibiotic residues, antibiotic resistance genes and zoonotic bacteria in soils fertilized with pig manure. Environ Sci Pollut Res Int 27(22):28016–28029. https://doi.org/10.1007/s11356-020-09119-y

Vetchapitak T, Misawa N (2019) Current status of campylobacter food poisoning in Japan. Food Saf (Tokyo) 7(3):61–73

Wagner AO, Gstraunthaler G, Illmer P (2008) Survival of bacterial pathogens during the thermophilic anaerobic digestion of biowaste: laboratory experiments and in situ validation. Anaerobe 14(3):181–183. https://doi.org/10.1016/j.anaerobe.2008.03.004

Watabe M, Rao JR, Stewart TA, Xu J, Millar BC, Xiao L, Lowery CJ, Dooley JS, Moore JE (2003) Prevalence of bacterial faecal pathogens in separated and unseparated stored pig slurry. Lett Appl Microbiol 36(4):208–212. https://doi.org/10.1046/j.1472-765x.2003.01293.x

Wilson DJ, Gabriel E, Leatherbarrow AJ, Cheesbrough J, Gee S, Bolton E, Fox A, Fearnhead P, Hart CA, Diggle PJ (2008) Tracing the source of campylobacteriosis. PLoS Genet 4(9):e1000203. https://doi.org/10.1371/journal.pgen.1000203

Zhao Y, Aarnink AJA, Koerkamp PWGG, Hagenaars TJ (2011) Detection of airborne campylobacter with three bioaerosol samplers for alarming bacteria transmission in broilers. Biol Eng 3(4):177–186. https://doi.org/10.13031/2013.37122

Zilbauer M, Dorrell N, Wren BW, Bajaj-Elliott M (2008) Campylobacter jejuni-mediated disease pathogenesis: an update. Trans R Soc Trop Med Hyg 102(2):123–129. https://doi.org/10.1016/j.trstmh.2007.09.019

Zweifel C, Scheu KD, Keel M, Renggli F, Stephan R (2008) Occurrence and genotypes of Campylobacter in broiler flocks, other farm animals, and the environment during several rearing periods on selected poultry farms. Int J Food Microbiol 125(2):182–187. https://doi.org/10.1016/j.ijfoodmicro.2008.03.038

Phage Biocontrol of *Campylobacter*: A One Health Approach

Sophie Kittler, Severin Steffan, Elisa Peh, and Madeleine Plötz

Contents

Abstract Human infections by *Campylobacter* species are among the most reported bacterial gastrointestinal diseases in the European Union and worldwide with severe outcomes in rare cases. Considering the transmission routes and farm animal reservoirs of these zoonotic pathogens, a comprehensive One Health approach will be necessary to reduce human infection rates. Bacteriophages are viruses that specifically infect certain bacterial genera, species, strains or isolates. Multiple studies have demonstrated the general capacity of phage treatments to reduce *Campylobacter* loads in the chicken intestine. However, phage treatments are not yet approved for extensive use in the agro-food industry in Europe. Technical inconvenience is mainly related to the efficacy of phages, depending on the optimal choice of phages and their

S. Kittler (✉) · S. Steffan · E. Peh · M. Plötz
Institute for Food Quality and Food Safety, University of Veterinary Medicine Hannover, Foundation, Bischofsholer Damm 15, 30173 Hannover, Germany
e-mail: sophie.kittler@tiho-hannover.de

S. Steffan
e-mail: Severin.Michael.Steffan@tiho-hannover.de

E. Peh
e-mail: Elisa.Peh@tiho-hannover.de

M. Plötz
e-mail: Madeleine.Ploetz@tiho-hannover.de

S. Backert (ed.), *Fighting Campylobacter Infections*, Current Topics in Microbiology and Immunology 431, https://doi.org/10.1007/978-3-030-65481-8_6

combination, as well as application route, concentration and timing. Additionally, regulatory uncertainties have been a major concern for investment in commercial phage-based products. This review addresses the question as to how phages can be put into practice and can help to solve the issue of human campylobacteriosis in a sustainable One Health approach. By compiling the reported findings from the literature in a standardized manner, we enabled inter-experimental comparisons to increase our understanding of phage infection in *Campylobacter* spp. and practical on-farm studies. Further, we address some of the hurdles that still must be overcome before this new methodology can be adapted on an industrial scale. We envisage that phage treatment can become an integrated and standardized part of a multi-hurdle anti-bacterial strategy in food production. The last part of this chapter deals with some of the issues raised by legal authorities, bringing together current knowledge on *Campylobacter*-specific phages and the biosafety requirements for approval of phage treatment in the food industry.

1 Introduction

Infections by *Campylobacter* species (spp.) are the most commonly reported gastrointestinal bacterial infections in the European Union (EU), with relative rare cases resulting in more serious outcomes (EFSA 2019). Although the disease is normally self-limiting, due to the high number of cases, a small fraction resulting in long-term immunological or neurological symptoms increases the overall public health burden of campylobacteriosis significantly. *Campylobacter* infections represent a major target for reducing the public health burden of intestinal infectious diseases worldwide (Newell et al. 2010). The *Campylobacter* species *C. jejuni, C. coli, C. lari* and *C. upsaliensis* are thermophilic and grow optimally at 42 °C. Importantly, *C. jejuni* and *C. coli* are most frequently associated with human gastrointestinal disease (Lee et al. 2016), but represent commensals in the intestinal tract of various warm-blooded animals. Both species colonize the chicken intestine at high concentrations of more than $\log_{10}7$ colony forming units (CFU) per gram cecal content without causing any symptoms of infection (Reich et al. 2008; Quinn 2011; Kittler et al. 2013). More than 45% of fresh-skinned broiler meat was found to be contaminated with *Campylobacter* spp. in 2018 in the EU (EFSA 2019).

It has been recognized for some time, based on different risk assessments, that a reduction of bacterial food contamination by a few $\log_{10}$-units might be sufficient to significantly improve public health, whereas complete elimination of the pathogen might not be feasible (Rosenquist et al. 2003; FDA 2003). Considering the complexity of interacting factors that impact the disease burden associated with *Campylobacter* spp., phage intervention is considered as a methodology to be incorporated in a combination of other measures taken to reduce the bacterial loads on meat. Phage-based biocontrol should be applied at the optimal stage of the food production chain with respect to the efficiency of bacterial reduction and its impact on public health. To evaluate presently available studies in regard to this question, here we discuss the Pro's and Con's of different settings. In accordance with the complexity of model systems, trials performed to reduce bacterial loads on meat (here collectively described as in vitro trials) will be discussed separately from trials where live chickens are treated with phages.

To reduce the risk for *Campylobacter* infections arising from broiler meat consumption in the EU, a limit of 1000 CFU of *Campylobacter* per gram neck skin of chilled broiler carcasses was set. If *Campylobacter* counts exceed this regulatory limit, improvements in slaughter hygiene and critical reviews of the process controls and biosecurity at farm level are mandatory (EG 2017). However, since *C. jejuni* and *C. coli* are well adapted to the avian and porcine gut, respectively, their control and eradication in poultry and pig farming represents an immense challenge for food production industries and food monitoring by state authorities (Lin 2009; Newell et al. 2010). Considering the transmission routes and reservoirs of thermophilic *Campylobacter* spp., a comprehensive One Health approach is vital for preventing human infections (Gölz et al. 2014).

Bacteriophages (phages) are viruses that specifically infect certain bacterial species or strains therein (Kutter 2009). Phages have been shown to be effective in reducing *Campylobacter* loads in the chicken intestine in various studies (Wagenaar et al. 2005; Loc Carrillo et al. 2005; El-Shibiny et al. 2009; Carvalho et al. 2010b, Fischer et al. 2013a; Kittler et al. 2013; Hammerl et al. 2014) and could potentially be used in combination with other measures in a multi-hurdle approach (Klein et al. 2015; Moye et al. 2018). A number of studies have effectively used phages to reduce *Campylobacter* numbers post-harvest (Goode et al. 2003; Atterbury et al. 2003a) and in biofilms (Siringan et al. 2011). The *Campylobacter*-specific phages used in these studies belong to the subfamily of Eucampyvirinae (family Myoviridae, whose members have icosahedral heads containing a genome of double-stranded DNA and contractile tails) as far as their taxonomic position has been established, to which two genera are currently assigned: *Firehammervirus* and *Fletchervirus* (Crippen et al. 2019). Unlike in bacterial nomenclature, the genus name does not typically appear in the virus species name: currently, five *Firehammervirus* species and more than 14 *Fletchervirus* species are known that are all named 'Campylobacter virus' (or 'Campylobacter phage') with a letter/number combination. In the past, some *Fletcherviruses* were named Cp8viruses, Cp8unalikeviruses or Group III phages, while *Firehammerviruses* were formerly known as Cp220viruses, Cp220likeviruses or Group II phages. The members of *Fletchervirus* and *Firehammervirus* differ in their host specificity. While the latter infect both *C. jejuni* and *C. coli* strains, *Fletchervirus* species specifically infect *C. jejuni* (Wagenaar et al. 2005; Loc Carrillo et al. 2005; Denou et al. 2009; Kittler et al. 2013; Jackel et al. 2019).

The specificity of bacteriophages offers the opportunity to target pathogenic bacteria without disrupting the benign microflora at the site of application (Abedon 2014; Galtier et al. 2016; Richards et al. 2019). One successful example is the *Listeria*-specific phage P100, which has been applied for more than ten years in food production settings and has shown to remain highly efficient in bacterial reduction (Moye et al. 2018). Nonetheless, many phages are not yet approved for extensive use in the agro-food industry in the EU, in contrast to the USA, Israel, Canada,

Switzerland, Australia and New Zealand that have already adopted appropriate regulations (Fernandez et al. 2018). Technical and legal issues were regarded as the main reasons for the hesitation in the EU (Fernandez et al. 2018). Technical issues are mainly related to differences in storage requirements and efficacy of different phages; for each application, the right choice of phage species or combination of species, their pharmaceutical formulation, the optimal infection route, concentration and the timing of application must be established, as general rules can hardly be defined. This provides regulatory hurdles and uncertainties that have been a major concern for industrial investments, to which the issue of intellectual property can be added. At present, most efforts were the result of start-up companies that have not yet been applied to large-scale industrial production.

This review summarizes the findings reported in the literature and further focuses on the questions that must be addressed in order to use phages in agricultural practices, as we believe they can contribute to solve the health issues of human campylobacteriosis in a sustainable One Health approach (Fig. 1). Reviews discussing the efficacy

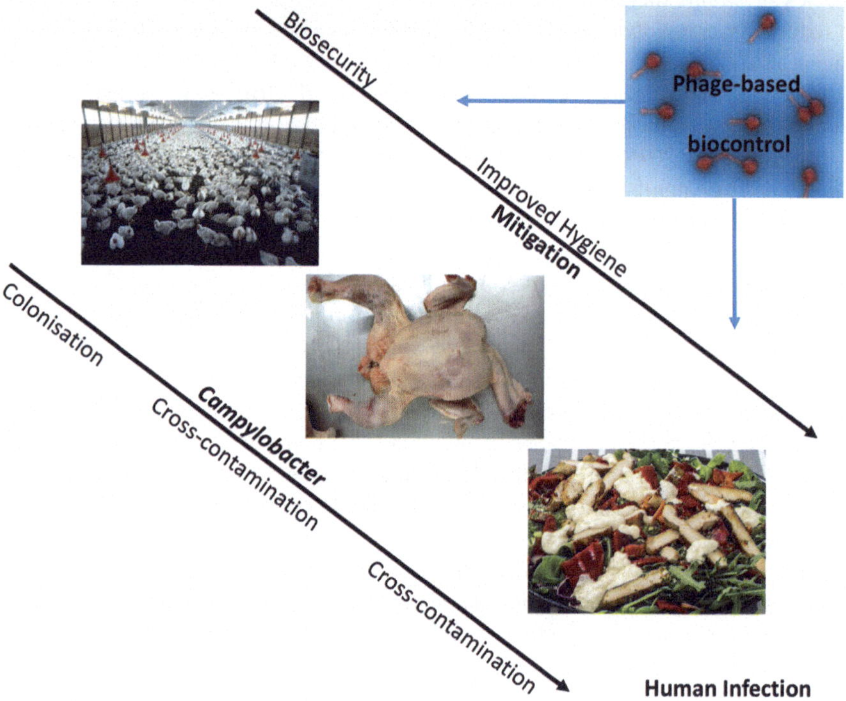

Fig. 1 ***Campylobacter*distribution and control measures along the food chain.** *Campylobacter* spp. colonization of broiler flocks starts when the animals are two to three weeks old. Cross-contamination of meat with *Campylobacter* occurs at slaughter and during food preparation. Improved biosecurity and hygiene measures could prevent some positive broiler flocks. Phage-based biocontrol of *Campylobacter* could reduce the level of bacterial contamination in the flocks and on food products in future applications

of phage applications in vitro and in vivo have been previously published, and these have already pointed out the need for a better understanding of the mode of action of the used bacteriophages (Connerton et al. 2011; Kittler et al. 2013). Multiple recent studies have addressed the biology of phage infection on a molecular basis (Holst Sørensen et al. 2012; Gencay et al. 2018; Crippen et al. 2019). Bringing together the knowledge of practical on-farm phage applications and the recent increased understanding of phage infection mechanisms in *Campylobacter* spp. will help to address issues of technical inconvenience and adapt novel application concepts. In the last part of this chapter, some of the issues raised by legal authorities, especially by EFSA, will be discussed, to compile present knowledge on *Campylobacter*-specific phages and the prerequisites for phage approval according to current knowledge (EFSA , 2012, 2009,2016).

2 *Campylobacter* Bacteriophages in a One Health Approach

Phages can be used at different stages of the poultry food production chain, with the ultimate goal to reduce bacterial load on food items at retail. Risk assessment models have suggested that interventions aiming at a reduction in *Campylobacter* loads directly on the meat might be most effective. Nevertheless, the public health benefits of controlling viable *Campylobacter* in the poultry flocks might be higher, since *Campylobacter* may also spread from farms to humans by pathways other than via contaminated broiler meat (FDA 2003; EFSA 2011). When considering phage applications in terms of a One Health approach, three aspects should be examined in combination to enable effectivity against human campylobacteriosis in the long term: (i) fundamental knowledge on the interaction between *Campylobacter* and the chicken-host must be considered, as this may provide new perspectives and possibilities for comprehensive solutions by selecting 'weak points' of the colonization or contamination process as targets for successful reduction measures; (ii) since the outcome of the disease in humans results from a complex interplay of factors, all of these must be considered to identify the optimal point(s) for intervention that will achieve maximum effects at clinical endpoints; and (iii) different health and safety prerequisites in veterinary and human medicine need to be considered, as well as the perspectives from all stakeholders (e.g., veterinary authorities, public health authorities, regulatory agencies, farmers, industry, human medical experts and policymakers) in order to facilitate the implementation of phage application on a regular basis.

A word of caution is needed here. Even if phage application, in combination with possible other intervention strategies, would be able to eradicate *Campylobacter* spp. from industrially produced poultry, this would only partially reduce campylobacteriosis cases in humans. According to current knowledge, chickens and turkeys (which can be heavily colonized with *Campylobacter* without showing symptoms) represent the major source for campylobacteriosis cases in humans. However, other sources, such as ruminants and diverse environmental sources also considerably contribute to direct or indirect transmission to humans (Baldvinsson et al. 2014; Nowaczek et al. 2019). While foodborne transmission is the most important route for the majority of human infections (Baldvinsson et al. 2014), differences may exist in pathogenicity between *C. jejuni* strains that reach humans from different sources (Nowaczek et al. 2019). Currently, risk intervention and infection models do not yet reflect the full complexity of infection routes, while mitigation strategies should consider all of the known routes for effective intervention. These subjects are further discussed in Chap. 2 of this book. All applied measures are ultimately targeted to prevent the spread of highly pathogenic *Campylobacter* strains to humans. However, such targeted reduction continues to represent an immense challenge, since the level of pathogenicity cannot be easily determined for *Campylobacter* (Dasti et al. 2010). The outcome of the disease varies from mild to severe gastroenteritis and the post-infectious sequelae such as reactive arthritis, Guillain-Barré syndrome, irritable bowel syndrome and inflammatory bowel disease all contribute to a considerate public health burden, as is further discussed in Chap. 2. The contribution of individual bacterial genotypes, that may be particularly found in certain reservoirs, or may be particularly responsible for the more severe cases of campylobacteriosis, to the overall disease burden should be investigated in a One Health strategy (Baldvinsson et al. 2014).

In the context of food safety and the stakeholders involved, phages fit many requirements for sustainable biocontrol measures. The practical feasibility of any intervention strategy in broiler production is limited by economic considerations, as cost constraints are very strong in this industry. Moreover, the logistics of broiler production are tightly planned and timed, so that any additional steps must be compatible with current production systems. Fortunately, phage production is inexpensive and easy to achieve at an industrial level (Moye et al. 2018). Furthermore, phages can be easily applied to flocks via drinking water (Kittler et al. 2013). Any intervention strategy must be acceptable to consumers, who may pay attention to criteria such as environmental friendliness or the relatively vague characteristic of an intervention being 'natural' (Román et al. 2017). Phages do not change organoleptic properties of the treated food products (Loc-Carrillo and Abedon 2011; Moye et al. 2018), so this would not hamper acceptance. However, although people are continuously being exposed to bacteriophages with no known side effects, information to the general public must be carefully phrased, e.g., using the term 'phage' instead of 'virus,' to avoid negative associations. So far, phage science has received quite a positive image in the press and in the general public, and the use of phages in food production is easy to explain with reference to food safety and the ubiquitous nature of phages.

3 Interactions of Bacteria with Their Bacteriophages and the Development of Resistance

Bacteriophages infect bacterial cells and use their metabolism to replicate. In the final step of an infection, they lyse their host, thus releasing phage progenies into the surrounding environment. During the infection process, phages can enter two different lifecycles: phages that follow a lytic lifecycle transform the bacterial metabolism and use it to produce proteins for phage production which are coded on the bacteriophage's genome. This results in release of newly formed phages and induces bacterial lysis. In contrast, temperate phages follow a lysogenic lifecycle and integrate their genome into the bacterial chromosome. The integrated phage genome is called a prophage and replicates as part of the bacterial genome, until the phage finally enters the lytic life cycle, initiated by unfavorable environmental conditions and complex communication systems. A recent study demonstrated that communication through small molecules is essential for a lysis-lysogeny decision in phages infecting *Bacillus* species (Erez et al. 2017). Since phages for biocontrol or therapeutic applications have to be strictly lytic for safety reasons, the lysogenic cycle is not a subject of this review. The infection process of lytic phages includes (i) attachment of the phage to the surface of the bacterial host cell by specific recognition between a receptor located on the surface of the bacterial cell and the phage's receptor binding protein (RBP) located at the tip of the phage's tail. A variety of molecules with different functions can act as receptors for phage binding, such as porins and other transport proteins, flagella, structural proteins, and enzymes that are surface exposed (Mondigler et al. 1995; Baldvinsson et al. 2014; Sørensen et al. 2015; Islam et al. 2019). (ii) After successful binding of the phage, which may include initial reversible adsorption followed by an irreversible adsorption step, a conformational change in the phage's baseplate finally provokes contraction of the phages outer envelope and injection of nucleic acid into the bacterial host cell. Subsequent production of structural proteins results in phage morphogenesis, which is directed by a complex system of phage-encoded proteins (reviewed in Aksyuk and Rossman 2011). Progeny release and final lysis of the bacterial host is achieved by a single lysis protein or by a holin–endolysin system (Young and Young, 1982; Wang et al. 2003). In *Campylobacter*-specific phages, so far only few mechanisms and proteins of the infection process have been identified. Capsular polysaccharide (CPS) moieties, a motile flagellum and glycosylated flagellin have all been implied as structures involved in phage adsorption (Javed et al. 2015; Baldvinsson et al. 2014; Gencay et al. 2018), and the O-methyl phosphoramidate (MeOPN) modification of CPS was identified to be a phage receptor used by different *Campylobacter*-specific phages (Sørensen et al. 2015). The sequence of phage vB_CjeM_los1 suggests that a holin–endolysin system induces bacterial lysis of *Campylobacter* cells during infection by that phage. More studies are necessary to correctly assign the function of current hypothetical proteins and elucidate mechanisms and proteins involved in phage infection of *Campylobacter* species (O'Sullivan et al. 2018).

Bacteria can block phage infection by three approaches, targeting different stages: avoidance of phage adherence, destruction of phage DNA and interference with phage replication. Initial phage adsorption can be prevented by modifying the involved phage receptors or extracellular matrix so that phage attachment to the bacterial host is prevented, as was demonstrated for *Escherichia coli* (Soundararajan et al. 2019). Exclusion of superinfection is mediated by prophages that prevent subsequent phage infections by various molecular mechanisms (Bondy-Denomy et al.

2016). Once a phage's genome is successfully transferred into the bacterial cell, restriction-modification systems can degrade the incoming DNA, while simultaneously shielding the own bacterial DNA from restriction by methylation. Bacteriophages can escape these defense systems by modification of their DNA, as has been observed in *Fletcherviruses* where certain guanine (G) residues were replaced by inosine or in *Firehammerviruses* that use the base 7-amido-7-deazaguanine instead of G (Crippen et al. 2019). A second defense line attacking phage DNA is the CRISPR/Cas system that represents a rudimentary bacterial immune system. CRISPRs (short for clustered regularly interspaced short palindromic repeats) and the CRISPR-associated (Cas) genes trigger cleavage in the targeted DNA that aborts a phage infection before replication occurs. The spacers are the targeting sequences and function to memorize and recognize structures of the phage genome; they become integrated in the bacterial genome as a result of a previous attack with that phage species (Pyenson and Marraffini, 2020). The CRISPR/Cas9 system is nearly universally present in *C. jejuni*, but it is often lacking in *C. coli* (Pearson et al. 2015). The third level of bacterial defense represents intracellular mechanisms such as antisense RNA targeting of primase that can block phage infection, as was shown for *Streptococcus thermophilus* (Sturino and Klaenhammer 2004).

Resistance against phages comes with a cost. The emergence of phage-resistant bacteria is often associated with phenotypic changes that are frequently linked to reduced fitness compared to non-resistant strains (Labrie et al. 2010). These changes can also reduce virulence or the ability of phage resistant bacteria to colonize the gut. In *Campylobacter*, resistance was mainly associated with changed surface structures or changes in flagellar function (Coward et al. 2006; Sørensen et al. 2011; Kittler et al. 2014; Lis and Connerton 2016). Some studies reported changes of the CRISPR-Cas system (Louwen and Baarlen, 2013; Hooton et al. 2015), and one study observed upregulation of genes with similarity to T4 superinfection exclusion upon phage infection in *Campylobacter* (Sacher et al. 2018).

Although bacteria can become resistant against the viruses that pray on them, the interaction between a bacterial population and its viral attacker eventually often reaches an equilibrium: Neither can the virus completely destroy its host, nor do the bacteria become fully resistant to the virus so that it would be eliminated from the population. Indeed, infection by a single phage of *Campylobacter* in a chicken gut can result in a sub-population of bacteria that are resistant to the virus in question, but this subpopulation does not completely take over. Rather, resistant and susceptible bacteria continue to coexist, so that the bacterial population does not develop long-term resistance. This was shown in a longitudinal field study (Connerton et al. 2004). Individual *C. jejuni* sub-populations that were resistant to killing by the phages could be isolated but did not dominate or outgrow the phage-sensitive *C. jejuni* cells; rather, the two coexisted in the chicken gut in the presence of phages. The use of phage cocktails can further reduce the risk of resistance to develop (Tanji et al. 2004; Pereira et al. 2016; Kittler et al. 2020).

4 Phage Treatment of *Campylobacter* in the Literature

The ability of bacteriophages to reduce the number of viable bacteria in which they replicate, despite their dependence on the bacterial host cells for their reproduction, is illustrated by phage infections that occur naturally in the avian gut. In an influential publication, Atterbury and colleagues (2005) showed that mean numbers of *Campylobacter* in chicken flocks that naturally contained bacteriophages were approximately 100 times lower compared to flocks in which *Campylobacter*-specific phages were absent. This observation initiated experimental phage treatment to reduce *Campylobacter* numbers, which subsequently reported comparable reductions (e.g., El-Shibiny et al. 2009; Fischer et al. 2013a). A number of publications have described experimental use of phages to reduce *Campylobacter* loads. In order to identify these, we performed a literature study, the results of which are briefly presented below.

A comprehensive literature search was performed in Pubmed with Boolean searches (AND/OR combinations of different search terms) followed by a snowball approach and complemented with our own knowledge. That way we identified 16 studies available from the literature that have tested the effect of phage application to reduce *Campylobacter* numbers in chickens. The outcomes of these 16 studies are summarized in Table 1. The studies represent either in vitro experiments, where meat was experimentally contaminated with *Campylobacter* and then treated with phages, or in vivo experiments, where phages were administered to chickens. One early study (Wagenaar et al. 2005) compared prophylactic application (phages were given prior to a *Campylobacter* challenge) with intervention treatment (*Campylobacter* was given prior to the phages), while all other in vivo studies evaluated intervention in *Campylobacter*-colonized chickens. Since phages need their bacterial hosts to replicate, intervention applications have a higher chance to reduce *Campylobacter* numbers. Only one study failed to obtain any significant reduction in bacterial numbers on chicken meat (Study number 4 in the table, Orquera et al. 2012), which was an in vitro study testing two bacteria/phage combinations at relatively low doses. All other studies recorded reductions, at various levels. A publication bias effect cannot be completely ruled out as negative findings (absence of any reduction) may not be published as frequently as positive findings.

The experimental conditions of the individual studies are briefly summarized in the table, although the reader is referred to the original publication for detailed information. If comparable details on settings or *Campylobacter* concentrations were not directly stated in the original article, these were assumed or calculated based on the presented data or, if possible, by analyzing the corresponding figures and graphs. If sources other than the original article were used for retrieving information, these sources were also cited in the table. Finally, even though the duration of *Campylobacter* reduction is important, this was not included in the collected data, in order to keep the table as simple as possible. Nevertheless, this aspect should also be considered when discussing results of phage intervention trials on *Campylobacter* reduction.

Table 1 Summary of in vivo and in vitro trials on the use of phages for controlling *Campylobacter* spp

Nr. and reference	*Campylobacter* ID[a]	Phage name, host range (HR, when specified), Genus	Experimental settings[b]	Significant effects on *Campylobacter* compared to control[c]
1 Goode et al. (2003)	C.j. C222	NCTC 12673 h: 8/11 *Fletchervirus* (HR taken from 13)	In vitro: chicken skin with (per 60 cm^2): 10^4 CFU at 4 °C. Phage was administered at MOI > 1	1.25 log_{10} reduction (determined at 1 dpa)
2 Atterbury et al. (2003a)	C.j. NCTC 12662	NCTC 12674 h: 8/11 *Fletchervirus*	In vitro: chicken skin with (per 2cm^2): (2a) 10^6 CFU at 4 °C and −20 °C, phage at MOI 10, 0.1, 0.01 (2b) 10^4 CFU at 4 °C and −20 °C, Phage at MOI 10^3, 10, 1	(2a) at MOI 10 max. 1.2 log_{10} reduction (3 dpa) (2b) at MOI 10^3 max. 1.3 log_{10} reduction (3 dpa), at lower MOIs no significant reduction (3 dpa) (both 2a and 2b)
3 Bigwood et al. (2008)	Field strain SGCSCFT	Cj6 Specific to strain SGCSCFT Genus not indicated	In vitro: Cooked and raw beef with (per 4cm^2): (3a) $4{\cdot}10^4$ CFU at 24 °C, phage at MOI 10^4, 10 (3b) $4{\cdot}10^4$ CFU at 5 °C, phage at MOI 10^4 (3c) <$4{\cdot}10^2$ CFU at 24 °C, phage at MOI 10^4 (3d) <$4{\cdot}10^2$ CFU at 5 °C, phage at MOI 10^4	(3a) at MOI 10^4, cooked: max. 2.9 log_{10} red. (6 hpa) and raw: max 1.5 log_{10} red. (24 hpa). At MOI 10, cooked: max. 3.7 log_{10} red. (3 hpa) and raw: 0.4 log_{10} red. (6 hpa) (3b) cooked: max. 2.9 log_{10} red. at 6 hpa and raw: 1.5 log_{10} red. at 24 hpa (3c) cooked: max. 0.7 log_{10} red. at 3 hpa and raw: 0.6 log_{10} red. at 24 hpa (3d) cooked: max. 0.4 log_{10} red. at 6 hpa and raw: 0.4 log_{10} red. at 24 hpa

(continued)

Table 1 (continued)

Nr. and reference	*Campylobacter* ID[a]	Phage name, host range (HR, when specified), Genus	Experimental settings[b]	Significant effects on *Campylobacter* compared to control[c]
4 Orquera et al. (2012)	C.j. NCTC 11168 or C.j. NCTC 12668	CP81 HR: 47/245 *Fletchervirus* or NCTC 12684, HR: see 6. *Firehammervirus*	In vitro: chicken meat with (per 10 g): (4a) 10^6 CFU 11168 at 4 °C, phage CP81 at MOI 10^2, 10 (4b) 10^6 CFU 12668 at 4 °C, phage 12684 at MOI 10^2, 10	No significant reduction
5 Zampara et al. (2017)	NCTC 12662	F356 or F357, F356 + F357, F379 + F380 + F381 F356 + F380 + F381 All *Fletchervirus*, HR: 3/8–4/8 (from Sørensen et al. 2015)	In vitro: chicken neck skin with (per $12cm^2$) 10^4 CFU at 5 °C, (5a) phage F356 at MOI 10^3 (5b) phage F357 at MOI 10^3 (5c) phage F356 + F357 at MOI 10^3 each (5d) phage F379 + F380 + F381 at MOI 10^3 each (5e) phage F356 + F380 + F381 at MOI 10^3 each (for all: incubated at 30% CO_2, 70% N_2)	(5c, combination) 0.7 log_{10} reduction (1 dpa) other experimental settings: no significant reduction

(continued)

Table 1 (continued)

Nr. and reference	*Campylobacter* ID[a]	Phage name, host range (HR, when specified), Genus	Experimental settings[b]	Significant effects on *Campylobacter* compared to control[c]
6 Firlieyanti et al. (2016)	Combination of C.j. HPC5 C.j. 81-176 C.j. CLB44 C.j. CLB68 C.j. CLB104	NCTC 12682 (Φ3), NCTC 12684 (Φ15), HR: 3/5 for both *Firehammervirus* (HR from Coward et al. 2006)	In vitro: chicken liver with (per 10 g): (6a) 10^5 CFU/g at 4 °C, phage 12682 at MOI 10^3 (6b) 10^5 CFU/g at 4 °C, phage 12684 at MOI 10^3 (6c) 10^3 CFU/g at 4 °C, phage 12682 at MOI 10^5 (6d) 10^3 CFU/g at 4 °C, phage 12684 at MOI 10^5	Between <0.5 $\log_{10}$ reduction and <1 $\log_{10}$ reduction, depending on the phage, MOI and C. strain
7 Thung et al. (2020)	C.j. CF84	CJ01 HR: 11/13 including *C. lari, Fletchervirus*	In vitro: mutton or chicken meat with (per 20 g): $5 \cdot 10^4$ CFU at 4 °C, phage at MOI 10^2	Mutton meat: max. 1.7 $\log_{10}$ reduction (2 dpa) Chicken meat: max. 1.68 $\log_{10}$ reduction (2 dpa)
8 Wagenaar et al. (2005)	C.j. C356	(8a) NCTC 12671, (8b) 12671 + NCTC 12669 *Fletchervirus*	In vivo,10 d old chicks (prophylactic) or 32 d old (therapeutic) (8a) Prophylactic 10^9–10^{10} PFU 4 d before C for 10 d (8b) Intervention 10^9–10^{10} PFU 5 d after C for 6 d (8c) Intervention 10^9–10^{11} PFU 1 d after C	(8a) Delayed colonization and overall 1–2 $\log_{10}$ reduction (8b) max. >3 $\log_{10}$ reduction (2 days after starting application) (8c) max. 1.5 $\log_{10}$ reduction (1 day after starting application)

(continued)

Table 1 (continued)

Nr. and reference	*Campylobacter* ID[a]	Phage name, host range (HR, when specified), Genus	Experimental settings[b]	Significant effects on *Campylobacter* compared to control[c]
9 Loc Carrillo et al. (2005)	C.j. HPC5 or C.j. GIIC8	CP8 HR: 84/130 (Atterburyet al. 2003b) or CP34 HR: 53/130 (Connerton et al. 2004) *Fletchervirus*	In vivo, chickens, intervention (9a) HPC5 with phage CP8 at 10^5, 10^7, 10^9 PFU (9b) HPC5 with phage CP34 at 10^5, 10^7, 10^9 PFU (9c) GIIC8 with phage CP8 at 10^7 PFU In all cases phage administered 5 d after C.	(9a) no significant reduction, at all MOIs (9b) at 10^5 PFU max. >2 $\log_{10}$ reduction (1 dpa) at 10^7 PFU max. 3.9 $\log_{10}$ reduction (1 dpa) at 10^9 PFU max. 1.4 $\log_{10}$ reduction (4 dpa) (9c) at 10^7 PFU max. 5.6 $\log_{10}$ reduction (1 dpa)
10 Scott et al. (2007a)	C.j. F2E3 (sensitive) C.j. F2E1 (resistant) C.j. F2C10 (sensitive)	CP30 *Fletchervirus* (Jackel et al. 2019)	In vivo, chickens, Intervention (10a) F2E3 (10b) F2E3 co-cultivated with F2E1 (10c) F2C10 (10d) F2C10 co-cultivated with F2E1 in all cases 10^7 PFU administered 4 d after C.	(10a) 1.6 $\log_{10}$ reduction of F2E4 (2 dpa) (10b) 3.3 $\log_{10}$ reduction of F2E3 (2 dpa), none for F2E1 (10c) 3.2 $\log_{10}$ reduction of F2C10 (2 dpa) (10d) 5.5 $\log_{10}$ reduction of F2C10 (2 dpa), none for F2E1 Resistant F2E1 was unable to compete with the sensitive strains in absence of phage; but in presence of phage F2E1 dominated when present

(continued)

Table 1 (continued)

Nr. and reference	*Campylobacter* ID[a]	Phage name, host range (HR, when specified), Genus	Experimental settings[b]	Significant effects on *Campylobacter* compared to control[c]
11 El-Shibiny et al. (2009)	C.j. HPC5 or C.c. OR12	CP220 HR: 18/53 *Firehammervirus*	In vivo, chickens, Intervention (11a) HPC5 with 10^5, 10^7, 10^9 PFU 5 d after C. (11b) OR12 with 10^5, 10^7, 10^9 PFU 5 d after C.	(11a) at 10^5 PFU max.1.7 $\log_{10}$ reduction (1 dpa) at 10^7 PFU max. 2.4 $\log_{10}$ reduction (2 dpa) at 10^9 PFU max. 2.1 $\log_{10}$ reduction (5 dpa) (11b) at 10^5 PFU max. 0.2 $\log_{10}$ reduction (4 dpa) at 10^7 PFU max. 0.7 $\log_{10}$ reduction (4 dpa) at 10^9 PFU max.1.9 $\log_{10}$ reduction (2 dpa)
12 Carvalho et al. (2010b)	(12a) C.j. 2140CD1 or (12b) C.c. A11	Cocktail of phiCcoIBB35, BB27, BB12 HR: 13/15 to 14/15 Carvalho et al. (2010a) *Firehammervirus*	In vivo, chickens, Intervention (12a) oral application at 10^6 PFU 7 d after 2140CD1 (12b) oral application at 10^6 PFU 7 d after A11 (12c) in feed at MOI 10^7 7 d after A11	(12a) max. 2.34 $\log_{10}$ reduction (4 dpa, feces) (12b) max. 1.69 $\log_{10}$ reduction (7 dpa, feces) (12c) max. 2.0 $\log_{10}$ reduction (2 dpa, feces)

(continued)

Table 1 (continued)

Nr. and reference	*Campylobacter* ID[a]	Phage name, host range (HR, when specified), Genus	Experimental settings[b]	Significant effects on *Campylobacter* compared to control[c]
13 Fischer et al. (2013a)	C.j. 1474–06	NCTC 12673, cocktail: 12672 + 12673 + 12674 + 12678 HR: 7/11–8/11 (Hirsch 2010; Sails et al. 1998) *Fletchervirus*	In vivo, chickens, Intervention (13a) 12673 at 10^7 PFU (13b) cocktail at MOI 10^7 PFU (total content) (13c) cocktail at MOI 10^7 PFU (total content) (independent replicate) in all cases administered 3 d after C.	(13a) max. 2.8 $\log_{10}$ reduction (21 dpa) (13b) max. 2.3 $\log_{10}$ reduction (3 dpa) (13c) max. 2.1 $\log_{10}$ reduction (21 dpa)
14 Kittler et al. (2013)	natural colonization: (14a) C.j. ST 4819 (14b) C.j. ST 51, C.j. ST 905 (14c) C.j. ST 4755, C.j. new ST	Cocktail as in 13a: NCTC 12672 + 12673 + 12674 + 12678	In vivo, chicken field trials 14a: via drinking water at $10^{7.5}$ PFU 2 d after C. detection in sample 14b: via drinking water at $10^{5.8}$ PFU 4 d after C. detection in sample 14c: via drinking water at $10^{7.6}$ PFU 2 d after C. detection in sample	(14a) max. 3.2 $\log_{10}$ reduction (6 dpa) (14b, 14c) no significant reduction
15 Hammerl et al. (2014)	C.j. 3871	CP14 HR: 51/245, *Fletchervirus* or CP14 + CP81 (see 4) or CP14 + CP68 with HR: 31/245, *Firehammervirus*	In vivo, chickens, Intervention (15a) Phage CP14 at $5{\cdot}10^8$ PFU 7 d after C. (15b) Phage CP14 and CP18 at $5{\cdot}10^8$ PFU for each, 7 d after C. (15c) Phage CP14 at $5{\cdot}10^8$ PFU 7 d after C. and 18 h later phage CP68 at $5{\cdot}10^8$ PFU	(15a) max. ~1 $\log_{10}$ reduction (3 dpa, faeces) (15b) no significant reduction (faeces) (15c) max. >1 $\log_{10}$reduction (2 dpa, faeces)

(continued)

Table 1 (continued)

Nr. and reference	*Campylobacter* ID[a]	Phage name, host range (HR, when specified), Genus	Experimental settings[b]	Significant effects on *Campylobacter* compared to control[c]
16 Richards et al. (2019)	C.j. HPC5	CP20 *Firehammervirus* and CP30A *Fletchervirus*	In vivo, chickens, Intervention cocktail at 10^7 PFU total content administered 4 d after C.	max. 2.4 $\log_{10}$ reduction (2 dpa)

[a]No strain designations but MLST Sequence types are given for the field trial listed as number 14. Abbreviations: C.j. *Campylobacter jejuni*; C.c. *Campylobacter coli*

[b]*Campylobacter* as oral application in in vivo trials except field trial number 14. Abbreviations: C.: *Campylobacter*; CFU, colony forming units; PFU, plaque forming units; MOI, multiplicity of infection; d, days; h, hours

[c]Mean reduction in cecal content unless stated otherwise. Abbreviations: dpa, days post phage application; hpa, hours post phage application; red, reduction

From Table 1, it can be seen that most in vitro studies were conducted with a multiplicity of infection (MOI of phage per bacteria) of between 0.01 and 10^6, whereby an MOI below 1 was not effective (Atterbury et al. 2003a). The in vitro studies that resulted in observed reductions reported reductions of *Campylobacter* numbers between 0.4 $\log_{10}$ and 3.7 $\log_{10}$, depending on the study. In vivo experiments were mostly carried out with doses of 10^5 to 10^{10} PFU/animal, and depending on the study, reductions of cecal content between 0.2 $\log_{10}$ and 5.6 $\log_{10}$ were reported (Table 1).

The efficacy of phage treatment in the eradication of *Campylobacter* mainly depends on the optimal choice, formulation, application route and concentration of the phages (Moye et al. 2018; Lewis and Hill 2019). Of these, the choice of the phages is of crucial importance. Various previous and recent studies address the choice of the phage by applying different phages (belonging to both genera) or testing these on different *Campylobacter* host strains. A phage with a broad host range would be able to infect multiple *Campylobacter* strains, and this would be advantageous in an agricultural setting. A broader host spectrum can further be achieved by the use of phage mixtures (phage cocktails). Additionally to achieving a broader host range, cocktails consisting of different phages have the advantage that they might be active in more variable environmental conditions and that they can prevent the development of resistance (Tanji et al. 2004; Fischer et al. 2013a). For this reason, most in vivo studies used cocktails composed of two to four phages, while only two studies used single phages. This contrasts with the in vitro studies of which only one evaluated a cocktail (Zampara et al. 2017, Table 1). Most studies stated that phages with a broad host range were chosen (Wagenaar et al. 2005; Loc Carrillo et al. 2005; Scott et al. 2007a; Carvalho et al. 2010b), while only two studies stated that the efficiency of bacterial lysis was considered for selection of the applied phages (Hammerl et al. 2014; Hirsch 2010). Thus, a broad host range was considered more important for a priori selection than evidenced lysis efficacy by bacterial titer reduction in in vitro cultures. This may be because those titers do not necessarily reflect the phage's efficacy for bacterial lysis in environmental or in vivo settings. More work is needed to better experimentally predict the lytic efficacy of phages under the conditions resembling those where they need to be most effective.

4.1 Experimental Phage Treatment of Contaminated Meat

Modeling data have suggested that a reduction in disease burden of 90% can be achieved if poultry meat at retail would have 100-fold fewer *Campylobacter* numbers, or if the intestinal load of chickens prior to slaughter were 1000-times lower (EFSA 2011). Thus, lower reductions on meat are necessary than those required when the chicken gut is the target of intervention. Figure 2 summarizes the reported reductions from the various studies summarized in Table 1. For reference, the target reduction is indicated. Of the in vitro studies for which we were able to extract the data, three experiments reached the threshold of at least a 2 log reduction, all from study number

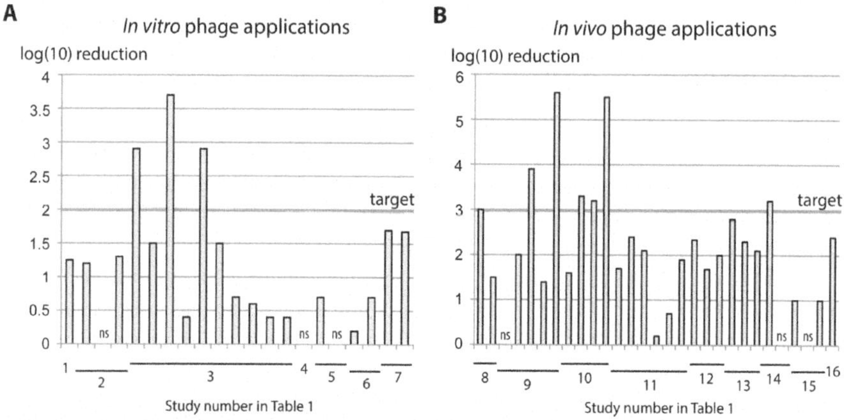

Fig. 2 Reported reduction of *Campylobacter* bacteria on meat following in vitro application (a) and in chickens following in vivo application (b) of phages. The studies from which the data were taken are numbered as in Table 1. The aimed minimum effectiveness is indicated as 'target.' Experiments without significant results are summarized by 'ns'

3 in Table 1 (Bigwood et al. 2008) (Fig. 2a). These successes were all obtained with a single phage/bacterial strain combination where the bacteria were present on cooked beef, while the same strain present on raw beef was less effectively reduced (Table 1). The second best results with chicken meat was reported for study number 7 (Thung et al. 2020) which nearly reached the threshold, again for a single phage/strain combination. Only one study tested the effectivity of phage combinations or cocktails applied to meat (study 5, Zampara et al. 2017) but none of their reported results reached a tenfold reduction for bacteria present on chicken neck skin (Table 1, Fig. 2). It should be pointed out that even phage treatment reaching an effectivity below the target threshold can still contribute to food safety and that it will be most effective when combined with other measures, such as improvements in slaughter hygiene and application of organic acids.

4.2 Experimental Phage Treatment of Colonized Chickens

Applying phages at the end of the food chain as an epidemiological end-point was considered desirable to prevent the development of *C. jejuni* resistance (Goode et al. 2003). However, various studies have since indicated that the development of resistances might not be a major obstacle for application of phage-based biocontrol in chickens, due to fitness costs of resistant isolates as stated above (Connerton et al. 2004; Kittler et al. 2014).

An advantage of achieving a reduction of *Campylobacter* numbers in chickens during their production is that this may prevent infections other than those directly related to food consumption, as it would reduce the overall population of *Campylobacter* in an area and may thus lower the burden from environmental sources as well (EFSA 2011). It is therefore quite reassuring that four of the so far conducted in vivo studies reported to reach the desired threshold of a 3 $\log_{10}$ reduction, at least in some of the tested conditions (Fig. 1b). While most studies used cocktails consisting of *Fletcherviruses* that mainly replicate in *C. jejuni*, four studies were identified that used *Firehammerviruses* (alone or in combination with *Fletcherviruses*), so that *C. coli* strains would also be targeted (Jackel et al. 2019) (Table 1). Finally, it is important to note that adverse effects of phage application on the health and growth characteristics of the birds were not observed in any of the studies reported here.

Most studies used phage doses of approximately 10^7 plaque-forming units (PFU) per bird, with a range between 10^5 and 10^{10} PFU. In contrast to the situation in the in vitro trials, the chances of bacteria and phages to meet may be increased by motility of the bacteria, their active growth and by the coprophagic nature of the birds. This may explain why no clear dose dependency was observed in those few trials that evaluated this (Loc Carrillo et al. 2005; Carvalho et al. 2010b). In various in vivo studies, the density of host bacteria at the time of phage application mainly relied on the state of colonization by *Campylobacter*. Thus, the period between oral inoculation or natural introduction of *Campylobacter* and phage application is specified in Table 1, and this varied from 1 to 21 days. Even in experiments conducted under almost similar conditions, the magnitude of an observed effect and its duration could vary considerably, clearly showing that the right choice of phages and factors regarding colonization of *Campylobacter* are essential for achieving reproducible reduction rates (Fischer et al. 2013a, Hammerl et al. 2014). Phage application at primary production can and should be combined with additional measures targeting *Campylobacter* at the subsequent production steps, in order to reach a maximum effect for public health and food safety.

5 Practical Hurdles and Open Questions that Need to Be Addressed

Methods for the isolation and characterization of *Campylobacter*-specific phages were reviewed recently and are not further treated here (Jackel et al. 2019). One in vivo study evaluated different routes for administering the phages to the chickens, either by oral gavage or via feed (Carvalho et al. 2010b). The results indicate that phage application via feed was more effective, and this would be far more practical than oral inoculation. However, applying certain substances via drinking water is more commonly used by the farmers. Administration via drinking water was tested in a field trial and proved to be an efficient and practically applicable route of administration (Kittler et al. 2014).

Phage suspensions need to remain stable during administration and passage through the chicken gut. In some of the in vivo studies we identified, protective buffering substances were added to protect the phages from the low pH of the stomach (Loc Carrillo et al. 2005; El-Shibiny et al. 2009; Carvalho et al. 2010b; Fischer et al. 2013a; Hammerl et al. 2014; Richards et al. 2019), while other studies used pure phage suspensions without additional buffers (Wagenaar et al. 2005; Scott et al. 2007a; Kittler et al. 2013). To date, convenient standard procedures and protocols that ensure stability during storage and application of bacteriophages in a farm environment and in the food industry are not yet available. Two recent studies investigated a lyophilization process and engineered spray-dried microparticles for the conservation of phages (Carrigy et al. 2019; Liang et al. 2020), which is a promising approach. The concentrations of the administered phages and of the bacteria being present were suggested to be crucial for the outcome of phage application (Abedon 2011a; Hagens and Loessner 2010). After all, the first step of phage infection depends on adsorption on the bacterial surface, which is the result of random contact between the phage and the receptor toward it is specific, with a given affinity. Although such random contact is mainly dependent on the density of bacteria and phages, Kasman and colleagues suggested that the measure how the ratio of phages and bacteria (the MOI) is expressed, might be generally inappropriate for conditions in which bacterial concentrations are below 10^7/mL, based on experimental results (Kasman et al. 2002). Taken together, the identification of a suitable phage or phage cocktail to reduce *Campylobacter* numbers in vivo under experimental conditions is only a first step. Reliable methods for preparation, storage and administration of the phages, at high titers, still need to be developed to ensure efficacy of phage application for biocontrol purposes.

Most of the in vitro studies selected single phages for application to laboratory-adapted *Campylobacter* strains, while only few used *Campylobacter* field strains for evaluating the lytic abilities of the phages towards bacteria residing on food (Bigwood et al. 2008; Firlieyanti et al. 2016; Thung et al. 2020). It is quite possible that *Campylobacter* cells present on a given food product may have adapted phenotypically to this matrix. Whether this would affect phage effectiveness is currently not known, but it was shown that the reduction in *Campylobacter* CFUs present on naturally contaminated food matrices can differ from reductions in artificially contaminated foods (Aidley et al. 2017). Therefore, further research on is needed to elucidate which bacterial factors impact the outcome of phage intervention and how such factors are expressed depending on the food matrix.

The low temperature at which meat is typically stored poses an interesting question as to how the method can be effective at all. All studies from Table 1 except one stored the treated samples at temperatures of 4–5 °C or at freezing temperatures, which is non-permissive for *Campylobacter* growth (Abedon 2014). It can be assumed that the bacterial metabolism is rather low, which questions how phages can reduce bacterial CFUs, as their reproduction depends on bacterial replication. This, and the identification of potential factors that might control the mode of action under these conditions, still remains largely open (Bigwood et al. 2008; Aidley et al. 2017). It has been observed that bacterial cells can lyse spontaneously following a large number of phages adsorbing to their cell membrane, without the need of phage replication; however, this is possible only at very high MOIs (Abedon 2011b). While most studies used *Campylobacter* inoculum sizes of 10^4 CFU, the volume size of the examined food matrix varied considerably among the reports. Further studies are needed to elucidate the role of the MOI in systems with low bacterial densities (Kasman et al. 2002; Hagens and Loessner 2010). Interestingly, in one study, it was reported that the efficacy of killing bacterial cells varied considerably at 5 °C for those phages that used capsular polysaccharides as receptors, while none of the phages using the flagellum for initial interaction were able to significantly reduce *Campylobacter* loads on chicken skin at this temperature (Zampara et al. 2017). The same study investigated samples after storage under anaerobic conditions, mimicking modified atmosphere packaging.

Several authors have highlighted the necessity of field trials as a next step to reduce *Campylobacter* in commercial broiler flocks with the help of phages (EFSA 2011; Connerton et al. 2011; Newell et al. 2011; Janez and Loc-Carrillo 2013). So far, most experiments were performed in a laboratory setting. Field trials are needed to examine the optimal choice of phages under production conditions, the most suitable application routes and doses, and the best timing, as all these factors affect the population dynamics of phages and their *Campylobacter*. A few field trial studies have already been carried out by our group with commercial broiler flocks in Germany, using a cocktail containing four phages, which had been previously tested in an experimental in vivo study resulting in reproducible reductions in *C. jejuni* loads by at least a 100-fold (Kittler et al. 2013; Fischer et al. 2013a). In these field trials, the cocktail was applied to a flock of *Campylobacter*-positive broilers via drinking water in three individual experiments, each with a non-treated control group. The used doses ranged from $\log_{10}$ 5.8 to 7.6 PFU per animal. The natural colonization by *C. jejuni* was confirmed prior to phage application, and a few days later, the phage cocktail was applied and fecal samples were examined for phage and *Campylobacter* concentrations during subsequent days, as were cecal samples at slaughter. In one trial, the reduction one day post phage application was so strong that *Campylobacter* levels decreased to below the detection limit and over 3 $\log_{10}$-units reductions were detected at slaughter compared to the control group. However, no significant reduction was observed in two other experimental groups at slaughter, indicating strong case-to-case variation. Nevertheless, a stagnation of *Campylobacter* colonization was observed in those birds that also occurred one day following phage application. It should be noted that the two trials that lacked reduction made use of phages that had not been replicated in the bacterial strains present in the naturally colonized birds; moreover, susceptibility testing had only been conducted after the phages had been applied due to time restrictions. Thus, a number of experimental modifications can be proposed that would suffice more field trials, and clearly, more work needs to be done before phages can be applied on a commercial basis to be beneficial for public health. The optimal timing of phage treatment targeted to maximize reduction at slaughter needs to be determined, and the reproducibility of the method needs to be demonstrated. We envisage that optimized cocktails that are most favorably adapted to apply to *Campylobacter* field strains can ultimately be defined, to be applied a few days prior to slaughter and possibly combined with other control measures. It is our opinion that this can reproducibly reduce *Campylobacter* numbers under commercial rearing conditions (Kittler et al. 2013).

6 Resistance of *Campylobacter* Phages

To develop valid and effective phage treatments to an industrial level, it is crucial to understand how the bacterial host responds to phage predation under practical conditions. An important environmental factor shaping the behavior of *Campylobacter* is the surrounding microbiota, which is present not only in the chicken gut, but also in and on the resulting food products (Abedon 2014). Therefore, it remains a challenging task to combine knowledge on molecular mechanisms of phage infection in *Campylobacter* under consideration of all factors at play in practical applications along the food production chains (Fernandez et al. 2018). Moreover, bacteria actively prevent phage infection by numerous intrinsic and extrinsic molecular mechanisms as described in section three (Hyman and Abedon 2010). The detailed examination of all these factors, including the possible interactions with the environmental microbiota, would be extremely time consuming and not expedient. However, we need to understand the relevant factors affecting effective and safe use of phages in the food environment. Thus, we aim here to connect the present knowledge on effects of gut microbiota on both bacterial phage resistance development and *Campylobacter* population dynamics.

Table 2 summarizes studies on the mechanisms by which *Campylobacter* bacteria combat phage infection. Currently known strategies of bacterial defense against phage attack include spontaneous mutations, restriction modification systems and adaptive immunity via CRISPR-Cas (Labrie et al. 2010; see also Chap. 10 of this book). In any case, the anti-viral defense systems impose enormous energy costs for the host bacterium (Oechslin 2018). In accordance with these general assumptions, resistance mechanisms by *Campylobacter* were reported to impose fitness costs that can reduce the competitiveness of bacteria in settings where phages are absent; when phages are then introduced, this effect possibly limits the dominance of resistant bacteria (Scott et al. 2007b; Kittler et al. 2014; Atterbury et al. 2005). When resistance develops, it does not linearly or exponentially increase over time, as one might assume; rather, phage resistance can be observed in unpredictable waves, and this should be monitored at different time points during the experiment (Fischer et al. 2013b; Bull et al. 2014; Kittler et al. 2014). Few studies examined resistances occurring during experiments performed on food products. Although the bacteria do not grow at refrigerating temperatures, it should be pointed out that *Campylobacter* maintains vital functions (Hazeleger et al. 1998). This implies that molecular changes in *Campylobacter* can still occur under these conditions, so that the occurrence of phage resistant isolates should be monitored even in refrigerated meat products. As already mentioned, the adsorption mechanism by which *Campylobacter* phages enter their host can be divided into flagellotropic bacteriophages that depend on a motile host with a functional flagellum and phages that depend on CPS for entry (Sørensen

Table 2 Overview of defense mechanisms of *Campylobacter* spp. against phages[a]

Campylobacter ID	Phage ID	Model	Resistance frequency	Molecular and phenotypical changes	Impact on bacterial characteristics	References
Field strains F2E1, F2E3, F2C10	Phages from same flock	Naturally colonized broiler flocks	Not specified	Horizontal gene transfer	Horizontal gene transfer was indicated to change of MLST ST and provoke loss of phage susceptibility Bacteriophage sensitive strains had a competitive advantage over the Bacteriophage insensitive strain in absence of phage but not in their presence	Connerton et al. (2004), Scott et al. (2007a)
Field strains	NCTC 12673, 12673, 12674, 12678	Field trial in commercial broiler flocks	Changing proportion of susceptible and non-susceptible subpopulation	Motility decreased, GGT	Reduced phage susceptibility was associated with reduced motility and GGT activity Phage susceptible *C. jejuni* subpopulation with increase in these factors did overgrow a non-susceptible subpopulation in the presence of phages	Kittler et al. (2014)

(continued)

Table 2 (continued)

Campylobacter ID	Phage ID	Model	Resistance frequency	Molecular and phenotypical changes	Impact on bacterial characteristics	References
HPC5	CP8, CP34	Chicken intestinal colonization	4% resistant isolates	Genomic rearrangements	Intra-genomic inversion between *Mu*-like prophage sequences, inverted genomic segments up to one third of genome In vivo three HPC5 variants observed: (i) resistant to virulent phages, (ii) inefficient in colonizing broilers, (iii) producing infectious CampMu phage particles Similar genotypes were not observed in vitro, see below Rapid phenotypic reversion when reintroduced in vivo	Loc Carrillo et al. (2005), Scott et al. (2007b)
2140CD1	phiCcoIBB35 phiCcoIBB27 phiCcoIBB12	Chicken intestinal colonization	13% resistant isolates	–	No decreased colonization, no phenotypic reversion in vivo	Carvalho et al. (2010b)

(continued)

Table 2 (continued)

Campylobacter ID	Phage ID	Model	Resistance frequency	Molecular and phenotypical changes	Impact on bacterial characteristics	References
NCTC 11168 NCTC12662	F336, F198, F287, F303, F326 F207	In vitro/chicken intestinal colonization	Not specified, 75% resistant isolates	Phase variable CPS MeOPN	Phase-variable MeOPN moiety, of the divers capsular polysaccharides was identified as a phage receptor Phase variable 3-*O*-Me and 6-*O*-Me groups of the NCTC 11168 CPS influence plaquing efficiency Resistant bacteria with lack or gain of MeOPN or 6-*O*-Me were recovered from chickens co-infected with F336 and NCTC 11168 Presence of phages selects for specific CPS variants in vivo CPS-dependent phages showed reduced plaque formation or could not infect NCTC12662R Resistance developed at high frequency due to phase variable gene expression of MeOPN modification of CPS Data indicate that CPS-dependent phages use diverse mechanisms for initial interaction with NCTC 12662	Sørensen et al. (2011); Holst Sorensen et al. (2012) Gencay et al. (2018)

(continued)

Table 2 (continued)

Campylobacter ID	Phage ID	Model	Resistance frequency	Molecular and phenotypical changes	Impact on bacterial characteristics	References
NCTC 11168	Phage F336	In vitro	Not specified	Phase-variable CPS genes *cj1421*, *cj1422*, *cj1426*	Dynamic switching was observed in the ON/OFF states of three phase variable genes upon phage F336 exposure Phage receptor loss was predominant but several events also prevented binding Dominant phage-resistant phasotype differed between cultures	Aidley et al. (2017)
NCTC 11168	16 phages of the NCTC typing-scheme	In vitro	Not specified	Capsular changes Reduced motility	Phage infection of different spontaneous mutants enables grouping of co-resistant phages into distinct groups Phages dependent on capsular polysaccharides and flagellotropic bacteriophages were identified Results for the transposon mutant cj0390 suggest that there may be overlap between groups	Coward et al. (2006)

(continued)

Table 2 (continued)

Campylobacter ID	Phage ID	Model	Resistance frequency	Molecular and phenotypical changes	Impact on bacterial characteristics	References
NCTC 12668 (*C. coli*)	NCTC 12684 *Firehammerviruses*	In vitro	NCTC 12668	ND	Resistant clones only acquired transient resistance, leading to cross protection against five other group II phages (NCTC 12675, 12683 and CP21, 7 and 68) Phage adsoption to resistant clones was impaired, motility or flaA sequence was not altered	Orquera et al. (2015)
Clinical strains and mutants	16 phages of the NCTC typing-scheme and 13 additional phages	In vitro	Depending on phage	Sialyltransferase Cst-II, CRISPR-Cas-system	First glimpse at potential role of subtype II-CRIPR-Cas system in bacteriophage defense Cst-II-positive *C. jejuni* harbored reduced CRISPR sizes or even absence A protective effect of lipooligosaccharide sialylation against phages was hypothesized, mechanism unknown	Louwen et al. (2013)

(continued)

Table 2 (continued)

Campylobacter ID	Phage ID	Model	Resistance frequency	Molecular and phenotypical changes	Impact on bacterial characteristics	References
NCTC 12662	CP8 CP30A *Fletchervirus*	In vitro	Not specified	CRISPR-like Cas4 protein	*Campylobacter* acquired new host-derived CRISPR spacers while being in association with bacteriophages harboring a Cas4-like protein Phages might use host DNA as an effective decoy to bacteriophage DNA Phage defense mechanism by sialylation might be abrogated by this mechanism CRISPR-mediated autoimmunity might impact the shape of evolution in *C. jejuni*	Hooton and Connerton (2015)

(continued)

Table 2 (continued)

Campylobacter ID	Phage ID	Model	Resistance frequency	Molecular and phenotypical changes	Impact on bacterial characteristics	References
NCTC 12662, 11168, HPC5 and others	CP8 CP30A *Fletchervirus*	In vitro/biofilms	7–10% in NCTC12662 after exposure to CP30A (Profound strain dependent differences in biofilms)	Carrier state life cycle Reduced motility	Equilibrium of phage sensitive and insensitive bacteria maintained in presence of phages, spontaneous phage production after biofilm disruption Cultures showing carrier state life cycles (CSLC) after phage exposure Phage resistant types showed growth phase dependent motility and were unable to colonize chickens and infect HCA-7 colonic epithelial cells The point mutation *flhF(T368A)* affects ability of FlhF to produce functional flagella *C. jejuni* carrying flhf(T368A) produce a flagellated subpopulation supporting phage replication At low phage titers, reversion to wild type motility may arise until phage density increases	Siringan et al. (2011, 2014) Jackel et al. (2019), Carrigy et al. (2019) Liang and Connerton (2018)

(continued)

Table 2 (continued)

Campylobacter ID	Phage ID	Model	Resistance frequency	Molecular and phenotypical changes	Impact on bacterial characteristics	References
HPC5	CP34 *Fletchervirus*	In vitro	91%	Reduced motility	Motility was impaired in resistant mutants	Scott et al. (2007b)
NCTC 12662	CP_F1 genus not specified	In vitro	Not specified	Reduced motility FlaB	In 78% of recovered resistant isolates attenuation or loss of motility Inactivation of the *flaB* gene resulted in reduced swarming motility and increased susceptibility of NCTC 12662 and in addition, increase in phage yields Plaques were clear and adsorption two-fold while burst size increased three-fold in mutant	Lis and Connerton (2016)

(continued)

Table 2 (continued)

Campylobacter ID	Phage ID	Model	Resistance frequency	Molecular and phenotypical changes	Impact on bacterial characteristics	References
NCTC 11168, 12658	F341 *Fletchervirus (Hansen)*	In vitro	Not specified	Reduced motility	Bacteriophage F341 does not infect mutants that have paralyzed or lack flagella F341 travels along the filament to reach the basal body of the bacterium The authors suggest the flagellum being used as a receptor and initial binding causing conformational change of the phage tail that enables DNA injection after binding to a secondary receptor	Baldvinsson et al. (2014)
NCTC 12662, 12658, RM1221	20 distinct phages	In vitro	Not specified	Reduced motility CPS	When isolating novel bacteriophages NCTC 12662 host strain selected for CP81-type phages dependent on the capsule for infection while RM1221 selected for CP220-type phages, requiring motility for infection	Sørensen et al. (2015)

(continued)

Table 2 (continued)

Campylobacter ID	Phage ID	Model	Resistance frequency	Molecular and phenotypical changes	Impact on bacterial characteristics	References
NCTC 11168	NCTC 12673	In vitro	Not specified	Transcriptome	The most highly upregulated gene was and uncharacterized operon Other significantly upregulated genes include those involved in oxidative stress defense and the *Campylobacter* multidrug efflux pump (CmeABC) Phage infectivity was altered by mutagenesis of the oxidative stress defense genes *katA*, *ahpC*,*sodB* and *cmeA* and *cmeB*	Sacher et al. (2018)

[a]Abbreviations: MLST multilocus sequence typing, ST sequence type, GGT gamma-glutamyl transpeptidase, *C. Campylobacter,* MeOPN *O*-methyl phosphoramidate, CPS capsular polysaccharides, CRISPR Clustered Regularly Interspaced Short Palindromic Repeats, DNA deoxyribonucleic acid

et al. 2015). Bacterial motility is important for chicken colonization by *C. jejuni*, and additionally, it is an important pathogenic determinant in humans (Grant et al. 1993; Lertsethtakarn et al. 2011, see also this chapter of this book). Several studies have investigated the association of motility and phage susceptibility, as summarized in Table 2.

We have shown previously that reduced phage susceptibility was associated with reduced *C. jejuni* motility and gamma-glutamyl transpeptidase (GGT) activity and that in presence of phage, the susceptible subpopulation of *C. jejuni* could outcompete the non-susceptible subpopulation with reduced fitness in field trials. Resistant isolates were not detected any more at the end of the trial (Kittler et al. 2014). Changes in bacterial motility and phage susceptibility occurred while phages and bacteria coexist (reviewed by Hooton and Connerton 2015). Flagellotropic phages were unable to infect non-motile bacteria lacking functional flagella (Baldvinsson et al. 2014). It is assumed that initial binding occurs between a flagellotropic phage and flagellin after which such phages enter the cell via a CPS-dependent mechanism. Additionally, the role of flagellin A and flagellin B in flagellotropic phage adsorption and infection was examined by several authors. One study described that inactivation of the minor flagellin encoded by the *flaB* gene resulted in an increased phage susceptibility and elevated phage CP_F1 yield, although these bacteria are less motile (Lis and Connerton 2016). This observation can at present not be fully explained.

Another study concentrated on phages that depend on lipooligosaccharide (LOS) for entry and examined changes in the Cst-II-generated ganglioside-like LOS structure that were associated with resistance against phage infection and significantly reduced CRISPR sizes (suggesting a less effective defense via spacer acquisition) in the tested *C. jejuni* model (Louwen et al. 2013). However, the underlying mechanism of the association between phage resistance changed LOS structures and a degenerated CRISPR-Cas system remains unclear, and this needs to be further investigated and elucidated. The CRISPR-Cas defense remains an interesting subject of research. In this context, it can be mentioned that a *C. jejuni* host-spacer acquisition by a conserved Cas-4 like protein of bacteriophages was activated during continuous association of bacteriophage and bacteria in the carrier state. This promoted the acquisition of spacers that produced crRNAs targeting ADP-heptose-lipooligosaccharide heptosyltransferase systems. These spacers would prevent the carbohydrate addition to growing LOS structures and as such could be responsible for the abovementioned changes in the LOS structures and thus abrogate the Cst-II generated changes of LOS structures that were suggested to be responsible for phage resistance (Louwen et al. 2013; Hooton and Connerton 2015). In addition, phase-variable expression of CPS has been reported that might point toward a mechanism found in the *Campylobacter* phage F336 (Sørensen et al. 2011; Holst Sørensen et al. 2012), providing synthesis of variable receptor binding proteins, as was previously reported for the *Campylobacter* phage CP220 (Timms et al. 2010). Finally, the relevance of different mechanisms might vary depending on the environment or food matrix used (Moye et al. 2018). Thus, general assumptions concerning the occurrence and mechanisms of phage resistance in *Campylobacter* should be handled with care.

7 Regulatory Aspects and Safety of *Campylobacter* Bacteriophages

Adverse effects of phage application to chickens, such as reduced feed intake or growth or increased mortality, have so far not been reported, to the best of our knowledge. However, as a result of phage application in chickens or on meat, ultimately phages would enter the food chain, and thus, considerations regarding the safety for consumers are needed to allow regulatory approval. Bacteriophages are already ingested in daily life and frequently encountered in the environment as a result of their natural existence, and such bacteriophages probably outnumber bacterial concentrations in most ecosystems (Comeau et al. 2008). However, some may consider that observation in itself is not sufficient to assume safety or allow regulation, as phage application results in human addition of a limited group of bacteriophages that will be ingested. Their safety toward humans must be demonstrated, and their beneficial effect of enhanced food safety by reducing *Campylobacter* numbers must be weighed into this.

The main hurdles for extensive use of phages in the agro-food sector might further be related to technical and legal inconveniences. Problems regarding the stability of phage preparations and stocks, absence of bacterial resistance over time, the possibility of unintended selection of particular *Campylobacter* strains as a result of selective pressure, and the possibility of phage-induced gene transfer all need to be considered. It is anticipated that these problems, together with the required scaling of the process, can eventually be solved (Bari et al. 2017; Fernandez et al. 2018). Biocontrol measure by means of phage application is increasingly accepted for targeting bacterial pathogens in various food products (Moye et al. 2018), although so far in the EU commercial phage products have not yet been approved by the responsible authorities. The prerequisites required for implementing phage-based products against *Campylobacter* in the European food industry have not yet been defined, but experience from other phage applications may pave the way here, and also identify potential hurdles. For example, the biosafety of the phage-based product Listex™ P100 against *Listeria monocytogenes* was confirmed by data from analyses of three parameters: (i) temperate phages that could potentially transfer certain virulence genes were absent in the bacterial propagation strain, (ii) P100 and its ingredients were precisely defined, and (iii) the mode of fabrication of the product is exactly described. The biosafety of Listex™ P100 was accepted because the phage in question is strictly lytic, it is strongly genus specific, and the likelihood of persistence in the environment is low in the absence of a susceptible host. The propagation procedures and purification steps were considered as safe due to propagation on an apathogenic *Listeria innocua* strain and the implemented HACCP-based control program. In the process of Listex™ P100 to be allowed in the EU, EFSA requested more studies using naturally contaminated food for evaluation of the efficacy of the product and asked for data showing the survival of this phage in processing wastewater (the wastewater produced during food production) and in the environment (EFSA 2012). If we take this as a road map for the development of effective, standardized and approved phage products for combating *Campylobacter*, the following considerations are likely to be important: (i) non-pathogenic standard strains of the species *C. jejuni* and *C. coli* are not available; (ii) no classical virulence factors or specific gene clusters of virulence are known that determines a given *Campylobacter* strain being pathogenic or not (Dasti et al. 2010); (iii) the host range of *Campylobacter* phages is relatively narrow, restricting the possibility to propagate the phages in other apathogenic bacterial species. Upon evaluation of all available data, EFSA considered Listex™ P100 as safe, also based on the absence of genes whose products were homologous to bacterial toxins or other virulence factors. Additionally, the host range of Listex™ P100 was limited to *Listeria*, ensuring containment of the treatment (EFSA 2016). Both requirements would be met by a number of *Campylobacter* phages that also do not contain genes required to display the lysogenic cycle and have a narrow host range limited to certain *Campylobacter* strains only (Jackel et al. 2019; Hobbs and Abedon 2016). However, a narrow host range may potentially limit the effectiveness against naturally occurring *Campylobacter* strains, which is something that needs to be investigated.

In conclusion, more studies are required for regulatory approval, which was already recognized a decade ago (Hagens and Loessner 2010) and has not been completely resolved since. Finally, in order to avoid mistakes that were previously made with antibiotic use and to implement phages as natural predators of pathogens in a sustainable way, the monitoring of phages and their bacterial hosts during application is strongly recommended, so that large-scale use of phages is regularly evaluated (Sommer et al. 2019). This might also clarify how the target bacteria ultimately respond to different environmental conditions and how this would affect their susceptibility to the used phages (Denes and Wiedmann 2014).

8 Concluding Remarks

The control of *Campylobacter* spp. along the food chain by using bacteriophages is no longer purely an academic exercise, as commercial application is now on the horizon, but more work is needed before it can be included as one of the measures taken to reduce the burden of *Campylobacter* spp. to public health. Although various studies have been conducted already, further studies are essential to adapt phage application techniques to the requirements of the food industry and to demonstrate reproducible efficiency of phages under the conditions of commercial production. Among other uncertainties, it needs to be established at which stage of the food chain phages can be most efficiently applied to maximize food safety. The application of phages to naturally contaminated chicken flocks and to potentially contaminated food products will be most effective when combined with other methods to reduce *Campylobacter*, in a multi-hurdle approach. In order to use phages in a sustainable manner and to avoid mistakes made in the past, the monitoring of phage resistance and eventual changes in bacterial characteristics should be performed on a regular basis. Ultimately, this One Health approach requires intensive cooperation of research groups, industry and stakeholders alike.

Acknowledgements This work was supported by the German Federal Ministry of Education and Research (BMBF) through the zoonoses research consortium PAC-*Campylobacter* (project IP5/01KI1725E).

References

Abedon S (2011) Phage therapy pharmacology: calculating phage dosing. Adv Appl Microbiol 77:1–40. https://doi.org/10.1016/B978-0-12-387044-5.00001-7

Abedon ST (2011b) Lysis from without. Bacteriophage 1:46–49. https://doi.org/10.4161/bact.1.1.13980

Abedon ST (2014) Phage therapy: eco-physiological pharmacology. Scientifica 2014:581639. https://doi.org/10.1155/2014/581639

Aidley J, Sørensen MCH, Bayliss CD, Brondsted L (2017) Phage exposure causes dynamic shifts in the expression states of specific phase-variable genes of *Campylobacter jejuni*. Microbiology 163:911–919. https://doi.org/10.1099/mic.0.000470

Aksyuk AA, Rossmann MG (2011) Bacteriophage assembly. Viruses 3:172–203. https://doi.org/10.3390/v3030172

Atterbury RJ, Connerton PL, Dodd CE, Rees CE, Connerton IF (2003) Application of host-specific bacteriophages to the surface of chicken skin leads to a reduction in recovery of *Campylobacter jejuni*. Appl Environ Microbiol 69:6302–6306. https://doi.org/10.1128/AEM.69.10.6302-6306.2003

Atterbury RJ, Connerton PL, Dodd CE, Rees CE, Connerton IF (2003) Isolation and characterization of *Campylobacter* bacteriophages from retail poultry. Appl Environ Microbiol 69:4511–4518. https://doi.org/10.1128/aem.69.8.4511-4518.2003

Atterbury RJ, Dillon E, Swift C, Connerton PL, Frost JA, Dodd CE, Rees CE, Connerton IF (2005) Correlation of *Campylobacter* bacteriophage with reduced presence of hosts in broiler chicken ceca. Appl Environ Microbiol 71:4885–4887. https://doi.org/10.1128/AEM.71.8.4885-4887.2005

Baldvinsson SB, Sørensen MC, Vegge CS, Clokie MR, Brondsted L (2014) *Campylobacter jejuni* motility is required for infection of the flagellotropic bacteriophage F341. Appl Environ Microbiol 80:7096–7106. https://doi.org/10.1128/AEM.02057-14

Bari SMN, Walker FC, Cater K, Aslan B, Hatoum-Aslan A (2017) Strategies for editing virulent staphylococcal phages using CRISPR-Cas10. ACS Synth Biol 6:2316–2325. https://doi.org/10.1021/acssynbio.7b00240

Bigwood T, Hudson JA, Billington C, Carey-Smith GV, Heinemann JA (2008) Phage inactivation of foodborne pathogens on cooked and raw meat. Food Microbiol 25:400–406. https://doi.org/10.1016/j.fm.2007.11.003

Bondy-Denomy J, Qian J, Westra ER, Buckling A, Guttman DS, Davidson AR, Maxwell KL (2016) Prophages mediate defense against phage infection through diverse mechanisms. ISME J 10:2854–2866. https://doi.org/10.1038/ismej.2016.79

Bull JJ, Vegge CS, Schmerer M, Chaudhry WN, Levin BR (2014) Phenotypic resistance and the dynamics of bacterial escape from phage control. PLoS ONE 9(4):e94690. https://doi.org/10.1371/journal.pone.0094690

Carrigy NB, Liang L, Wang H, Kariuki S, Nagel TE, Connerton IF, Vehring R (2019) Spray-dried anti-*Campylobacter* bacteriophage CP30A powder suitable for global distribution without cold chain infrastructure. Int J Pharm 569:118601. https://doi.org/10.1016/j.ijpharm.2019.118601

Carrigy NB, Liang L, Wang H, Kariuki S, Nagel TE, Connerton IF, Vehring R (2020) Trileucine and Pullulan improve anti-*Campylobacter* bacteriophage stability in engineered spray-dried microparticles. Ann Biomed Eng 48:1169–1180. https://doi.org/10.1007/s10439-019-02435-6

Carvalho C, Susano M, Fernandes E, Santos S, Gannon B, Nicolau A, Gibbs P, Teixeira P, Azeredo J (2010) Method for bacteriophage isolation against target *Campylobacter* strains. Lett Appl Microbiol 50:192–197. https://doi.org/10.1111/j.1472-765X.2009.02774.x

Carvalho CM, Gannon BW, Halfhide DE, Santos SB, Hayes CM, Roe JM, Azeredo J (2010) The in vivo efficacy of two administration routes of a phage cocktail to reduce numbers of *Campylobacter coli* and *Campylobacter jejuni* in chickens. BMC Microbiol 10:232. https://doi.org/10.1186/1471-2180-10-232

Comeau AM, Hatfull GF, Krisch HM, Lindell D, Mann NH, Prangishvili D (2008) Exploring the prokaryotic virosphere. Res Microbiol 159:306–313. https://doi.org/10.1016/j.resmic.2008.05.001

Connerton PL, Loc Carrillo CM, Swift C, Dillon E, Scott A, Rees CE, Dodd CE, Frost J, Connerton IF (2004) Longitudinal study of *Campylobacter jejuni* bacteriophages and their hosts from broiler chickens. Appl. Environ. Microbiol. 70:3877–3883. https://doi.org/10.1128/AEM.70.7.3877-3883.2004

Connerton PL, Timms AR, Connerton IF (2011) *Campylobacter* bacteriophages and bacteriophage therapy. J. Appl Microbiol 111:255–265. https://doi.org/10.1111/j.1365-2672.2011.05012.x

Coward C, Grant AJ, Swift C, Philp J, Towler R, Heydarian M, Frost JA, Maskell DJ (2006) Phase-variable surface structures are required for infection of *Campylobacter jejuni* by bacteriophages. Appl Environ Microbiol 72:4638–4647. https://doi.org/10.1128/AEM.00184-06

Crippen CS, Lee YJ, Hutinet G, Shajahan A, Sacher JC, Azadi P, de Crecy-Lagard V, Weigele PR, Szymanski CM (2019) Deoxyinosine and 7-deaza-2-deoxyguanosine as carriers of genetic information in the DNA of *Campylobacter* viruses. J Virol 93(23):e01111–19. https://doi.org/ARTNe01111-1910.1128/JVI.01111-19

Dasti JI, Tareen AM, Lugert R, Zautner AE, Gross U (2010) *Campylobacter jejuni*: a brief overview on pathogenicity-associated factors and disease-mediating mechanisms. Int J Med Microbiol 300:205–211. https://doi.org/10.1016/j.ijmm.2009.07.002

Denes T, Wiedmann M (2014) Environmental responses and phage susceptibility in foodborne pathogens: implications for improving applications in food safety. Curr Opin Biotechnol 26:45–49. https://doi.org/10.1016/j.copbio.2013.09.001

Denou E, Bruttin A, Barretto C, Ngom-Bru C, Brussow H, Zuber S (2009) T4 phages against *Escherichia coli* diarrhea: potential and problems. Virology 388:21–30. https://doi.org/10.1016/j.virol.2009.03.009

EFSA (2009) Scientific opinion on "The use and mode of action of bacteriophages in food production". EFSA J 2(26):1076. https://doi.org/10.2903/j.efsa.2009.1076

EFSA (2011) Scientific opinion on *Campylobacter* in broiler meat production: control options and performance objectives and/or targets at different stages of the food chain. EFSA J 9(4):2105. https://doi.org/10.2903/j.efsa.2011.2105

EFSA (2012) Scientific opinion on the evaluation of the safety and efficacy of ListexTM P100 for the removal of *Listeria monocytogenes* surface contamination of raw fish. EFSA J 10(3):2615. https://doi.org/10.2903/j.efsa.2012.2615

EFSA (2016) Evaluation of the safety and efficacy of ListexTMP100 forreduction of pathogens on different ready-to-eat (RTE) food products. EFSA J 14(8):4565. https://doi.org/10.2903/j.efsa.2016.4565

EFSA (2019) The European union one health 2018 zoonoses report. EFSA J 17(12):5926. https://doi.org/10.2903/j.efsa.2019.5926

EG (2017) Commission Regulation (EU) 2017/1495 amending Regulation (EC) No 2073/2005 as regards *Campylobacter* in broiler carcases. Available from https://eur-lex.europa.eu/eli/reg/2017/1495/oj

El-Shibiny A, Scott A, Timms A, Metawea Y, Connerton P, Connerton I (2009) Application of a group II *Campylobacter* bacteriophage to reduce strains of *Campylobacter jejuni* and *Campylobacter coli* colonizing broiler chickens. J Food Prot 72:733–740. https://doi.org/10.4315/0362-028x-72.4.733

Erez Z, Steinberger-Levy I, Shamir M, Doron S, Stokar-Avihail A, Peleg Y, Melamed S, Leavitt A, Savidor A, Albeck S, Amitai G, Sorek R (2017) Communication between viruses guides lysis-lysogeny decisions. Nature 541:488–493. https://doi.org/10.1038/nature21049

FDA (2003) Quantitative assessment of relative risk to public health from foodborne *Listeria monocytogenes* among selected categories of ready to eat foods. US food and drug administration center for food safety and applied nutrition. Available fromhttps://www.fda.gov/food/cfsan-risk-safety-assessments/quantitative-assessment-relative-risk-public-health-foodborne-listeria-monocytogenes-among-selected

Fernandez L, Gutierrez D, Rodriguez A, Garcia P (2018) Application of bacteriophages in the agro-food sector: a long way toward approval. Front Cell Infect Microbiol 8:296. https://doi.org/10.3389/fcimb.2018.00296

Firlieyanti AS, Connerton PL, Connerton IF (2016) Campylobacters and their bacteriophages from chicken liver: the prospect for phage biocontrol. Int J Food Microbiol 237:121–127. https://doi.org/10.1016/j.ijfoodmicro.2016.08.026

Fischer S, Kittler S, Klein G, Glunder G (2013) Impact of a single phage and a phage cocktail application in broilers on reduction of *Campylobacter jejuni* and development of resistance. PLoS ONE 8(10):e78543. https://doi.org/10.1371/journal.pone.0078543

Fischer S, Kittler S, Klein G, Glünder G (2013) Impact of a single phage and a phage cocktail application in broilers on reduction of *Campylobacter jejuni* and development of resistance. PLoS ONE 8(10):e78543. https://doi.org/10.1371/journal.pone.0078543

Galtier M, De Sordi L, Maura D, Arachchi H, Volant S, Dillies MA, Debarbieux L (2016) Bacteriophages to reduce gut carriage of antibiotic resistant uropathogens with low impact on microbiota composition. Environ Microbiol 18(7):2237–2245. https://doi.org/10.1111/1462-2920.13284

Gencay YE, Sørensen MCH, Wenzel CQ, Szymanski CM, Brøndsted L (2018) Phase variable expression of a single phage receptor in *Campylobacter jejuni* NCTC12662 influences sensitivity toward several diverse CPS-dependent phages. Front Microbiol 9:82. https://doi.org/10.3389/fmicb.2018.00082

Gölz G, Rosner B, Hofreuter D, Josenhans C, Kreienbrock L, Lowenstein A, Schielke A, Stark K, Suerbaum S, Wieler LH, Alter T (2014) Relevance of Campylobacter to public health—the need for a one health approach. Int J Med Microbiol 304:817–823. https://doi.org/10.1016/j.ijmm.2014.08.015

Goode D, Allen VM, Barrow PA (2003) Reduction of experimental *Salmonella* and *Campylobacter* contamination of chicken skin by application of lytic bacteriophages. Appl Environ Microbiol 69:5032–5036. https://doi.org/10.1128/aem.69.8.5032-5036.2003

Grant CC, Konkel ME, Cieplak W Jr, Tompkins LS (1993) Role of flagella in adherence, internalization, and translocation of *Campylobacter jejuni* in nonpolarized and polarized epithelial cell cultures. Infect Immun 61:1764–1771

Hagens S, Loessner MJ (2010) Bacteriophage for biocontrol of foodborne pathogens: calculations and considerations. Curr Pharm Biotechnol 11:58–68. https://doi.org/10.2174/138920110790725429

Hammerl JA, Jackel C, Alter T, Janzcyk P, Stingl K, Knuver MT, Hertwig S (2014) Reduction of Campylobacter jejuni in broiler chicken by successive application of group II and group III phages. PLoS ONE 9(12):e114785. https://doi.org/10.1371/journal.pone.0114785

Hazeleger WC, Wouters JA, Rombouts FM, Abee T (1998) Physiological activity of *Campylobacter jejuni* far below the minimal growth temperature. Appl Environ Microbiol 64:3917–3922

Hirsch K (2010) Entwicklung von Minimierungsstrategien für *Campylobacter* im Geflügel durch Anwendung von Bakteriophagen. Doctoral thesis, University of Veterinary Medicine Hannover, Foundation, Hannover

Hobbs Z, Abedon ST (2016) Diversity of phage infection types and associated terminology: the problem with 'Lytic or lysogenic'. FEMS Microbiol Lett 363(7). https://doi.org/10.1093/femsle/fnw047

Holst Sørensen MC, van Alphen LB, Fodor C, Crowley SM, Christensen BB, Szymanski CM, Brondsted L (2012) Phase variable expression of capsular polysaccharide modifications allows *Campylobacter jejuni* to avoid bacteriophage infection in chickens. Front Cell Infect Microbiol 2:11. https://doi.org/10.3389/fcimb.2012.00011

Hooton SPT, Connerton IF (2015) *Campylobacter jejuni* acquire new host-derived CRISPR spacers when in association with bacteriophages harboring a CRISPR-like Cas4 protein. Front Microbiol 5:744. https://doi.org/10.3389/Fmicb.2014.00744

Hyman P, Abedon ST (2010) Bacteriophage host range and bacterial resistance. Adv Appl Microbiol 70:217–248. https://doi.org/10.1016/S0065-2164(10)70007-1

Islam MZ, Fokine A, Mahalingam M, Zhang Z, Garcia-Doval C, van Raaij MJ, Rossmann MG, Rao VB (2019) Molecular anatomy of the receptor binding module of a bacteriophage long tail fiber. PLoS Pathog. 15(12):e1008193. https://doi.org/10.1371/journal.ppat.1008193

Jackel C, Hammerl JA, Hertwig S (2019) *Campylobacter* phage isolation and characterization: what we have learned so far. Methods Protoc 2(1):18. https://doi.org/10.3390/mps2010018

Janez N, Loc-Carrillo C (2013) Use of phages to control *Campylobacter* spp. J Microbiol Methods 95:68–75. https://doi.org/10.1016/j.mimet.2013.06.024

Javed MA, van Alphen LB, Sacher J, Ding W, Kelly J, Nargang C, Smith DF, Cummings RD, Szymanski CM (2015) A receptor-binding protein of *Campylobacter jejuni* bacteriophage NCTC 12673 recognizes flagellin glycosylated with acetamidino-modified pseudaminic acid. Mol Microbiol. 95:101–115. https://doi.org/10.1111/mmi.12849

Kasman LM, Kasman A, Westwater C, Dolan J, Schmidt MG, Norris JS (2002) Overcoming the phage replication threshold: a mathematical model with implications for phage therapy. J Virol 76(11):5557–5564. https://doi.org/10.1128/jvi.76.11.5557-5564.2002

Kittler S, Fischer S, Abdulmawjood A, Glünder G, Klein G (2013) Effect of bacteriophage application on *Campylobacter jejuni* loads in commercial broiler flocks. Appl Environ Microbiol 79:7525–7533. https://doi.org/10.1128/AEM.02703-13

Kittler S, Fischer S, Abdulmawjood A, Glünder G, Klein G (2014) Colonisation of a phage susceptible *Campylobacter jejuni* population in two phage positive broiler flocks. PLoS ONE 9(4):e94782. https://doi.org/10.1371/journal.pone.0094782

Kittler S, Wittmann J, Mengden RALP, Klein G, Rohde C, Lehnherr H (2017) The use of bacteriophages as one-health approach to reduce multidrug-resistant bacteria. Sus Chem Pharm 5:80–83. https://doi.org/10.1016/j.scp.2016.06.001

Kittler S, Mengden R, Korf IHE, Bierbrodt A, Wittmann J, Plötz M, Jung A, Lehnherr T, Rohde C, Lehnherr H, Klein G, Kehrenberg C (2020) Impact of bacteriophage-supplemented drinking water on the *E. coli* population in the chicken gut. Pathogens 9(4):293. https://doi.org/10.3390/pathogens9040293

Klein G, Jansen W, Kittler S, Reich F (2015) Mitigation strategies for *Campylobacter* spp. in broiler at pre-harvest and harvest level. Berl Munch Tierarztl Wochenschr 128(3–4):132–140

Kutter E (2009) Phage host range and efficiency of plating. Methods Mol Biol 501:141–149. https://doi.org/10.1007/978-1-60327-164-6_14

Labrie SJ, Samson JE, Moineau S (2010) Bacteriophage resistance mechanisms. Nat Rev Microbiol 8:317–327. https://doi.org/10.1038/nrmicro2315

Lee S, Lee J, Ha J, Choi Y, Kim S, Lee H, Yoon Y, Choi KH (2016) Clinical relevance of infections with zoonotic and human oral species of *Campylobacter*. J Microbiol 54:459–467. https://doi.org/10.1007/s12275-016-6254-x

Lertsethtakarn P, Ottemann KM, Hendrixson DR (2011) Motility and chemotaxis in *Campylobacter* and *Helicobacter*. Annu Rev Microbiol 65:389–410. https://doi.org/10.1146/annurev-micro-090110-102908

Lewis R, Hill C (2019) Overcoming barriers to phage application in food and feed. Curr Opin Biotechnol 61:38–44. https://doi.org/10.1016/j.copbio.2019.09.018

Liang L, Carrigy NB, Kariuki S, Muturi P, Onsare R, Nagel T, Vehring R, Connerton PL, Connerton IF (2020) Development of a lyophilization process for *Campylobacter* bacteriophage storage and transport. Microorganisms 8(2):282. https://doi.org/10.3390/microorganisms8020282

Liang L, Connerton IF (2018) FlhF(T368A) modulates motility in the bacteriophage carrier state of *Campylobacter jejuni*. Mol Microbiol 110:616–633. https://doi.org/10.1111/mmi.14120

Lin J (2009) Novel approaches for *Campylobacter* control in poultry. Foodborne Pathog Dis 6:755–765. https://doi.org/10.1089/fpd.2008.0247

Lis L, Connerton IF (2016) The minor flagellin of *Campylobacter jejuni* (FlaB) confers defensive properties against bacteriophage infection. Front Microbiol 7:1908. https://doi.org/10.3389/fmicb.2016.01908

Loc-Carrillo C, Abedon ST (2011) Pros and cons of phage therapy. Bacteriophage 1:111–114. https://doi.org/10.4161/bact.1.2.14590

Loc Carrillo C, Atterbury RJ, el-Shibiny A, Connerton PL, Dillon E, Scott A, Connerton IF (2005) Bacteriophage therapy to reduce *Campylobacter jejuni* colonization of broiler chickens. Appl Environ Microbiol 71:6554-6563. https://doi.org/10.1128/AEM.71.11.6554-6563.2005

Louwen R, van Baarlen P (2013) Are bacteriophage defence and virulence two sides of the same coin in *Campylobacter jejuni*? Biochem Soc Trans 41:1475–1481. https://doi.org/10.1042/BST20130127

Louwen R, Horst-Kreft D, de Boer AG, van der Graaf L, de Knegt G, Hamersma M, Heikema AP, Timms AR, Jacobs BC, Wagenaar JA, Endtz HP, van der Oost J, Wells JM, Nieuwenhuis EE, van Vliet AH, Willemsen PT, van Baarlen P, van Belkum A (2013) A novel link between *Campylobacter jejuni* bacteriophage defence, virulence and Guillain-Barre syndrome. Eur J Clin Microbiol Inf Dis 32:207–226. https://doi.org/10.1007/s10096-012-1733-4

Mondigler M, Vögele RT, Heller KJ (1995) Overproduced and purified receptor binding protein pb5 of bacteriophage T5 binds to the T5 receptor protein FhuA. FEMS Microbiol Lett 130:293–300. https://doi.org/10.1111/j.1574-6968.1995.tb07734.x

Moye ZD, Woolston J, Sulakvelidze A (2018) Bacteriophage applications for food production and processing. Viruses 10(4):205. https://doi.org/10.3390/v10040205

Newell DG, Elvers KT, Dopfer D, Hansson I, Jones P, James S, Gittins J, Stern NJ, Davies R, Connerton I, Pearson D, Salvat G, Allen VM (2011) Biosecurity-based interventions and strategies to reduce *Campylobacter* spp. on poultry farms. Appl Environ Microbiol 77:8605–8614. https://doi.org/10.1128/AEM.01090-10

Newell DG, Koopmans M, Verhoef L, Duizer E, Aidara-Kane A, Sprong H, Opsteegh M, Langelaar M, Threfall J, Scheutz F, van der Giessen J, Kruse H (2010) Food-borne diseases—the challenges of 20 years ago still persist while new ones continue to emerge. Int J Food Microbiol 139(Suppl 1):S3-15. https://doi.org/10.1016/j.ijfoodmicro.2010.01.021

Nowaczek A, Urban-Chmiel R, Dec M, Puchalski A, Stepien-Pysniak D, Marek A, Pyzik E (2019) *Campylobacter* spp. and bacteriophages from broiler chickens: characterization of antibiotic susceptibility profiles and lytic bacteriophages. Microbiol Open 8:e784. https://doi.org/10.1002/mbo3.784

Oechslin F (2018) Resistance development to bacteriophages occurring during bacteriophage therapy. Viruses 10(7):351. https://doi.org/10.3390/v10070351

Orquera S, Golz G, Hertwig S, Hammerl J, Sparborth D, Joldic A, Alter T (2012) Control of *Campylobacter* spp. and *Yersinia enterocolitica* by virulent bacteriophages. J Mol Genet Med 6:273–278. https://doi.org/10.4172/1747-0862.1000049

Orquera S, Hertwig S, Alter T, Hammerl JA, Jirova A, Golz G (2015) Development of transient phage resistance in *Campylobacter coli* against the group II phage CP84. Berl Munch Tierarztl Wochenschr 12(3–4):141–147

O'Sullivan L, Lucid A, Neve H, Franz CMAP, Bolton D, McAuliffe O, Paul Ross R, Coffey A (2018) Comparative genomics of Cp8viruses with special reference to *Campylobacter* phage vB_CjeM_los1, isolated from a slaughterhouse in Ireland. Arch Virol 163:2139–2154. https://doi.org/10.1007/s00705-018-3845-3

Pearson BM, Louwen R, van Baarlen P, van Vliet AH (2015) Differential distribution of type II CRISPR-Cas systems in agricultural and nonagricultural *Campylobacter coli* and *Campylobacter jejuni* isolates correlates with lack of shared environments. Genome Biol Evol 7(9):2663–2679. https://doi.org/10.1093/gbe/evv174

Pereira C, Moreirinha C, Lewicka M, Almeida P, Clemente C, Cunha Â, Delgadillo I, Romalde JL, Nunes ML, Almeida A (2016) Bacteriophages with potential to inactivate *Salmonella Typhimurium*: use of single phage suspensions and phage cocktails. Virus Res 220:179–192. https://doi.org/10.1016/j.virusres.2016.04.020

Pyenson NC, Marraffini LA (2020) Co-evolution within structured bacterial communities results in multiple expansion of CRISPR loci and enhanced immunity. Elife 9:e53078. https://doi.org/10.7554/eLife.53078

Quinn PJ (2011) Veterinary Microbiology and Microbial Disease. Campylobacter and Helicobacter species. Blackwell Publishing Ltd., West Sussex, UK

Reich F, Atanassova V, Haunhorst E, Klein G (2008) The effects of *Campylobacter* numbers in caeca on the contamination of broiler carcasses with *Campylobacter*. Int J Food Microbiol 127:116–120. https://doi.org/10.1016/j.ijfoodmicro.2008.06.018

Richards PJ, Connerton PL, Connerton IF (2019) Phage biocontrol of *Campylobacter jejuni* in chickens does not produce collateral effects on the gut microbiota. Front Microbiol 10:476. https://doi.org/10.3389/fmicb.2019.00476

Román S, Sánchez-Siles LM, Siegrist M (2017) The importance of food naturalness for consumers: results of a systematic review. Trends Food Sci Technol 67:44–57. https://doi.org/10.1016/j.tifs.2017.06.010

Rosenquist H, Nielsen NL, Sommer HM, Norrung B, Christensen BB (2003) Quantitative risk assessment of human campylobacteriosis associated with thermophilic *Campylobacter* species in chickens. Int J Food Microbiol 83:87–103. https://doi.org/10.1016/s0168-1605(02)00317-3

Sacher JC, Flint A, Butcher J, Blasdel B, Reynolds HM, Lavigne R, Stintzi A, Szymanski CM (2018) Transcriptomic analysis of the *Campylobacter jejuni* response to T4-like phage NCTC 12673 infection. Viruses 10(6):322. https://doi.org/10.3390/v10060332

Sails AD, Wareing DR, Bolton FJ, Fox AJ, Curry A (1998) Characterisation of 16 *Campylobacter jejuni* and *C. coli* typing bacteriophages. J Med Microbiol 47:123–128. https://doi.org/10.1099/00222615-47-2-123

Scott AE, Timms AR, Connerton PL, El-Shibiny A, Connerton IF (2007) Bacteriophage influence *Campylobacter jejuni* types populating broiler chickens. Environ Microbiol 9:2341–2353. https://doi.org/10.1111/j.1462-2920.2007.01351.x

Scott AE, Timms AR, Connerton PL, Loc Carrillo C, Adzfa Radzum K, Connerton IF (2007) Genome dynamics of *Campylobacter jejuni* in response to bacteriophage predation. PLoS Path 3(8):e119. https://doi.org/10.1371/journal.ppat.0030119

Siringan P, Connerton PL, Cummings NJ, Connerton IF (2014) Alternative bacteriophage life cycles: the carrier state of *Campylobacter jejuni*. Open Biol 4:130200. https://doi.org/10.1098/rsob.130200

Siringan P, Connerton PL, Payne RJ, Connerton IF (2011) Bacteriophage-mediated dispersal of *Campylobacter jejuni* biofilms. Appl Environ Microbiol 77:3320–3326. https://doi.org/10.1128/AEM.02704-10

Sommer J, Trautner C, Witte AK, Fister S, Schoder D, Rossmanith P, Mester PJ (2019) Don't shut the stable door after the phage has bolted-the importance of bacteriophage inactivation in food environments. Viruses 11(5):468. https://doi.org/10.3390/v11050468

Sørensen MC, Gencay YE, Birk T, Baldvinsson SB, Jäckel C, Hammerl JA, Vegge CS, Neve H, Brøndsted L (2015) Primary isolation strain determines both phage type and receptors recognised by *Campylobacter jejuni* bacteriophages. PLoS ONE 10(1):e0116287. https://doi.org/10.1371/journal.pone.0116287

Sørensen MC, van Alphen LB, Harboe A, Li J, Christensen BB, Szymanski CM, BrøndstedL, (2011) Bacteriophage F336 recognizes the capsular phosphoramidate modification of *Campylobacter jejuni* NCTC11168. J Bacteriol 193:6742–6749. https://doi.org/10.1128/JB.05276-11

Soundararajan M, von Bünau R, Oelschlaeger TA (2019) K5 capsule and lipopolysaccharide are important in resistance to T4 phage attack in probiotic *E. coli* Strain Nissle 1917. Front Microbiol 10:2783. https://doi.org/10.3389/fmicb.2019.02783

Sturino JM, Klaenhammer TR (2004) Antisense RNA targeting of primase interferes with bacteriophage replication in *Streptococcus thermophilus*. Appl Environ Microbiol 70:1735–1743. https://doi.org/10.1128/aem.70.3.1735-1743.2004

Tanji Y, Shimada T, Yoichi M, Miyanaga K, Hori K, Unno H (2004) Toward rational control of *Escherichia coli* O157:H7 by a phage cocktail. Appl Microbiol Biotechnol 64:270–274. https://doi.org/10.1007/s00253-003-1438-9

Thung TY, Lee E, Mahyudin NA, Radzi CWJWM, Mazlan N, Tan CW, Radu S (2020) Partial characterization and in vitro evaluation of a lytic bacteriophage for biocontrol of *Campylobacter jejuni* in mutton and chicken meat. J Food Safety 40:e12770. https://doi.org/10.1111/jfs.12770

Timms AR, Cambray-Young J, Scott AE, Petty NK, Connerton PL, Clarke L, Seeger K, Quail M, Cummings N, Maskell DJ, Thomson NR, Connerton IF (2010) Evidence for a lineage of virulent bacteriophages that target *Campylobacter*. BMC Genom 11:214. https://doi.org/10.1186/1471-2164-11-214

Wagenaar JA, Van Bergen MA, Mueller MA, Wassenaar TM, Carlton RM (2005) Phage therapy reduces *Campylobacter jejuni* colonization in broilers. Vet Microbiol 109:275–283. https://doi.org/10.1016/j.vetmic.2005.06.002

Wang IN, Deaton J, Young R (2003) Sizing the holin lesion with an endolysin-beta-galactosidase fusion. J Bacteriol 185:779–787. https://doi.org/10.1128/jb.185.3.779-787.2003

Young KD, Young R (1982) Lytic action of cloned phi X174 gene E. J Virol. 44:993–1002. https://doi.org/10.1128/JVI.44.3.993-1002.1982

Zampara A, Sørensen MC, Elsser-Gravensen A, Brøndsted L (2017) Significance of phage-host interactions for biocontrol of *Campylobacter jejuni* in food. Food Control 73:1169–1175. https://doi.org/10.1016/j.foodcont.2016.10.033

Campylobacter Virulence Factors and Molecular Host–Pathogen Interactions

Nicole Tegtmeyer, Irshad Sharafutdinov, Aileen Harrer, Delara Soltan Esmaeili, Bodo Linz, and Steffen Backert

Contents

Abstract *Campylobacter jejuni* and *Campylobacter coli* can be frequently isolated from poultry and poultry-derived products, and in combination these two species

---

N. Tegtmeyer · I. Sharafutdinov · A. Harrer · D. Soltan Esmaeili · B. Linz · S. Backert (✉)
Department of Biology, Division of Microbiology, Friedrich Alexander University Erlangen-Nuremberg, Staudtstrasse 5, 91058 Erlangen, Germany
e-mail: steffen.backert@fau.de

N. Tegtmeyer
e-mail: nicole.tegtmeyer@fau.de

I. Sharafutdinov
e-mail: irshad.sharafutdinov@fau.de

A. Harrer
e-mail: Aileen.Harrer@anatomie.med.uni-giessen.de

D. Soltan Esmaeili
e-mail: delara.esmaeili@fau.de

B. Linz
e-mail: bodo.linz@fau.de

S. Backert (ed.), *Fighting Campylobacter Infections*, Current Topics in Microbiology and Immunology 431, https://doi.org/10.1007/978-3-030-65481-8_7

cause a large portion of human bacterial gastroenteritis cases. While birds are typically colonized by these *Campylobacter* species without clinical symptoms, in humans they cause (foodborne) infections at high frequencies, estimated to cost billions of dollars worldwide every year. The clinical outcome of *Campylobacter* infections comprises malaise, diarrhea, abdominal pain and fever. Symptoms may continue for up to two weeks and are generally self-limiting, though occasionally the disease can be more severe or result in post-infection sequelae. The virulence properties of these pathogens have been best-characterized for *C. jejuni,* and their actions are reviewed here. Various virulence-associated bacterial determinants include the flagellum, numerous flagellar secreted factors, protein adhesins, cytolethal distending toxin (CDT), lipooligosaccharide (LOS), serine protease HtrA and others. These factors are involved in several pathogenicity-linked properties that can be divided into bacterial chemotaxis, motility, attachment, invasion, survival, cellular transmigration and spread to deeper tissue. All of these steps require intimate interactions between bacteria and host cells (including immune cells), enabled by the collection of bacterial and host factors that have already been identified. The assortment of pathogenicity-associated factors now recognized for *C. jejuni*, their function and the proposed host cell factors that are involved in crucial steps leading to disease are discussed in detail.

1 Introduction

Zoonotic infections by bacterial, viral and parasitic microbes represent a significant health burden to humans (Cunningham et al. 2017; Logue et al. 2017; Plowright et al. 2017). Among these, a number of important foodborne pathogens are responsible for high degrees of morbidity and mortality worldwide. Examples from the bacterial kingdom include *Campylobacter* species, which are often detected in the natural environment such as water surface habitats as well as in the gastrointestinal tract of certain birds and mammals, where they colonize asymptomatically as commensals (Young et al. 2007; Burnham and Hendrixson 2018). In addition, the presence of *Campylobacter* species, in particular *C. jejuni* and *C. coli*, in poultry flocks and other farm animals such as dairy cows provides a high zoonotic potential. By means of multilocus sequence typing (MLST), it was demonstrated that certain genetic *C. jejuni* variants can better survive in the environment than others and are frequently found in natural soil and water reservoirs, especially during the warm summer seasons (Epps et al. 2013; Bronowski et al. 2014). From such sources, the bacteria can transfer to new hosts, *e.g.*, to poultry flocks through rodents, flies or direct contact (Jorgensen et al. 2011; Bronowski et al. 2014). Even though *Campylobacter* spp. represent fastidious microaerophilic bacteria, they are well adapted to persistence in natural ecosystems, for instance through biofilm formation, aerotolerance mechanisms and starvation strategies (Gölz et al. 2012; Tram et al. 2020). Multiple surveys have demonstrated that the majority of commercial poultry flocks become colonized with *C. jejuni* or *C. coli* within about 2–4 weeks after hatching (Potturi-Venkata et al.

2007; van Gerwe et al. 2009). This colonization of chicks most frequently proceeds through horizontal transfer from the environment, rather than by vertical transmission from mother hens. Cross-transmission from other *Campylobacter*-positive flocks present on the same farm or from previous flocks can be prevented by strict hygiene procedures and fumigation (Herman et al. 2003; Wedderkopp et al. 2003; Bronowski et al. 2014). Taken together, although we now have a better understanding of the transmission routes by which *Campylobacter* enters the farm environment, it appears that more studies are required to find ways to effectively combat this pathogen on farms (see also Chaps. 4 and 5 of this book).

The major transmission route of *C. jejuni* (unless specifically stated otherwise, all reference to *C. jejuni* in this chapter also applies to *C. coli*) to humans mostly proceeds via the handling and consumption of contaminated poultry meat, raw milk, cross-contamination to other food products and, less frequently, by contact with freshwater or consumption of well water (Kaakoush et al. 2015). Occasionally, close contact to infected animals such as pets, particularly young dogs with diarrhea, can be sources of infection by *C. jejuni* (Campagnolo et al. 2018; Bronowski et al. 2014). Thus, the main infection route toward humans occurs by a fecal-to-oral pathway. Nevertheless, human-to-human spread is relatively uncommon. However, upon ingestion *Campylobacter* enters the gastrointestinal tract and colonizes the jejunal mucosa of the intestine by successfully competing with the intestinal microbiota (Masanta et al. 2013). The overall prevalence of *C. jejuni* infections is very high and represents a major fraction of all bacteria-caused gastroenteritis cases: It was estimated to account for about 400–500 million human incidences across the entire planet annually (Friedman et al. 2000). Infection by *C. jejuni* can lead to watery or bloody diarrheal disease, which can vary from non-inflammatory and self-limiting to a severe and inflammatory nature, resulting in significant medical and socioeconomic consequences (Nachamkin et al. 2008; Oyarzabal and Backert 2012). In a small subset of persons, the infection can be accompanied by more serious complications such as bacteraemia, or may result in the development of reactive arthritis and Reiter's syndrome or the neurological sequelae Guillain–Barré and Miller Fisher syndrome (Smith 2002; Yuki and Koga 2006). The rate of human infections by zoonotic foodborne *C. jejuni* has been progressively growing over the years, which creates a significant public health burden worldwide (Kaakoush et al. 2015).

Genetic typing tools and whole-genome sequencing have clearly demonstrated that *C. jejuni* isolates obtained from chicken meat can result in human campylobacteriosis (see Chap. 3 of this book). Nevertheless, the collective genetic diversity among *Campylobacter* isolates suggests that other transmission routes also exist, and that highly mutable sequences in multiple genetic loci across the chromosome appear to have an important function for bacterial adaptation in a new host (Sheppard and Maiden 2015). For example, comparative genomics and phenotypic analyses identified the emergence of specific *C. jejuni* lineages to become cattle specialists, which coincided with the enormous rise in the global cattle population in recent decades (Mourkas et al. 2020). Genome sequencing and other approaches have demonstrated the presence of various virulence factors, some of which have been well characterized in recent years. Once the bacteria have reached the human intestine, *C. jejuni* has been shown to interact with the gut epithelium as well as with immune cells (as summarized in a model in Fig. 1). A major difference between the infection of chicken and human hosts is the markedly higher capability of *C. jejuni* to invade human vs. avian epithelial cells (Young et al. 2007; Ó Cróinín and Backert 2012; Burnham and Hendrixson 2018). These findings imply that *C. jejuni* adhesion to and invasion into the epithelium is most likely associated with disease outcome. Consequently, it can be assumed that pinpointing factors involved in bacteria-host interactions are crucial to understand *C. jejuni* pathogenesis and for the development of new antimicrobial therapies. This chapter reviews in detail the infection strategy by *C. jejuni* toward the human host and the interplay of bacterial factors with epithelial as well as immune cells, which is important for the development of gut disease.

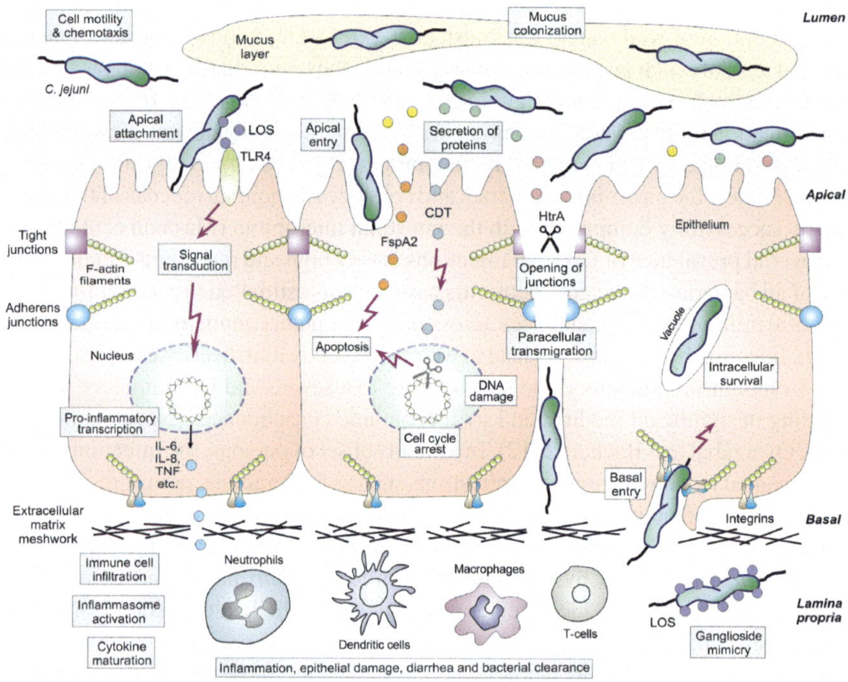

Fig. 1 Model for crucial steps and *C. jejuni* mechanisms during infection in the human intestine. The intestinal epithelium functions as a tight physical barrier and serves as sensor of microbial infections such as *C. jejuni*. Various indicated surface-exposed and secreted bacterial factors are proposed to enable colonization of the mucus layer, adhere to epithelial cells, open the tight and adherens junctions, allow cell invasion as well as trafficking and survival in intracellular vacuoles. In addition, factors produced by *C. jejuni* can trigger nuclear responses such as cell cycle arrest, DNA damage, apoptosis and pro-inflammatory cytokine production. The latter leads to the infiltration of various immune cell types to the sites of infection. Production of reactive oxygen species and other anti-bacterial responses enhance cell damage, leading to campylobacteriosis and eventually to bacterial clearance. Furthermore, modified lipooligosaccharide (LOS) structures of *C. jejuni* mimic human gangliosides, which can lead to auto-antibody production and neural disorders in some patients

2 Bacterial Virulence Factors and Epithelial Cell Responses

2.1 *Specialized Metabolism and Enteric Life Style*

It is well established that *C. jejuni* does not utilize sugar metabolites as a carbon source, which was shown to be due to absence of the glycolytic enzyme phosphofructokinase (Parkhill et al. 2000; Velayudhan and Kelly 2002). Instead, *C. jejuni* growth depends on the presence of single amino acids or keto acids, either supplied by the host or by the residual gut microbiota (Lee and Newell 2006; Hofreuter 2014). The amino acids present in the chicken gut were quantified, and the most abundant ones were those that *C. jejuni* depends on for its metabolism, demonstrating its adaptation to this host (Parsons et al. 1983). Early *C. jejuni* growth experiments in vitro have shown that aspartate, serine, proline and glutamate are favorably utilized as nutritive substances (Leach et al. 1997; Elharrif and Mégraud 1986; Leon-Kempis et al. 2006; Velayudhan et al. 2004; Guccione et al. 2008). The bacteria contain dedicated membrane transporters for certain amino acids that are essential for serine metabolism. By means of mutagenesis, it was demonstrated that these transporter systems are required for colonization in the intestine of chickens (Hendrixson and DiRita 2004; Velayudhan et al. 2004; Ribardo and Hendrixson 2011). Furthermore, it appears that natural *C. jejuni* strains display a high genetic diversity and sometimes utilize specific pathways to metabolize a given amino acid (Hofreuter 2014; Gao et al. 2017). In addition, specific genetic polymorphisms have been associated with varying capabilities to acquire nutrients and establish colonization in mice. For example, *C. jejuni* strains expressing a certain γ-glutamyltranspeptidase (GGT) enzyme were also able to metabolize glutathione and glutamine, leading to elevated colonization rates in the murine gut (Hofreuter et al. 2008; Floch et al. 2014). Similarly, when an asparaginase enzyme was present with a sec-mediated secretion motif, *C. jejuni* could acquire asparagine from the environment, which not only improved colonization of the intestinal tract, but also of the liver of mice (Hofreuter et al. 2008). Finally, a recent comprehensive in vitro analysis of a *C. jejuni* transposon mutant library was performed for bacterial multiplication in distinct broth media and subsequently for the capability to colonize mice, and this was merged with isotopolog profiling experiments and metabolic flow studies (Gao et al. 2017). This work identified that *C. jejuni* can consume various metabolic end products from the gut microbiome, including acetate and carbon dioxide in the form of hydrogen carbonate, and particularly single amino acids as mentioned above, as well as oligo-peptides made available from food and degraded host proteins. Furthermore, it was observed that single amino acids and di-peptides are present in the intestinal mucus layer in considerable quantities. However, certain required amino acids may not be available in abundances high enough to support *C. jejuni* growth, as demonstrated by auxotrophic mutations that prevented the production of serine or aromatic and branched amino acids and led to the inability to colonize mice (Gao et al. 2017). It seems that *C. jejuni* surmounts these metabolic substrate constraints by utilizing the tricarboxylic acid cycle, the non-oxidative pentose phosphate pathway and gluconeogenesis, which can collectively promote growth in vitro and in vivo (Gao et al. 2017). Together, this detailed study has pinpointed multiple routes of a highly specialized metabolism and life style of *C. jejuni* to achieve bacterial fitness in the gut.

2.2 Campylobacter *Motility and Chemotaxis*

Motility is a characteristic property of *C. jejuni* and is essential for effective colonization in the avian, murine or human host (Wassenaar et al. 1993; Guerry 2007; Chang and Miller 2006; Artymovich et al. 2013; Black et al. 1988; Schmidt et al. 2019). The bacteria are motile by means of two flagella, one on each end of the spirally-shaped bacterial cell body. The flagellum functions as a propeller driven by a rotating motor, which enables the bacterium to swim by a corkscrew-like mechanism (Purcell 1997; Karim et al. 1998; Shigematsu et al. 1998; Cohen et al. 2020). The *C. jejuni* flagellar structure resembles that of *Salmonella* or *E. coli,* but *C. jejuni* has three additional disk structures making up the motor, called the basal disk (mainly consisting of FlgP), the medial disk (composed of PflA) and the proximal disk (formed by PflB and MotAB stator units) (Chen et al. 2011; Beeby et al. 2016). These disk structures are located in the periplasm, where they surround the flagellar rod (Beeby et al. 2016). In addition, the *C. jejuni* flagellum has an MS ring and a C ring of higher complexity that contributes to enhanced activity of the flagellar type III secretion system (Henderson et al. 2020), which will be further described in Sect. 2.6 below. Connected to this structure is the so-called surface hook (FlgE) to which the flexible extracellular flagellar filament (composed of FlaA and FlaB) is attached (Guerry et al. 1991; Hendrixson and DiRita 2003; Chen et al. 2011; Beeby et al. 2016). In addition to a function in bacterial motility, the flagellum may also be used for the secretion of proteins into the extracellular space (Konkel et al. 1999; Poly et al. 2007; Christensen et al. 2009; Barrero-Tobon and Hendrixson 2012; Faber et al. 2015). It was also reported that the flagellum can play a role in the adhesion of *C. jejuni* to certain host cells (Yao et al. 1994). In line with that observation, it was described that the secreted flagellin-like protein FlaC binds to the host cell, may contribute to *C. jejuni* invasion and as such may play a role by modulation of the host immune response (Song et al. 2004). The flagellar filament undergoes O-linked glycosylation, which enables it to colonize chickens (Howard et al. 2009). Protein glycosylation is also important for the correct assembly of the filament's building blocks, flagellin (FlaA and FlaB). To this end, it appears that glycosylation allows the flagellin subunits to interact with each other (Goon et al. 2003; Guerry et al. 2006; Kreutzberger et al. 2020). A two-component signal transduction system (FlgS and FlgR) regulates transcription of many flagellar genes essential for flagellar biosynthesis, some of which are produced with Sigma 54 (Hendrixson and DiRita 2003; Wösten et al. 2004). They are controlled by supercoiling of chromosomal DNA (Shortt et al. 2016) as well as by phase variation and phosphorylation, which is unique in this bacterium (Hendrixson 2006, 2008).

By means of chemotactic sensors, *C. jejuni* is able to senses metabolic concentration gradients, such as those represented by certain components surrounding the mucosa of the gut (Korolik 2019). For the human intestine, these are mainly aspartate, asparagine and lactate, while in the chicken intestine L-fucose is present (Vegge et al. 2009; Hartley-Tassell et al. 2010; Rahman et al. 2014; Dwivedi et al. 2016). Thus, chemotaxis plays an important role both in commensal and pathogenic microbe–host interactions. Through a broad genome sequence analysis among multiple *C. jejuni* strains, a number of orthologous chemotaxis genes including *cheA, cheW, cheV, cheY, cheR* and *cheB* were identified (Marchant et al. 2002). It was shown that CheY acts as a response regulator and interacts with the flagellar motor to affect the rotation direction to turn either clockwise or counterclockwise. Therefore, it is a particularly important factor for flagellar function (Yao et al. 1997). A recent publication demonstrated that CheY has no effect on the speed of rotation, because speed was not affected by deletion of *cheY* (Cohen et al. 2020). Deletion of one of the above mentioned chemotaxis components leads to a colonization defect as shown in ferret and mouse models (Yao et al. 1997; Chang and Miller 2006). As a result, *C. jejuni* is no longer able to induce the associated disease during infection. Together, it can be concluded that motility in combination with the chemotaxis cascade is important for colonization and proper interaction of *C. jejuni* with its host, either asymptomatic as in chickens or symptomatic as in humans (Korolik 2019).

2.3 *CDT Toxin Production*

The cytolethal distending toxin (CDT) is the only known toxin found in most, though not all *C. jejuni* strains (Mortensen et al. 2011). In contrast to other diarrheal pathogens, *C. jejuni* does not encode other toxins, which makes the CDT unique for this bacterium (Lai et al. 2016). When cultured host cells are exposed to CDT, it leads to a characteristically enlarged cell surface and to cell death, which gave the toxin its name. This was originally shown for several sensitive cell lines upon infection with *C. jejuni* (Johnson and Lior 1988). CDT is a holotoxin comprising of three subunits, CdtA, CdtB and CdtC. Of these, CdtB exhibits enzymatic Dnase activity that ultimately results in cell-cycle arrest and cell death, while CdtA and CdtC are responsible for the translocation of CdtB across the target cell membrane (Lara-Tejero and Galán 2001). Subunits CdtA, CdtB and CdtC are located on the bacterial cell surface, where they assist in binding to the host cell (Guerra et al. 2011); however, the exact mechanism of CdtA/CdtC-assisted CdtB translocation remains controversial. Similarities to the B chain of ricin toxin, which is important for the receptor-induced endocytosis of ricin, have been demonstrated (Lara-Tejero and Galán 2001). It has also been shown that the CDT holotoxin is either directly secreted into the extracellular space or packaged into outer membrane vesicles (OMVs) that are continuously being shed from the bacteria (Lindmark et al. 2009). OMVs are commonly formed by Gram-negative bacteria and fulfil a number of tasks such as the delivery of toxins (Wai et al. 2003). So far it remains unclear, however, how the OMVs carrying CDT bind to target cells, or whether the OMVs enter the cell. It appears that the holotoxin, when packed in OMVs, can enter the host cell by membrane fusion (DiRienzo 2014). Thus, it remains to be determined how exactly CdtB is translocated into the host cell.

Once CdtB has reached the cytoplasm of the target cell, it is transported to the endoplasmic reticulum with the help of the Golgi apparatus to get into the cell nucleus (Heywood et al. 2005). Both actin and the microtubulin systems assist in this transport of CdtB into the nucleus (Méndez-Olvera et al. 2016). Inside the cell nucleus, CdtB induces DNA double-strand breaks, which lead to arrest of the cell cycle in the G2/M phase (Whitehouse et al. 1998). This mechanism results in the activation of DNA repair mechanisms and blocking of the nuclear CDC2 kinase via phosphorylation, which is responsible for entering mitosis. This ultimately leads to apoptosis of the cell (Pickett and Whitehouse 1999; Guerra et al. 2005). In some human cell lines (Hela, Caco-2), cell enlargement is indeed observed during infection with *C. jejuni*, leading to cell death (Johnson and Lior 1988; Elmi et al. 2016). Further studies showed that CDT is able to trigger the secretion of IL-8 as an immune response, which leads to the recruitment of macrophages and neutrophils to the infected site (Hickey et al. 2000; Purdy et al. 2000; Zheng et al. 2008). This leads to a massive infiltration of immune cells into the infected tissue, which is a typical histopathological hallmark of campylobacteriosis. In support of this, a Δ*cdt* knockout mutant was still able to colonize immuno-suppressed mice but was no longer able to induce symptoms or systemic infection (Purdy et al. 2000; Fox et al. 2004). Thus, these results clearly show that CDT plays a role in the pathogenicity of *C. jejuni* and can therefore be counted as virulence factor. Nevertheless, human symptomatic infections are sometimes caused by strains defective in CDT production (Mortensen et al. 2011).

2.4 Serine Protease HtrA and Epithelial Barrier Disruption

The trypsin-like serine protease HtrA (high-temperature requirement A) is a highly conserved enzyme found in both prokaryotes and eukaryotes and was first described in *Escherichia coli* (Lipinska et al. 1989). That species typically contains three *htrA* orthologous genes, called *degQ*, *degP* and *degS*, while other bacteria such as *C. jejuni* encode only one *htrA* gene copy, whose product is most similar to DegQ. HtrA has a domain-like structure and consists of a signal peptide required for secretion, a protease domain and two PDZ domains. The bi-functional protein acts as a chaperone and also has protease activity. The protease domain contains a catalytic triad which is composed of the amino acids histidine (His), aspartate (Asp) and serine (Ser). Following removal of the signal peptide, the protein reaches the periplasm, where it forms proteolytically active multimers that carry out protein quality control functions (Clausen et al. 2002; Kim and Kim 2005; Krojer et al. 2010). Cryo-electron microscopy of *C. jejuni* HtrA revealed that it forms a dodecamer, built of four trimers (Zarzecka et al. 2020). Biochemical studies disclosed proteolytically active hexamers, dodecamers and even larger oligomers with a remarkable stability compared to previously investigated HtrA orthologs in other bacteria (Zarzecka et al.

2020). It was further shown that HtrA protects the *C. jejuni* bacteria against the enrichment of denatured or non-properly folded proteins in the periplasm under stress conditions (Brøndsted et al. 2005; Bæk et al. 2011a). Indeed, a *C. jejuni* Δ*htr*A deletion mutant exhibited reduced growth compared to the wild-type bacteria. This was attributed to the chaperone activity of the protein, since a protease-inactive S197A mutant showed no impairment in terms of growth (that single amino acid mutation does not affect the chaperonin activity of HtrA). Similar behavior was also observed in response to oxidative stress (Brøndsted et al. 2005; Bæk et al. 2011a). Further studies demonstrated that the Δ*htr*A deletion mutant had defects to adhere and invade into host cells. The defective phenotype was restored after genetic complementation with the wild-type *htrA* gene (Bæk et al. 2011a; Boehm et al. 2012, 2013, 2015). However, the ability to adhere to the host cell did not require protease activity of HtrA (Bæk et al. 2011b). Altogether, these findings suggest that the chaperone part of HtrA has an important role in the adherence and invasion process during *C. jejuni* infection (Konkel et al. 2001).

It has been demonstrated that HtrA of *C. jejuni* is also secreted into the extracellular space, which suggests stress response and survival is not its only function (Boehm et al. 2012, 2018; Backert et al. 2018). A plausible hypothesis is that the HtrA protein is released as a soluble enzyme and/or packaged as cargo into OMVs (Boehm et al. 2012; Elmi et al. 2012, 2016; Yoon 2016). The number of HtrA molecules secreted per *C. jejuni* cell has been quantified and is considerably high: On average, about 4000–5000 HtrA molecules can be secreted by a single bacterium during 2 h of culture in liquid broth (Neddermann and Backert 2019). When the protein is secreted into the extracellular environment during infection, HtrA comes in direct contact with host cell surface proteins, where it may exhibit protease activity, as has been experimentally demonstrated. The first described HtrA target was the adherens junction and tumor suppressor protein E-cadherin, which HtrA cleaves into various fragments (Boehm et al. 2012). Another recently discovered target protein which is cleaved by *C. jejuni* HtrA is the tight junction protein occludin (Harrer et al. 2019). These cleavage events lead to the temporary opening of cell-to-cell junctions in the epithelium, allowing *C. jejuni* to transmigrate between two neighboring cells (Fig. 1). Interestingly, it was observed that such a temporary opening of the junctions has no major impact on the transepithelial electrical resistance (TER), which demonstrates the overall tightness of an epithelial layer (Boehm et al. 2012; Harrer et al. 2019). It was also shown that both the Δ*htrA* and an S197A mutants were unable to transmigrate across a cell monolayer, despite being fully motile, which clearly demonstrated the importance of the HtrA protease of *C. jejuni* for crossing the epithelial cell monolayer. Thus, our hypothesis is that HtrA has a dual function for the pathogen: (i) intracellular protein quality control and (ii) extracellular cleavage of host cell junctional proteins to establish a proper infection (Boehm et al. 2012, 2013; Backert et al. 2018).

In two different mouse models (one using IL-$10^{-/-}$ knockout mice and the other using infant wild-type mice), the impact of *C. jejuni* HtrA on the infection process has been monitored (Boehm et al. 2018). When IL-$10^{-/-}$ knockout mice were infected with wild-type *C. jejuni*, strong changes in the crypt architecture were observed, accompanied by a strong infiltration of immune cells into the tissue. These features were diminished in mice infected with the Δ*htrA* mutant (Heimesaat et al. 2014a). Infection of infant mice with wild-type and Δ*htrA* mutant *C. jejuni* resulted in similar observations, and in this model presence or absence of HtrA did not affect the colonization rates of *C. jejuni* in the intestine (Heimesaat et al. 2014b). Taken together, these observations clearly identified HtrA as a virulence factor of *C. jejuni*. We therefore consider HtrA an ideal candidate for future development of new antimicrobial drugs. The development of new drugs against this foodborne pathogen is very important, as *Campylobacter*-mediated enteritis is highly frequent in a number of countries (Gölz et al. 2014). Efforts to develop an inhibitor against HtrA have been conducted in vitro, mostly targeting the proteins of *E. coli* or *Helicobacter pylori* HtrA (Hauske et al. 2009; Perna et al. 2014, 2015; Schmidt et al. 2016; Tegtmeyer et al. 2016). These attempts had shown some success; however, the identified inhibitors are still somewhat unspecific and can also inhibit HtrAs of other bacteria, including commensals. Thus, further research must be carried out in this area to narrow down the specificity of HtrA inhibitors toward the proteins of pathogenic bacteria only.

2.5 *Outer Membrane Adhesins and Host Cell Binding*

C. jejuni invasion into host epithelial cells is first initiated through cell adherence provided by various adhesion proteins (called adhesins), which recognize and bind to specific host cell receptors resulting in stable attachment. The binding activity of adhesins is widely accepted to be fundamental for the effective interaction of a given bacterium with host cells and is a necessary prerequisite for subsequent invasion (Hermans et al. 2011; Backert et al. 2013). Several major *C. jejuni* adhesins have been shown to result in attachment to the host cell (Fig. 2), with CadF (*Campylobacter* adhesin to fibronectin) believed to play a key role in this process. Other major adhesins include fibronectin like protein A (FlpA), *jejuni* lipoprotein A (JlpA), major outer membrane protein (MOMP), *Campylobacter* autotransporter protein A (CapA), a 95 kDa outer membrane protein (p95) and periplasmic binding protein 1 (PEB1) (Ó Cróinín and Backert 2012). CadF is a 37-kDa outer membrane protein that mediates bacterial attachment to the host cells through the extracellular matrix protein fibronectin (Konkel et al. 1997; Schmidt et al. 2019; Krause-Gruszczynska et al. 2007a). The CadF/fibronectin interaction during *C. jejuni* colonization was mostly studied using INT-407 cells as a model for infection. Originally thought to be derived from normal embryonic intestinal tissue, INT-407 was subsequently found to have been established via HeLa cell contamination, which should be taken into account when working with these cells (Nelson-Rees and Flandermeyer 1976; Neimark 2015). Disruption of the *cadF* gene resulted in reduction of bacterial adherence to the INT-407 cells (Monteville and Konkel 2002; Monteville et al. 2003; Krause-Gruszczynska et al. 2007b). CadF deficiency further rendered the bacteria less capable to colonize chickens in comparison with wild-type *C. jejuni* (Ziprin et al. 1999). It was proposed that CadF is responsible for the uptake of *C. jejuni* at

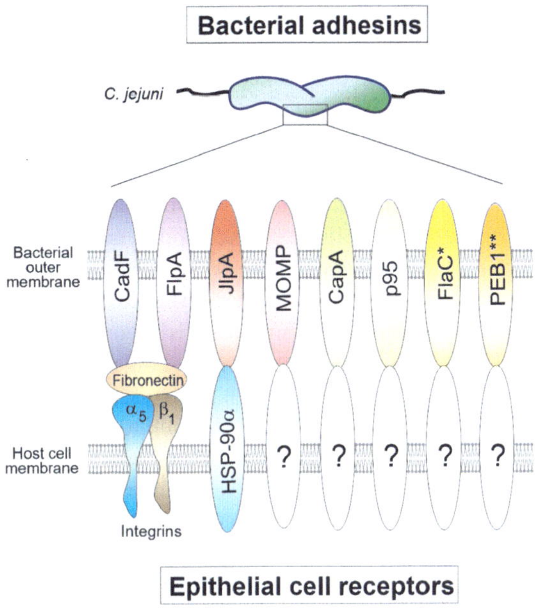

Fig. 2 Major surface-exposed adhesins and involved host cell receptors described for *C. jejuni*. At least eight membrane-anchored bacterial proteins have been described to act as binding factors to host target cells. For three of these factors, the corresponding host cell interaction partner has been identified. For the others, the host receptor is still unknown and these are labeled with question marks. FlaC* is secreted via the flagellar T3SS and PEB1** has been described as an aspartate/glutamate-binding protein of an ABC transporter. For more details on these adhesins, see text

the basolateral area of host cells, where fibronectin is linked to the integrin-based receptor complex (Monteville and Konkel 2002). However, in the IL-$10^{-/-}$ mouse model observations revealed that *C. jejuni* flagellin A and B, but not cell adhesion mediated by CadF, are essential for inducing murine campylobacteriosis (Schmidt et al. 2019).

Another binding factor of *C. jejuni* acting through interaction with fibronectin is the 46-kDa protein FlpA. Deletion of *flpA* resulted in a significant reduction of bacterial attachment to INT407 cells, while in wild-type *C. jejuni* the FlpA protein interacted with cellular fibronectin in a dose-dependent manner (Konkel et al. 2010; Eucker and Konkel 2012). This interaction was studied in more detail, revealing that the FlpA/fibronectin interaction is mediated by the fibronectin-binding linear motif in domain-2 of FlpA with the gelatin-binding domain of fibronectin (Larson et al. 2013). Altogether, it seems that both CadF and FlpA proteins enable attachment of *C. jejuni* to host cells via fibronectin in a cooperative manner.

A third adhesin is the constitutively expressed JlpA, a 43-kDa surface lipoprotein that was shown to interact with intestinal heat shock protein 90α. Mutation in the *jlpA* gene led to reduced adherence of *C. jejuni* to cultured HEp-2 cells by 18–19.4% (Jin et al. 2001, 2003). In addition, pretreatment of HEp-2 cells with purified JlpA decreased *C. jejuni* adhesion in a dose-dependent manner (Jin et al. 2001), suggesting saturation of a receptor. In an interesting study, recombinant JlpA was expressed in *Lactococcus lactis* and when these bacteria were fed to chickens, IgA antibodies were raised against the protein that was present in chicken feces. When these antibodies were used for pretreatment of either human INT407 or primary chicken embryo intestinal cells, it resulted in a significant reduction in bacterial adherence and invasion of *C. jejuni* (Gorain et al. 2020), again suggesting the interaction between JlpA and its targets can be blocked, this time with JlpA-directed antibodies.

Similarly, inactivation of the *peb1A* gene, which encodes a 28-kDa protein PEB1, decreased adherence of *C. jejuni* to cultured HeLa cells, and prevented colonization of mice (Pei et al. 1998); however, the exact adhesion mechanism of this protein remains unclear. Later studies showed that PEB1 had less influence on adherence of *C. jejuni* to T84 cells or chicken epithelial cells (Novik et al. 2010; Flanagan et al. 2009). PEB1 actually functions as a periplasmic binding protein as part of an aspartate/glutamate ABC transporter system, which is required for optimal microaerobic growth on dicarboxylic amino acids (Leon-Kempis et al. 2006). This could explain why PEB1-deficient bacteria are less able to colonize a given host.

The MOMP protein, which is also known as PorA, has also been suggested to contribute to the adherence of *C. jejuni* to host cells, although its major role seems to be the transport of nutrients and other small molecules. Nevertheless, a recent study showed that the transcription terminator of the *porA* gene enhances the expression level of MOMP by stabilizing its mRNA and therefore influences the virulence of *C. jejuni* (Dai et al. 2019). Some of the proposed *C. jejuni* adhesins are controversial in the literature, and their contribution to the bacterial adhesive properties might be indirect, as may apply to, among others, PEB1, MOMP, CapA and p95 (Ó Cróinín and Backert 2012). Interestingly, novel genes regulating adhesion factors remain to be discovered, as illustrated by the recent report of a new two-component signal transduction system which is involved in regulating adhesion (Xi et al. 2020). These genes (*cj1492c* and *cj1507c*) encode a histidine kinase and a transcriptional regulator, respectively, and when inactivated, this impaired motility, adherence and invasion, and fewer bacteria survived intracellularly. The gene pair has been renamed BumR and BumS and has also been found to be involved in directing a response to butyrate (Luethy et al. 2015; Goodman et al., 2020). Together, these findings illustrate that adhesion is a key prerequisite for the successful colonization of host by *C. jejuni*.

2.6 The Flagellum as a Specialized Type III Secretion System

C. jejuni does not possess classical type III or type IV secretion systems (T3SS or T4SS) to inject effector molecules into host cells; however, T3SS functions were found to be provided by the flagellum, which has been demonstrated to export effector proteins that can control bacteria–host interactions as discussed below (Barrero-Tobon and Hendrixson 2012; Christensen et al. 2009; Young et al. 1999; Ziprin et al. 1999). The flagellar filament of *C. jejuni* consists of two glycosylated structural flagellins, FlaA and FlaB, as described above. One of the proteins secreted through the flagellar filament is the non-structural protein FlaC, which is implemented in host cell invasion. The flagellar proteins FliS and FliW in *C. jejuni* assist in the secretion of FlaC protein (Radomska et al. 2017). The FliS protein (a flagellar chaperone) preferentially binds to the glycosylated flagellins and is essential for flagellar assembly;

it also directs FlaC toward the flagella for its secretion, while FliW mainly acts as sensor of intracellular FlaA/FlaB flagellin levels. The FlaC protein is thus secreted from the flagellar apparatus of *C. jejuni* cells and plays an important role in entry to epithelial cells (Song et al. 2004). Moreover, FlaC has been shown to directly interact with toll-like receptor 5 (TLR5), resulting in p38 activation (Faber et al. 2015). Preincubation with FlaC modulated the immune responses of chicken and human macrophage-like cells toward the bacterial TLR4 agonist lipopolysaccharide (LPS) by promoting cross-tolerance with subsequent reduction of interleukin-1β (IL-1β) expression (Faber et al. 2015). Consequently, the flagellum is a complex machinery that not only renders *C. jejuni* motile, but also enables protein secretion and administration into the host cell, providing a crucial step in the process leading to host invasion (Burnham and Hendrixson 2018). In this context, it is noteworthy that the flagellum is evolutionarily related to T3SSs used by, for instance, pathogenic *Salmonella* or *Yersinia* species, and was therefore named flagellar T3SS (fT3SS).

FlaC is not the only protein secreted by the *C. jejuni* fT3SS. There are two other distinct groups of proteins secreted this way, described as (i) flagellar co-expressed determinants (FedA-D) and (ii) *Campylobacter* invasion antigens (CiaA-I) (Konkel et al. 1999; Eucker and Konkel 2012; Burnham and Hendrixson 2018). The Fed proteins were found to be important in commensal colonization of chickens, while FedA is also involved in invasion of human intestinal cells (Barrero-Tobon and Hendrixson 2012). However, the individual functions of Fed proteins remain largely unknown (Burnham and Hendrixson 2018). The Cia proteins have been reported to influence *C. jejuni* interaction with human intestinal cells; however, their mechanisms of action during adhesion and invasion are also relatively unclear. One of the best-characterized Cia members, the 73-kDa protein CiaB, appears to be necessary for the secretion process itself, and is required for maximal invasion of *C. jejuni* into host target cells (Konkel et al. 1999). *C. jejuni* strain F38011 with a deleted Δ*ciaB* gene exhibited significantly lower invasion capacity into human cells, along with reduced colonization in chickens (Ziprin et al. 1999). However, inactivation of *ciaB* in strain 81–176 did not influence invasion capacity toward cultured intestinal epithelial cells (Novik et al. 2010). These conflicting outcomes may be related to strain differences or differences in experimental procedures. A gene screening of *C. jejuni* strain NCTC 11168 revealed at least 42 proteins with putative fT3SS amino-terminal sequences directing their export through the flagellum (Christensen et al. 2009). CiaC is an example and was reported to be essential for maximal invasion of epithelial cells, which took place through the recruitment and activation of small Rho GTPase member Rac1, while *ciaC*-deficient *C. jejuni* resulted in significant decrease of Rac1 activation (Eucker and Konkel 2012). Another Cia member, CiaI, may have a function in intracellular survival in human cells (Buelow et al. 2011) and/or colonization in chickens (Barrero-Tobon and Hendrixson 2012) and is discussed below. Taken together, it appears that the flagellar export machinery fT3SS represents a crucial secretory apparatus, which enables *C. jejuni* to invade and manipulate host cells.

2.7 *Bacterial Factors and Signaling Involved in Host Cell Invasion*

A major disease-associated feature of *C. jejuni* is its capability to invade host tissues, which is believed to represent a primary mechanism of pathogenesis associated with host tissue damage. A number of molecular players of the bacterium and host cell involved in invasion have been discovered (Fig. 3). High-resolution electron microscopy of infected epithelial cells revealed that *C. jejuni* can induce membrane rearrangements upon contact with a host membrane, which eventually leads to cell invasion (Boehm et al. 2011; Krause-Gruszczynska et al. 2011). According to multiple studies, there are two major strategies utilized by *C. jejuni* to enter cultured cells, either microtubule-dependent or actin-filament-dependent mechanisms, which

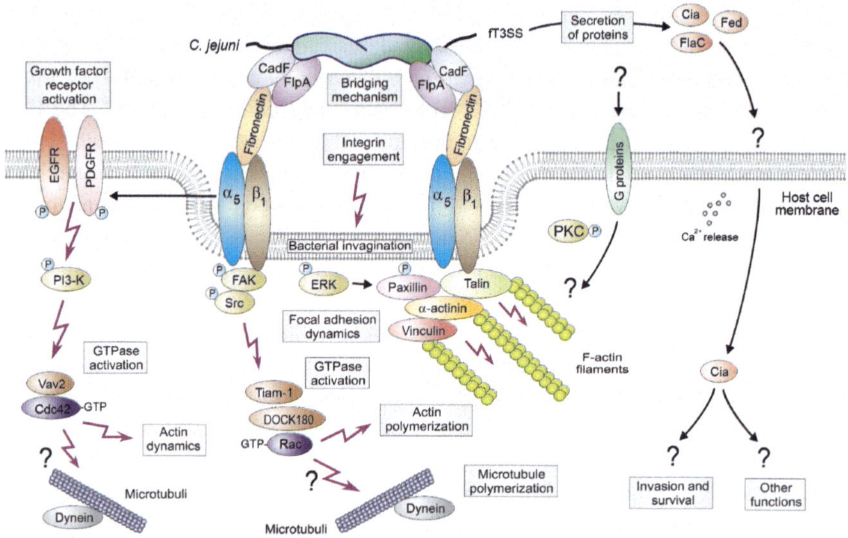

Fig. 3 Molecular signal transduction model for *C. jejuni*-induced events leading to bacterial invasion of the human intestinal epithelium. *C. jejuni* express two fibronectin-binding proteins, CadF and FlpA, which mediate attachment of the bacteria to integrin-based focal adhesion structures. In this way, integrin receptors are activated leading to various indicated signaling events. For example, the growth factor receptors EGFR and PDGFR are activated by phosphorylation, which leads to stimulation of PI3-kinase (PI3-K) and the guanine exchange factor Vav-2, activating the small Rho GTPase Cdc42. In addition, integrin engagement leads to autophosphorylation of the cytoplasmic kinases FAK and Src as well as paxillin, which in turn activate the guanine exchange factors Tiam-1 and Dock180, stimulating small Rho GTPase Rac1. Both Cdc42 and Rac1 can then trigger microtubule and F-actin polymerization/reorganization events leading to the membrane engulfment and subsequent uptake of *C. jejuni* into the host cells. Activation of additional host receptors such as G proteins by yet unknown bacterial factor(s) also appears to contribute to bacterial host cell entry. Finally, various secreted effector proteins of the *C. jejuni* flagellum (fT3SS) are proposed to function in bacterial attachment and invasion, but their exact mechanisms are yet unclear. For more details, see text

seem to vary between *C. jejuni* strains (Ó Cróinín and Backert 2012). While the detailed signaling mechanism of the microtubule-dependent invasion pathway is unclear, members of the small Rho GTPase proteins are known as key regulators, which trigger actin polymerization, membrane ruffling and bacterial internalization (Stradal and Schelhaas 2018). Various experiments reported that entry of *C. jejuni* into intestinal epithelial cells depends on the activation of the small Rho GTPase members Cdc42 and Rac1 (Krause-Gruszczynska et al. 2007b; Krause-Gruszczynska et al. 2011; Boehm et al. 2011; Eucker and Konkel 2012). *C. jejuni* binds to fibronectin by means of its CadF and FlpA adhesins, which may act together, and in turn drive the phosphorylation of EGF receptor by activation of the integrin receptor (Eucker and Konkel 2012). Upon attachment to the host cell surface, CadF invokes two signaling cascades that lead to either Cdc42 or Rac1 activation, both resulting in F-actin-dependent engulfment and uptake of *C. jejuni*. This process involves a wide range of intermediate factors initiated by binding of CadF to fibronectin, followed by signal cascades involving PDGF and EGF receptors, integrins, cytosolic kinases and guanine exchange factors. The activation of Cdc42 and Rac1, respectively, takes place by the following proposed pathways. The difference of these two routes is the step following integrin/focal adhesion kinase (FAK) interaction: CadF → fibronectin → β1-integrin → FAK/Src → PDGFR/EGFR → PI3-kinase → Vav2 → Cdc42 (Krause-Gruszczynska et al 2011) (left side of Fig. 3); or CadF → fibronectin → β1-integrin → FAK → Tiam-1/DOCK180 → Rac1 (Krause-Gruszczynska et al. 2011; Boehm et al. 2011) (middle part of Fig. 3). The importance of all these host factors for efficient bacterial uptake was demonstrated by *C. jejuni* infection of fibroblasts derived from fibronectin$^{-/-}$, integrin-$\beta_1{}^{-/-}$, FAK$^{-/-}$ and Src$^{-/-}$/Yes$^{-/-}$/Fyn$^{-/-}$ (SYF) triple knockout mice, each of which resulted in invasion failure (Krause-Gruszczynska et al. 2011; Boehm et al. 2011). Whether the activation of Cdc42 and Rac1 are involved in microtubule-dependent uptake of *C. jejuni* has not yet been investigated. In addition, various studies using inhibitors and gentamycin protection assays suggested that heterotrimeric G proteins, the mitogen-activated kinases ERK and p38, protein kinase C (PKC), phosphatidylinositol-3-kinase (PI3K) and Ca^{2+} release from host cytoplasmic compartments all play a role in *C. jejuni* host cell entry (Wooldridge et al. 1996; Biswas et al. 2000; Hu et al. 2005; 2006), but the involved bacterial factors are not yet clear and need further investigation.

It should be mentioned that many of the above in vitro studies were performed with non-polarized host cells that lack proper cell-to-cell junctions, so that the basolateral fibronectin/integrin complex is easily accessible, a situation that vastly differs from an intact intestinal epithelium encountered in vivo. By means of polarized intestinal Caco-2 cells, it was demonstrated that the serine protease HtrA enables disruption of the cellular tight and adherens junctions as discussed above, which facilitates bacterial invasion of epithelial cells at the basolateral site (Harrer et al. 2019). Taken together, invasion of the epithelium by *C. jejuni* is a complex process involving dozens of bacterial effector molecules as well as a range of host receptors and proteins. While fibronectin-mediated effects of the proteins CadF and FlpA are well described, the molecular mechanisms of action of other proteins, such as adhesins PEB1, MOMP, p95 or CapA, remain to be elucidated.

2.8 *Intracellular Survival and Trafficking of* Campylobacter

Invasion of *C. jejuni* into the gut epithelium has been examined in detail; however, there are only a handful of studies investigating the fate and persistence of the bacteria once they are inside the host cells (Pesci et al. 1994; Gaynor et al. 2005; Naikare et al. 2006; Buelow et al. 2011; Bouwman et al. 2013). Gentamicin protection assays and electron microscopy studies have shown that *C. jejuni* can survive for up to 1–3 days in intestinal epithelial cell lines in vitro. In particular, intracellular *C. jejuni* were observed in a membrane-enclosed compartment in the cellular cytoplasm that was named CCV (short for *Campylobacter*-containing vacuole) (Watson and Galán 2008). These findings initiated investigations toward the *C. jejuni* factors that are involved in intracellular survival and trafficking. The first *C. jejuni* gene reported to contribute to survival within epithelial cells was *sodB*, encoding a superoxide dismutase catalyzing the breakdown of superoxide radicals, which represents an important defense mechanism against oxidative damage (Pesci et al. 1994). Another intracellular bacterial survival factor is *spoT*, a gene encoding a bifunctional ppGpp synthetase/ pyrophosphohydrolase (Gaynor et al. 2005). Microarray expression studies showed that SpoT regulates the so-called stringent stress response by *C. jejuni*. This response appears to be important for bacterial survival in the stationary phase as well as persistence during changing O_2 or CO_2 concentrations. In addition, the stringent response was necessary for various pathogenicity-related phenotypes including *C. jejuni* viability inside the cultured intestinal epithelial cells (Gaynor et al. 2005). Other genes essential to intracellular survival include *aspA* (aspartate ammonia-lyase) and *aspB* (aspartate aminotransferase) as demonstrated by their inactivation, which decreased intracellular survival, probably due to decreased viability and/or unidentified consequences of the bacterial physiology (Novik et al. 2010). However, in such studies it is difficult to differentiate factors specifically needed from intracellular survival, as mutation of many housekeeping enzymes would result in impaired intracellular survival as well. In another study, the gene of the fT3SS-delivered protein CiaI was inactivated, which compromised invasion and reduced intracellular survival levels (Buelow et al. 2011). However, the latter function is not yet fully clear because other studies supported the view that CiaI may exhibit a different function in establishing commensalism during colonization of chicken (Barrero-Tobon and Hendrixson 2011). Thus, more work is required to characterize the *C. jejuni* factors facilitating intracellular survival functions.

It appears that LOS enhances *C. jejuni* attachment and endocytosis into intestinal epithelial cells (Louwen et al. 2008, 2012). Survival of *C. jejuni* inside CCVs, a compartment in which other pathogens can be killed, may be due to a mechanism evading their maturation to a typical lysosome (Watson and Galán 2008). It has been reported that the CCV diverges from the conventional "canonical" endocytic route (Fig. 4). Interestingly, experimental recovery of intracellular *C. jejuni* from CCVs was only possible by culturing the bacteria under oxygen-limiting conditions (Watson and Galán 2008). This implies that the bacteria undergo crucial physiological adaptations inside the CCVs that may be irreversible. Furthermore, the CCVs were shown

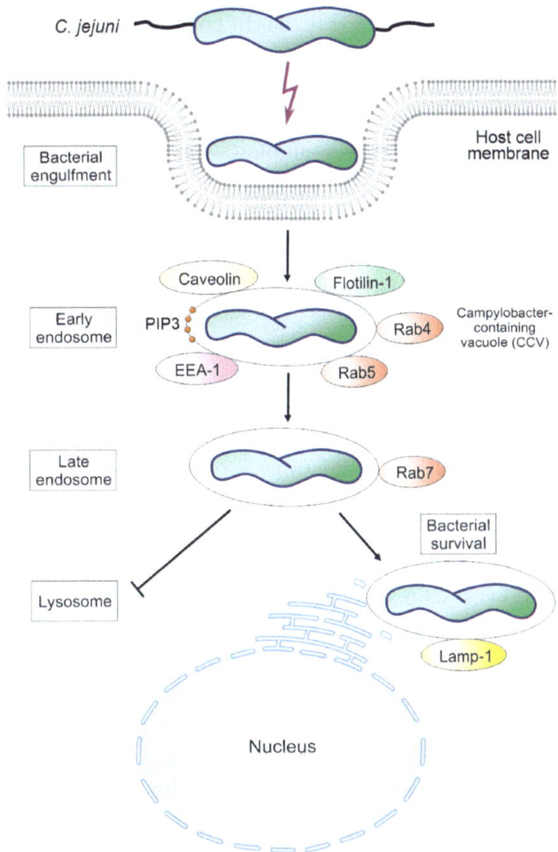

Fig. 4 Model for the establishing and maturation of *C. jejuni*-containing vacuoles (CCVs) in the cytoplasm of infected epithelial cells. *C. jejuni* enters the intestinal epithelium via mechanisms described in Fig. 3. After internalization, a so-called CCV is formed by membrane engulfment of invading *C. jejuni*. This CCV transiently recruits various marker molecules of the endocytic cascade including flotilin-1, Rab4, Rab5, Rab7, Caveolin and phosphatidylinositol 3,4,5 trisphosphate (PIP3). This leads to trafficking and intracellular survival of the bacteria in this compartment, inhibiting the canonical endocytic route toward lysosome development. In particular, intracellular *C. jejuni* inhibit the fusion of the late endosome to form lysosomes, and thus avoid bacterial killing. The CCVs also contain the marker protein Lamp-1 and localize in close proximity to the host Golgi apparatus near the nucleus. For more details, see text

to interact with endosomal compartments, and since they can be stained for the early endosomal marker EEA-1 (Early Endosome Antigen 1) as well as for two trafficking GTPases (called Rab4 and Rab5) (Watson and Galán 2008; Louwen et al. 2012); the presence of these markers on the outside of CCVs can be assumed. Nevertheless, this early event obviously appears only temporary and does not continue along the canonical cascade of endocytosis and is evidently different from conventional lysosome formation. However, the CCVs stained positive for the well-known late endosomal marker protein Lamp-1 (Fig. 4). In contrast, the CCVs did not contain signals for cathepsin B, a well-known lysosomal marker protease, and it is also not stainable for specific other endocytic tracer proteins (Watson and Galán 2008). Altogether, the recruitment of Lamp-1, which happens at a very early stage of CCV development appears to progress by a unique *C. jejuni*-triggered signaling cascade and does not lead to fusion with lysosomes. And this pathway does not demand the functional expression of the GTPases Rab5 and Rab7, despite their presence in the CCVs. Thus, further analyses are necessary to clarify in better detail how *C. jejuni* hijacks endocytic compartments for trafficking and to cause disease in humans.

3 Bacterial Virulence Factors and Immune Cell Responses

Innate immunity identifies multiple microbes through the action of various immune receptors. The recognition of *C. jejuni* by such receptors has raised much consideration because their activities could explain the immune pathology of this pathogen (Phongsisay 2016). Below we discuss important receptors binding to *C. jejuni* factors and examine their downstream signaling cascades and resulting immune responses. Numerous pro- and anti-inflammatory pathways that are triggered by *C. jejuni* and control the infection, respectively, are highlighted in Figs. 5 and 6. In vitro and in vivo studies have demonstrated that cytokines including TNF-α, IL-1β, IL-6, IL-8, IFN-β, IFN-γ and others are induced by *C. jejuni*, which not only leads to a pronounced inflammatory response (Phongsisay 2016), but also disturbs the intestinal epithelial barrier function in various ways (Bücker et al. 2018; for details see Chapter 8 in this book).

3.1 Interaction with Toll-Like Receptors

Toll-like receptors (TLRs) are key players in the activation of innate immunity. These transmembrane proteins are typically expressed on the surface of macrophages and dendritic cells but can also be found on intestinal epithelial cells. TLRs comprise specific patter-recognition receptors that recognize structurally conserved components of pathogenic microbes, including fungi, bacteria and viruses (Trinchieri and Sher 2007). For example, TLR2 recognizes peptidoglycan and lipoteichoic acid from Gram-positive bacteria (Schwandner et al. 1999), TLR4 recognizes LPS of Gram-negative bacteria (Poltorak et al. 1998), while TLR5 commonly recognizes bacterial flagellin (Andersen-Nissen et al. 2005). In vitro studies using mouse dendritic cells revealed that contact with *C. jejuni* activates their signal transduction protein MyD88 via the receptors TLR2 and TLR4 (Rathinam et al. 2009). Specifically, it is the contact of TLR4 with specific bacterial glycoconjugates that stimulates this signal cascade. At least five different *C. jejuni* glycoconjugates were found to trigger this response, three of which may be components of low-molecular-weight LOS. A

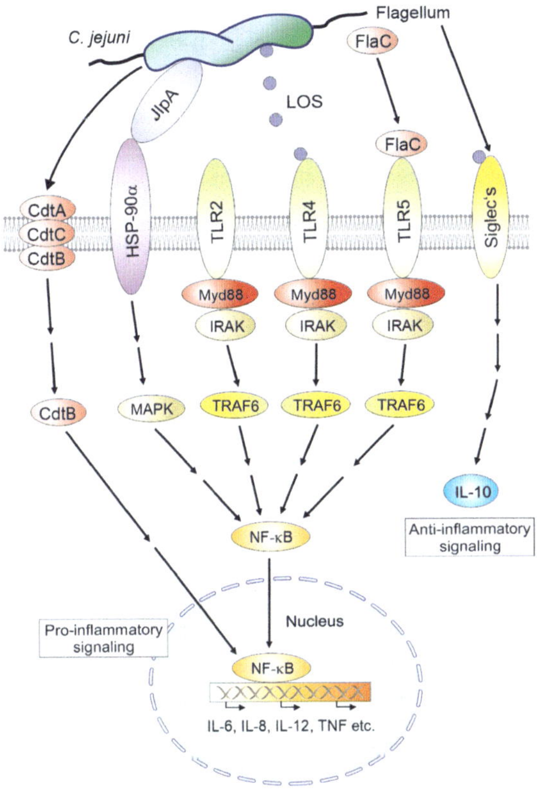

Fig. 5 ***C. jejuni*** **targets various pro-inflammatory and anti-inflammatory immune signaling pathways by interaction of bacterial factors with indicated receptor molecules.** *C. jejuni* passes the mucus layer in the human gut and interacts with the intestinal epithelial cells, triggering nuclear responses such as cytokine expression including IL-6, IL-8, IL-12 or TNF. *C. jejuni*-induced cytokine secretion can be stimulated by CDT toxin, adhesin JlpA, LOS and functional flagella. The pro-inflammatory signal transduction proceeds via the activation of host cell receptors HSP-90α, toll-like receptors (TLR2, TLR4 and TLR5), intracellular signaling proteins and transcription factor NF-κB; while engagement of Siglec receptors by *C. jejuni* stimulate the production of anti-inflammatory cytokine IL-10

fourth glycoconjugate appears as a ladder-like band with a molecular size between 30 and 50 kDa on SDS–polyacrylamide gels, while the fifth is a large protein of approximately 150 kDa (Phongsisay et al. 2015). However, the exact *C. jejuni* factor(s) that trigger TLR2 receptor activation remain to be identified. In addition, *C. jejuni* flagellin FlaA is a very poor stimulator of TLR5 (Watson and Galán 2005). Instead, cellular contact to the *C. jejuni* flagellar factor FlaC activates TLR5 (Faber et al. 2015). Upon stimulation, TRL2, TLR4 and TLR5 activate MyD88, which in turn mediates an inflammatory response via the MyD88-NF-κB signaling pathway that involves IRAK, TRAF6 and possibly other mediators, and results in the release of

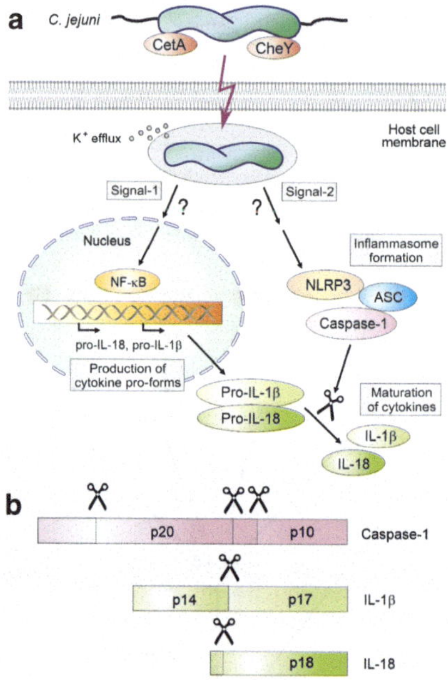

◂**Fig. 6 Signal transduction events leading to inflammasome activation by *C. jejuni*. a** *C. jejuni* stimulates NLRP3 inflammasome formation, typically assembling in infected human dendritic cells, macrophages or monocytes, which finally leads to bacterial clearance. The canonical pathway of inflammasome activation comprises two signals coming from the bacterium, signal-1 and signal-2. Signal-1 leads to activation of transcription factor NF-κB and mRNA production of NLRP3 and the pro-forms of IL-1β and IL-18. Engagement of the NLRP3 protein through signal-2 and adaptor protein ASC induces the recruitment and cleavage of autoproteolytic pro-caspase-1. Activated caspase-1 then cleaves interleukin pro-forms leading to the production of mature IL-1β and IL-18 cytokines. It was described that energy taxis protein CetA and CheY-controlled host cell invasion by *C. jejuni* play an important role, but the actual bacterial factors representing signal-1 and signal-2, respectively, are yet unknown and labeled with question marks. **b** Autoproteolytic processing events in pro-caspase-1 leading to the activation of caspase-1 and maturation steps of IL-1β and IL-18 pro-forms through caspase-1 are shown. For more details, please see text

pro-inflammatory cytokines such as IL-6, IL-8 and IL-12. In addition to MyD88-NF-κB, TRL4 also induces TIR-domain-containing adapter-inducing interferon-β (TRIF)-mediated phosphorylation of Interleukin release factor 3 (IRF3), which stimulates IFN-β secretion from the challenged cells (Rathinam et al. 2009). All these responses are part of immune stimulation that eventually recruits immune cells. Taken together, recognition of *C. jejuni* biomolecules by a variety of TLRs activates pro-inflammatory responses (Fig. 5) and finally results in clearance of the pathogen.

3.2 Role of Siglec Receptors

Sialic acid-binding immunoglobulin-like lectins (Siglecs) are transmembrane proteins comprising 15 members in humans and that are expressed by various immune cell types. The N-terminus is exposed to the extracellular space and binds to ligands containing sialic acid components (von Gunten and Bochner 2008). Due to their specificity, several of the known Siglecs can bind to a variety of pathogens, including Group B *Streptococcus* and *Trypanosoma cruzi*, resulting in the production of interleukins (Crocker et al. 2007). Systematic analyses of the binding capacity of 10 different Siglecs revealed that *C. jejuni* LOS interacted with Siglec7 (Avril et al. 2006). As expected, this binding was only observed for sialylated, but not with unsialylated LOS components. In addition, sialylated LOS, particularly α2,3-sialylated LOS, was shown to interact with the soluble Siglec-1/Sn (Heikema et al. 2010), while Siglec7 preferably binds to α2,8-sialylated LOS (Bax et al. 2011). Furthermore, modified LOS structures of *C. jejuni* mimic human gangliosides, which can lead to auto-antibody production and neural disorders in some patients (Yuki et al. 2004). However, in addition to Siglec1/Sn and Siglec7, *C. jejuni* is also recognized by Siglec10, which was reported to bind both live *C. jejuni* bacteria and purified flagella, suggesting that activation of Siglec10 might be mediated by the flagella. In contrast to TLRs that activate a pro-inflammatory response, flagellin-Siglec10 contact increased the expression of IL-10 (Stephenson et al. 2014) and thus anti-inflammatory signaling (Fig. 5). Further studies are required to confirm these findings in in vivo models of infection.

3.3 Activation of the NLRP3 Inflammasome

Inflammasomes represent cytosolic multiprotein complexes of the host innate immune system. A wide variety of pathogens, including yeasts, bacteria and viruses, induce inflammasome-dependent production of IL-1β (Schroder and Tschopp 2010). Upregulation of pro-IL-1β transcription and subsequent secretion of IL-1β by mouse macrophages was also observed following infection with *C. jejuni*, which suggested processing of the pro-cytokine by the inflammasome (Bouwman et al. 2014). There are several inflammasome types, e.g., NLRP1, NLRP3 and NLRC4 (Schroder and Tschopp 2010), of which *C. jejuni* activates NLRP3 (Bouwman et al. 2014). The NLRP3 inflammasome is triggered by two bacterial signals (signal-1 and signal-2, respectively) and assembles with pro-caspase-1 and adaptor protein ASC, which stimulates maturation and formation of active caspase-1 (Fig. 6a). Active caspase-1 then conventionally processes pro-IL-1β and pro-IL-18 into the respective mature cytokines (Fig. 6b). An analysis of a series of *C. jejuni* mutants defective for a variety of known virulence-associated factors, including LOS, flagella and adhesins, attempted to identify *C. jejuni* components involved in triggering this inflammasome response (Bouwman et al. 2014). Only two of the tested mutants, resulted in decreased IL-1β production, namely the deletion of the chemotaxis protein CheY and of the energy taxis protein CetA. This suggests that there is a direct correlation between the number of intracellular *C. jejuni* bacteria (which is regulated by at least CheY and CetA) and NLRP3 inflammasome activation. Yet, all engineered deletion mutants still induced mature IL-1β secretion, albeit at slightly lower levels, suggesting that multiple components may activate this response. It is suggested that contact of macrophages with *C. jejuni* induces a transmembrane ion flux (e.g., K^+-efflux), as seen for other pathogens (Muñoz-Planillo et al. 2013), but the exact signals that trigger the increased transcription of pro-IL-1β and possibly pro-IL-18, as well as the formation of the NLRP3 inflammasome, remain to be elucidated.

4 Concluding Remarks

C. jejuni is a fascinating microorganism which colonizes the chicken intestinal tract as a commensal but is associated with various diseases when infecting humans. Interestingly, by comparison to other enteric pathogens, *C. jejuni* comprises only a relatively small array of known virulence factors. While other pathogens like *Salmonella, Yersinia* or *Listeria* species exhibit a broad collection of "weapons" including various toxins, multiple secretion systems and effector molecules, *C. jejuni* has a relatively limited number of identified disease-associated factors, most notably CDT, LOS, HtrA, CadF and fT3SS. As reviewed here, some of these are involved in adhesion to and invasion into host epithelial cells, where the bacteria can survive and even spread into neighboring tissue. Although all details are still not fully clear, most data support an invasion mechanism that depends on the fibronectin-integrin-β_1 receptor and the small Rho GTPases Rac1 and Cdc42, which appear to dominate *C. jejuni* host cell entry (Fig. 3). However, the process of cell invasion would involve molecular dissection of the mechanic forces activated by host cells, which then trigger bacterial engulfment, allow entry and result in membrane closing behind the invading *C. jejuni*. Future studies should examine how this works in detail. It is also important to investigate in more detail the mechanisms by which *C. jejuni* survives and spreads intracellularly, or how the bacteria cause infection of other organs in the human body that can include the spleen, mesenteric lymph nodes and the liver (Burnham and Hendrixson 2018; Backert et al. 2013). It also appears that *C. jejuni* survives inside the macrophages using its catalase enzyme KatA (Day et al. 2000), a phenomenon which deserves further investigation. Furthermore, for many of the proposed *C. jejuni* virulence proteins, a general knowledge of how they function mechanistically in vivo is still missing. Also, the actual advantage of some bacterial factors such as the CDT toxin for the bacterium is still unclear. It might be that the DNA damage triggered by its DNase activity not only initiates apoptosis but also modulates the host immune response, operating as an immunoregulatory factor, which helps the bacteria producing CDT to establish a suitable niche in its host. In this context, the process of apoptotic cell death induced by *C. jejuni* is more complicated than originally envisaged. Besides CDT described above (Pickett and Whitehouse 1999; Guerra et al.

2005), the serine protease HtrA (Heimesaat et al. 2014b) and the fT3SS substrate FspA2 (Poly et al. 2007) were also reported to stimulate apoptosis in epithelial cells, which can be counteracted by the *C. jejuni*-induced ubiquitin-editing enzyme A20 that negatively regulates apoptotic cell death (Lim et al. 2017). Thus, the interplay of apoptotic and anti-apoptotic signal pathways modulated by *C. jejuni* are still not fully understood, deserving further investigation. Lastly, it must be mentioned that there are controversies and contradictions in various reports describing the functional mechanistics and importance of particular *C. jejuni* virulence factors, such as the Cia proteins or some proposed adhesins. Strain-dependent differences, variation in experimental procedures, and a disconnect between in vitro work, in vivo (mouse) models, and the human situation (where individual immunological variation may be of crucial importance to clinical outcome) further murken the waters. These important issues also need to be addressed and clarified in future studies. Thus, it occurs that *C. jejuni* will continue to be an attractive and gratifying research subject with high importance to public health.

Acknowledgements We thank Drs. David Hendrixson (UT Southwestern, Dallas, Texas/USA) and Tadhg Ó Cróinín (University College Dublin, Dublin/Ireland) for helpful comments on the manuscript. This work is supported by the German Federal Ministry of Education and Research (BMBF) in frame of the zoonoses research consortium PAC-*Campylobacter* to S.B. (project IP9/01KI1725E).

References

Andersen-Nissen E, Smith KD, Strobe KL, Barrett SL, Cookson BT, Logan SM, Aderem A (2005) Evasion of Toll-like receptor 5 by flagellated bacteria. Proc Natl Acad Sci USA 102:9247–9252. https://doi.org/10.1073/pnas.0502040102

Artymovich K, Kim JS, Linz JE, Hall DF, Kelley LE, Kalbach HL, Kathariou S, Gaymer J, Paschke B (2013) A "successful allele" at *Campylobacter jejuni* contingency locus Cj0170 regulates motility; "successful alleles" at locus Cj0045 are strongly associated with mouse colonization. Food Microbiol 34(2):425–430. https://doi.org/10.1016/j.fm.2013.01.007

Avril T, Wagner ER, Willison HJ, Crocker PR (2006) Sialic acid-binding immunoglobulin-like lectin 7 mediates selective recognition of sialylated glycans expressed on *Campylobacter jejuni* lipooligosaccharides. Infect Immun 74:4133–4141. https://doi.org/10.1128/IAI.02094-05

Backert S, Bernegger S, Skorko-Glonek J, Wessler S (2018) Extracellular HtrA serine proteases: an emerging new strategy in bacterial pathogenesis. Cell Microbiol 20:e12845. https://doi.org/10.1111/cmi.12845

Backert S, Boehm M, Wessler S, Tegtmeyer N (2013) Transmigration route of *Campylobacter jejuni* across polarized intestinal epithelial cells: paracellular, transcellular or both? Cell Commun Signal 11:72. https://doi.org/10.1186/1478-811X-11-72

Bæk KT, Vegge CS, Brøndsted L (2011) HtrA chaperone activity contributes to host cell binding in *Campylobacter jejuni*. Gut Pathog 3:13. https://doi.org/10.1186/1757-4749-3-13.10.1186/1757-4749-3-13

Bæk KT, Vegge CS, Skorko-Glonek J, Brøndsted L (2011) Different contributions of HtrA protease and chaperone activities to *Campylobacter jejuni* stress tolerance and physiology. Appl Environ Microbiol 77:57–66. https://doi.org/10.1128/aem.01603-10

Barrero-Tobon AM, Hendrixson DR (2012) Identification and analysis of flagellar coexpressed determinants (Feds) of *Campylobacter jejuni* involved in colonization. Mol Microbiol 84:352–369. https://doi.org/10.1111/j.1365-2958.2012.08027.x

Bax M, Kuijf ML, Heikema AP, van Rijs W, Bruijns SC, Garcia-Vallejo JJ, Crocker PR, Jacobs BC, van Vliet SJ, van Kooyk Y (2011) *Campylobacter jejuni* lipooligosaccharides modulate dendritic cell-mediated T cell polarization in a sialic acid linkage-dependent manner. Infect Immun 79:2681–2689. https://doi.org/10.1128/IAI.00009-11

Beeby M, Ribardo DA, Brennan CA, Ruby EG, Jensen GJ, Hendrixson DR (2016) Diverse high-torque bacterial flagellar motors assemble wider stator rings using a conserved protein scaffold. Proc Natl Acad Sci USA 113:E1917-1926. https://doi.org/10.1073/pnas.1518952113

Biswas D, Itoh K, Sasakawa C (2000) Uptake pathways of clinical and healthy animal isolates of *Campylobacter jejuni* into INT-407 cells. FEMS Immunol Med Microbiol 29(3):203–211. https://doi.org/10.1111/j.1574-695X.2000.tb01524.x

Black RE, Levine MM, Clements ML, Hughes TP, Blaser MJ (1988) Experimental *Campylobacter jejuni* infection in humans. J Infect Dis 157(3):472–479. https://doi.org/10.1093/infdis/157.3.472

Boehm M, Haenel I, Hoy B, Brøndsted L, Smith TG, Hoover T, Wessler S, Tegtmeyer N (2013) Extracellular secretion of protease HtrA from *Campylobacter jejuni* is highly efficient and independent of its protease activity and flagellum. Eur J Microbiol Immunol (Bp) 3:163–173. https://doi.org/10.1556/EuJMI.3.2013.3.3

Boehm M, Hoy B, Rohde M, Tegtmeyer N, Bæk KT, Oyarzabal OA, Brøndsted L, Wessler S, Backert S (2012) Rapid paracellular transmigration of *Campylobacter jejuni* across polarized epithelial cells without affecting TER: role of proteolytic-active HtrA cleaving E-cadherin but not fibronectin. Gut Pathog 4:3. https://doi.org/10.1186/1757-4749-4-3

Boehm M, Krause-Gruszczynska M, Rohde M, Tegtmeyer N, Takahashi S, Oyarzabal OA, Backert S (2011) Major host factors involved in epithelial cell invasion of *Campylobacter jejuni*: role of fibronectin, integrin beta1, FAK, Tiam-1, DOCK180 in activating Rho GTPase Rac1. Front. In Cell Infect. Microbiol 1:17. https://doi.org/10.3389/fcimb.2011.00017

Boehm M, Lind J, Backert S, Tegtmeyer N (2015) *Campylobacter jejuni* serine protease HtrA plays an important role in heat tolerance, oxygen resistance, host cell adhesion, invasion, and transmigration. Eur J Microbiol Immunol (Bp) 5:68–80. https://doi.org/10.1556/eujmi-d-15-00003

Boehm M, Simson D, Escher U, Schmidt AM, Bereswill S, Tegtmeyer N, Backert S, Heimesaat MM (2018) Function of serine protease HtrA in the lifecycle of the foodborne pathogen *Campylobacter jejuni*. Eur J Microbiol Immunol (Bp) 8:70–77. https://doi.org/10.1556/1886.2018.00011

Bouwman LI, de Zoete MR, Bleumink-Pluym NM, Flavell RA, van Putten JP (2014) Inflammasome activation by *Campylobacter jejuni*. J Immunol 193:4548–4557. https://doi.org/10.4049/jimmunol.1400648

Bouwman LI, Niewold P, van Putten JP (2013) Basolateral invasion and trafficking of *Campylobacter jejuni* in polarized epithelial cells. PLoS ONE 8(1):e54759. https://doi.org/10.1371/journal.pone.0054759

Brøndsted L, Andersen MT, Parker M, Jorgensen K, Ingmer H (2005) The HtrA protease of *Campylobacter jejuni* is required for heat and oxygen tolerance and for optimal interaction with human epithelial cells. Appl Environ Microbiol 71:3205–3212. https://doi.org/10.1128/aem.71.6.3205-3212.2005

Bronowski C, James CE, Winstanley C, Bronowski C (2014) Role of environmental survival in transmission of *Campylobacter jejuni*. FEMS Microbiol Lett 356(1):8–19. https://doi.org/10.1111/1574-6968.12488

Bücker R, Krug SM, Moos V, Bojarski C, Schweiger MR, Kerick M, Fromm A, Janßen S, Fromm M, Hering NA, Siegmund B, Schneider T, Barmeyer C, Schulzke JD (2018) *Campylobacter jejuni* impairs sodium transport and epithelial barrier function via cytokine release in human colon. Mucosal Immunol 11(2):474–485. https://doi.org/10.1038/mi.2017.66

Buelow DR, Christentsen JE, Neal-McKinney JM, Konkel ME (2011) *Campylobacter jejuni* survival within human epithelial cells is enhanced by the secreted protein CiaI. Mol Microbiol 80:1296–1312. https://doi.org/10.1111/j.1365-2958.2011.07645.x

Burnham PM, Hendrixson DR (2018) *Campylobacter jejuni*: collective components promoting a successful enteric lifestyle. Nat Rev Microbiol 16(9):551–565. https://doi.org/10.1038/s41579-018-0037-9

Campagnolo ER, Philipp LM, Long JM, Hanshaw NL (2018) Pet-associated Campylobacteriosis: a persisting public health concern. Zoonoses Public Health 65(3):304–311. https://doi.org/10.1111/zph.12389

Chang C, Miller JF (2006) *Campylobacter jejuni* colonization of mice with limited enteric flora. Infect Immun 74:5261–5271. https://doi.org/10.1128/iai.01094-05

Chen S, Beeby M, Murphy GE, Leadbetter JR, Hendrixson DR, Briegel A, Li Z, Shi J, Tocheva EI, Muller A, Dobro MJ, Jensen GJ (2011) Structural diversity of bacterial flagellar motors. Embo J 30:2972–2981. https://doi.org/10.1038/emboj.2011.186

Christensen JE, Pacheco SA, Konkel ME (2009) Identification of a *Campylobacter jejuni*—secreted protein required for maximal invasion of host cells. Mol Microbiol 73:650–662. https://doi.org/10.1111/j.1365-2958.2009.06797.x

Clausen T, Southan C, Ehrmann M (2002) The HtrA family of proteases: implications for protein composition and cell fate. Mol Cell 10:443–455. https://doi.org/10.1016/s1097-2765(02)00658-5

Cohen EJ, Nakane D, Kabata Y, Hendrixson DR, Nishizaka T, Beeby M (2020) *Campylobacter jejuni* motility integrates specialized cell shape, flagellar filament, and motor, to coordinate action of its opposed flagella. PLoS Pathog 2;16(7):e1008620. https://doi.org/10.1371/journal.ppat.1008620

Crocker PR, Paulson JC, Varki A (2007) Siglecs and their roles in the immune system. Nat Rev Immunol 7:255–266. https://doi.org/10.1038/nri2056

Cunningham AA, Daszak P, Wood JLN (2017) One Health, emerging infectious diseases and wildlife: two decades of progress? Philos Trans R Soc Lond B Biol Sci 372(1725):20160167. https://doi.org/10.1098/rstb.2016.0167

Dai L, Wu ZW, Xu CY, Sahin O, Yaeger M, Plummer PJ, Zhang QJ (2019) The rho-independent transcription terminator for the porA gene enhances expression of the major outer membrane protein and *Campylobacter jejuni* virulence in abortion induction. Infect Immun 87(12):e00687-e719. https://doi.org/10.1128/IAI.00687-19

Day WA Jr, Sajecki JL, Pitts TM, Joens LA (2000) Role of catalase in *Campylobacter jejuni* intracellular survival. Infect Immun 68:6337–6345. https://doi.org/10.1128/iai.68.11.6337-6345.2000

DiRienzo JM (2014) Uptake and processing of the cytolethal distending toxin by mammalian cells. Toxins (Basel) 6:3098–3116. https://doi.org/10.3390/toxins6113098

Dwivedi R, Nothaft H, Garber J, Xin Kin L, Stahl M, Flint A, van Vliet AH, Stintzi A, Szymanski CM (2016) L-fucose influences chemotaxis and biofilm formation in *Campylobacter jejuni*. Mol Microbiol 101:575–589. https://doi.org/10.1111/mmi.13409

Elharrif Z, Mégraud F (1986) Characterization of thermophilic *Campylobacter*: carbon-substrate utilization tests. Curr Microbiol 13:117–122

Elmi A, Nasher F, Jagatia H, Gundogdu O, Bajaj-Elliott M, Wren B, Dorrell N (2016) *Campylobacter jejuni* outer membrane vesicle-associated proteolytic activity promotes bacterial invasion by mediating cleavage of intestinal epithelial cell E-cadherin and occludin. Cell Microbiol 18:561–572. https://doi.org/10.1111/cmi.12534

Elmi A, Watson E, Sandu P, Gundogdu O, Mills DC, Inglis NF, Manson E, Imrie L, Bajaj-Elliott M, Wren BW, Smith DG, Dorrell N (2012) *Campylobacter jejuni* outer membrane vesicles play an important role in bacterial interactions with human intestinal epithelial cells. Infect Immun 80:4089–4098. https://doi.org/10.1128/IAI.00161-12

Epps SV, Harvey RB, Hume ME, Phillips TD, Anderson RC, Nisbet DJ (2013) Foodborne *Campylobacter*: infections, metabolism, pathogenesis and reservoirs. Int J Environ Res Public Health 10(12):6292–6304. https://doi.org/10.3390/ijerph10126292

Eucker TP, Konkel ME (2012) The cooperative action of bacterial fibronectin-binding proteins and secreted proteins promote maximal *Campylobacter jejuni* invasion of host cells by stimulating membrane ruffling. Cell Microbiol 14:226–238. https://doi.org/10.1111/j.1462-5822.2011.01714.x

Faber E, Gripp E, Maurischat S, Kaspers B, Tedin K, Menz S, Zuraw A, Kershaw O, Yang I, Rautenschlein S, Josenhans C (2015) Novel Immunomodulatory Flagellin-Like Protein FlaC in *Campylobacter jejuni* and Other *Campylobacterales*. mSphere 1(1):e00028-15. https://doi.org/10.1128/mSphere.00028-15

Flanagan RC, Neal-McKinney JM, Dhillon AS, Miller WG, Konkel ME (2009) Examination of *Campylobacter jejuni* putative adhesins leads to the identification of a new protein, designated FlpA, required for chicken colonization. Infect Immunol 77:2399–23407. https://doi.org/10.1128/IAI.01266-08

Floch P, Pey V, Castroviejo M, Dupuy JW, Bonneu M, de la Guardia AH, Pitard V, Mégraud F, Lehours P (2014) Role of *Campylobacter Jejuni* gamma-glutamyl transpeptidase on epithelial cell apoptosis and lymphocyte proliferation. Gut Pathog 6:20. https://doi.org/10.1186/1757-4749-6-20

Fox JG, Rogers AB, Whary MT, Ge Z, Taylor NS, Xu S, Horwitz BH, Erdman SE (2004) Gastroenteritis in NF-kappaB-deficient mice is produced with wild-type *Camplyobacter jejuni* but not with *C. jejuni* lacking cytolethal distending toxin despite persistent colonization with both strains. Infect Immun 72:1116–1125. https://doi.org/10.1128/iai.72.2.1116-1125.2004

Friedman CR, Neimann J, Wegener HC, Tauxe RV (2000) Epidemiology of *Campylobacter jejuni* infections in the United States and other industrialized nations. In: Nachamkin I, Blaser MJ (eds) Campylobacter. ASM Press, Washington, DC, pp 121–138

Gao B, Vorwerk H, Huber C, Lara-Tejero M, Mohr J, Goodman AL, Eisenreich W, Galán JE, Hofreuter D (2017) Metabolic and fitness determinants for in vitro growth and intestinal colonization of the bacterial pathogen *Campylobacter jejuni*. PLoS Biol 15(5):e2001390. https://doi.org/10.1371/journal.pbio.2001390

Gaynor EC, Wells DH, MacKichan JK, Falkow S (2005) The *Campylobacter jejuni* stringent response controls specific stress survival and virulence-associated phenotypes. Mol Microbiol 56:8–27. https://doi.org/10.1111/j.1365-2958.2005.04525.x

Gölz G, Rosner B, Hofreuter D, Josenhans C, Kreienbrock L, Lowenstein A, Schielke A, Stark K, Suerbaum S, Wieler LH, Alter T (2014) Relevance of *Campylobacter* to public health—the need for a one health approach. Int J Med Microbiol 304:817–823. https://doi.org/10.1016/j.ijmm.2014.08.015

Gölz G, Sharbati S, Backert S, Alter T (2012) Quorum sensing dependent phenotypes and their molecular mechanisms in *Campylobacterales*. Eur J Microbiol Immunol (Bp) 2(1):50–60. https://doi.org/10.1556/EuJMI.2.2012.1.8

Goodman KN, Powers MJ, Crofts AA, Trent MS, Hendrixson DR (2020) *Campylobacter jejuni* BumSR directs a response to butyrate via sensor phosphatase activity to impact transcription and colonization. Proc Natl Acad Sci USA 117(21):11715–11726. https://doi.org/10.1073/pnas.1922719117

Goon S, Kelly JF, Logan SM, Ewing CP, Guerry P (2003) Pseudaminic acid, the major modification on *Campylobacter* flagellin, is synthesized via the Cj1293 gene. Mol Microbiol 50:659–671. https://doi.org/10.1046/j.1365-2958.2003.03725.x

Gorain C, Singh A, Bhattacharyya S, Kundu A, Lahiri A, Gupta S, Mallick AI (2020) Mucosal delivery of live *Lactococcus lactis* expressing functionally active JlpA antigen induces potent local immune response and prevent enteric colonization of *Campylobacter jejuni* in chickens. Vaccine 38:1630–1642. https://doi.org/10.1016/j.vaccine.2019.12.064

Guccione E, del Rocio L-K, Pearson BM, Hitchin E, Mulholland F, van Diemen PM, Stevens MP, Kelly DJ (2008) Amino acid-dependent growth of *Campylobacter jejuni*: key roles for aspartase (AspA) under microaerobic and oxygen-limited conditions and identification of AspB (Cj0762), essential for growth on glutamate. Mol Microbiol. 69(1):77–93. https://doi.org/10.1111/j.1365-2958.2008.06263.x.10.1016/j.vaccine.2019.12.064

Guerra L, Teter K, Lilley BN, Stenerlow B, Holmes RK, Ploegh HL, Sandvig K, Thelestam M, Frisan T (2005) Cellular internalization of cytolethal distending toxin: a new end to a known pathway. Cell Microbiol 7:921–934. https://doi.org/10.1111/j.1462-5822.2005.00520.x

Guerra L, Cortes-Bratti X, Guidi R, Frisan T (2011) The biology of the cytolethal distending toxins. Toxins (Basel) 3(3):172–190. https://doi.org/10.3390/toxins3030172

Guerry P (2007) *Campylobacter* flagella: not just for motility. Trends Microbiol 15(10):456–461. https://doi.org/10.1016/j.tim.2007.09.006

Guerry P, Alm RA, Power ME, Logan SM, Trust TJ (1991) Role of two flagellin genes in *Campylobacter* motility. J Bacteriol 173(15):4757–4764. https://doi.org/10.1128/jb.173.15.4757-4764.1991

Guerry P, Ewing CP, Schirm M, Lorenzo M, Kelly J, Pattarini D, Majam G, Thibault P, Logan S (2006) Changes in flagellin glycosylation affect *Campylobacter* autoagglutination and virulence. Mol Microbiol 60:299–311. https://doi.org/10.1111/j.1365-2958.2006.05100.x

Harrer A, Bucker R, Boehm M, Zarzecka U, Tegtmeyer N, Sticht H, Schulzke JD, Backert S (2019) *Campylobacter jejuni* enters gut epithelial cells and impairs intestinal barrier function through cleavage of occludin by serine protease HtrA. Gut Pathog 11:4. https://doi.org/10.1186/s13099-019-0283-z

Hartley-Tassell LE, Shewell LK, Day CJ, Wilson JC, Sandhu R, Ketley JM, Korolik V (2010) Identification and characterization of the aspartate chemosensory receptor of *Campylobacter jejuni*. Mol Microbiol 75:710–730. https://doi.org/10.1111/j.1365-2958.2009.07010.x

Hauske P, Meltzer M, Ottmann C, Krojer T, Clausen T, Ehrmann M, Kaiser M (2009) Selectivity profiling of DegP substrates and inhibitors. Bioorg Med Chem 17:2920–2924. https://doi.org/10.1016/j.bmc.2009.01.073

Heikema AP, Bergman MP, Richards H, Crocker PR, Gilbert M, Samsom JN, van Wamel WJB, Endtz HP, van Belkum A (2010) Characterization of the specific interaction between sialoadhesin and sialylated *Campylobacter jejuni* lipooligosaccharides. Infect Immun 78:3237–3246. https://doi.org/10.1128/IAI.01273-09

Heimesaat MM, Alutis M, Grundmann U, Fischer A, Tegtmeyer N, Böhm M, Kuhl AA, Göbel UB, Backert S, Bereswill S (2014) The role of serine protease HtrA in acute ulcerative enterocolitis and extra-intestinal immune responses during *Campylobacter jejuni* infection of gnotobiotic IL-10 deficient mice. Front Cell Infect Microbiol 4:77. https://doi.org/10.3389/fcimb.2014.00077

Heimesaat MM, Fischer A, Alutis M, Grundmann U, Boehm M, Tegtmeyer N, Gobel UB, Kuhl AA, Bereswill S, Backert S (2014) The impact of serine protease HtrA in apoptosis, intestinal immune responses and extra-intestinal histopathology during *Campylobacter jejuni* infection of infant mice. Gut Pathog 6:16. https://doi.org/10.1186/1757-4749-6-16

Henderson LD, Matthews-Palmer TRS, Gulbronson CJ, Ribardo DA, Beeby M, Hendrixson DR. (2020) Diversification of *Campylobacter jejuni* Flagellar C-Ring composition impacts its structure and function in motility, flagellar assembly, and cellular processes. mBio 11(1):e02286-19. https://doi.org/10.1128/mBio.02286-19

Hendrixson DR, DiRita VJ (2003) Transcription of sigma54-dependent but not sigma28-dependent flagellar genes in *Campylobacter jejuni* is associated with formation of the flagellar secretory apparatus. Mol Microbiol 50(2):687–702. https://doi.org/10.1046/j.1365-2958.2003.03731.x

Hendrixson DR, DiRita VJ (2004) Identification of *Campylobacter jejuni* genes involved in commensal colonization of the chick gastrointestinal tract. Mol Microbiol 52:471–484. https://doi.org/10.1111/j.1365-2958.2004.03988.x

Hendrixson DR (2006) A phase-variable mechanism controlling the *Campylobacter jejuni* FlgR response regulator influences commensalism. Mol Microbiol 61:1646–1659. https://doi.org/10.1111/j.1365-2958.2006.05336.x

Hendrixson DR (2008) Restoration of flagellar biosynthesis by varied mutational events in *Campylobacter jejuni*. Mol Microbiol 70:519–536. https://doi.org/10.1111/j.1365-2958.2008.06428.x

Herman L, Heyndrickx M, Grijspeerdt K, Vandekerchove D, Rollier I, De Zutter L (2003) Routes for *Campylobacter* contamination of poultry meat: epidemiological study from hatchery to slaughter. Epidemiol Infect 131:1169–1180. https://doi.org/10.1017/s0950268803001183

Hermans D, Van Deun K, Martel A, Van Immerseel F, Messens W, Heyndrickx M, Haesebrouck F, Pasmans F (2011) Colonization factors of *Campylobacter jejuni* in the chicken gut. Vet Res 42:14. https://doi.org/10.1186/1297-9716-42-82

Heywood W, Henderson B, Nair SP (2005) Cytolethal distending toxin: creating a gap in the cell cycle. J Med Microbiol 54:207–216. https://doi.org/10.1099/jmm.0.45694-0

Hickey TE, McVeigh AL, Scott DA, Michielutti RE, Bixby A, Carroll SA, Bourgeois AL, Guerry P (2000) *Campylobacter jejuni* cytolethal distending toxin mediates release of interleukin-8 from intestinal epithelial cells. Infect Immun 68:6535–6541. https://doi.org/10.1128/iai.68.12.6535-6541.2000

Hofreuter D, Novik V, Galán JE (2008) Metabolic diversity in *Campylobacter jejuni* enhances specific tissue colonization. Cell Host Microbe 4:425–433. https://doi.org/10.1016/j.chom.2008.10.002

Hofreuter D (2014) Defining the metabolic requirements for the growth and colonization capacity of *Campylobacter jejuni*. Front Cell Infect Microbiol 4:137. https://doi.org/10.3389/fcimb.2014.00137

Howard SL, Jagannathan A, Soo EC, Hui JP, Aubry AJ, Ahmed I, Karlyshev A, Kelly JF, Jones MA, Stevens MP, Logan SM, Wren BW (2009) *Campylobacter jejuni* glycosylation island important in cell charge, legionaminic acid biosynthesis, and colonization of chickens. Infect Immun 77(6):2544–2556. https://doi.org/10.1128/IAI.01425-08

Hu L, Raybourne RB, Kopecko DJ (2005) Ca^{2+} release from host intracellular stores and related signal transduction during *Campylobacter jejuni* 81–176 internalization into human intestinal cells. Microbiology (Reading) 151(Pt9):3097–3105. https://doi.org/10.1099/mic.0.27866-0

Hu L, McDaniel JP, Kopecko DJ (2006) Signal transduction events involved in human epithelial cell invasion by *Campylobacter jejuni* 81–176. Microb Pathog. 40(3):91–100. https://doi.org/10.1016/j.micpath.2005.11.004

Jin S, Joe A, Lynett J, Hani EK, Sherman P, Chan VL (2001) JlpA, a novel surface-exposed lipoprotein specific to *Campylobacter jejuni*, mediates adherence to host epithelial cells. Mol Microbiol 39:1225–1236. https://doi.org/10.1111/j.1365-2958.2001.02294.x

Jin S, Song YC, Emili A, Sherman PM, Chan VL (2003) JlpA of *Campylobacter jejuni* interacts with surface-exposed heat shock protein 90 and triggers signalling pathways leading to the activation of NF-kabbaB and p38 MAP kinase in epithelial cells. Cell Microbiol 5:165–174. https://doi.org/10.1046/j.1462-5822.2003.00265.x

Johnson WM, Lior H (1988) A new heat-labile cytolethal distending toxin (CLDT) produced by *Campylobacter* spp. Microb Pathog 4:115–126. https://doi.org/10.1016/0882-4010(88)90053-8

Jorgensen F, Ellis-Iversen J, Rushton S, Bull SA, Harris SA, Bryan SJ, Gonzalez A, Humphrey TJ (2011) Influence of season and geography on *Campylobacter jejuni* and *C. coli* subtypes in housed broiler flocks reared in Great Britain. Appl Environ Microbiol 77:3741–3748. https://doi.org/10.1128/AEM.02444-10

Kaakoush NO, Castaño-Rodríguez N, Mitchell HM, Man SM (2015) Global epidemiology of *Campylobacter* infection. Clin Microbiol Rev 28(3):687–720. https://doi.org/10.1128/CMR.00006-15

Karim QN, Logan RP, Puels J, Karnholz A, Worku ML (1998) Measurement of motility of *Helicobacter pylori, Campylobacter jejuni,* and *Escherichia coli* by real time computer tracking using the Hobson BacTracker. J Clin Pathol 51:623–628. https://doi.org/10.1136/jcp.51.8.623

Kim DY, Kim KK (2005) Structure and function of HtrA family proteins, the key players in protein quality control. J Biochem Mol Biol 38:266–274. https://doi.org/10.5483/bmbrep.2005.38.3.266

Konkel ME, Garvis SG, Tipton SL, Anderson DE Jr, Cieplak W Jr (1997) Identification and molecular cloning of a gene encoding a fibronectin-binding protein (CadF) from *Campylobacter jejuni*. Mol. Microbiol 24:953–963. https://doi.org/10.1046/j.1365-2958.1997.4031771.x

Konkel ME, Kim BJ, Rivera-Amill V, Garvis SG (1999) Bacterial secreted proteins are required for the internalization of *Campylobacter jejuni* into cultured mammalian cells. Mol Microbiol 32:691–701. https://doi.org/10.1046/j.1365-2958.1999.01376.x

Konkel ME, Monteville MR, Rivera-Amill V, Joens LA (2001) The pathogenesis of *Campylobacter jejuni*-mediated enteritis. Curr Issues Intest Microbiol 2:55–71

Konkel ME, Larson CL, Flanagan RC (2010) *Campylobacter jejuni* FlpA binds fibronectin and is required for maximal host cell adherence. J Bacteriol 192:68–76. https://doi.org/10.1128/JB.00969-09

Korolik V (2019) The role of chemotaxis during *Campylobacter jejuni* colonisation and pathogenesis. Curr Opin Microbiol 47:32–37. https://doi.org/10.1016/j.mib.2018.11.001

Krause-Gruszczynska M, Rohde M, Hartig R, Genth H, Schmidt G, Keo T, Koenig W, Miller WG, Konkel ME, Backert S (2007) Role of the small Rho GTPases Rac1 and Cdc42 in host cell invasion of *Campylobacter jejuni*. Cell Microbiol 9:2431–2444. https://doi.org/10.1111/j.1462-5822.2007.00971.x

Krause-Gruszczynska M, van Alphen LB, Oyarzabal OA, Alter T, Hänel I, Schliephake A, König W, van Putten JP, Konkel ME, Backert S (2007) Expression patterns and role of the CadF protein in *Campylobacter jejuni* and *Campylobacter coli*. FEMS Microbiol Lett 274(1):9–16. https://doi.org/10.1111/j.1574-6968.2007.00802.x

Krause-Gruszczynska M, Boehm M, Rohde M, Tegtmeyer N, Takahashi S, Buday L, Oyarzabal OA, Backert S (2011) The signaling pathway of *Campylobacter jejuni*-induced Cdc42 activation: role of fibronectin, integrin beta1, tyrosine kinases and guanine exchange factor Vav2. Cell Commun Signal 9:32. https://doi.org/10.1186/1478-811X-9-32

Kreutzberger MAB, Ewing C, Poly F, Wang F, Egelman EH (2020) Atomic structure of the *Campylobacter jejuni* flagellar filament reveals how ε Proteobacteria escaped Toll-like receptor 5 surveillance. Proc Natl Acad Sci U S a 117(29):16985–16991. https://doi.org/10.1073/pnas.2010996117

Krojer T, Sawa J, Huber R, Clausen T (2010) HtrA proteases have a conserved activation mechanism that can be triggered by distinct molecular cues. Nat Struct Mol Biol 17:844–852. https://doi.org/10.1038/nsmb.1840

Lai CK, Chen YA, Lin CJ, Lin HJ, Kao MC, Huang MZ, Lin YH, Chiang-Ni C, Chen CJ, Lo UG, Lin LC, Lin H, Hsieh JT, Lai CH (2016) Molecular Mechanisms and potential clinical applications of *Campylobacter jejuni* cytolethal distending toxin. Front Cell Infect Microbiol 6:9. https://doi.org/10.3389/fcimb.2016.00009

Lara-Tejero M, Galán JE (2001) CdtA, CdtB, and CdtC form a tripartite complex that is required for cytolethal distending toxin activity. Infect Immun 69:4358–4365. https://doi.org/10.1128/iai.69.7.4358-4365.2001

Larson CL, Samuelson DR, Eucker TP, O'Loughlin JL, Konkel ME (2013) The fibronectin-binding motif within FlpA facilitates *Campylobacter jejuni* adherence to host cell and activation of host cell signaling. Emerg Microbes Infect 2(10):e65. https://doi.org/10.1038/emi.2013.65

Leach S, Harvey P, Wali R (1997) Changes with growth rate in the membrane lipid composition of and amino acid utilization by continuous cultures of *Campylobacter jejuni*. J Appl Microbiol 82:631–640. https://doi.org/10.1111/j.1365-2672.1997.tb02873.x

Lee MD, Newell DG (2006) *Campylobacter* in poultry: filling an ecological niche. Avian Dis 50:1–9. https://doi.org/10.1637/7474-111605R.1

Leon-Kempis MR, Guccione E, Mulholland F, Williamson MP, Kelly DJ (2006) The *Campylobacter jejuni* PEB1a adhesin is an aspartate/glutamate-binding protein of an ABC transporter essential for microaerobic growth on dicarboxylic amino acids. Mol Microbiol 60:1262–1275. https://doi.org/10.1111/j.1365-2958.2006.05168.x

Lim MCC, Maubach G, Sokolova O, Feige MH, Diezko R, Buchbinder J, Backert S, Schlüter D, Lavrik IN, Naumann M (2017) Pathogen-induced ubiquitin-editing enzyme A20 bifunctionally shuts off NF-κB and caspase-8-dependent apoptotic cell death. Cell Death Differ 24(9):1621–1631. https://doi.org/10.1038/cdd.2017.89

Lindmark B, Rompikuntal PK, Vaitkevicius K, Song T, Mizunoe Y, Uhlin BE, Guerry P, Wai SN (2009) Outer membrane vesicle-mediated release of cytolethal distending toxin (CDT) from *Campylobacter jejuni*. BMC Microbiol 9:220. https://doi.org/10.1186/1471-2180-9-220

Lipinska B, Fayet O, Baird L, Georgopoulos C (1989) Identification, characterization, and mapping of the *Escherichia coli htrA* gene, whose product is essential for bacterial growth only at elevated temperatures. J Bacteriol 171:1574–1584. https://doi.org/10.1128/jb.171.3.1574-1584.1989

Logue CM, Barbieri NL, Nielsen DW (2017) Pathogens of food animals: sources, characteristics, human risk, and methods of detection. Adv Food Nutr Res 82:277–365. https://doi.org/10.1016/bs.afnr.2016.12.009

Louwen R, Heikema A, van Belkum A, Ott A, Gilbert M, Ang W, Endtz HP, Bergman MP, Nieuwenhuis EE (2008) The sialylated lipooligosaccharide outer core in *Campylobacter jejuni* is an important determinant for epithelial cell invasion. Infect Immunol 76:4431–4438. https://doi.org/10.1128/IAI.00321-08

Louwen R, Nieuwenhuis EE, van Marrewijk L, Horst-Kreft D, de Ruiter L, Heikema AP, van Wamel WJ, Wagenaar JA, Endtz HP, Samsom J, van Baarlen P, Akhmanova A, van Belkum A (2012) *Campylobacter jejuni* translocation across intestinal epithelial cells is facilitated by ganglioside-like lipooligosaccharide structures. Infect Immun 80(9):3307–3318. https://doi.org/10.1128/IAI.06270-11

Luethy PM, Huynh S, Parker CT, Hendrixson DR (2015) Analysis of the activity and regulon of the two-component regulatory system composed by Cjj81176_1484 and Cjj81176_1483 of *Campylobacter jejuni*. J Bacteriol 197(9):1592–1605. https://doi.org/10.1128/JB.02564-14

Marchant J, Wren B, Ketley J (2002) Exploiting genome sequence: predictions for mechanisms of *Campylobacter* chemotaxis. Trends Microbiol 10(4):155–159. https://doi.org/10.1016/s0966-842x(02)02323-5

Masanta WO, Heimesaat MM, Bereswill S, Tareen AM, Lugert R, Groß U, Zautner AE (2013) Modification of intestinal microbiota and its consequences for innate immune response in the pathogenesis of campylobacteriosis. Clin Dev Immunol 2013:526860. https://doi.org/10.1155/2013/526860

Méndez-Olvera ET, Bustos-Martinez JA, Lopez-Vidal Y, Verdugo-Rodriguez A, Martinez-Gomez D (2016) Cytolethal distending toxin from *Campylobacter jejuni* requires the cytoskeleton for toxic activity. Jundishapur J Microbiol 9:e35591. https://doi.org/10.5812/jjm.35591

Monteville MR, Konkel ME (2002) Fibronectin-facilitated invasion of T84-eukaryotic cells by *Campylobacter jejuni* occurs preferentially at the basolateral cell surface. Infect Immun 70:6665–6671. https://doi.org/10.1128/iai.70.12.6665-6671.2002

Monteville MR, Yoon JE, Konkel ME (2003) Maximal adherence and invasion of INT 407 cells by *Campylobacter jejuni* requires the CadF outer-membrane protein and microfilament reorganization. Microbiology 149:153–165. https://doi.org/10.1099/mic.0.25820-0

Mortensen NP, Schiellerup P, Boisen N, Klein BM, Locht H, Abuoun M, Newell D, Krogfelt KA (2011) The role of *Campylobacter jejuni* cytolethal distending toxin in gastroenteritis: toxin detection, antibody production, and clinical outcome. Apmis 119:626–634. https://doi.org/10.1111/j.1600-0463.2011.02781.x

Mourkas E, Taylor AJ, Méric G, Bayliss SC, Pascoe B, Mageiros L, Calland JK, Hitchings MD, Ridley A, Vidal A, Forbes KJ, Strachan NJC, Parker CT, Parkhill J, Jolley KA, Cody AJ, Maiden MCJ, Kelly DJ, Sheppard SK (2020) Agricultural intensification and the evolution of host specialism in the enteric pathogen *Campylobacter jejuni*. Proc Natl Acad Sci USA 117(20):11018–11028. https://doi.org/10.1073/pnas.1917168117

Muñoz-Planillo R, Kuffa P, Martínez-Colón G, Smith BL, Rajendiran TM, Núñez G (2013) K(+) efflux is the common trigger of NLRP3 inflammasome activation by bacterial toxins and particulate matter. Immunity 38:1142–1153. https://doi.org/10.1016/j.immuni.2013.05.016

Nachamkin I, Szymanski CM, Blaser MJ (2008) Campylobacter, 3rd edn. ASM Press, Washington

Naikare H, Palyada K, Panciera R, Marlow D, Stintzi A (2006) Major Role for FeoB in *Campylobacter Jejuni* ferrous iron acquisition, gut colonization, and intracellular survival. Infect Immun 74(10):5433–5444. https://doi.org/10.1128/IAI.00052-06

Neddermann M, Backert S (2019) Quantification of serine protease HtrA molecules secreted by the foodborne pathogen *Campylobacter jejuni*. Gut Pathog 11:14. https://doi.org/10.1186/s13099-019-0295-8

Neimark J (2015) Line of attack. Science 347(6225):938–940. https://doi.org/10.1126/science.347.6225.938

Nelson-Rees WA, Flandermeyer RR (1976) HeLa cultures defined. Science 191(4222):96–98. https://doi.org/10.1126/science.1246601

Novik V, Hofreuter D, Galan JE (2010) Identification of *Campylobacter jejuni* genes involved in its interaction with epithelial cells. Infect Immun 78(8):3540–3553. https://doi.org/10.1128/IAI.00109-10

Ó Cróinín T, Backert S (2012) Host epithelial cell invasion by *Campylobacter jejuni*: trigger or zipper mechanism? Front Cell Infect Microbiol 2:25. https://doi.org/10.3389/fcimb.2012.00025

Oyarzabal OA, Backert S (2012) Microbial food safety: an introduction (food science text series). Springer, Heidelberg

Parkhill J, Wren B, Mungall K, Ketley J, Churcher C, Basham D, Chillingworth T, Davies R, Feltwell T, Holroyd S, Jagels K, Karlyshev AV, Moule S, Pallen MJ, Penn CW, Quail MA, Rajandream MA, Rutherford KM, van Vliet AH, Whitehead S, Barrell BG (2000) The genome sequence of the food-borne pathogen *Campylobacter jejuni* reveals hypervariable sequences. Nature 403:665–668. https://doi.org/10.1038/35001088

Parsons CM, Potter LM, Brown RD Jr (1983) Effects of dietary carbohydrate and of intestinal microflora on excretion of endogenous amino acids by poultry. Poult Sci. 62(3):483–489. https://doi.org/10.3382/ps.0620483

Pei ZH, Burucoa C, Grignon B, Baqar S, Huang XZ, Kopecko DJ, Bourgeois AL, Fauchère JL, Blaser MJ (1998) Mutation in the peb1A locus of *Campylobacter jejuni* reduces interactions with epithelial cells and intestinal colonization of mice. Infect Immun 66:938–943

Perna AM, Reisen F, Schmidt TP, Geppert T, Pillong M, Weisel M, Hoy B, Simister PC, Feller SM, Wessler S, Schneider G (2014) Inhibiting *Helicobacter pylori* HtrA protease by addressing a computationally predicted allosteric ligand binding site. Chem Sci 5:3583–3590. https://doi.org/10.1039/c4sc01443j

Perna AM, Rodrigues T, Schmidt TP, Bohm M, Stutz K, Reker D, Pfeiffer B, Altmann KH, Backert S, Wessler S, Schneider G (2015) Fragment-based De Novo design reveals a small-molecule inhibitor of *Helicobacter pylori* HtrA. Angew Chem Int Ed Engl 54:10244–10248. https://doi.org/10.1002/anie.201504035

Pesci EC, Cottle DL, Pickett CL (1994) Genetic, enzymatic, and pathogenic studies of the iron superoxide dismutase of *Campylobacter jejuni*. Infect Immun 62(7):2687–2694. https://doi.org/10.1128/IAI.62.7.2687-2694.1994

Phongsisay V, Hara H, Fujimoto S (2015) Toll-like receptors recognize distinct proteinase-resistant glycoconjugates in *Campylobacter jejuni* and *Escherichia coli*. Mol Immunol 64:195–203. https://doi.org/10.1016/j.molimm.2014.11.020

Phongsisay V (2016) The immunobiology of *Campylobacter jejuni*: innate immunity and autoimmune diseases. Immunobiol. 221(4):535–543. https://doi.org/10.1016/j.imbio.2015.12.005

Pickett CL, Whitehouse CA (1999) The cytolethal distending toxin family. Trends Microbiol 7:292–297. https://doi.org/10.1016/s0966-842x(99)01537-1

Plowright RK, Parrish CR, McCallum H, Hudson PJ, Ko AI, Graham AL, Lloyd-Smith JO (2017) Pathways to zoonotic spillover. Nat Rev Microbiol 15(8):502–510. https://doi.org/10.1038/nrmicro.2017.45

Poltorak A, He X, Smirnova I, Liu MY, Van Huffel C, Du X, Birdwell D, Alejos E, Silva M, Galanos C, Freudenberg M, Ricciardi-Castagnoli P, Layton B, Beutler B (1998) Defective LPS signaling in C3H/HeJ and C57BL/10ScCr mice: mutations in Tlr4 gene. Science 282(5396):2085–2208. https://doi.org/10.1126/science.282.5396.2085

Poly F, Ewing C, Goon S, Hickey TE, Rockabrand D, Majam G, Lee L, Phan J, Savarino NJ, Guerry P (2007) Heterogeneity of a *Campylobacter jejuni* protein that is secreted through the flagellar filament. Infect Immun 75:3859–3867. https://doi.org/10.1128/IAI.00159-07

Potturi-Venkata LP, Backert S, Vieira SL, Oyarzabal OA (2007) Evaluation of logistic processing to reduce cross-contamination of commercial broiler carcasses with *Campylobacter* spp. J Food Prot 70:2549–2554. https://doi.org/10.4315/0362-028x-70.11.2549

Purcell EM (1997) The efficiency of propulsion by a rotating flagellum. Proc Natl Acad Sci USA 94:11307–11311. https://doi.org/10.1073/pnas.94.21.11307

Purdy D, Buswell CM, Hodgson AE, Mc AK, Henderson I, Leach SA (2000) Characterisation of cytolethal distending toxin (CDT) mutants of *Campylobacter jejuni*. J Med Microbiol 49:473–479. https://doi.org/10.1099/0022-1317-49-5-473

Radomska KA, Wosten M, Ordonez SR, Wagenaar JA, van Putten JPM (2017) Importance of *Campylobacter jejuni* FliS and FliW in flagella biogenesis and flagellin secretion. Front Microbiol 8:1060. https://doi.org/10.3389/fmicb.2017.01060

Rahman H, King RM, Shewell LK, Semchenko EA, Hartley-Tassell LE, Wilson JC, Day CJ, Korolik V (2014) Characterisation of a multi-ligand binding chemoreceptor CcmL (Tlp3) of *Campylobacter jejuni*. PLoS Pathog 10:e1003822. https://doi.org/10.1371/journal.ppat.1003822

Rathinam VA, Appledorn DM, Hoag KA, Amalfitano A, Mansfield LS (2009) *Campylobacter jejuni*-induced activation of dendritic cells involves cooperative signaling through Toll-like receptor 4 (TLR4)-MyD88 and TLR4-TRIF axes. Infect Immun 77:2499–2507. https://doi.org/10.1128/IAI.01562-08

Ribardo DA, Hendrixson DR (2011) Analysis of the LIV system of *Campylobacter jejuni* reveals alternative roles for LivJ and LivK in commensalism beyond branched-chain amino acid transport. J Bacteriol 193(22):6233–6243. https://doi.org/10.1128/JB.05473-11

Schmidt AM, Escher U, Mousavi S, Tegtmeyer N, Boehm M, Backert S, Bereswill S, Heimesaat MM (2019) Immunopathological properties of the *Campylobacter jejuni* flagellins and the adhesin CadF as assessed in a clinical murine infection model. Gut Pathogens 11:24. https://doi.org/10.1186/s13099-019-0306-9

Schmidt TP, Perna AM, Fugmann T, Bohm M, Jan H, Haller S, Gotz C, Tegtmeyer N, Hoy B, Rau TT, Neri D, Backert S, Schneider G, Wessler S (2016) Identification of E-cadherin signature motifs functioning as cleavage sites for *Helicobacter pylori* HtrA. Sci Rep 6:23264. https://doi.org/10.1038/srep23264

Schroder K, Tschopp J (2010) The inflammasomes. Cell 140:821–832. https://doi.org/10.1016/j.cell.2010.01.040

Schwandner R, Dziarski R, Wesche H, Rothe M, Kirschning CJ (1999) Peptidoglycan- and lipoteichoic acid-induced cell activation is mediated by toll-like receptor 2. J Biol Chem 274(25):17406–17409. https://doi.org/10.1074/jbc.274.25.17406

Sheppard SK, Maiden MC (2015) The evolution of *Campylobacter jejuni* and *Campylobacter coli*. Cold Spring Harb Perspect Biol 7(8):a018119:22. https://doi.org/10.1101/cshperspect.a018119

Shigematsu M, Umeda A, Fujimoto S, Amako K (1998) Spirochaete-like swimming mode of *Campylobacter jejuni* in a viscous environment. J Med Microbiol 47:521–526. https://doi.org/10.1099/00222615-47-6-521

Shortt C, Scanlan E, Hilliard A, Cotroneo CE, Bourke B, Ó Cróinín T (2016) DNA supercoiling regulates the motility of *Campylobacter jejuni* and is altered by growth in the presence of chicken mucus. mBio 7(5):e01227-16. https://doi.org/10.1128/mBio.01227-16

Smith JL (2002) *Campylobacter jejuni* infection during pregnancy: long-term consequences of associated bacteremia, Guillain-Barré syndrome, and reactive arthritist. J Food Prot 65(4):696–708. https://doi.org/10.4315/0362-028x-65.4.696

Song YC, Jin S, Louie H, Ng D, Lau R, Zhang Y, Weerasekera R, Al Rashid S, Ward LA, Der SD, Chan VL (2004) FlaC, a protein of *Campylobacter jejuni* TGH9011 (ATCC43431) secreted through the flagellar apparatus, binds epithelial cells and influences cell invasion. Mol Microbiol 53:541–553. https://doi.org/10.1111/j.1365-2958.2004.04175.x

Stephenson HN, Mills DC, Jones H, Milioris E, Copland A, Dorrell N, Wren BW, Crocker PR, Escors D, Bajaj-Elliott M (2014) Pseudaminic acid on *Campylobacter jejuni* flagella modulates dendritic cell IL-10 expression via Siglec-10 receptor: a novel flagellin-host interaction. J Infect Dis 210(9):1487–1498. https://doi.org/10.1093/infdis/jiu287

Stradal TEB, Schelhaas M (2018) Actin dynamics in host-pathogen interaction. FEBS Lett 592(22):3658–3669. https://doi.org/10.1002/1873-3468.13173

Tegtmeyer N, Moodley Y, Yamaoka Y, Pernitzsch SR, Schmidt V, Traverso FR, Schmidt TP, Rad R, Yeoh KG, Bow H, Torres J, Gerhard M, Schneider G, Wessler S, Backert S (2016) Characterisation of worldwide *Helicobacter pylori* strains reveals genetic conservation and essentiality of serine protease HtrA. Mol Microbiol 99(5):925–944. https://doi.org/10.1111/mmi.13276

Tram G, Day CJ, Korolik V (2020) Bridging the gap: a role for *Campylobacter jejuni* biofilms. Microorganisms 8(3):452. https://doi.org/10.3390/microorganisms8030452

Trinchieri G, Sher A (2007) Cooperation of Toll-like receptor signals in innate immune defense. Nat Rev Immunol 7:179–190. https://doi.org/10.1038/nri2038

van Gerwe T, Miflin JK, Templeton JM, Bouma A, Wagenaar JA, Jacobs-Reitsma WF, Stegeman A, Klinkenberg D (2009) Quantifying transmission of *Campylobacter jejuni* in commercial broiler flocks. Appl Environ Microbiol 75:625–628. https://doi.org/10.1128/AEM.01912-08

Vegge CS, Brøndsted L, Li YP, Bang DD, Ingmer H (2009) Energy taxis drives *Campylobacter jejuni* toward the most favorable conditions for growth. Appl Environ Microbiol 75:5308–5314. https://doi.org/10.1128/aem.00287-09

Velayudhan J, Kelly DJ (2002) Analysis of gluconeogenic and anaplerotic enzymes in *Campylobacter jejuni*: an essential role for phosphoenolpyruvate carboxykinase. Microbiology 148(3):685–694. https://doi.org/10.1099/00221287-148-3-685

Velayudhan J, Jones MA, Barrow PA, Kelly DJ (2004) L-serine catabolism via an oxygen-labile L-serine dehydratase is essential for colonization of the avian gut by *Campylobacter jejuni*. Infect Immun 72:260–268. https://doi.org/10.1128/iai.72.1.260-268.2004

von Gunten S, Bochner BS (2008) Basic and clinical immunology of Siglecs. Ann N Y Acad Sci 1143:61–82. https://doi.org/10.1196/annals.1443.011

Wai SN, Lindmark B, Soderblom T, Takade A, Westermark M, Oscarsson J, Jass J, Richter-Dahlfors A, Mizunoe Y, Uhlin BE (2003) Vesicle-mediated export and assembly of pore-forming oligomers of the enterobacterial ClyA cytotoxin. Cell 115:25–35. https://doi.org/10.1016/s0092-8674(03)00754-2

Wassenaar TM, van der Zeijst BA, Ayling R, Newell DG (1993) Colonization of chicks by motility mutants of *Campylobacter jejuni* demonstrates the importance of flagellin A expression. J Gen Microbiol 139(6):1171–1175. https://doi.org/10.1099/00221287-139-6-1171

Watson RO, Galán JE (2005) Signal transduction in *Campylobacter jejuni*-induced cytokine production. Cell Microbiol 7:655–665. https://doi.org/10.1111/j.1462-5822.2004.00498.x

Watson RO, Galán JE (2008) *Campylobacter jejuni* survives within epithelial cells by avoiding delivery to lysosomes. PLoS Pathog 4:e14. https://doi.org/10.1371/journal.ppat.0040014

Wedderkopp A, Nielsen EM, Pedersen K (2003) Distribution of *Campylobacter jejuni* Penner serotypes in broiler flocks 1998–2000 in a small Danish community with special reference to serotype 4-complex. Epidemiol Infect 131:915–921. https://doi.org/10.1017/s0950268803008975

Whitehouse CA, Balbo PB, Pesci EC, Cottle DL, Mirabito PM, Pickett CL (1998) *Campylobacter jejuni* cytolethal distending toxin causes a G2-phase cell cycle block. Infect Immun 66:1934–1940

Wooldridge KG, Williams PH, Ketley JM (1996) Host signal transduction and endocytosis of *Campylobacter jejuni*. Microb Pathog 21(4):299–305. https://doi.org/10.1006/mpat.1996.0063

Wösten MM, Wagenaar JA, van Putten JP (2004) The FlgS/FlgR two-component signal transduction system regulates the *fla* regulon in *Campylobacter jejuni*. J Biol Chem. 279(16):16214–16222. https://doi.org/10.1074/jbc.M400357200

Xi D, Alter T, Einspanier R, Sharbati S, Golz G (2020) *Campylobacter jejuni* genes Cj1492c and Cj1507c are involved in host cell adhesion and invasion. Gut Pathogens 12. https://doi.org/10.1186/s13099-020-00347-8

Yao R, Burr DH, Doig P, Trust TJ, Niu H, Guerry P (1994) Isolation of motile and non-motile insertional mutants of *Campylobacter jejuni*: the role of motility in adherence and invasion of eukaryotic cells. Mol Microbiol 14:883–893. https://doi.org/10.1111/j.1365-2958.1994.tb01324.x

Yao R, Burr DH, Guerry P (1997) CheY-mediated modulation of *Campylobacter jejuni* virulence. Mol Microbiol 23:1021–1031. https://doi.org/10.1046/j.1365-2958.1997.2861650.x

Yoon H (2016) Bacterial outer membrane vesicles as a delivery system for virulence regulation. J Microbiol Biotechnol 26:1343–1347. https://doi.org/10.4014/jmb.1604.04080

Young GM, Schmiel DH, Miller VL (1999) A new pathway for the secretion of virulence factors by bacteria: The flagellar export apparatus functions as a protein-secretion system. Proc Natl Acad Sci USA 96:6456–6461. https://doi.org/10.1073/pnas.96.11.6456

Young KT, Davis LM, Dirita VJ (2007) *Campylobacter jejuni*: molecular biology and pathogenesis. Nat Rev Microbiol 5:665–679. https://doi.org/10.1038/nrmicro1718

Yuki N, Susuki K, Koga M, Nishimoto Y, Odaka M, Hirata K, Taguchi K, Miyatake T, Furukawa K, Kobata T, Yamada M (2004) Carbohydrate mimicry between human ganglioside GM1 and *Campylobacter jejuni* lipooligosaccharide causes Guillain-Barré syndrome. Proc Natl Acad Sci USA 101(31):11404–11409. https://doi.org/10.1073/pnas.0402391101

Yuki N, Koga M (2006) Bacterial infections in Guillain-Barré and Fisher syndromes. Curr Opin Neurol 19(5):451–457. https://doi.org/10.1097/01.wco.0000245367.36576.e9

Zarzecka U, Grinzato A, Kandiah E, Cysewski D, Berto P, Skorko-Glonek J, Zanotti G, Backert S (2020) Functional analysis and cryo-electron microscopy of *Campylobacter jejuni* serine protease HtrA. Gut Microbes (in Press). https://doi.org/10.1080/19490976.2020.1810532

Zheng J, Meng J, Zhao S, Singh R, Song W (2008) *Campylobacter*-induced interleukin-8 secretion in polarized human intestinal epithelial cells requires *Campylobacter*-secreted cytolethal distending toxin- and Toll-like receptor-mediated activation of NF-kappaB. Infect Immun 76:4498–4508. https://doi.org/10.1128/iai.01317-07

Ziprin RL, Young CR, Stanker LH, Hume ME, Konkel ME (1999) The absence of cecal colonization of chicks by a mutant of *Campylobacter jejuni* not expressing bacterial fibronectin-binding protein. Avian Dis 43:586–589

Diarrheal Mechanisms and the Role of Intestinal Barrier Dysfunction in *Campylobacter* Infections

Fábia Daniela Lobo de Sá, Jörg-Dieter Schulzke, and Roland Bücker

Contents

Abstract *Campylobacter* enteritis is the most common cause of foodborne bacterial diarrhea in humans. Although various studies have been performed to clarify the pathomechanism in *Campylobacter* infection, the mechanism itself and bacterial virulence factors are yet not completely understood. The purpose of this chapter is to (i) give an overview on *Campylobacter*-induced diarrheal mechanisms, (ii) illustrate underlying barrier defects, (iii) explain the role of the mucosal immune response and (iv) weigh preventive and therapeutic approaches. Our present knowledge of pathogenetic and diarrheal mechanisms of *Campylobacter jejuni* is explained in the first part of this chapter. In the second part, the molecular basis for the *Campylobacter*-induced barrier dysfunction is compared with that of other species in the *Campylobacter* genus. The bacteria are capable of overcoming the intestinal epithelial barrier. The invasion into the intestinal mucosa is the initial step of the infection, followed by a second step, the epithelial barrier impairment. The extent

F. D. Lobo de Sá · J.-D. Schulzke · R. Bücker (✉)
Institute of Clinical Physiology/Nutritional Medicine, Medical Department, Division of Gastroenterology, Infectiology, Rheumatology, Charité - University Medicine Berlin, Campus Benjamin Franklin, Hindenburgdamm 30, 12203 Berlin, Germany
e-mail: roland-felix.buecker@charite.de

F. D. Lobo de Sá
e-mail: fabia.lobo-da-fonseca@charite.de

J.-D. Schulzke
e-mail: joerg.schulzke@charite.de

S. Backert (ed.), *Fighting Campylobacter Infections*, Current Topics in Microbiology and Immunology 431, https://doi.org/10.1007/978-3-030-65481-8_8

of the impairment depends on various factors, including *tight junction* dysregulation and epithelial apoptosis. The disturbed intestinal epithelium leads to a loss of water and solutes, the *leak flux* type of diarrhea, and facilitates the uptake of harmful antigens, the *leaky gut* phenomenon. The barrier dysfunction is accompanied by increased pro-inflammatory cytokine secretion, which is partially responsible for the dysfunction. Moreover, cytokines also mediate ion channel dysregulation (e.g., epithelial sodium channel, ENaC), leading to another diarrheal mechanism, which is sodium malabsorption. Future perspectives of *Campylobacter* research are the clarification of molecular pathomechanisms and the characterization of therapeutic and preventive compounds to combat and prevent *Campylobacter* infections.

1 Introduction

First descriptions of the *Campylobacter*-induced colitis in humans were published in the beginning of the 1980s, when standardized cultivation of campylobacters became routine (Blaser et al. 1980). Confirming the causality between *Campylobacter* and the enteritis, histological analysis of experimental infected mouse models revealed acute mucosal inflammation with focal epithelial lesions together with neutrophils infiltrating the crypt epithelium (cryptitis) of the small intestine (Blaser et al. 1983) and in the colon of experimentally infected macaques with diarrheal outcome (Russell et al. 1989). The invasion of the gastrointestinal mucosa is a main virulence characteristic of *C. jejuni* (van Spreeuwel et al. 1985). In *C. jejuni*-infected humans, it was proven that the bacteria induce diarrhea, and it was described that the acute inflammatory response is a hallmark of the infection (Black et al. 1988). The pathohistological classification of the *Campylobacter* enteritis revealed a picture with invasion of campylobacters into colonic epithelial cells, goblet cells and into the lamina propria together with a massive infiltration of immune cells and a marked distortion of the crypt architecture (van Spreeuwel et al. 1985).

Different enteropathogens invade the intestinal epithelium and in this manner disrupt the epithelial barrier. Often an involvement of *tight junction* (TJ) proteins, especially claudins, is detected during this process. Since the invasion ratio of *C. jejuni* migrating into host cells in vitro was low (<1% of the applied bacteria invade the epithelial monolayer) (Blaser and Reller 1981) and the bacteria were intracellularly eliminated (De Melo et al. 1989), bacterial toxins have been originally proposed as important virulence factors for the *Campylobacter* diarrhea. Cytotoxicity and enterotoxicity were made responsible as pathomechanisms in the intestine, but the existence of an enterotoxin could not be confirmed. Bacterial pathomechanisms instead include, e.g., the secretion of different virulence factors or toxins, but also bacterial motility-dependent virulence mechanisms like adherence, invasion and transmigration via the transcellular and paracellular pathway, which were found to be substantial for the *Campylobacter* pathogenesis (Backert et al. 2013). However, the pathogenetic principles of the *Campylobacter* enteritis comprise a combination of several processes in the intestinal epithelium and subepithelium.

2 Pathogenetic Principles and Diarrheal Mechanisms in campylobacteriosis

In the etiological classification, the *Campylobacter* diarrhea is an infectious inflammatory disease. The infection can develop from acute diarrhea (<2 weeks) to chronic diarrhea (>2 weeks) (Spiller et al. 2000; Nielsen et al. 2012). In general, every diarrhea, etiology independently, can be described as an imbalance of intestinal absorption and secretion and is driven by osmotic forces. From the pathophysiological view, there are five ways diarrhea can occur. The pathogenetic principles were (i) high osmotic pressure of not-observed luminal solutes (osmotic diarrhea), (ii) increased gut motility, with increased transit time of intestinal content and reduced absorption of nutrients (motility-related diarrhea), (iii) lack of absorptive transporters or decreased surface area (malabsorptive diarrhea), (iv) increased active ion secretion (secretory diarrhea) or (v) paracellular loss of water and solutes into the lumen (leak flux diarrhea). The classification of the diarrheal mechanisms of infectious diarrhea, based on the pathophysiology, can be assigned to three main categories: the secretory diarrhea, the malabsorptive diarrhea and the leak flux diarrhea. For the *Campylobacter* infection, all three mechanisms were experimentally proposed and are at least partially confirmed as discussed below.

Secretory Diarrhea

Since *Campylobacter* patients often exhibit watery diarrhea, it was proposed early that an enterotoxin could be responsible, like shown for other enteropathogens as, e.g., the cholera toxin (CT) (Fig. 1). In a rat ileal loop model, *C. jejuni* could induce cholera-like enterotoxigenecity with intraluminal fluid accumulation (Ruiz-Palacios et al. 1983). Enterotoxic and non-enterotoxic *C. jejuni* and *C. coli* strains and the *Campylobacter jejuni* enterotoxins (here abbreviated CJT) elongated Chinese hamster ovary (CHO) cells by an increase in intracellular cAMP (reviewed in Wassenaar 1997). CHO cells were chosen, because they are insensitive to the *C. jejuni* cytolethal distending toxin (CDT). The enterotoxicity of campylobacters could be functionally shown (McCardell et al. 1984), but the existence of a CJT or a cytotonic toxin was questioned by most researchers. Contradictory evidence was shown for the CJT, as sonicates of *C. jejuni* and *C. coli* were non-enterotoxic toward CHO cells (Wadström et al. 1983). The controversy over a structural CJT protein continued as an enterotoxin was not identified in hundreds of sequenced *C. jejuni* genomes and all attempts to clone CJT failed.

For the *Campylobacter* infection, no induction of an active chloride (Cl^-) secretion as known for CT could be detected. CT commonly activates the cystic fibrosis transmembrane conductance regulator (CFTR), which is a Cl^- channel transporting Cl^- into the lumen. Thus, the CJT effect seems to be less relevant for the watery part of diarrhea. By measurement of the short-circuit current (Isc) in Ussing chambers in the mucosa from colon biopsies of *C. jejuni*-infected patients, the induction of active anion secretion could be excluded (Bücker et al. 2018). No change in basal Isc was observed. After stimulation of electrogenic Cl^- secretion with prostaglandin

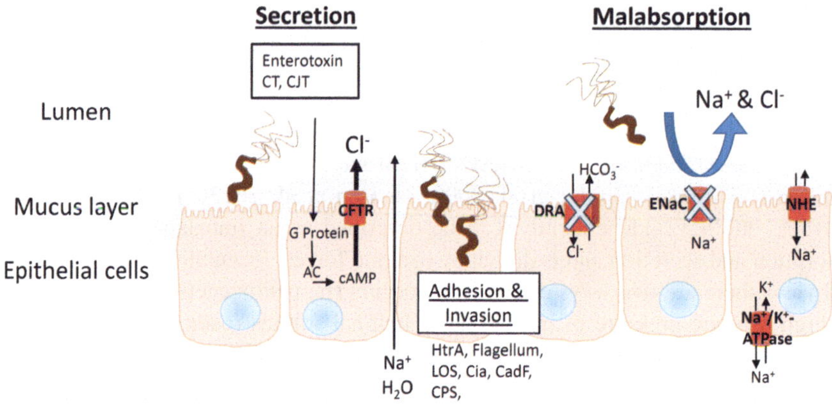

Fig. 1 Proposed secretory and malabsorptive diarrheal mechanisms in *Campylobacter jejuni*infection. Bacterial enterotoxin secretion as well as adhesion was associated with a secretory type of diarrhea. CFTR might have only a minor influence in *Campylobacter* diarrhea; no induction of active Cl^- secretion has been identified. In contrast, *Campylobacter* infection decreases DRA and ENaC activity, while effects on Na^+/K^+-ATPase and NHE regulation are questionable. Abbreviations: AC: Adenylyl cyclase; CadF: *Campylobacter* adhesin to fibronectin; cAMP: cyclic adenosine monophosphate; Cia: *Campylobacter* invasion antigens; CFTR: cystic fibrosis transmembrane conductance regulator; CPS: capsular polysaccharide; CT: cholera toxin; CJT: *Campylobacter jejuni* enterotoxins; DRA: downregulated-in-adenoma exchanger; ENaC: epithelial sodium (Na^+) channel; HtrA: high-temperature requirement A serine protease; NHE: Sodium–hydrogen antiporter; LOS: lipooligosaccharide

2 (PGE_2) and theophylline (via cAMP-dependent activation) or with the cholinergic agonist carbachol, the Isc change was lower than in controls, pointing even to a diminished Cl^- channel activity (Bücker et al. 2018). As the maximum anion transport capacity was decreased, the reason for this could be a downregulation of transporters or channels. RNA expression analysis revealed that in the mucosa of *C. jejuni*-infected patients, the CFTR was downregulated (Bücker et al. 2018). Similar effects were shown also in vitro in T84 cells with suppressed Cl^- secretion via CFTR activity after *C. jejuni* infection (Negoro et al. 2014). Moreover, as the secretion takes place in the crypts—that are the most affected regions in *Campylobacter* diarrhea (the cryptitis)—a loss of the overall secretory capacity is inevitable. Thus, this diarrheal mechanism alone could not explain the *Campylobacter* enteritis. But a dysregulation of other ion transporters present in the mucosal surface like DRA (downregulated-in-adenoma, *SLC26A3*) or the epithelial sodium channel (ENaC) could contribute as well to diarrhea via a malabsorptive mechanism for ions and water (Fig. 1).

Malabsorptive Diarrhea

A common example for malabsorptive diarrhea is the decreased expression of GLUT5 in fructose malabsorption, leading to fructose accumulation in the small intestine and subsequently to uncleaved fructose passage into the colon. In this way, the fructose is fermented leading to gas production, abdominal pain and diarrhea. In campylobacteriosis, an increase in gas production and transporter downregulation in the small intestine was not reported yet. However, ion transporter and channel downregulation could be demonstrated for the large intestine. In the large intestine, epithelial surface cells are responsible for electrogenic Na^+ absorption, whereas the ion secretion takes place in the crypts (Welsh et al. 1982). In inflammatory bowel disease (IBD), defects in the surface epithelium of the colon are involved in disturbed electrolyte transport (Sandle et al. 1990).

The experimental infection of the colonocyte Caco-2 cell model revealed that *C. jejuni* inhibits absorptive transport functions (MacCallum et al. 2005). Here, fluid accumulation and cell dome formation were characterized indicating malabsorption as mechanism in *Campylobacter* diarrhea (MacCallum et al. 2005). As outlined above, CFTR is compromised during *C. jejuni* infection. In addition, the expression of the anion exchanger DRA was found to be decreased in the infected human colon mucosa (Bücker et al. 2018). This electroneutral transporter for Cl^- absorption was expression-regulated by lowered mRNA levels. A comparable mRNA expression regulation was found for the regulatory subunits (β and γ) of the ENaC. The decreased ENaC activity in the colon of *Campylobacter* patients led to lower Na^+ absorption (Bücker et al. 2018). Therefore, Na^+ malabsorption as one diarrheal mechanism was introduced into the literature. This is in part the consequence of a direct interaction of *C. jejuni* with the epithelial cells as well as due to the suppression of ENaC by pro-inflammatory cytokines. The malabsorptive diarrheal mechanism by β- and γ-ENaC downregulation as well as by phosphorylation of the extracellular signal-regulated kinase (ERK) was also shown for *C. concisus* (Nattramilarasu et al. 2020). Thus, for *C. jejuni* and *C. concisus* sodium malabsorption together with barrier dysfunction in the colon (*leak flux*) were shown as diarrheal mechanisms (Nielsen et al. 2011; Bücker et al. 2018; Nattramilarasu et al. 2020). As another hint for malabsorptive mechanismsm Na^+/K^+-ATPase inhibition was demonstrated after *C. jejuni* infection (Kanwar et al. 1994). Na^+/H^--antiporter (NHE) inhibition was observed in various infectious diarrhea (reviewed in Gurney et al. 2017), but was not shown in *Campylobacter* infections.

Leak Flux Diarrhea

Leak flux diarrhea is defined as a consequence of a passive passage of solutes and water into the lumen via the disruption of tight junctions (TJs) and/or the induction of epithelial damage (Schulzke et al. 2009). When the paracellular leak pathway is opened, epithelial barrier function is compromised. The leakiness of the epithelium may involve also macromolecules with a distinct molecule size and flux ratio. As one pathogenetic principle, the term "leak flux mechanism" was introduced to describe the epithelial barrier dysfunction mechanism (Gitter et al. 2000a). It contributes also as diarrheal mechanism to HIV enteropathy, celiac disease and ulcerative colitis (UC) (Stockmann et al. 1998; Schulzke et al. 1998; Schmitz et al. 1999a).

In experimental investigations where the TJ is affected and/or cell damage (epithelial cell death) occurs, the leak flux mechanism can explain the functional epithelial dysregulation, induced by cytokines like TNF-α (Gitter et al. 2000a), or bacteria like *Yersinia enterocolitica* in vitro or in vivo (Hering et al. 2011, 2016). Even the well-studied bacterial pore-forming toxin aerolysin from *Aeromonas hydrophila*, with cytotoxic and enterotoxic properties (Epple et al. 2004), contributes to a leak flux mechanism by a rapid TJ redistribution (Bücker et al. 2011). Also *Arcobacter butzleri* showed a leak flux mechanism in vitro, which could explain the watery diarrhea (Bücker et al. 2009). A leak flux mechanism is also present as a pathogenetic principle in infections with *C. concisus*, *C. fetus* and *C. jejuni* (Nielsen et al. 2011; Bücker et al. 2017, 2018). The watery type of *Campylobacter* diarrhea can be explained by the three diarrheal mechanisms; secretion, malabsorption and leak flux. The escalating pathomechanisms leading to the bloody part of the diarrhea are the consequence of the epithelial barrier dysfunction together with the mucosal immune response toward the bacteria.

2.1 Barrier Dysfunction, Leak Flux and Leaky Gut

The balanced epithelial barrier function in the gut is maintained by the expression pattern and distribution of distinct TJ proteins, the claudins. Depending on the intestinal segment (proximal or distal), or the differentiation side in the mucosa (crypt, surface or villus), the composition of barrier maintaining together with channel-forming claudins (located only bicellularly) is unique in the respective sides of the epithelia. An expression change or the subcellular redistribution of claudins is the molecular and structural correlate for the barrier dysfunction and the leak flux type of diarrhea.

Adjacent cells are connected in the paracellular space at the most apical part with TJ proteins, composing TJ strands, which include 26 different claudins in humans, TAMPs (occludin, tricellulin, MarvelD3) and junctional adhesion molecules (JAMs) (Fig. 2). The TJ proteins create the polarization of the epithelial layer into an apical and basolateral compartment (fence function) and a limitation in the electrolyte transport between the cells (gate function) (Mandel et al. 1993). The first identified TJ protein was occludin (Furuse et al. 1993) followed by claudins (Furuse et al. 1998) and tricellulin (Ikenouchi et al. 2005). The regulation of the epithelial barrier function and its impairment can be classified into three mechanisms. Two are TJ-dependent mechanisms: the pore pathway and the leak pathway (Shen et al. 2011). The third is a TJ-independent mechanism: the unrestricted pathway by epithelial damage (France and Turner 2017). The disruption of the epithelial barrier by pathogens like *Campylobacter* comprises at least (i) the induction of epithelial apoptosis by *C. jejuni*, which is barrier-relevant and (ii) that the barrier-maintaining claudins are downregulated and redistributed off the TJ domain in the human colon mucosa (Butkevych et al. 2020; Bücker et al. 2018). These two cellular mechanisms can explain the leak flux type of *Campylobacter* diarrhea (Fig. 3).

Pore Pathway

The pore pathway describes a high-capacity paracellular pathway with charge and size selectivity, for molecule diameters ranging from approx. 5 to 10 Å (Van Itallie et al. 2008; Yu et al. 2009; Krug et al. 2012). The concept of mucosal leakiness was

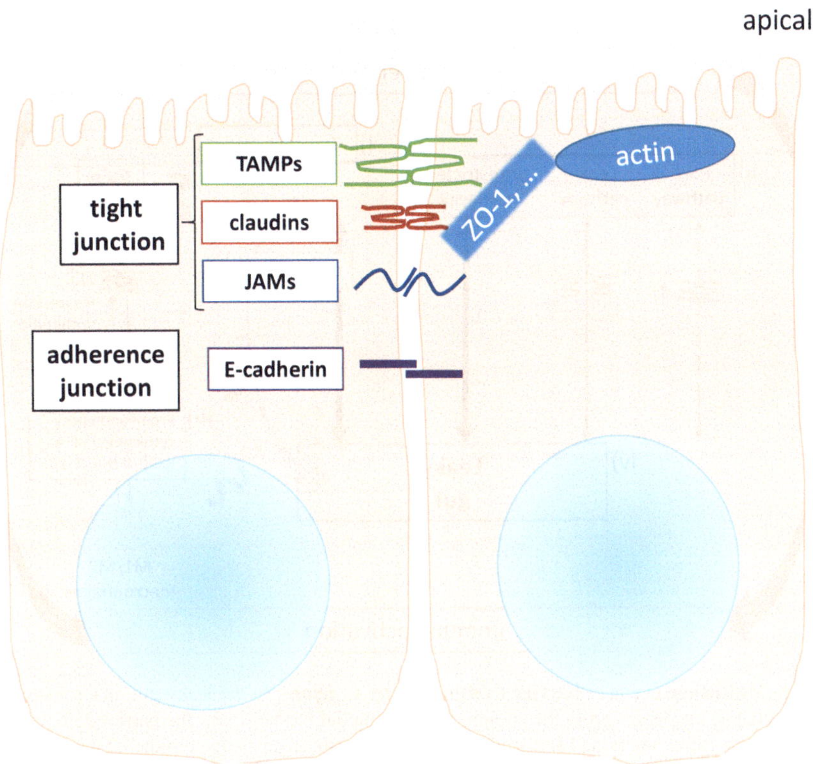

Fig. 2 Scheme of the tight junction proteins. Neighboring cells are connected at the most apical part with transmembrane tight junction proteins, including 27 different claudins (barrier-forming claudins in mammals, e.g., claudin-1, -3, -4, -5, -8 or channel-forming claudins, e.g., claudin-2 and -15), the tight junction-associated Marvel proteins (TAMPs) family members occludin, tricellulin and MarvelD3 and the junctional adhesion molecules (JAMs). Tight junction proteins are linked to peripheral scaffolding proteins such as zonula occludens proteins (ZO-1, -2, -3) and connected to the actin cytoskeleton. Further basolateral adherence junction proteins (E-cadherin) belong also to the junctional complex

first mentioned as explanation for the anion flux dysregulation in the large intestine of UC patients (Edmonds and Pilcher 1973). For the inflamed colon in UC and the defective anion transport, the leakiness was proposed as pathomechanism (Sandle et al. 1990). Here, a pore pathway should be sufficient to explain the paracellular anion flux, which was discovered later. The paracellular convective transport (solvent drag phenomenon) of ions and water through the intercellular spaces was shown to have its molecular correlate in channel-forming claudins (e.g., claudin-2, -15). Claudin-2 was described as anion channel and more important for the paracellular transport mechanism, it was identified as paracellular water channel (Amasheh et al. 2002; Rosenthal et al. 2010).

In IBD, claudin-2 is upregulated upon the inflammatory condition (Heller et al. 2005; Zeissig et al. 2007). Claudin-2-dependent paracellular water flux is thought

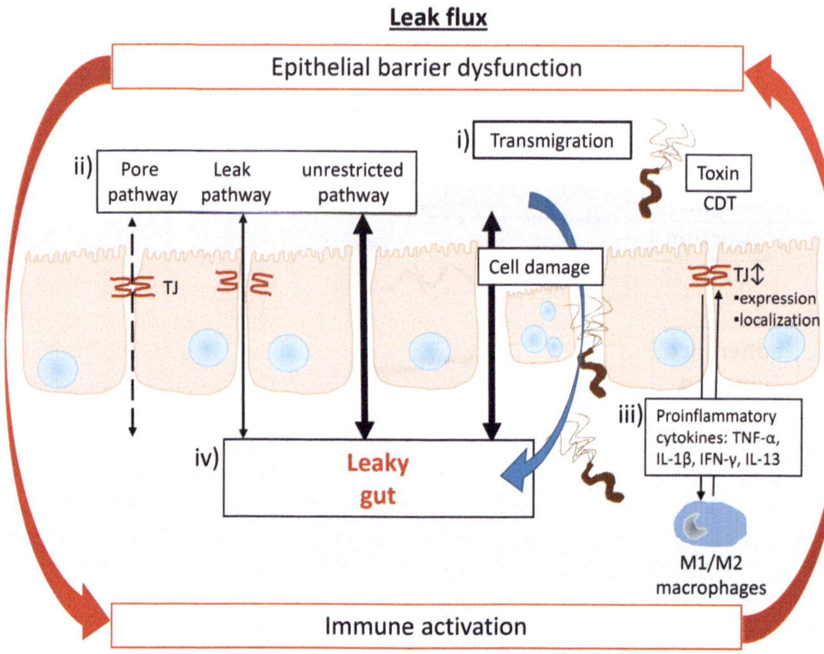

Fig. 3 Multistage event of barrier dysfunction by *C. jejuni*. In the leaky gut, macromolecules with antigenic properties cross the compromised epithelial barrier. Here, the barrier defects of the leak pathway (tight junction (TJ) disruption) and the unrestricted passage (epithelial damage) build the basis for the antigen influx and the concomitant immune activation. (i) Barrier passage of *C. jejuni* (by paracellular transmigration and/or transcellular translocation via transcytosis) without any measurable changes in permeability for water and solutes in electrophysiological investigations. (ii) Opening of the leak pathway and the unrestricted pathway (direct epithelial cell interaction). (iii) Induction of subepithelial immune cell cytokine release, leading to opening of the pore pathway, leak pathway and the unrestricted pathway (indirect immune cell–epithelial cell interaction) contributing to leak flux diarrhea. (iv) Mucosal cytokine storm with potentiated barrier dysfunction by antigen influx (leaky gut phenomenon)

to rinse off noxious agents from the colon mucosa. The paracellular opening of the pore pathway may contribute to the protection of the epithelium, but can also contribute to watery diarrhea. The contribution of the claudin-2 increase to the diarrhea was first shown for UC, but is important also for Crohn's disease and lymphocytic colitis (Heller et al. 2005; Zeissig et al. 2007; Barmeyer et al. 2017). This claudin-2-dependent relevance for the pore pathway was amended by the discovery of other ion-selective channel claudins that may contribute to a barrier dysfunction or malabsorption when compromised (Tamura et al. 2011).

In inflammatory *Campylobacter* diarrhea, no protein expression increase of channel-forming claudins in the colon has been observed so far (only *CLDN2* mRNA increased; Bücker et al. 2018). However, in a *C. jejuni*-infected immune cell–epithelial cell co-culture, the claudin-2 protein expression was increased (Butkevych et al. 2020). Thus, the induction of the pore pathway by claudin-2 is likely another diarrheal pathomechanism of *C. jejuni* or still a defense reaction of the host, with subepithelial immune cell involvement via cytokines. That the subepithelial immune system is activated and how it compromises in turn the epithelial barrier is described with the *leaky gut* concept. However, this comprises the leak pathway as well as the unrestricted passage and bacterial transcytosis.

Leak Pathway

The leak pathway describes a low-capacity paracellular pathway that allows passage of large solutes without regard to charge with a paracellular flux of molecules with diameters up to 63 Å (~47 kDa) (Buschmann et al. 2013). The molecular basis for the leak pathway is displayed by the barrier-maintaining TJ proteins. For the increased permeability of small macromolecules, the disruption of claudins is the main target. Equally important, the TJ protein tricellulin seals the paracellular space against macromolecules between three or four adjacent epithelial cells (Krug et al. 2009a). In the context of the leak pathway, the tricellular passage of small macromolecules could increase when tricellulin is downregulated. This was shown for UC and is regulated via IL-13 receptor α2 (Krug et al. 2018).

In general, a downregulation of barrier-maintaining claudins and tricellulin can occur at the protein or gene expression level. However, a rapid retraction of these proteins from the TJ strands into intracellular compartments was shown by myosin light-chain kinase (MLCK)-dependent cytoskeletal redistribution or endocytosis processes. The interference with the assembly and disassembly of TJ strands can be followed by confocal or electron microscopy. The consequence of retracted claudins was shown, e.g., for Crohn's disease in freeze fracture electron microscopy (FFEM) by discontinuous TJ strands with strand breaks (Zeissig et al. 2007). For the *Campylobacter* infection, these kinds of FFEM analyses on TJ strand formation are not published yet. The accessory intracellular TJ protein ZO-1 has primary structural importance and is not directly involved in the sealing function. However, the ZO-1 actin-binding domain is responsible for the regulation of the leak pathway through the perijunctional actomyosin cytoskeleton (Van Itallie et al. 2009; Yu et al. 2010). The perijunctional cytoskeleton can be dysregulated via MLCK activation (Yu et al. 2010). This MLCK dependence was also observed in aerolysin-mediated barrier dysfunction and was regulated by intracellular Ca^{2+} signaling (Bücker et al. 2011). Inhibition of MLCK prevented the aerolysin-driven TJ redistribution (Bücker et al. 2011). T cell-mediated barrier dysfunction via MLCK could also be prevented by inhibitors and was suggested as a therapeutic approach in immune-mediated gastrointestinal diseases (Clayburgh et al. 2005).

In *C. jejuni*-infected T84 cells, ZO-1 distribution or expression was not affected (Chen et al. 2006). However, occludin was redistributed off the TJ domain to intracellular domains of the cells, similar to previous observations in Caco-2 cells (Chen et al. 2006; MacCallum et al. 2005). Occludin was reported as a molecular target of *C. jejuni*. The contribution of occludin to the barrier function of the leak pathway covers more the regulatory part, but not a direct sealing function for small solutes, since occludin knockout mice did not exhibit intestinal barrier impairment for ions as indicated by impedance spectroscopy (Schulzke et al. 2005). Furthermore, occludin downregulation has been proposed to trigger tricellulin redistribution off the tricellular into the bicellular TJ, as a result of which the tricellular TJ could become more permeable for macromolecules, while this does not significantly influence ionic permeability. However, the direct cleavage of occludin by the secreted serine proteinase HtrA from *C. jejuni* was shown to be a major virulence mechanism for the bacterial transmigration between the host cells (Boehm et al. 2018; Harrer et al. 2019, see also Chap. 7 of this book).

In general, pathogens can induce an opening of the leak pathway by altering claudin gene expression, posttranslational and posttranscriptional regulation as well as claudin redistribution, altogether are considered to contribute to leak flux diarrhea. Responsible for the opened leak pathway in *C. jejuni* infection are several claudins. In the *C. jejuni*-infected human colon mucosa, mRNA and protein downregulation for barrier-forming claudin-3, -4, -5 and -8 were shown (Bücker et al. 2018). Moreover, a subcellular redistribution of claudin-1, -5 and -8 off the TJ to intracellular compartments was observed. Altogether, it has to be assumed that this leads to a disruption of the TJ allowing even larger molecules to pass (i.e., opening of the leak pathway) (Bücker et al. 2018). Consequently, a strong reduction of the sealing TJ proteins should be relevant for the leak pathway. At this point, we have to admit that a clear experimental assignment of a distinct claudin alteration to either the pore or the leak pathway or both is not possible. Nevertheless, it may be reasonable to assume that a low-grade reduction of sealing claudins (e.g., claudin-8) may rather contribute to the pore pathway only, whereas a high-grade reduction of several TJ proteins as here in campylobacteriosis may affect also the leak pathway. Moreover, also the epithelial damage induction in campylobacteriosis contributes to the barrier dysfunction (see unrestricted passage below).

Unrestricted Passage

When the selectivity of the TJ is no longer given by an opened intracellular space of detached cells or by single-cell loss after epithelial cell death (apoptosis, necrosis, necroptosis, etc.), water, solutes and large molecules can flow out of the tissue into the lumen via the unrestricted passage. The efflux from the serosal to the mucosal side by epithelial damage substantially contributes to diarrhea, to watery diarrhea but especially to bloody diarrhea. Cytotoxicity of bacteria is the primary cause of epithelial damage. A number of bacterial toxins act on host cell death mechanisms. For *C. jejuni,* the CDT is well-studied (reviewed in Lai et al. 2016), but also *C. jejuni* strains lacking CDT exhibit cytotoxic effects (reviewed in Wassenaar, 1997; Purdy et al. 2000).

In response to the *C. jejuni* infection, several cytokines were secreted. Some cytokines (i.e., TNF-α) have effects on epithelial cell death. Thus, a direct induction of epithelial apoptosis by *C. jejuni* and/or its virulence factors accompanied by an indirect induction of apoptosis by cytokines have to be assumed. Another possibility is the direct cytotoxic action of immune cells (i.e., cytotoxic T cells) inducing epithelial cell death, as proposed for norovirus infection (Troeger et al. 2009). It is known that apoptotic defects induce epithelial single-cell lesions, which are rapidly closed by an actomyosin constriction: the "purse-string" mechanism (Florian et al.

2002; Günzel et al. 2006). As long as no toxins inhibit the restitution of single-cell lesions, like shown for aerolysin or *E. coli* α-hemolysin, the purse-string mechanisms prevents from the flux of water and solutes (Bücker et al. 2011; Günzel et al. 2006). Whether or not *C. jejuni* interferes with the restitutional purse-string mechanism is not yet clear and should be investigated in future.

The opening of the paracellular space by the described cellular mechanism for an unrestricted passage opens this route in both directions; therefore, large macromolecules can overcome the epithelial barrier from the mucosal to the serosal side. The epithelial damage opens the paracellular route for the entry of antigens, proteins, toxins, lipopolysaccharides and even whole bacteria. Due to its selectivity, the pore pathway cannot serve as pathway for antigens. However, the leak and the unrestricted pathway substantially contribute to antigen influx and trigger the *leaky gut* phenomenon.

Transcytotic Pathway

Another important mechanism of bacterial invasion and antigen uptake is epithelial transcytosis. Different studies demonstrated that *C. jejuni* transmigrates within the first hours of infection (Blaser and Reller 1981; Everest et al. 1992; Konkel et al. 1992; Harvey et al. 1999; Hu et al. 2008; Watson and Galán 2008), indicating a translocation of living bacteria through the epithelial cell layer without affecting the integrity of the intestinal barrier (Monteville and Konkel 2002; Hu et al. 2008; Boehm et al. 2012). But *C. jejuni* is also able to potentiate transcytosis of antigens and non-invasive commensal bacteria via M cells (Kalischuk et al. 2010). The regulation of transcytosis is cell type specific and depends also on the bacterial adhesion and motility abilities (Grant et al. 1993; Hatayama et al. 2018; Simson et al. 2020). For more details, see Chap. 7 of this book.

Leaky Gut concept

The triad of epithelial barrier dysfunction, the passage of noxious agents into the mucosa and the overreaction of the immune response in the subepithelium are delineated by the leaky gut concept. The basis for the leaky gut hypothesis was introduced by Fine and co-workers in the 1950s in a series of experiments in models of hemorrhagic shock and other animal shock models, where toxins coming from the intestinal lumen into the organism aggravated the outcome of shock (Jacob et al. 1954; Schweinburg et al. 1954). Apart from the experimental hypothesis, the leaky gut concept was found to be present in several intestinal diseases and described as increased uptake of macromolecules and a loss of the barrier against bacterial compounds or even viable bacteria (translocation) (Fink 1990). The term "leaky gut" was first used for an increased intestinal permeability to toxic luminal compounds in alcoholic patients (Bjarnason et al. 1984). The leaky gut concept does not describe the outflow of water and solutes by leak flux or unrestricted efflux, but rather the influx of antigens and noxious agents or even whole bacteria into the organism. The passage of antigens from the lumen into the organism can occur via the disrupted TJ (the leak pathway) and/or via focal epithelial cell leaks (the unrestricted pathway). The involvement of focal leaks in the pathogenesis of hemolytic *E. coli* was shown to be a pivotal epithelial pathway for the immune activation and initiation of the intestinal inflammation and an experimental demonstration of the leaky gut concept (Bücker et al. 2014). In turn, the mucosal immune response also influences mucosal integrity, so that the epithelial leakage is manifested and aggravated, which leads again to influx of antigens, a self-reinforcing downward spiral (Fig. 3).

2.2 *Immune Cell Response and Barrier Function*

In intestinal inflammatory conditions as present in several gastrointestinal diseases including the Campylobacteriosis, mucosal cytokines influence epithelial barrier function. *Campylobacter* itself as well as its virulence factors attack the epithelial layer and in parallel induce an immune response in the subepithelium. The resulting direct barrier defects are accompanied by an increased secretion of cytokines, chemokines and other inflammatory and neurogastrointestinal mediators like serotonin (Spiller et al. 2000), PGE_2 (Schmitz et al. 1996) or lactate dehydrogenase (LDH) (Beltinger et al. 2008), having in turn an effect on the intestinal physiology.

In *C. jejuni*-infected patients, the colon mucosa increasingly released pro-inflammatory cytokines such as TNF-α, INF-γ, IL-13 and IL-1β (Bücker et al. 2018). Various in vitro and in vivo studies have proven that TNF-α, IFN-γ, IL-1β, IL-6 and IL-13 disturb the intestinal epithelial barrier function by different pathways and are therefore barrier relevant (Schmitz et al. 1999b; Adams et al. 1993; Barmeyer et al. 2004; Wang et al. 2007; Heller et al. 2005; Hu and Hickey 2005). Cytokines regulate also the expression and localization of TJ proteins. The *C. jejuni*-induced mucosal immune response is therefore substantial for the barrier disturbance and contributes to a leak flux type of diarrhea.

The pro-inflammatory cytokine TNF-α is of central importance in a wide range of infectious inflammatory diseases. Different intestinal epithelial cells, when challenged with TNF-α, showed impaired intestinal barrier function, by reduction of the transepithelial electrical resistance (TER) and increased permeabilities for ions and macromolecules (Schmitz et al. 1999b; Gitter et al. 2000a; Mankertz et al. 2000; Amasheh et al. 2010; Gitter et al. 2000b; Luettig et al. 2016; Hatayama et al. 2018). Moreover, TNF-α induced a loss of K^+ and Cl^- by a subepithelial PGE_2-mediated pathway (Schmitz et al. 1996) and led to epithelial leaks by single-cell apoptosis (Gitter et al. 2000b) contributing to leak flux diarrhea (Schulzke et al. 2006).

INF-γ is another main player in intestinal inflammation and influences barrier function as measured by decreased TER in vitro (Madara and Stafford 1989; Adams et al. 1993; Mankertz et al. 2000), which was accompanied by increased Na^+ and mannitol permeability (Madara and Strafford 1989). IFN-γ diminished active anion secretion (Colgan et al. 1994; Besançon et al. 1994) as well as downregulated ZO-1 and occludin (Sugi et al. 2001), redistributed TJ proteins off the TJ domain (Bruewer et al. 2003) and cleaved claudin-2, likewise shown for TNF-α (Mankertz et al. 2009). Furthermore, a proteolytic cleavage of occludin was characterized before and was assumed to be a result of epithelial apoptosis (Bojarski et al. 2004). Interestingly, epithelial apoptosis was more pronounced upon TNF-α as opposed to IFN-γ exposure (Schulzke et al. 2006). However, INF-γ together with *C. jejuni* synergistically enhanced the intestinal epithelial loss and disrupted occludin (Rees et al. 2008). For further details on cytokine regulation, see Chap. 9 of this book.

C. jejuni infection leads also to higher secretion of the chemokine IL-8, an initial event in acute inflammation triggering the inflammatory response (Hickey et al. 1999; Borrmann et al. 2007; Beltinger et al. 2008; Zheng et al. 2008). IL-8 secretion is associated either with adhesion and invasion abilities of viable bacteria (Hickey et al. 1999), with the flagellum (Zheng et al. 2008), or with CDT (Hickey et al. 2000; Zheng et al. 2008). *C. jejuni* infection activates ERK, as well as p38 MAP kinase, and NF-κB via IκBα degradation and DNA binding, evoking IL-8 secretion (Mellits et al. 2002; Watson and Galán 2008). IL-1β and IL-6 affect the intestinal barrier function in different gastrointestinal disorders (Barmeyer et al. 2004; Wang et al. 2007). Also, IL-13 impaired the epithelial barrier function (Heller et al. 2005; Schulzke et al. 2009), whereas IL-13 increased the apoptotic ratio and induced claudin-2 expression (Heller et al. 2005). More recent studies from our group could delineate that the immune response during *Campylobacter* infection has an important impact on barrier function (Lobo de Sá et al. 2019; Butkevych et al. 2020).

3 Barrier Dysfunction in *Campylobacter* Infection

The epithelial barrier function can be electrophysiologically characterized by TER measurements. This method was developed by Hans Ussing (Ussing 1949). The Ussing chamber is also suitable for measurements of the I_{SC} (active ion transport), and it serves also for flux measurements with small macromolecules. Another method for TER measurements is the transwell system with chopstick electrodes and an Ohm voltmeter, which was mainly used in *Campylobacter* research. In Ussing chambers, more sophisticated resistance measurements of mucosal tissue are possible. The impedance measurements (one-path impedance spectroscopy) can distinguish between epithelial (R^{epi}) and subepithelial (R^{sub}) resistance contributions to the total TER (Gitter et al. 1998). The two-path impedance spectroscopy can distinguish para- and transcellular resistances (Krug et al. 2009b).

3.1 Barrier Defects by C. jejuni, C. coli, C. fetus, C. concisus *and Related Bacteria*

As the most common cause of bacterial gastroenteritis, *C. jejuni* is the main species for campylobacteriosis. Also *C. coli*, *C. concisus* and *C. fetus* are causing human gastroenteritis. Other non-jejuni campylobacters like *C. lari*, *C. rectus*, *C. sputorum*, *C. hyointestinalis*, *C. upsaliensis* and further species are associated to enteritis, with uncharacterized pathogenicity. The barrier defects caused by different *Campylobacter* species share similarities, but every species and even different strains develop own pathogenic outcomes with differences in the severity of barrier disturbance, host specificity and intensity of immune activation. For more information on host specificity, see Chap. 4 of this book.

C. jejuni was shown to have effects on barrier function in the human intestinal tissue, which occur in the following sequence; (i) invasion (as particle barrier permeation), (ii) claudin-2 upregulation (pore pathway as part of the leak flux diarrhea), (iii) claudin-4, -5, -8 downregulation with TJ barrier disruption (opened leak pathway for antigen influx and leak flux diarrhea), (iv) epithelial apoptosis induction (opening of unrestricted pathway) and (v) cytokine release (accelerating the leaky gut phenomenon). The released cytokines in turn target on claudin-2 upregulation, claudin-5 and -8 downregulation and apoptosis induction.

The functional assessment of the barrier function in *Campylobacter* infection initiated with work using the Caco-2 cell model showing unchanged TER in the first six hours of infection (Konkel et al. 1992). First slight reductions in TER could be observed 8 h post-infection with different *C. jejuni* strains in vitro, although *C. jejuni* translocates (Brás and Ketley 1999). This time-dependently TER drop in vitro was accompanied by occludin rearrangements days after *C. jejuni* infection (Brás and Ketley 1999; MacCallum et al. 2005; Chen et al. 2006). For the *C. jejuni* serine proteinase HtrA, the cleavage of occludin and the adherens junction protein E-cadherin was shown together with increased cell invasion (Hoy et al. 2012; Boehm et al. 2012; Harrer et al. 2019). Thus, together with the junction protein cleavage, the secreted HtrA represents an important virulence factor for the paracellular transmigration of *C. jejuni* through the epithelium (Backert et al. 2018). First results on changes in barrier-forming TJ proteins after *C. jejuni* infection were obtained with canine intestinal epithelial cells. Claudin-4 distribution was observed, and the permeability for 3 kDa dextran was increased after *C. jejuni* infection (Lamb-Rosteski et al. 2008). Animal models with clinical similarity to human campylobacteriosis have been developed (Murphy et al. 2011; Otto et al. 2012; Haag et al. 2012). But only recently, a functional measurement of the epithelial barrier dysfunction in the colon of *C. jejuni*-infected IL-$10^{-/-}$ mice could be shown by an increased fluorescein flux (Lobo de Sá et al. 2019). The in vivo intestinal permeability test with the measurements of mannitol/lactulose ratio in *C. jejuni* patients revealed an increased permeability for these small macromolecules (Spiller et al. 2000). Early barrier defects of *C. jejuni* infection in vitro were demonstrated to be predominantly based on apoptosis induction in the first 22 h after infection (Butkevych et al. 2020). Thus, the evidence for *C. jejuni*-induced barrier defects is strong but the connection of cell death mechanisms and TJ changes to the barrier dysfunction has only insufficiently been characterized so far.

In human colon biopsies from *C. jejuni*-infected patients, TJ expression was changed with an upregulation of claudin-1, a trend to an increase of claudin-2 as well as a downregulation of claudin-3, -4, -5 and -8 (Bücker et al. 2018). Moreover, claudin-1, -5 and -8 were redistributed off the TJ as shown by confocal microscopy.

Furthermore, the induction of epithelial apoptosis and focal leak formation was functionally characterized by one-path impedance spectroscopy, revealing a decrease in epithelial resistance and an increase in fluorescein permeability (Bücker et al. 2018). The flux of 4 kDa dextran was not significantly different, pointing to a leak pathway type in the *C. jejuni*-induced barrier defect. In human mucosal specimens, RNA-seq analysis revealed mRNA expression changes of epithelial transporters and channels as ENaC, as well as mRNA of the abovementioned TJ proteins, confirming that the barrier dysfunction is caused by TJ expression gene regulation (Bücker et al. 2018). The included pathway analysis of the RNA-seq by Ingenuity Pathway Analysis confirmed affected signaling pathways (e.g., IFN-, TNF-pathways), and even more important, the analysis predicted potential protective substances against campylobacteriosis, which are in part addressed in Sect. 3.2 below. Furthermore, the immune system involvement in barrier disruption was experimentally confirmed in an adapted immune cell–epithelial cell co-culture system. Here, the basal presence of *C. jejuni* and the cytokine release (TNF-α, IL-1β IL-6) from mucosal macrophages show substantial impact on barrier dysfunction, without any direct bacterial contact to the epithelium, highlighting the immune-mediated part of the leak flux mechanism (Lobo de Sá et al. 2019).

The closest relative to *C. jejuni* is the zoonotic pathogen *C. coli*, often isolated from human diarrheic stool samples. Experimentally infected human HT-29/B6 epithelial cell monolayers showed an impaired barrier function 48 h after infection indicated by a decreased TER due to a reduction in claudin-3 and claudin-4 protein expression (Bücker et al. 2017). Like *C. jejuni* or *C. fetus*, the *C. coli* infection in HT-29/B6 cells caused an upregulation of claudin-1, the so-called claudin-1 paradox (Butkevych et al. 2020; Bücker et al. 2018, 2017). Claudin-1 paradox means that the upregulation of the expression of barrier-forming claudin-1 does not increase the epithelial resistance, since it is intracellularly located and not inserted into the TJ. Since *C. coli* predominantly colonizes swine, with occasionally diarrheal outcome in piglets, *C. coli* infection was tracked throughout the organism in a pig colonization trial by oral inoculation. Here, the bacteria translocated through the intestinal mucosa into lymphoid organs, without any obvious pathologies in the intestinal epithelium (Bratz et al. 2013). However, barrier-breaking abilities of *C. coli* were shown for human T84 cells as well as porcine IPEC-1 epithelial cells. In both cell lines, the infection with *C. coli* resulted in a decreased TER 48 h post-infection, accompanied by a redistribution of occludin and an increase in epithelial cell death (Murphy et al. 2011). Moreover, release of IL-8 was measured from infected cells as well as the increase of the inflammatory subunit of p65 of transcription factor NF-κB (Murphy et al. 2011).

As the *Campylobacter* species with the most severe barrier-breaking capabilities, *C. fetus* is known to overcome several barriers, from the intestinal to the placental barrier. *C. fetus* is mostly isolated from the intestines of humans, sheep and cattle as well as other animals as swine, poultry or reptiles (Harvey and Greenwood 1985). In animals and also in humans, the *C. fetus* infection can result in gastroenteritis, septicemia and abortion (Steinkraus and Wright 1994). First functional experiments showed that *C. fetus* did neither reduce TER nor enhanced passage of FITC-inulin in the first hours after infection in vitro (Baker and Graham 2010). The known *C. fetus* virulence factors are comparable to that of *C. jejuni*, but the barrier defect induction occurs earlier (Bücker et al. 2017). *C. fetus* infected HT-29/B6 monolayers showed a reduced transcellular and paracellular resistance in two-path impedance measurements (Krug et al. 2009b). The results of the barrier analysis point more to an opened unrestricted pathway, by membrane damage and focal leak induction as well as to an activation of anion secretion already 24 h post-infection (Bücker et al. 2017). In contrast to *C. jejuni*, *C. fetus* did not reduce the expression of barrier-forming claudins. Only claudin-4 cleavage and upregulation of claudin-1 were observed (Bücker et al. 2017, 2018; Butkevych et al. 2020). Due to the association with apoptotic cell death in the epithelium, the claudin-1 paradox could be assigned to apoptotic events. In addition to cell death induction and focal leak formation, a high ratio of bacterial translocation was observed, indicating the leaky gut phenomenon for the *C. fetus* infection (Bücker et al. 2017).

C. concisus, the only non-zoonotic *Campylobacter* species, is mostly found in the human oral cavity or intestine. The bacteria cause diarrhea and enteritis and display a specific risk for immunocompromised persons (Aabenhus et al. 2002; Nielsen et al. 2012). The prevalence of *C. concisus* is estimated to be as common as that of *C. jejuni*, but the severity of infection differs (Nielsen et al. 2012). A clinical observation study demonstrated that *C. concisus* provokes a milder course of gastroenteritis but with a longer period of diarrhea in comparison with *C. jejuni* (Nielsen et al. 2012). The pathogenic potential of *C. concisus* together with disturbed epithelial barrier integrity was shown in vitro in different cell lines by a decrease of ZO-1 and occludin expression accompanied by higher IL-8 secretion (Man et al. 2010). Barrier-breaking properties of different clinical *C. concisus* isolates from oral or fecal source were shown to decrease TER in the same time course (Nielsen et al. 2011; Nattramilarasu et al. 2020), which was comparable to that of *C. jejuni* or *C. coli* (Bücker et al. 2017). Moreover, the *C. concisus* infection resulted in increased fluorescein permeability (332 Da), a fivefold increase of apoptotic events and a reduced expression of claudin-5 indicating a leak flux mechanism (Nielsen et al. 2011). A recent study demonstrated that *C. concisus* reduced the amiloride-sensitive short-circuit current, indicating an inhibition of the ENaC-dependent Na^+ transport as an additional malabsorptive diarrheal mechanism (Nattramilarasu et al. 2020). The ENaC dysregulation was mediated via ERK activation and mRNA downregulation of the β- and γ-ENaC subunit (Nattramilarasu et al. 2020). Furthermore, *C. concisus* reduced the claudin-8 expression and induced its redistribution (Nattramilarasu et al. 2020). Interestingly, in this experimental model only claudin-8 was affected, which has relevance in the paracellular sealing function toward Na^+ (Amasheh et al. 2009). As the paracellular back leakage of Na^+ into the lumen is accelerated by claudin-8 dysregulation, the ENaC dysfunction together with claudin-8 disruption could potentiate the loss of Na^+ ions into the lumen (Nattramilarasu et al. 2020).

Whereas *Campylobacter* spp. with clinical relevance (*C. jejuni*, *C. coli*, *C. fetus* and *C. concisus*) present with watery and bloody diarrhea (Lindblom et al. 1995; Aabenhus et al. 2002; Graham 2002; Inglis et al. 2011; Nielsen et al. 2012) further *Campylobacter* spp. such as *C. showae, C. hominis, C. sputorum* and *C. insulaenigrae* are associated with watery diarrhea (Inglis et al. 2011; Lindblom et al.

1995; Chua et al. 2007) and *C. lari*, *C. upsaliensis* with watery and bloody diarrhea (Broczyk et al. 1987; Patton et al. 1989), but the information on the pathomechanisms of these species is sparse. It is conceivable that diarrhea-inducing *Campylobacter* spp. might have barrier-weakening properties. Descriptions on bacteremia cases or extra-intestinal manifestations can lead to the assumption that these species translocate and may have barrier-breaking properties (Patton et al. 1989; Lindblom et al. 1995; Chiu et al. 1995; Macuch and Tanner, 2000). For *C. showae,* epithelial adhesion and for *C. upsaliensis* and *C. rectus* epithelial invasion was experimentally shown (Man et al 2010; Mooney et al. 2003; Arce et al. 2010). Cytotoxicity or apoptosis induction was reported for *C. lari* or *C. upsaliensis*, respectively (Johnson and Lior, 1986; Trott et al. 2001). For *C. rectus*, a drop in TER comparable to *C. jejuni* was shown in HT-29/B6 cells (Bücker et al. 2017). For the species *C. hyointestinalis, C. helveticus, C. mucosalis, C. hyoilei, C. nitrofigilis* and *C. lanienae,* no information on barrier dysfunction is available.

A related bacterium which causes barrier defects and induces mainly watery diarrhea is *Arcobacter butzleri*. HT-29/B6 cells infected with *A. butzleri* showed threefold higher apoptotic events as well as a time- and dose-dependent drop in TER and an increase of permeability for different tracer molecules (Bücker et al. 2009). Moreover, claudin-1, -5 and -8 were redistributed off the TJ, while claudin-2, -3 and -4 were unchanged (Bücker et al. 2009). These barrier compromising events represent a leak flux mechanism fitting to the watery type of diarrhea (Bücker et al. 2009).

3.2 Protective Approaches

Various barrier-protective and/or anti-inflammatory compounds could be used in the treatment of *Campylobacter* infections. Beside oral rehydration solution (ORS), antibiotics are used just in severe cases. A WHO recommendation is to supplement ORS with the essential micronutrient zinc in the treatment of diarrhea. Several studies demonstrated the beneficial effects of zinc in infectious diarrhea, inflammation and intestinal epithelial wound healing. The epithelial cell restitution by zinc was demonstrated in vitro (Cario et al. 2000). Zinc inhibits also the expression of pro-inflammatory cytokines in plasma samples of volunteers and in a cell model (Prasad et al. 2004). Zinc oxide nanoparticles acted even bactericidal against *C. jejuni* (Xie et al. 2011). Zinc treatment improved the intestinal barrier in other bacterial-induced epithelial dysfunctions, such as α-hemolysin-positive *E. coli* (Wiegand et al. 2017; Bücker et al. 2020).

Phytochemical interventions with secondary plant compounds could also reduce the *Campylobacter*-induced pathologies by strengthening the epithelial barrier function. Some therapeutic approaches have been successfully applied against *Campylobacter* infections or could be promising, because of its barrier-protective properties. Such a barrier improving compound is the flavonoid quercetin from onions, kale and apples (Amasheh et al. 2008). In Caco-2 cells, quercetin upregulated the claudin-4 expression (Amasheh et al. 2008) leading to a decrease in paracellular electrical conductance (Van Itallie et al. 2001). Also the growth of different *C. jejuni* strains was inhibited in the presence of quercetin (Campana et al. 2009).

In traditional Chinese medicine, berberine is used for the treatment of infectious diarrhea, gastroenteritis and abdominal pain (Zhou and Mineshita 2000). Berberine showed anti-inflammatory properties in TNF-α- and IFN-γ-induced barrier dysfunction in vitro (Amasheh et al. 2010) and reduced the IL-8 production in vivo (Zhou and Mineshita 2000). Berberine showed also antimicrobial effects against *E. coli* and in high concentrations against *B. subtilis* (Kong et al. 2012). No data about antimicrobial activity of berberine against *Campylobacter* are available so far.

The polyphenol resveratrol is the main compound in the skin of red grapes and thus in red wine. In vivo it was shown, that resveratrol obviates bacterial translocation of *Toxoplasma gondii* into the systemic circulation, diminished inflammation (Bereswill et al. 2010), preserved barrier function in a DSS mouse model (Mayangsari and Suzuki 2018) and reduced effects of oxidative stress (Cao et al. 2019). Resveratrol also modulated and inhibited the growth and invasion capability of *Campylobacter* into different intestinal epithelial cell lines (Klančnik et al. 2017).

Vitamin D (VD) is a liposoluble hormone, which plays an essential role in bone metabolism, in immune regulation and also in gastrointestinal functions as well as the gastrointestinal microbiome (Ooi et al. 2013). 1,25-dihydroxyvitamin D3 (1,25(OH)2D3; calcitriol) is the active form of VD. The biological activity is controlled through the VD receptor (VDR) (Barbalho et al. 2016). The protective influence of VD on the intestinal epithelial barrier was demonstrated in VD-deficient IL-$10^{-/-}$ mice in experimental IBD (Zhu et al. 2005). $VDR^{-/-}$ mice were more susceptible to DSS colitis and bacteria translocated from the gut into the systemic circulation (Froicu and Cantorna 2007). In contrary, VD inhibits DSS colitis development (Ooi et al. 2013). An overexpression of VDR leads to an improved integrity of the epithelial barrier in vivo (Zhu et al 2005). VD-treated Caco-2 cells were protected against ethanol-induced barrier impairment but also against TNF-α-induced barrier disturbance (Chen et al. 2015a; 2015b). The *C. jejuni*-induced cytotoxic effects on HT-29/B6 cells were diminished by VD in a dose-dependent manner and partially recovered the decreased TER (Bücker et al. 2018). In vivo the translocation of *C. jejuni* to extraintestinal compartments as well as the intestinal and extraintestinal inflammation was reduced in VD-treated mice (Mousavi et al. 2019).

Another VDR ligand, the polyphenol curcumin from the turmeric root of the plant *Curcuma longa* has been used for centuries in traditional medicine against gastrointestinal and digestive disorders. Curcumin reduced the motility and inhibited the growth of different bacteria (Mahady et al. 2002; Gunes et al. 2016). In concentrations above 87 μM, curcumin was effective against *C. jejuni* growth in vitro (Lobo de Sá et al. 2019). Anti-inflammatory properties of curcumin were also demonstrated in LPS-induced barrier dysfunction with a reduced IL-1β secretion (Wang et al. 2017). Barrier-protective abilities were shown in vitro in Caco-2 cells, with an increased gene expression of claudin-4 (Watari et al. 2012). Curcumin protected *C. jejuni*-infected co-cultures from the disruption of the epithelial barrier, indicated by an ameliorated TER and permeability for fluorescein with similar results in an in vivo model (Lobo de Sá et al. 2019). Furthermore, the in vitro studies enlightened the anti-apoptotic and anti-inflammatory properties of curcumin, as well as its inhibition of the redistribution of claudin-4 and claudin-8 (Lobo de Sá et al. 2019). The induction of apoptosis is one of the main pathomechanisms beginning in the early phase of the *C. jejuni* infection, also responsible for leak flux diarrhea (Lobo de Sá et al. 2019; Butkevych et al. 2020), which could be inhibited by curcumin treatment (Lobo de Sá et al. 2019). Hence, compounds with anti-apoptotic properties could find their deployments in the treatment of *Campylobacter* infections (Butkevych et al. 2020). Taken together, added to the feed of livestock, the above presented substances may represent an easy and effective strategy to prevent the colonization of *C. jejuni* for the future. Consequently, a reduced bacterial load would lead to a reduced incidence of campylobacteriosis in humans.

4 Concluding Remarks

C. jejuni adhere, invade and transmigrate through the intestinal epithelial layer and thus provoke direct barrier disturbance. The bacteria arrive in the lamina propria together with antigens from the lumen and reach underlying immune cells, inducing an immune response with boosted secretion of inflammatory and barrier-relevant mediators, such as TNF-α, INF-γ, IL-1β and IL-13. This mucosal cytokine storm in turn affects the barrier indirectly (leaky gut concept). The understanding of the immune response in *Campylobacter* infections is well advanced as more detailed data related to cytokines in association with barrier function are now available. Although not all mechanisms are clarified, *Campylobacter* infections affect the TJ disruption-dependent leak pathway as well as the unrestricted pathway by barrier-relevant cytotoxicity. Thus, *Campylobacter* induces epithelial apoptosis, epithelial leaks, expression changes of TJ proteins and also subcellular redistributions of TJ proteins off the TJ domain. Moreover, it was proven that *Campylobacter* itself and the induced immune response disturb the transport of ions and molecules. The *Campylobacter*-induced barrier dysfunction leads to the development of a leak flux type of diarrhea. Moreover, a sodium malabsorption was identified as another diarrheal mechanism in the colon. Further elucidation of the pathomechanisms behind the *Campylobacter* infection will facilitate the prevention of and combat against *Campylobacter* infections, in order to identify targets for potential protective and therapeutic compounds. Consequently, *Campylobacter* colonization in animals and hence infections in humans can be reduced with anti-inflammatory, anti-apoptotic or barrier-protective substances. Promising candidates to fight *Campylobacter* infections in human and veterinary medicine are micronutrients such as zinc and vitamins, but also phytopharmaceuticals such as quercetin, berberine, resveratrol, curcumin or combinations of these compounds, which should be considered as therapeutic regimens.

Acknowledgements This work is supported by the German Federal Ministry of Education and Research (BMBF) in frame of the zoonoses research consortium PAC-*Campylobacter* to R.B. and J.D.S. (project IP8 / 01KI1725D).

References

Aabenhus R, Permin H, On SL, Andersen LP (2002) Prevalence of *Campylobacter concisus* in diarrhoea of immunocompromised patients. Scand J Infect Dis 34:248–252. https://doi.org/10.1080/00365540110080566

Adams RB, Planchon SM, Roche JK (1993) IFN-gamma modulation of epithelial barrier function. Time course, reversibility, and site of cytokine binding. J Immunol 150:2356–2363

Amasheh M, Fromm A, Krug SM, Amasheh S, Andres S, Zeitz M, Fromm M, Schulzke JD (2010) TNFalpha-induced and berberine-antagonized tight junction barrier impairment via tyrosine kinase, Akt and NFkappaB signaling. J Cell Sci 123:4145–4155. https://doi.org/10.1242/jcs.070896

Amasheh M, Schlichter S, Amasheh S, Mankertz J, Zeitz M, Fromm M, Schulzke JD (2008) Quercetin enhances epithelial barrier function and increases claudin-4 expression in Caco-2 cells. J Nutr 138:1067–1073. https://doi.org/10.1093/jn/138.6.1067

Amasheh S, Meiri N, Gitter AH, Schöneberg T, Mankertz J, Schulzke JD, Fromm M (2002) Claudin-2 expression induces cation-selective channels in tight junctions of epithelial cells. J Cell Sci 115:4969–4976. https://doi.org/10.1242/jcs.00165

Amasheh S, Milatz S, Krug SM, Bergs M, Amasheh M, Schulzke JD, Fromm M (2009) Na+ absorption defends from paracellular back-leakage by claudin-8 upregulation. Biochem Biophys Res Commun 378:45–50. https://doi.org/10.1016/j.bbrc.2008.10.164

Arce RM, Diaz PI, Barros SP, Galloway P, Bobetsis Y, Threadgill D, Offenbacher S (2010) Characterization of the invasive and inflammatory traits of oral *Campylobacter rectus* in a murine model of fetoplacental growth restriction and in trophoblast cultures. J Reprod Immunol 84:145–153. https://doi.org/10.1016/j.jri.2009.11.003

Backert S, Boehm M, Wessler S, Tegtmeyer N (2013) Transmigration route of *Campylobacter jejuni* across polarized intestinal epithelial cells: paracellular, transcellular or both? Cell Commun Signal 11:72. https://doi.org/10.1186/1478-811X-11-72

Backert S, Bernegger S, Skorko-Glonek J, Wessler S (2018) Extracellular HtrA serine proteases: an emerging new strategy in bacterial pathogenesis. Cell Microbiol 20:e12845. https://doi.org/10.1111/cmi.12845

Baker NT, Graham LL (2010) *Campylobacter fetus* translocation across Caco-2 cell monolayers. Microb Pathog 49:260–272. https://doi.org/10.1016/j.micpath.2010.06.008

Barbalho SM, Bechara MD, de Alvares GR, Quesada K, Gasparini RG, de Cássio Alves de Carvalho A, Fiorini AM, (2016) Reflections About inflammatory bowel disease and vitamins A and D. J Med Food 19:1105–1110. https://doi.org/10.1089/jmf.2016.0101

Barmeyer C, Amasheh S, Tavalali S, Mankertz J, Zeitz M, Fromm M, Schulzke JD (2004) IL-1beta and TNFalpha regulate sodium absorption in rat distal colon. Biochem Biophys Res Commun. 317:500–507. https://doi.org/10.1016/j.bbrc.2004.03.072

Barmeyer C, Erko I, Awad K, Fromm A, Bojarski C, Meissner S, Loddenkemper C, Kerick M, Siegmund B, Fromm M, Schweiger MR, Schulzke JD (2017) Epithelial barrier dysfunction in lymphocytic colitis through cytokine-dependent internalization of claudin-5 and -8. J Gastroenterol 52:1090–1100. https://doi.org/10.1007/s00535-017-1309-2

Beltinger J, del Buono J, Skelly MM, Thornley J, Spiller RC, Stack WA, Hawkey CJ (2008) Disruption of colonic barrier function and induction of mediator release by strains of *Campylobacter jejuni* that invade epithelial cells. World J Gastroenterol 14:7345–7352. https://doi.org/10.3748/wjg.14.7345

Bereswill S, Muñoz M, Fischer A, Plickert R, Haag LM, Otto B, Kühl AA, Loddenkemper C, Göbel UB, Heimesaat MM (2010) Anti-inflammatory effects of resveratrol, curcumin and simvastatin in acute small intestinal inflammation. PLoS ONE 5:e15099. https://doi.org/10.1371/journal.pone.0015099

Besançon F, Przewlocki G, Baró I, Hongre AS, Escande D, Edelman A (1994) Interferon-gamma downregulates CFTR gene expression in epithelial cells. Am J Physiol 267:C1398–C1404. https://doi.org/10.1152/ajpcell.1994.267.5.C1398

Bjarnason I, Peters TJ, Wise RJ (1984) The leaky gut of alcoholism: possible route of entry for toxic compounds. Lancet 1:179–182. https://doi.org/10.1016/s0140-6736(84)92109-3

Black RE, Levine MM, Clements ML, Hughes TP, Blaser MJ (1988) Experimental *Campylobacter jejuni* infection in humans. J Infect Dis 157:472–479. https://doi.org/10.1093/infdis/157.3.472

Blaser MJ, Duncan DJ, Warren GH, Wang WL (1983) Experimental *Campylobacter jejuni* infection of adult mice. Infect Immun 39:908–916. https://doi.org/10.1128/IAI.39.2.908-916.1983

Blaser MJ, Parsons RB, Wang WL (1980) Acute colitis caused by *Campylobacter fetus* ss. jejuni. Gastroenterology 78:448–453

Blaser MJ, Reller LB (1981) *Campylobacter* enteritis. N Engl J Med 305:1444–1452. https://doi.org/10.1056/NEJM198112103052404

Boehm M, Hoy B, Rohde M, Tegtmeyer N, Bæk KT, Oyarzabal OA, Brøndsted L, Wessler S, Backert S (2012) Rapid paracellular transmigration of *Campylobacter jejuni* across polarized epithelial cells without affecting TER: role of proteolytic-active HtrA cleaving E-cadherin but not fibronectin. Gut Pathog 4:3. https://doi.org/10.1186/1757-4749-4-3

Boehm M, Simson D, Escher U, Schmidt AM, Bereswill S, Tegtmeyer N, Backert S, Heimesaat MM (2018) Function of serine protease HtrA in the lifecycle of the foodborne pathogen *Campylobacter jejuni*. Eur J Microbiol Immunol (Bp) 8:70–77. https://doi.org/10.1556/1886.2018.00011

Bojarski C, Weiske J, Schöneberg T, Schröder W, Mankertz J, Schulzke JD, Florian P, Fromm M, Tauber R, Huber O (2004) The specific fates of tight junction proteins in apoptotic epithelial cells. J Cell Sci 117:2097–2107. https://doi.org/10.1242/jcs.01071

Borrmann E, Berndt A, Hänel I, Köhler H (2007) *Campylobacter*-induced interleukin-8 responses in human intestinal epithelial cells and primary intestinal chick cells. Vet Microbiol 124:115–124. https://doi.org/10.1016/j.vetmic.2007.04.041

Brás AM, Ketley JM (1999) Transcellular translocation of *Campylobacter jejuni* across polarised epithelial monolayers. FEMS Microbiol Lett 179:209–215. https://doi.org/10.1111/j.1574-6968.1999.tb08729.x

Bratz K, Bücker R, Gölz G, Zakrzewski SS, Janczyk P, Nöckler K, Alter T (2013) Experimental infection of weaned piglets with *Campylobacter coli*—excretion and translocation in a pig colonisation trial. Vet Microbiol 162:136–143. https://doi.org/10.1016/j.vetmic.2012.08.016

Broczyk A, Thompson S, Smith D, Lior H (1987) Water-borne outbreak of *Campylobacter laridis*-associated gastroenteritis. Lancet 1:164–165. https://doi.org/10.1016/s0140-6736(87)92003-4

Bruewer M, Luegering A, Kucharzik T, Parkos CA, Madara JL, Hopkins AM, Nusrat A (2003) Proinflammatory cytokines disrupt epithelial barrier function by apoptosis-independent mechanisms. J Immunol 171:6164–6172. https://doi.org/10.4049/jimmunol.171.11.6164

Bücker R, Krug SM, Fromm A, Nielsen HL, Fromm M, Nielsen H, Schulzke JD (2017) *Campylobacter fetus* impairs barrier function in HT-29/B6 cells through focal tight junction alterations and leaks. Ann N Y Acad Sci. 1405:189–201. https://doi.org/10.1111/nyas.13406

Bücker R, Krug SM, Moos V, Bojarski C, Schweiger MR, Kerick M, Fromm A, Janßen S, Fromm M, Hering NA, Siegmund B, Schneider T, Barmeyer C, Schulzke JD (2018) *Campylobacter jejuni* impairs sodium transport and epithelial barrier function via cytokine release in human colon. Mucosal Immunol 11:474–485. https://doi.org/10.1038/mi.2017.78

Bücker R, Krug SM, Rosenthal R, Günzel D, Fromm A, Zeitz M, Chakraborty T, Fromm M, Epple HJ, Schulzke JD (2011) Aerolysin from *Aeromonas hydrophila* perturbs tight junction integrity and cell lesion repair in intestinal epithelial HT-29/B6 cells. J Infect Dis 204:1283–1292. https://doi.org/10.1093/infdis/jir504

Bücker R, Schulz E, Günzel D, Bojarski C, Lee IF, John LJ, Wiegand S, Janßen T, Wieler LH, Dobrindt U, Beutin L, Ewers C, Fromm M, Siegmund B, Troeger H, Schulzke JD (2014) α-Haemolysin of *Escherichia coli* in IBD: a potentiator of inflammatory activity in the colon. Gut 63:1893–1901. https://doi.org/10.1136/gutjnl-2013-306099

Bücker R, Troeger H, Kleer J, Fromm M, Schulzke JD (2009) *Arcobacter butzleri* induces barrier dysfunction in intestinal HT-29/B6 cells. J Infect Dis 200:756–64

Bücker R, Zakrzewski SS, Wiegand S, Pieper R, Fromm A, Fromm M, Günzel D, Schulzke JD (2020) Zinc prevents intestinal epithelial barrier dysfunction induced by alpha-hemolysin-producing *Escherichia coli* 536 infection in porcine colon. Vet Microbiol 243:108632. https://doi.org/10.1016/j.vetmic.2020.108632

Buschmann MM, Shen L, Rajapakse H, Raleigh DR, Wang Y, Wang Y, Lingaraju A, Zha J, Abbott E, McAuley EM, Breskin LA, Wu L, Anderson K, Turner JR, Weber CR (2013) Occludin OCEL-domain interactions are required for maintenance and regulation of the tight junction barrier to macromolecular flux. Mol Biol Cell 24:3056–3068. https://doi.org/10.1091/mbc.E12-09-0688

Butkevych E, Lobo De Sá FD, Nattramilarasu PK, Bücker R (2020) Epithelial apoptosis and subepithelial immune responses in *Campylobacter jejuni*-induced barrier disruption. Front Microbiol 11:344. https://doi.org/10.3389/fmicb.2020.00344

Campana R, Patrone V, Franzini IT, Diamantini G, Vittoria E, Baffone W (2009) Antimicrobial activity of two propolis samples against human *Campylobacter jejuni*. J Med Food 12:1050–1056. https://doi.org/10.1089/jmf.2008.0173

Cao S, Shen Z, Wang C, Zhang Q, Hong Q, He Y, Hu C (2019) Resveratrol improves intestinal barrier function, alleviates mitochondrial dysfunction and induces mitophagy in diquat challenged piglets. Food Funct 10:344–354. https://doi.org/10.1039/c8fo02091d

Cario E, Jung S, Harder D'Heureuse J, Schulte C, Sturm A, Wiedenmann B, Goebell H, Dignass AU (2000) Effects of exogenous zinc supplementation on intestinal epithelial repair in vitro. Eur J Clin Invest 30:419–428. https://doi.org/10.1046/j.1365-2362.2000.00618.x

Chen ML, Ge Z, Fox JG, Schauer DB (2006) Disruption of tight junctions and induction of proinflammatory cytokine responses in colonic epithelial cells by *Campylobacter jejuni*. Infect Immun 74:6581–6589. https://doi.org/10.1128/IAI.00958-06

Chen S, Zhu J, Chen G, Zuo S, Zhang J, Chen Z, Wang X, Li J, Liu Y, Wang P (2015) 1,25-Dihydroxyvitamin D3 preserves intestinal epithelial barrier function from TNF-α induced injury via suppression of NF-κB p65 mediated MLCK-P-MLC signaling pathway. Biochem Biophys Res Commun 460:873–878. https://doi.org/10.1016/j.bbrc.2015.03.125

Chen SW, Ma YY, Zhu J, Zuo S, Zhang JL, Chen ZY, Chen GW, Wang X, Pan YS, Liu YC, Wang PY (2015) Protective effect of 1,25-dihydroxyvitamin D3 on ethanol-induced intestinal barrier injury both in vitro and in vivo. Toxicol Lett 237:79–88. https://doi.org/10.1016/j.toxlet.2015.06.006

Chiu CH, Kuo CY, Ou JT (1995) Chronic diarrhea and bacteremia caused by *Campylobacter lari* in a neonate. Clin Infect Dis 21:700–701. https://doi.org/10.1093/clinids/21.3.700

Chua K, Gürtler V, Montgomery J, Fraenkel M, Mayall BC, Grayson ML (2007) *Campylobacter insulaenigrae* causing septicaemia and enteritis. J Med Microbiol 56:1565–1567. https://doi.org/10.1099/jmm.0.47366-0

Clayburgh DR, Barrett TA, Tang Y, Meddings JB, Van Eldik LJ, Watterson DM, Clarke LL, Mrsny RJ, Turner JR (2005) Epithelial myosin light chain kinase-dependent barrier dysfunction mediates T cell activation-induced diarrhea in vivo. J Clin Invest 115:2702–2715. https://doi.org/10.1172/JCI24970

Colgan SP, Parkos CA, Matthews JB, D'Andrea L, Awtrey CS, Lichtman AH, Delp-Archer C, Madara JL (1994) Interferon-gamma induces a cell surface phenotype switch on T84 intestinal epithelial cells. Am J Physiol 267:C402–C410. https://doi.org/10.1152/ajpcell.1994.267.2.C402

De Melo MA, Gabbiani G, Pechère JC (1989) Cellular events and intracellular survival of *Campylobacter jejuni* during infection of HEp-2 cells. Infect Immun 57:2214–2222. https://doi.org/10.1128/IAI.57.7.2214-2222.1989

Edmonds CJ, Pilcher D (1973) Electrical potential difference and sodium and potassium fluxes across rectal mucosa in ulcerative colitis. Gut 14:784–789. https://doi.org/10.1136/gut.14.10.784

Epple HJ, Mankertz J, Ignatius R, Liesenfeld O, Fromm M, Zeitz M, Chakraborty T, Schulzke JD (2004) *Aeromonas hydrophila* beta-hemolysin induces active chloride secretion in colon epithelial cells (HT-29/B6). Infect Immun 72:4848–4858. https://doi.org/10.1128/IAI.72.8.4848-4858.2004

Everest PH, Goossens H, Butzler JP, Lloyd D, Knutton S, Ketley JM, Williams PH (1992) Differentiated Caco-2 cells as a model for enteric invasion by *Campylobacter jejuni* and *C. coli*. J Med Microbiol 37:319–325. https://doi.org/10.1099/00222615-37-5-319

Fink MP (1990) Leaky gut hypothesis: a historical perspective. Crit Care Med 18:579–580

Florian P, Schöneberg T, Schulzke JD, Fromm M, Gitter AH (2002) Single-cell epithelial defects close rapidly by an actinomyosin purse string mechanism with functional tight junctions. J Physiol 545:485–499. https://doi.org/10.1113/jphysiol.2002.031161

France MM, Turner JR (2017) The mucosal barrier at a glance. J Cell Sci 130:307–314. https://doi.org/10.1242/jcs.193482

Froicu M, Cantorna MT (2007) Vitamin D and the vitamin D receptor are critical for control of the innate immune response to colonic injury. BMC Immunol 8:5. https://doi.org/10.1186/1471-2172-8-5

Furuse M, Fujita K, Hiiragi T, Fujimoto K, Tsukita S (1998) Claudin-1 and -2: novel integral membrane proteins localizing at tight junctions with no sequence similarity to occludin. J Cell Biol 141:1539–1550. https://doi.org/10.1083/jcb.141.7.1539

Furuse M, Hirase T, Itoh M, Nagafuchi A, Yonemura S, Tsukita S, Tsukita S (1993) Occludin: a novel integral membrane protein localizing at tight junctions. J Cell Biol 123:1777–1788. https://doi.org/10.1083/jcb.123.6.1777

Gitter AH, Bendfeldt K, Schmitz H, Schulzke JD, Bentzel CJ, Fromm M (2000) Epithelial barrier defects in HT-29/B6 colonic cell monolayers induced by tumor necrosis factor-alpha. Ann N Y Acad Sci 915:193–203. https://doi.org/10.1111/j.1749-6632.2000.tb05242.x

Gitter AH, Bendfeldt K, Schulzke JD, Fromm M (2000) Leaks in the epithelial barrier caused by spontaneous and TNF-alpha-induced single-cell apoptosis. FASEB 14:1749–1753. https://doi.org/10.1096/fj.99-0898com

Gitter AH, Fromm M, Schulzke JD (1998) Impedance analysis for the determination of epithelial and subepithelial resistance in intestinal tissues. J Biochem Biophys Methods 37:35–46. https://doi.org/10.1016/s0165-022x(98)00016-5

Graham LL (2002) *Campylobacter fetus* adheres to and enters INT 407 cells. Can J Microbiol 48:995–1007. https://doi.org/10.1139/w02-096

Grant CC, Konkel ME, Cieplak W Jr, Tompkins LS (1993) Role of flagella in adherence, internalization, and translocation of *Campylobacter jejuni* in nonpolarized and polarized epithelial cell cultures. Infect Immun 61:1764–1771. https://doi.org/10.1128/IAI.61.5.1764-1771.1993

Gunes H, Gulen D, Mutlu R, Gumus A, Tas T, Topkaya AE (2016) Antibacterial effects of curcumin: an in vitro minimum inhibitory concentration study. Toxicol Ind Health 32:246–250. https://doi.org/10.1177/0748233713498458

Günzel D, Florian P, Richter JF, Troeger H, Schulzke JD, Fromm M, Gitter AH (2006) Restitution of single-cell defects in the mouse colon epithelium differs from that of cultured cells. Am J Physiol Regul Integr Comp Physiol 290:R1496–R1507. https://doi.org/10.1152/ajpregu.00470.2005

Gurney MA, Laubitz D, Ghishan FK, Kiela PR (2017) Pathophysiology of Intestinal Na+/H+ Exchange. Cell Mol Gastroenterol Hepatol 3:27–40. https://doi.org/10.1016/j.jcmgh.2016.09.010

Haag LM, Fischer A, Otto B, Plickert R, Kühl AA, Göbel UB, Bereswill S, Heimesaat MM (2012) *Campylobacter jejuni* induces acute enterocolitis in gnotobiotic IL-10-/- mice via Toll-like-receptor-2 and -4 signaling. PLoS ONE 7:e40761. https://doi.org/10.1371/journal.pone.0040761

Harrer A, Bücker R, Boehm M, Zarzecka U, Tegtmeyer N, Sticht H, Schulzke JD, Backert S (2019) *Campylobacter jejuni* enters gut epithelial cells and impairs intestinal barrier function through cleavage of occludin by serine protease HtrA. Gut Pathog 11:4. https://doi.org/10.1186/s13099-019-0283-z

Harvey P, Battle T, Leach S (1999) Different invasion phenotypes of *Campylobacter* isolates in Caco-2 cell monolayers. J Med Microbiol 48:461–469. https://doi.org/10.1099/00222615-48-5-461

Harvey S, Greenwood JR (1985) Isolation of *Campylobacter fetus* from a pet turtle. J Clin Microbiol 21:260–261. https://doi.org/10.1128/JCM.21.2.260-261.1985

Hatayama S, Shimohata T, Amano S, Kido J, Nguyen AQ, Sato Y, Kanda Y, Tentaku A, Fukushima S, Nakahashi M, Uebanso T, Mawatari K, Takahashi A (2018) Cellular tight junctions prevent effective *Campylobacter jejuni* invasion and inflammatory barrier disruption promoting bacterial invasion from lateral membrane in polarized intestinal epithelial cells. Front Cell Infect Microbiol 8:15. https://doi.org/10.3389/fcimb.2018.00015

Heller F, Florian P, Bojarski C, Richter J, Christ M, Hillenbrand B, Mankertz J, Gitter AH, Bürgel N, Fromm M, Zeitz M, Fuss I, Strober W, Schulzke JD (2005) Interleukin-13 is the key effector Th2 cytokine in ulcerative colitis that affects epithelial tight junctions, apoptosis, and cell restitution. Gastroenterology 129:550–564. https://doi.org/10.1016/j.gastro.2005.05.002

Hering NA, Fromm A, Kikhney J, Lee IF, Moter A, Schulzke JD, Bücker R (2016) *Yersinia enterocolitica* affects intestinal barrier function in the colon. J Infect Dis 213:1157–1162. https://doi.org/10.1093/infdis/jiv571

Hering NA, Richter JF, Krug SM, Günzel D, Fromm A, Bohn E, Rosenthal R, Bücker R, Fromm M, Troeger H, Schulzke JD (2011) *Yersinia enterocolitica* induces epithelial barrier dysfunction through regional tight junction changes in colonic HT-29/B6 cell monolayers. Lab Invest 91:310–324. https://doi.org/10.1038/labinvest.2010.180

Hickey TE, Baqar S, Bourgeois AL, Ewing CP, Guerry P (1999) *Campylobacter jejuni*-stimulated secretion of interleukin-8 by INT407 cells. Infect Immun 67:88–93. https://doi.org/10.1128/IAI.67.1.88-93.1999

Hickey TE, McVeigh AL, Scott DA, Michielutti RE, Bixby A, Carroll SA, Bourgeois AL, Guerry P (2000) *Campylobacter jejuni* cytolethal distending toxin mediates release of interleukin-8 from intestinal epithelial cells. Infect Immun 68:6535–6541. https://doi.org/10.1128/iai.68.12.6535-6541.2000

Hoy B, Geppert T, Boehm M, Reisen F, Plattner P, Gadermaier G, Sewald N, Ferreira F, Briza P, Schneider G, Backert S, Wessler S (2012) Distinct roles of secreted HtrA proteases from gram-negative pathogens in cleaving the junctional protein and tumor suppressor E-cadherin. J Biol Chem 287:10115–10120. https://doi.org/10.1074/jbc.C111.333419

Hu L, Hickey TE (2005) *Campylobacter jejuni* induces secretion of proinflammatory chemokines from human intestinal epithelial cells. Infect Immun 73:4437–4440. https://doi.org/10.1128/IAI.73.7.4437-4440.2005

Hu L, Tall BD, Curtis SK, Kopecko DJ (2008) Enhanced microscopic definition of *Campylobacter jejuni* 81–176 adherence to, invasion of, translocation across, and exocytosis from polarized human intestinal Caco-2 cells. Infect Immun 76:5294–5304. https://doi.org/10.1128/IAI.01408-07

Ikenouchi J, Furuse M, Furuse K, Sasaki H, Tsukita S, Tsukita S (2005) Tricellulin constitutes a novel barrier at tricellular contacts of epithelial cells. J Cell Biol 171:939–945. https://doi.org/10.1083/jcb.200510043

Inglis GD, Boras VF, Houde A (2011) Enteric campylobacteria and RNA viruses associated with healthy and diarrheic humans in the Chinook health region of southwestern Alberta, Canada. Clin Microbiol 49:209–219. https://doi.org/10.1128/JCM.01220-10

Jacob S, Weizel H, Gordon E, Korman H, Schweinburg F, Frank H, Fine J (1954) Bacterial action in development of irreversibility to transfusion in hemorrhagic shock in the dog. Am J Physiol 179:523–531. https://doi.org/10.1152/ajplegacy.1954.179.3.523

Johnson WM, Lior H (1986) Cytotoxic and cytotonic factors produced by *Campylobacter jejuni*, *Campylobacter coli*, and *Campylobacter laridis*. J Clin Microbiol 24:275–281. https://doi.org/10.1128/JCM.24.2.275-281.1986

Kalischuk LD, Leggett F, Inglis GD (2010) *Campylobacter jejuni* induces transcytosis of commensal bacteria across the intestinal epithelium through M-like cells. Gut Pathog 2:14. https://doi.org/10.1186/1757-4749-2-14

Kanwar RK, Ganguly NK, Kanwar JR, Kumar L, Walia BN (1994) Impairment of Na+, K(+)-ATPase activity following enterotoxigenic*Campylobacter jejuni* infection: changes in Na^+, Cl^- and 3-O-methyl-D-glucose transport*in vitro*, in rat ileum. FEMS Microbiol Lett 124:381–385. https://doi.org/10.1111/j.1574-6968.1994.tb07312.x

Klančnik A, Šikić Pogačar M, Trošt K, Tušek Žnidarič M, Mozetič Vodopivec B, Smole Možina S (2017) Anti-campylobacter activity of resveratrol and an extract from waste Pinot noir grape skins and seeds, and resistance of *Camp. jejuni* planktonic and biofilm cells, mediated via the CmeABC efflux pump. J Appl Microbiol 122:65–77. https://doi.org/10.1111/jam.13315

Kong WJ, Xing XY, Xiao XH, Zhao YL, Wei JH, Wang JB, Yang RC, Yang MH (2012) Effect of berberine on *Escherichia coli, Bacillus subtilis*, and their mixtures as determined by isothermal microcalorimetry. Appl Microbiol Biotechnol 96:503–510. https://doi.org/10.1007/s00253-012-4302-y

Konkel ME, Mead DJ, Hayes SF, Cieplak W Jr (1992) Translocation of *Campylobacter jejuni* across human polarized epithelial cell monolayer cultures. J Infect Dis 166:308–315. https://doi.org/10.1093/infdis/166.2.308

Krug SM, Amasheh S, Richter JF, Milatz S, Günzel D, Westphal JK, Huber O, Schulzke JD, Fromm M (2009) Tricellulin forms a barrier to macromolecules in tricellular tight junctions without affecting ion permeability. Mol Biol Cell 20:3713–3724. https://doi.org/10.1091/mbc.E09-01-0080

Krug SM, Bojarski C, Fromm A, Lee IM, Dames P, Richter JF, Turner JR, Fromm M, Schulzke JD (2018) Tricellulin is regulated via interleukin-13-receptor α2, affects macromolecule uptake, and is decreased in ulcerative colitis. Mucosal Immunol 11:345–356. https://doi.org/10.1038/mi.2017.52

Krug SM, Fromm M, Günzel D (2009) Two-path impedance spectroscopy for measuring paracellular and transcellular epithelial resistance. Biophys J 97:2202–2211. https://doi.org/10.1016/j.bpj.2009.08.003

Krug SM, Günzel D, Conrad MP, Rosenthal R, Fromm A, Amasheh S, Schulzke JD, Fromm M (2012) Claudin-17 forms tight junction channels with distinct anion selectivity. Cell Mol Life Sci 69:2765–2778. https://doi.org/10.1007/s00018-012-0949-x

Lai CK, Chen YA, Lin CJ, Lin HJ, Kao MC, Huang MZ, Lin YH, Chiang-Ni C, Chen CJ, Lo UG, Lin LC, Lin H, Hsieh JT, Lai CH (2016) Molecular mechanisms and potential clinical applications of *Campylobacter jejuni* cytolethal distending toxin. Front Cell Infect Microbiol 6:9. https://doi.org/10.3389/fcimb.2016.00009

Lamb-Rosteski JM, Kalischuk LD, Inglis GD, Buret AG (2008) Epidermal growth factor inhibits *Campylobacter jejuni*-induced claudin-4 disruption, loss of epithelial barrier function, and *Escherichia coli* translocation. Infect Immun 76:3390–3398. https://doi.org/10.1128/IAI.01698-07

Lindblom GB, Sjogren E, Hansson-Westerberg J, Kaijser B (1995) *Campylobacter upsaliensis, C. Sputorum sputorum* and *C. Concisus* as common causes of diarrhoea in Swedish children. Scand. J Infect Dis 27:187–188. https://doi.org/10.3109/00365549509019006

Lobo de Sá FD, Butkevych E, Nattramilarasu PK, Fromm A, Mousavi S, Moos V, Golz JC, Stingl K, Kittler S, Seinige D, Kehrenberg C, Heimesaat MM, Bereswill S, Schulzke JD, Bücker R (2019) Curcumin Mitigates immune-induced epithelial barrier dysfunction by *Campylobacter jejuni*. Int J Mol Sci 20:pii: E4830. https://doi.org/10.3390/ijms20194830

Luettig J, Rosenthal R, Lee IM, Krug SM, Schulzke JD (2016) The ginger component 6-shogaol prevents TNF-α-induced barrier loss via inhibition of PI3K/Akt and NF-κB signaling. Mol Nutr Food Res 60:2576–2586. https://doi.org/10.1002/mnfr.201600274

MacCallum A, Hardy SP, Everest PH (2005) *Campylobacter jejuni* inhibits the absorptive transport functions of Caco-2 cells and disrupts cellular tight junctions. Microbiology 151:2451–2458. https://doi.org/10.1099/mic.0.27950-0

Macuch PJ, Tanner AC (2000) *Campylobacter* species in health, gingivitis, and periodontitis. J Dent Res 79:785–792. https://doi.org/10.1177/00220345000790021301

Madara JL, Stafford J (1989) Interferon-gamma directly affects barrier function of cultured intestinal epithelial monolayers. J Clin Invest 83:724–727. https://doi.org/10.1172/JCI113938

Mahady GB, Pendland SL, Yun G, Lu ZZ (2002) Turmeric (Curcuma longa) and curcumin inhibit the growth of *Helicobacter pylori*, a group 1 carcinogen. Anticancer Res 22:4179–4181

Man SM, Kaakoush NO, Leach ST, Nahidi L, Lu HK, Norman J, Day AS, Zhang L, Mitchell HM (2010) Host attachment, invasion, and stimulation of proinflammatory cytokines by *Campylobacter concisus* and other non-*Campylobacter jejuni* campylobacter species. J Infect Dis 202:1855–1865. https://doi.org/10.1086/657316

Mandel LJ, Bacallao R, Zampighi G (1993) Uncoupling of the molecular 'fence' and paracellular 'gate' functions in epithelial tight junctions. Nature 361:552–555. https://doi.org/10.1038/361552a0

Mankertz J, Amasheh M, Krug SM, Fromm A, Amasheh S, Hillenbrand B, Tavalali S, Fromm M, Schulzke JD (2009) TNFalpha up-regulates claudin-2 expression in epithelial HT-29/B6 cells via phosphatidylinositol-3-kinase signaling. Cell Tissue Res 336:67–77. https://doi.org/10.1007/s00441-009-0751-8

Mankertz J, Tavalali S, Schmitz H, Mankertz A, Riecken EO, Fromm M, Schulzke JD (2000) Expression from the human occludin promoter is affected by tumor necrosis factor alpha and interferon gamma. J Cell Sci 113:2085–2090

Mayangsari Y, Suzuki T (2018) Resveratrol Ameliorates intestinal barrier defects and inflammation in colitic mice and intestinal cells. J Agric Food Chem 66:12666–12674. https://doi.org/10.1021/acs.jafc.8b04138

McCardell BA, Madden JM, Lee EC (1984) *Campylobacter jejuni* and *Campylobacter coli* production of a cytotonic toxin immunologically similar to cholera toxin. J Food Prot 47:943–949. https://doi.org/10.4315/0362-028X-47.12.943

Mellits KH, Mullen J, Wand M, Armbruster G, Patel A, Connerton PL, Skelly M, Connerton IF (2002) Activation of the transcription factor NF-kappaB by *Campylobacter jejuni*. Microbiology 148:2753–2763. https://doi.org/10.1099/00221287-148-9-2753

Monteville MR, Konkel ME (2002) Fibronectin-facilitated invasion of T84 eukaryotic cells by *Campylobacter jejuni* occurs preferentially at the basolateral cell surface. Infect Immun 70:6665–6671. https://doi.org/10.1128/iai.70.12.6665-6671.2002

Mooney A, Byrne C, Clyne M, Johnson-Henry K, Sherman P, Bourke B (2003) Invasion of human epithelial cells by *Campylobacter upsaliensis*. Cell Microbiol 5:835–847. https://doi.org/10.1046/j.1462-5822.2003.00325.x

Mousavi S, Lobo de Sá FD, Schulzke JD, Bücker R, Bereswill S, Heimesaat MM (2019) Vitamin D in acute campylobacteriosis-results from an intervention study applying a clinical *Campylobacter jejuni* induced enterocolitis model. Front Immunol 10:2094. https://doi.org/10.3389/fimmu.2019.02094

Murphy H, Cogan T, Hughes R, Humphrey T (2011) Porcine intestinal epithelial responses to *Campylobacter* infection. Comp Immunol Microbiol Infect Dis 34:489–495. https://doi.org/10.1016/j.cimid.2011.08.005

Nattramilarasu PK, Bücker R, Lobo de Sá FD, Fromm A, Nagel O, Lee IM, Butkevych E, Mousavi S, Genger C, Kløve S, Heimesaat MM, Bereswill S, Schweiger MR, Nielsen HL, Troeger H, Schulzke JD (2020) *Campylobacter concisus* impairs sodium absorption in colonic epithelium via ENaC dysfunction and claudin-8 disruption. Int J Mol Sci 21:pii:E373. https://doi.org/10.3390/ijms21020373

Negoro S, Shimohata T, Hatayama S, Sato Y, Matsumoto M, Iba H, Aihara M, Uebanso T, Hamada Y, Nishikawa Y, Yamasaki S, Mawatari K, Takahashi A (2014) *Campylobacter jejuni* infection suppressed Cl^- secretion induced by CFTR activation in T-84 cells. J Infect Chemother 20:682–688. https://doi.org/10.1016/j.jiac.2014.07.007

Nielsen HL, Engberg J, Ejlertsen T, Bücker R, Nielsen H (2012) Short-term and medium-term clinical outcomes of *Campylobacter concisus* infection. Clin Microbiol Infect 18:E459–E465. https://doi.org/10.1111/j.1469-0691.2012.03990.x

Nielsen HL, Nielsen H, Ejlertsen T, Engberg J, Günzel D, Zeitz M, Hering NA, Fromm M, Schulzke JD, Bücker R (2011) Oral and fecal *Campylobacter concisus* strains perturb barrier function by apoptosis induction in HT-29/B6 intestinal epithelial cells. PLoS ONE 6:e23858. https://doi.org/10.1371/journal.pone.0023858

Ooi JH, Li Y, Rogers CJ, Cantorna MT (2013) Vitamin D regulates the gut microbiome and protects mice from dextran sodium sulfate-induced colitis. J Nutr 143:1679–1686. https://doi.org/10.3945/jn.113.180794

Otto B, Haag LM, Fischer A, Plickert R, Kühl AA, Göbel UB, Heimesaat MM, Bereswill S (2012) *Campylobacter jejuni* induces extra-intestinal immune responses via Toll-like-receptor-4 signaling in conventional IL-10 deficient mice with chronic colitis. Eur J Microbiol Immunol (Bp) 2:210–219. https://doi.org/10.1556/EuJMI.2.2012.3.7

Patton CM, Shaffer N, Edmonds P, Barrett TJ, Lambert MA, Baker C, Perlman DM, Brenner DJ (1989) Human disease associated with "*Campylobacter upsaliensis*" (catalase-negative or weakly positive *Campylobacter* species) in the United States. J Clin Microbiol 27:66–73. https://doi.org/10.1128/JCM.27.1.66-73.1989

Prasad AS, Bao B, Beck FW, Kucuk O, Sarkar FH (2004) Antioxidant effect of zinc in humans. Free Radic Biol Med 37:1182–1190. https://doi.org/10.1016/j.freeradbiomed.2004.07.007

Purdy D, Buswell CM, Hodgson AE, McAlpine K, Henderson I, Leach SA (2000) Characterisation of cytolethal distending toxin (CDT) mutants of *Campylobacter jejuni*. J Med Microbiol 49:473–479. https://doi.org/10.1099/0022-1317-49-5-473

Rees LE, Cogan TA, Dodson AL, Birchall MA, Bailey M, Humphrey TJ (2008) *Campylobacter* and IFNgamma interact to cause a rapid loss of epithelial barrier integrity. Inflamm Bowel Dis 14:303–309. https://doi.org/10.1002/ibd.20325

Rosenthal R, Milatz S, Krug SM, Oelrich B, Schulzke JD, Amasheh S, Günzel D, Fromm M (2010) Claudin-2, a component of the tight junction, forms a paracellular water channel. J Cell Sci 123:1913–1921. https://doi.org/10.1242/jcs.060665

Ruiz-Palacios GM, Torres J, Torres NI, Escamilla E, Ruiz-Palacios BR, Tamayo J (1983) Cholera-like enterotoxin produced by *Campylobacter jejuni*. Characterization and clinical significance. Lancet 2:250–253. https://doi.org/10.1016/s0140-6736(83)90234-9

Russell RG, Blaser MJ, Sarmiento JI, Fox J (1989) Experimental *Campylobacter jejuni* infection in *Macaca nemestrina*. Infect Immun 57:1438–1444. https://doi.org/10.1128/IAI.57.5.1438-1444.1989

Sandle GI, Higgs N, Crowe P, Marsh MN, Venkatesan S, Peters TJ (1990) Cellular basis for defective electrolyte transport in inflamed human colon. Gastroenterology 99:97–105. https://doi.org/10.1016/0016-5085(90)91235-x

Schmitz H, Barmeyer C, Fromm M, Runkel N, Foss HD, Bentzel CJ, Riecken EO, Schulzke JD (1999) Altered tight junction structure contributes to the impaired epithelial barrier function in ulcerative colitis. Gastroenterology 116:301–309. https://doi.org/10.1016/s0016-5085(99)70126-5

Schmitz H, Fromm M, Bentzel CJ, Scholz P, Detjen K, Mankertz J, Bode H, Epple HJ, Riecken EO, Schulzke JD (1999) Tumor necrosis factor-alpha (TNFalpha) regulates the epithelial barrier in the human intestinal cell line HT-29/B6. J Cell Sci 112:137–146

Schmitz H, Fromm M, Bode H, Scholz P, Riecken EO, Schulzke JD (1996) Tumor necrosis factor-alpha induces Cl^{2212} and K^{+} secretion in human distal colon driven by prostaglandin E2. Am J Physiol 271:669–674. https://doi.org/10.1152/ajpgi.1996.271.4.G669

Schulzke JD, Bentzel CJ, Schulzke I, Riecken EO, Fromm M (1998) Epithelial tight junction structure in the jejunum of children with acute and treated celiac sprue. Pediatr Res 43:435–441. https://doi.org/10.1203/00006450-199804000-00001

Schulzke JD, Bojarski C, Zeissig S, Heller F, Gitter AH, Fromm M (2006) Disrupted barrier function through epithelial cell apoptosis. Ann N Y Acad Sci 1072:288–299. https://doi.org/10.1196/annals.1326.027

Schulzke JD, Gitter AH, Mankertz J, Spiegel S, Seidler U, Amasheh S, Saitou M, Tsukita S, Fromm M (2005) Epithelial transport and barrier function in occludin-deficient mice. Biochim Biophys Acta 1669:34–42. https://doi.org/10.1016/j.bbamem.2005.01.008

Schulzke JD, Ploeger S, Amasheh M, Fromm A, Zeissig S, Troeger H, Richter J, Bojarski C, Schumann M, Fromm M (2009) Epithelial tight junctions in intestinal inflammation. Ann N Y Acad Sci 1165:294–300. https://doi.org/10.1111/j.1749-6632.2009.04062.x

Schweinburg FB, Frank HA, Fine J (1954) Bacterial factor in experimental hemorrhagic shock; evidence for development of a bacterial factor which accounts for irreversibility to transfusion and for the loss of the normal capacity to destroy bacteria. Am J Physiol 179:532–540. https://doi.org/10.1152/ajplegacy.1954.179.3.532

Shen L, Weber CR, Raleigh DR, Yu D, Turner JR (2011) Tight junction pore and leak pathways: a dynamic duo. Annu Rev Physiol 73:283–309. https://doi.org/10.1146/annurev-physiol-012110-142150

Simson D, Boehm M, Backert S (2020) HtrA-dependent adherence and invasion of *Campylobacter jejuni* in human vs avian cells. Lett Appl Microbiol 70:326–330. https://doi.org/10.1111/lam.13277

Spiller RC, Jenkins D, Thornley JP, Hebden JM, Wright T, Skinner M, Neal KR (2000) Increased rectal mucosal enteroendocrine cells, T lymphocytes, and increased gut permeability following acute *Campylobacter* enteritis and in post-dysenteric irritable bowel syndrome. Gut 47:804–811. https://doi.org/10.1136/gut.47.6.804

Steinkraus GE, Wright BD (1994) Septic abortion with intact fetal membranes caused by *Campylobacter fetus* subsp. *fetus*. J Clin Microbiol 32:1608–1609. https://doi.org/10.1128/JCM.32.6.1608-1609.1994

Stockmann M, Fromm M, Riecken E, Schulzke JD (1998) Nonmalabsorptive mechanisms of diarrhea in HIV infection. Pathobiology 66:165–169. https://doi.org/10.1159/000028016

Sugi K, Musch MW, Field M, Chang EB (2001) Inhibition of Na^+, K^+-ATPase by interferon gamma down-regulates intestinal epithelial transport and barrier function. Gastroenterology 120:1393–403. https://doi.org/10.1053/gast.2001.24045

Tamura A, Hayashi H, Imasato M, Yamazaki Y, Hagiwara A, Wada M, Noda T, Watanabe M, Suzuki Y, Tsukita S (2011) Loss of claudin-15, but not claudin-2, causes Na+ deficiency and glucose malabsorption in mouse small intestine. Gastroenterology 140:913–923. https://doi.org/10.1053/j.gastro.2010.08.006

Troeger H, Loddenkemper C, Schneider T, Schreier E, Epple HJ, Zeitz M, Fromm M, Schulzke JD (2009) Structural and functional changes of the duodenum in human norovirus infection. Gut 58:1070–1077. https://doi.org/10.1136/gut.2008.160150

Trott S, Bauer R, Knackmuss HJ, Stolz A (2001) *Campylobacter upsaliensis* exerts a cytolethal distending toxin effect on HeLa cells and T lymphocytes. Microbiology 147:1815–1824. https://doi.org/10.1099/00221287-147-7-1815

Ussing HH (1949) Distinction by means of tracers between active transport and diffusion. Acta Physiol Scand 19:43–56. https://doi.org/10.1111/j.1748-1716.1949.tb00633.x

Van Itallie C, Rahner C, Anderson JM (2001) Regulated expression of claudin-4 decreases paracellular conductance through a selective decrease in sodium permeability. J Clin Invest 107:1319–1327. https://doi.org/10.1172/JCI12464

Van Itallie CM, Fanning AS, Bridges A, Anderson JM (2009) ZO-1 stabilizes the tight junction solute barrier through coupling to the perijunctional cytoskeleton. Mol Biol Cell 20:3930–3940. https://doi.org/10.1091/mbc.E09-04-0320

Van Itallie CM, Holmes J, Bridges A, Gookin JL, Coccaro MR, Proctor W, Colegio OR, Anderson JM (2008) The density of small tight junction pores varies among cell types and is increased by expression of claudin-2. J Cell Sci 121:298–305. https://doi.org/10.1242/jcs.021485

van Spreeuwel JP, Duursma GC, Meijer CJ, Bax R, Rosekrans PC, Lindeman J (1985) *Campylobacter colitis*: histological immunohistochemical and ultrastructural findings. Gut 26:945–951. https://doi.org/10.1136/gut.26.9.945

Wadström T, Baloda SB, Krovacek K, Faris A, Bengtson S, Walder M (1983) Swedish isolates of *Campylobacter jejuni/coli* do not produce cytotonic or cytotoxic enterotoxins. Lancet 2:911. https://doi.org/10.1016/s0140-6736(83)90893-0

Wang J, Ghosh SS, Ghosh S (2017) Curcumin improves intestinal barrier function: modulation of intracellular signaling, and organization of tight junctions. Am J Physiol Cell Physiol 312:C438–C445. https://doi.org/10.1152/ajpcell.00235.2016

Wang L, Srinivasan S, Theiss AL, Merlin D, Sitaraman SV (2007) Interleukin-6 induces keratin expression in intestinal epithelial cells: potential role of keratin-8 in interleukin-6-induced barrier function alterations. J Biol Chem 282:8219–8227. https://doi.org/10.1074/jbc.M604068200

Wassenaar TM (1997) Toxin production by *Campylobacter* spp. Clin Microbiol Rev 10:466–476. https://doi.org/10.1128/CMR.10.3.466-476.1997

Watari A, Yagi K, Kondoh M (2012) A simple reporter assay for screening Claudin-4 modulators. Biochem Biophys Res Commun 426:454–460. https://doi.org/10.1016/j.bbrc.2012.08.083

Watson RO, Galán JE (2008) *Campylobacter jejuni* survives within epithelial cells by avoiding delivery to lysosomes. PLoS Pathog 4:e14. https://doi.org/10.1371/journal.ppat.0040014

Welsh MJ, Smith PL, Fromm M, Frizzell RA (1982) Crypts are the site of intestinal fluid and electrolyte secretion. Science 218:1219–1221. https://doi.org/10.1126/science.6293054

Wiegand S, Zakrzewski SS, Eichner M, Schulz E, Günzel D, Pieper R, Rosenthal R, Barmeyer C, Bleich A, Dobrindt U, Schulzke JD, Bücker R (2017) Zinc treatment is efficient against *Escherichia coli* α-haemolysin-induced intestinal leakage in mice. Sci Rep 7:45649. https://doi.org/10.1038/srep45649

Xie Y, He Y, Irwin PL, Jin T, Shi X (2011) Antibacterial activity and mechanism of action of zinc oxide nanoparticles against *Campylobacter jejuni*. Appl Environ Microbiol 77:2325–2331. https://doi.org/10.1128/AEM.02149-10

Yu AS, Cheng MH, Angelow S, Günzel D, Kanzawa SA, Schneeberger EE, Fromm M, Coalson RD (2009) Molecular basis for cation selectivity in claudin-2-based paracellular pores: identification of an electrostatic interaction site. J Gen Physiol 133:111–127. https://doi.org/10.1085/jgp.200810154

Yu D, Marchiando AM, Weber CR, Raleigh DR, Wang Y, Shen L, Turner JR (2010) MLCK-dependent exchange and actin binding region-dependent anchoring of ZO-1 regulate tight junction barrier function. Proc Natl Acad Sci USA 107:8237–8241. https://doi.org/10.1073/pnas.0908869107

Zeissig S, Bürgel N, Günzel D, Richter JF, Mankertz J, Wahnschaffe U, Kroesen AJ, Zeitz M, Fromm M, Schulzke JD (2007) Changes in expression and distribution of claudin-2, -5 and -8 lead to discontinuous tight junctions and barrier dysfunction in active Crohn's disease. Gut 56:61–72. https://doi.org/10.1136/gut.2006.094375

Zheng J, Meng J, Zhao S, Singh R, Song W (2008) *Campylobacter*-induced interleukin-8 secretion in polarized human intestinal epithelial cells requires *Campylobacter*-secreted cytolethal distending toxin- and Toll-like receptor-mediated activation of NF-kappaB. Infect Immun 76:4498–4508. https://doi.org/10.1128/IAI.01317-07

Zhou H, Mineshita S (2000) The effect of berberine chloride on experimental colitis in rats in vivo and in vitro. J Pharmacol Exp Ther 294:822–829

Zhu Y, Mahon BD, Froicu M, Cantorna MT (2005) Calcium and 1 alpha,25-dihydroxyvitamin D3 target the TNF-alpha pathway to suppress experimental inflammatory bowel disease. Eur J Immunol 35:217–224. https://doi.org/10.1002/eji.200425491

Murine Models for the Investigation of Colonization Resistance and Innate Immune Responses in *Campylobacter Jejuni* Infections

Soraya Mousavi, Stefan Bereswill, and Markus M. Heimesaat

Contents

Abstract Human infections with the food-borne pathogen *Campylobacter jejuni* are progressively increasing worldwide and constitute a significant socioeconomic burden to mankind. Intestinal campylobacteriosis in humans is characterized by bloody diarrhea, fever, abdominal pain, and severe malaise. Some individuals develop chronic post-infectious sequelae including neurological and autoimmune diseases such as reactive arthritis and Guillain-Barré syndrome. Studies unraveling the molecular mechanisms underlying campylobacteriosis and post-infectious sequelae have

S. Mousavi · S. Bereswill · M. M. Heimesaat (✉)
Institute of Microbiology, Infectious Diseases and Immunology, Gastrointestinal Microbiology Research Group, Charité-University Medicine Berlin, Corporate Member of Free University Berlin, Humboldt-University of Berlin, Berlin Institute of Health, Berlin, Germany
e-mail: markus.heimesaat@charite.de

S. Mousavi
e-mail: soraya.mousavi@charite.de

S. Bereswill
e-mail: stefan.bereswill@charite.de

S. Backert (ed.), *Fighting Campylobacter Infections*, Current Topics in Microbiology and Immunology 431, https://doi.org/10.1007/978-3-030-65481-8_9

been hampered by the scarcity of appropriate experimental in vivo models. Particularly, conventional laboratory mice are protected from *C. jejuni* infection due to the physiological colonization resistance exerted by the murine gut microbiota composition. Additionally, as compared to humans, mice are up to 10,000 times more resistant to *C. jejuni* lipooligosaccharide (LOS) constituting a major pathogenicity factor responsible for the immunopathological host responses during campylobacteriosis. In this chapter, we summarize the recent progress that has been made in overcoming these fundamental obstacles in *Campylobacter* research in mice. Modification of the murine host-specific gut microbiota composition and sensitization of the mice to *C. jejuni* LOS by deletion of genes encoding interleukin-10 or a single IL-1 receptor-related molecule as well as by dietary zinc depletion have yielded reliable murine infection models resembling key features of human campylobacteriosis. These substantial improvements pave the way for a better understanding of the molecular mechanisms underlying pathogen–host interactions. The ongoing validation and standardization of these novel murine infection models will provide the basis for the development of innovative treatment and prevention strategies to combat human campylobacteriosis and collateral damages of *C. jejuni* infections.

1 Introduction

The Gram-negative, spirally curved, microaerophilic, and highly motile *Campylobacter* bacteria live as part of the commensal gut microbiota in many avian species such as chicken and turkey (Bolton 2015; Masanta et al. 2013). Hence, farm animals, especially poultry, are the primary origin of human infections (Stahl and Vallance 2015; Ellström et al. 2016). In the European Union, the species *Campylobacter jejuni* and *C. coli* are the most commonly reported bacterial food-borne pathogens causing human infections with an increasing number of cases (EFSA 2019). Major clinical symptoms of *Campylobacter* infection such as fever, bloody diarrhea, and abdominal cramps are mostly self-limiting and resolve within one to two weeks (Mousavi et al. 2020c; Cróinín and Backert 2012; Kist and Bereswill 2001; Boehm et al. 2018). However, in rare instances the infection can result in the development of post-infectious complications such as reactive arthritis (RA), Guillain-Barré syndrome (GBS) and Miller Fisher syndrome (MFS) (Backert et al. 2017; Masanta et al. 2013; Dorrell and Wren 2007; Dasti et al. 2010; Kist and Bereswill 2001). Additionally, it has been shown that the development of several intestinal inflammatory morbidities, including inflammatory bowel disease (IBD), irritable bowel syndrome (IBS), and celiac disease, is associated with *C. jejuni* infection (Gradel et al. 2009).

The complex gut microbiota causes colonization resistance (CR) which protects mice against *C. jejuni* by nutrient depletion and production of bacteriocins (Mathur et al. 2015; Hooper and Macpherson 2010; Bereswill et al. 2011; Fiebiger et al. 2016; Ducarmon et al. 2019; Kim et al. 2017). The direct killing of *C. jejuni* by the commensal murine microbiota maintains a strict CR which has limited the availability of suitable murine infection models for the investigation of campylobacteriosis for a long time (Heimesaat and Bereswill 2015). In addition, mice are highly resistant against inflammation induced by lipopolysaccharide (LPS) derived from the cell walls of Gram-negative bacteria including the lipooligosaccharide (LOS) of *Campylobacter* (Warren et al. 2010). Together with the enormous LPS and LOS resistance of birds, this provided first evidence that murine and avian models of infection displaying clinical aspects of human campylobacteriosis can be generated by the combination of measures to overcome CR with modifications rendering the animals susceptible to LOS of *C. jejuni*.

Following depletion of the gut microbiota by antibiotic treatment, for instance, *C. jejuni* can stably establish within the gastrointestinal tract of mice at high loads following peroral challenge. This also holds true for secondary abiotic mice that are re-associated with a complex microbiota derived from human as opposed to murine donors, for infant mice, and for mice with high intestinal loads of enterobacterial commensals such as *Escherichia coli* during intestinal inflammation or following enteral enterobacterial feeding. Notably, in case *C. jejuni* can overcome the CR, mice become infected and display several pro-inflammatory features of human campylobacteriosis, but do not develop bloody diarrhea (Bereswill et al. 2011; Haag et al. 2012a, 2012c). The full symptom complex of campylobacteriosis only establishes if the LOS resistance of mice is abrogated by zinc deficiency, for instance (Giallourou et al. 2018)). In order to establish murine models for more severe campylobacteriosis, researchers took advantage of mice lacking interleukin-10 (IL-10; Mansfield et al. 2007) or single Ig IL-1 receptor related molecule (SIGIRR; Stahl et al. 2014) that were infected with *C. jejuni* after antimicrobial pretreatment (Haag et al. 2012b; Heimesaat et al. 2017; Heimesaat et al. 2014c). These excellent novel clinical models resemble different severe courses of human campylobacteriosis. The present chapter focuses on the major recent progress made in the generation of clinically reliable murine *C. jejuni* infection models and aims at summarizing the most important findings in this field of research.

2 Factors Affecting *Campylobacter* Colonization and Infection

Remarkably, mice with a human intestinal microbiota have been shown to be less resistant against *C. jejuni* colonization as compared to conventionally colonized mice (Heimesaat and Bereswill 2015). Moreover, the commensal *E. coli* loads within the human intestinal microbiota are associated with an increased risk for acquiring campylobacteriosis (Dicksved et al. 2014). Together with the finding that the murine CR can be abrogated by feeding live *E. coli* to mice (Haag et al. 2012c), this suggests that the host-specific composition of the gut microbiota plays a pivotal role in mediating CR against pathogens and in particular against *C. jejuni*. In further support, several investigators including our group have shown that modification or even eradication of the murine gut microbiota by antibiotic treatment facilitated *C. jejuni* colonization and resulted in the induction of pro-inflammatory immune responses (Bereswill et al. 2011; Yrios and Balish 1986a, c; Haag et al. 2012c; Mansfield et al. 2007; Masanta et al. 2013). This will be discussed in more detail below.

The bacterial diversity and density differ along the gastrointestinal tract and increase from the proximal (i.e., the stomach) to the very distal part (i.e., the rectum) (O'Hara and Shanahan 2006). The colon is colonized by up to 10^{11} bacteria per milliliter luminal content (O'Hara and Shanahan 2006; Palmer et al. 2007; Jandhyala et al. 2015; Fiebiger et al. 2016). The colonic microbiota consists of approximately 90% obligate anaerobic bacteria, mostly *Bacteroides*, *Bifidobacteria*, *Eubacteria*, *Fusobacteria*, *Clostridiales*, and *Lactobacillales*, and less commonly of enterococci and *Enterobacteriaceae* (Guarner and Malagelada 2003; Ley et al. 2005, 2008). Besides these commensal bacteria, the human intestines may be invaded by pathogens such as *C. jejuni*, *Salmonella enterica*, *Vibrio cholera*, and others (HMPC 2012). In order to successfully establish within the intestinal tract and induce disease, the flagellum and the helical shape of the highly motile *C. jejuni*, for instance, are essential for intestinal colonization and infection (Ducarmon et al. 2019; Ferrero and Lee 1988; Szymanski et al. 1995; Stahl et al. 2016). It has been recently shown that a non-motile deletion mutant of *C. jejuni*, lacking both flagella genes *flaA* and *flaB*, did colonize the colon at high concentrations but did not induce intestinal inflammation (Schmidt et al. 2019a). In addition, *C. jejuni* strains, which lost the helical shape due to a lack of the peptidoglycan genes *pgp1* and *pgp2*, showed significantly reduced bacterial colonization properties and less induced inflammation within the murine intestinal tract (Stahl et al. 2016).

Results from a recent study suggest that a high bacterial diversity of the gut microbiota hampers the colonization properties of *C. jejuni* in the human gastrointestinal tract (Kampmann et al. 2016). In the healthy vertebrate host, the gut microbiota exerts beneficial effects by supplying nutrients and energy and supports the development of the immune system (Nishida et al. 2018; Jandhyala et al. 2015; Masanta et al. 2013). In addition to the utilization of physical and nutritional niches in the intestines by competitively colonizing members of the commensal microbiota, all these factors together contribute to an effective physiological CR against invading pathogenic microorganisms (Vollaard and Clasener 1994; Sekirov et al. 2010; Buffie and Pamer 2013). The metabolization of dietary compounds by the gastrointestinal microbiota further leads to the formation of several metabolic products such as short-chain fatty acids (SCFAs) and bile salts that could be beneficial for both, the vertebrate host and the commensal bacteria, but might be potentially harmful to pathogens (Masanta et al. 2013; Ducarmon et al. 2019). The fermentation of indigestible carbohydrates by *Firmicutes* and *Bacteroidetes* results in the production of SCFAs such as acetate, propionate, and butyrate (Marchesi et al. 2016; Sun et al. 2017). Butyrate, for instance, serves as an energy source for colonic epithelial cells (Pomare et al. 1985) and plays a significant role in the differentiation and expansion of colonic regulatory T cells (Atarashi et al. 2013). Besides these beneficial properties for the host cells, SCFAs can also disrupt pathogenic growth by affecting intracellular pH and in consequence, compromising bacterial metabolisms (Ducarmon et al. 2019). In non-ionized forms, SCFAs are able to diffuse across the bacterial membrane and reduce their intracellular pH (Ducarmon et al. 2019; Cummings et al. 1987; Repaske and Adler 1981). Remarkably, a study applying a *C. jejuni* strain with mutations in the enzymes phosphotransacetylase (Δ*pta*) and acetate kinase (Δ*ackA*) that are involved in acetogenesis revealed that the expression of genes encoding for distinct catabolic enzymes and transport systems required for in vivo growth was modulated by SCFAs in the lower intestinal tract and reduced by the organic acid lactate in the upper intestinal compartments (Luethy et al. 2017). These results indicate that SCFAs may also interfere with colonization properties of *C. jejuni* in distinct parts of the intestinal tract (Luethy et al. 2017; Ducarmon et al. 2019). Furthermore, the exact roles of bile acids and salts in *C. jejuni* colonization and survival remain still unclear. In vitro studies demonstrated that deoxycholate induced *C. jejuni* virulence gene expression (Malik-Kale et al. 2008; Dasti et al. 2010). A *C. jejuni* strain deficient in the efflux pump *cmeABC* exhibited a decreased resistance to various bile salts such as cholic acid in vitro and further lost its colonization ability in the avian intestine (Lin et al. 2003). Remarkably, the depletion of deoxycholate-producing commensal bacteria upon clindamycin treatment promoted *C jejuni*-induced colitis in specific-pathogen-free (SPF) IL-10 deficient mice (Sun et al. 2018).

Even though the gastrointestinal tract harbors a multitude of different commensal microbial species which are noninvasive and do not produce exotoxins, the mucosal immune system is able to detect invading pathogens such as *C. jejuni* by pattern recognition receptors including Toll-like receptors (TLRs), which identify specific molecular structures of microorganisms such as LOS, LPS, and flagellin, for instance (Ulevitch 1999; Medzhitov 2007; Medzhitov and Janeway Jr 2000; Purchiaroni et al. 2013). The *C. jejuni* LOS acts as a potent TLR4 agonist and has been suggested to play an essential role in the immunopathogenesis of campylobacteriosis (Dorrell and Wren 2007; Kuijf et al. 2010; Mousavi et al. 2020a). The activation of TLRs by bacterial molecules including LOS induces an inflammatory chain reaction resulting in the secretion of pro-inflammatory cytokines such as IL-6 and tumor necrosis factor (TNF)-α, of chemokines, and of antimicrobial peptides including cathelicidins and C-type lectins (Medzhitov 2007; Rakoff-Nahoum et al. 2004). However, the susceptibility of vertebrates to TLR4 ligands is species-specific. In contrast to humans, birds and rodents are highly resistant against LPS/LOS, for instance. Hence, the intensity of the TLR4 signaling cascade induced by *C. jejuni* has a great impact on both, the colonization capacity and the outcome of campylobacteriosis, which both provide the basis for the development of efficient animal models for *C. jejuni* infection.

3 Susceptibility to Lipooligosaccharide Determines Host Specificity of *Campylobacter* Colonization Versus Infection

The bacterial LOS is essential for the intestinal colonization, cellular invasion, and translocation of *C. jejuni* across the intestinal epithelial barrier in human cell systems (Bolton 2015; Louwen et al. 2008; Perera et al. 2007; Muller et al. 2007; Naito et al. 2010; Louwen et al. 2012). Furthermore, LOS also plays a significant role as a permeability barrier against hydrophobic antibiotic and phenolic antimicrobial compounds (Vaara 1992; Oh and Jeon 2015). The inactivation of the *C. jejuni waaF* gene resulted in truncation of the LOS inner core (Oldfield et al. 2002), causing a reduction of the minimal inhibitory concentrations of phenolic antimicrobials, such as *p*-coumaric acid, epigallocatechin gallate, and hesperidin (Oh and Jeon 2015). Interestingly, results from a study with mice that had been infected with *C. jejuni ΔwaaF, ΔlgtF*, and parental strains revealed that LOS is involved in bacterial colonization properties (Naito et al. 2010).

The surface LPS of most Gram-negative bacteria comprises lipid A and an oligosaccharide core with an O-antigen. In *C. jejuni,* the LOS consists of the hydrophobic lipid A anchor and an oligosaccharide with a conserved inner and a variable outer core. In LOS, however, the characteristic O-antigen of bacterial LPS is missing (Golec 2007; Louwen et al. 2008; Naito et al. 2010; Karlyshev et al. 2005). The LOS biosynthesis genes in *C. jejuni* are localized at a hypervariable locus with approximately 20 identified classes (Houliston et al. 2011), which explains the variations in *C. jejuni*-associated pathologies (Gilbert et al. 2002; Guerry et al. 2002). Additionally, the horizontal gene transfer between different *C. jejuni* strains may lead to variations in LOS synthesis genes. Molecular analyses revealed that the *C. jejuni* GB11 strain (isolated from a GBS patient) was genetically similar to the completely sequenced *C. jejuni* NCTC 11168 strain. However, the expression of the LOS genes was shown to strongly vary between the respective strains. Interestingly, an identical LOS locus in both, *C. jejuni* ATCC 43446 and GB11 strains could be found (Gilbert et al. 2004). Taken together, the modulation of the LOS expression is an important strategy for avoiding host defenses and for possibly adapting to microenvironments (Karlyshev et al. 2005).

The specific binding of the lipid A moiety of sugar-containing molecules from the bacterial cell wall such as LPS/LOS to pattern recognition receptors (including TLR4) on the plasma membrane or in the cytosol of specialized host cells is the first molecular event required for the activation of immune responses and subsequent inflammation (Mogensen 2009). Others have extensively reviewed the TLR signaling pathways and their activation (Jungi et al. 2011; O'Neill et al. 2003; Vaure and Liu 2014; Cochet and Peri 2017). Briefly, TLR4 induces intracellular signaling through at least two major pathways: (I) the Toll–interleukin receptor (TIR)-domain-containing interferon-β (TRIF), TRIF-related adapter molecule (TRAM), TRIF–TRAM pathway, which upregulates genes encoding type I interferons (IFNs), and activates TNF-α production and secretion; and (II) the TIR-domain-containing adapter protein (TIRAP), myeloid differentiation primary response 88 (MyD88), TIRAP–MyD88 pathway, which regulates early nuclear factor kappa-light-chain-enhancer of activated B cell (NF-κB) activation and related inflammatory cytokine production such as IL-12, which is responsible for the majority of the LPS/LOS responses (Vaure and Liu 2014; de Zoete et al. 2010; Rathinam et al. 2009; van Mourik et al. 2010; see also Chap. 6 of this book). It has been shown that the immune responses induced by the LPS/LOS of different Gram-negative bacteria upon TLR4 activation differ remarkably (Stephenson et al. 2013). While the LPS of *Helicobacter pylori* (evolutionary related to *C. jejuni*) shows low reactivity (Pachathundikandi et al. 2015; Pachathundikandi et al. 2019), the lipid A moiety of *C. jejuni* LOS acts as a highly potent TLR4 agonist (Kuijf et al. 2010). Hence, species-specific differences in TLR expression, in LOS/LPS recognition, and thus in TLR4 activation determine the host specificity of *C. jejuni* infection versus colonization in humans, poultry, and mice as outlined in more detail below.

3.1 Chickens

As a natural host, chickens are frequently colonized with *C. jejuni* at high loads of 10^6–10^8 bacterial cells per g intestinal content indicating that the microbiota of the birds does not establish a CR against *C. jejuni* (Ridley et al. 2008). In line with this, it has been suggested that the colonization of *C. jejuni* in poultry is mainly determined by the avian immunity, while the host-specific microbiota composition is less relevant (Hermans et al. 2012). It is of note that avian hosts are completely non-responsive to *C. jejuni* colonization. Even if live *C. jejuni* bacteria can invade extra-intestinal avian organs such as the thymus, spleen, and liver, chickens usually do not display overt *C. jejuni*-induced clinical signs of inflammation (Hermans et al. 2012; Beery et al. 1988; Zhang and Sahin 2020; Young et al. 1999). This has led to the hypothesis that the anergy of the avian immune system to *C. jejuni* is the cause for benign commensal colonization versus in contrast to highly acute inflammatory responses seen in humans (Young et al. 2007).

Various investigations revealed the adhesive and invasive properties of *C. jejuni* within chicken epithelial cells as well as recruitment of innate immune cells such as monocytes/macrophages, natural killer cells, dendritic cells, and neutrophils to the intestines (Van Deun et al. 2008; Larson et al. 2008). These could subsequently induce activation of TLRs expressed on peripheral blood lymphocytes, for instance (Kannaki et al. 2010), which promoted an inflammatory response against *C. jejuni*. The chicken receptor complex TLR4 and myeloid differentiation protein-2 (MD-2) as well as the cell surface chicken-specific chTLR2 recognize LOS and lipopeptides of *C. jejuni* (Hermans et al. 2012). Additionally, the chicken-specific chTLR21 interacts with *C. jejuni* CpG DNA, sharing similarities with the interaction between CpG DNA and TLR9 in mammals (de Zoete et al. 2010). The activation of these TLRs results in an innate immune response leading to the production of pro-inflammatory cytokines and chemokines (de Zoete et al. 2010; Keestra et al. 2010; Hermans et al. 2012). Given that birds do not develop overt disease symptoms during *Campylobacter* colonization, the molecular mechanisms underlying the anergy of chickens and other avian hosts to *C. jejuni* await further investigation.

In mammalian TLR4 signaling, a MyD88-independent pathway is also involved, which is associated with the induction of a late phase of the pro-inflammatory transcription factor NF-κB (Yamamoto et al. 2003). This pathway leads to the activation of natural killer cells and maturation of dendritic cells. The absence of this signaling in chickens, however, may contribute to their resistance against TLR4 agonists such as LOS (Hermans et al. 2012). Remarkably, the low susceptibility of avian species toward TLR4 ligands was further underlined by the intravenous injection of a high dose of purified *E. coli* LPS (i.e., 577 mg per kg body weight) into chickens, which induced only rather mild clinical signs, from which the animals recovered within two days (Adler and DaMassa 1979). On the contrary, a much lower LPS dose of 45.5 mg per kg body weight was lethal to mice within only four hours (Adler and DaMassa 1979). Taken together, the LOS/LPS resistance of chickens might provide the molecular basis for the complete absence of inflammatory responses upon *C. jejuni* colonization (Adler and DaMassa 1979; van Dijk et al. 2012).

3.2 Mice

Remarkably, the experimental LPS/LOS doses applied to mice (1–25 mg/kg) have been shown to be about 100–10,000 times higher than the concentrations that are needed to induce severe systemic inflammation including septic shock in humans (Warren et al. 2010). These observations underline the fact that the LPS/LOS tolerance of mice renders the murine host highly resistant against bacterial LPS and LOS (Warren et al. 2010; Nomura et al. 2000). In the late 1990s, researchers identified the TLR4 as the major LPS/LOS recognition receptor in mice (Poltorak et al. 1998; Qureshi et al. 1999). Murine lymphoid cells such as naïve B cells and T cells and myeloid cells including macrophages, monocytes, and granulocytes (with the highest levels) express TLR4 (Ketloy et al. 2008; Peng 2005). In contrast to human plasmacytoid dendritic cells (pDCs), murine pDCs express TLR4 (Ketloy et al. 2008). Thus, the differences of LOS/LPS tolerance in mice and men are not fully clear and need further investigation. However, in contrast to humans, mice produce a soluble "TLR4-like" structure which may contribute to the murine LOS/LPS resistance by extracellular binding of the ligands without initiation of the inflammatory signaling cascades. This soluble mouse (sm) TLR4 is a partially secreted 20-kDa protein expressed from a alternatively spliced TLR4 mRNA (Iwami et al. 2000). In murine macrophages, binding of smTLR4 to the TLR4/cluster of differentiation 14 complex resulted in decreased LPS-mediated TNF-α secretion and NF-κB activation indicating that smTLR4 may function as a negative feedback mechanism to reduce the excessive LPS/LOS responses by inhibiting the TLR4/MD-2 interaction in mice (Iwami et al. 2000; Vaure and Liu 2014).

Despite the high similarity between human and murine TLR4/MD-2 signaling, MD-2 displays various species-related differences, leading to species-specific LPS/LOS tolerance. The human, but not murine MD-2 can bind LPS/LOS in the absence of TLR4 and induce a pro-inflammatory immune response. Murine MD-2, however, can only support TLR4-dependent cell activation by LPS/LOS when MD-2 and TLR4 are co-expressed in the same cell (Vašl et al. 2009). As already mentioned, mice have been shown to be up to 10,000-fold more resistant to LOS than humans (Vaure and Liu 2014; Heimesaat and Bereswill 2015). This is not only due to the strong CR mediated by the murine microbiota, but also due to the LOS recognition pattern of the murine TLR4/MD-2 receptor complex (Hajjar et al. 2002; Vaure and Liu 2014). For instance, solely murine TLR4/MD-2 is able to recognize Taxol (an anti-tumor drug with no structural similarity to lipid A) and lipid$_{IVA}$ (an intermediate in the biosynthetic pathway of lipid A and major component of *Yersinia pestis*) (Kawasaki et al. 2000; Vogel et al. 1984).

3.3 Humans

As mentioned before, the main difference between human and murine TLR4 activation is most probably due to different LPS/LOS recognition patterns (Vaure and Liu 2014). Human as opposed to murine TLR4 can differentiate between the hexa- and penta-acylated forms of LPS produced by *Pseudomonas aeruginosa*, for instance (Hajjar et al. 2002). Moreover, the TLR4 expression levels upon LPS challenge differ between humans and mice. While LPS exposure reduced the TLR4 expression levels in murine peritoneal macrophages and neutrophils (Nomura et al. 2000), human macrophages and monocytes displayed increased TLR4 mRNA levels (Muzio et al. 2000). Both human and murine cells of myeloid origin exhibit high levels of TLR4 expression (Vaure and Liu 2014). In contrast to murine lymphoid cells, TLR4 mRNAs were undetectable in human counterparts (Hornung et al. 2002; Ketloy et al. 2008).

The expression of TLR4 is low but detectable in human and in non-human primate epithelial cells (Imaeda et al. 2002; Weindl et al. 2007). Interestingly, in IBD patients the TLR4 expression was shown to be up-regulated in colonic epithelial cells, which may explain the higher susceptibility of these patients to LOS/LPS as compared to healthy individuals (Cario and Podolsky 2000). Remarkably, recent studies revealed that *C. jejuni*-infected individuals were at higher risk to develop post-infectious intestinal morbidities such as IBD, IBS, or celiac disease (Gradel et al. 2009). Results from various clinical studies further revealed that the hyper-activation of the innate immune system by sialylated *C. jejuni* LOS not only triggers the development of severe forms of campylobacteriosis but also increases the risk for development of post-infectious neurological sequelae such as GBS and MFS (Mortensen et al. 2009; Gilbert et al. 2004). Recently, *C. jejuni* strains of the classes A, B, and C received increasing attention due to their arsenal of genes encoding for the sialic acid-processing enzymes and glycosyltransferases, which are necessary for the synthesis of ganglioside-like glycans (Gilbert et al. 2000; Dorrell and Wren 2007; Habib et al. 2009). These *C. jejuni* strains produce a sialylated form of LOS, which mimic distinct surface structures of human gangliosides (Godschalk et al. 2007). After infection with *C. jejuni*, antibodies directed against LOS might thus react with human peripheral nerves and induce autoimmune disorders resulting in the onset of GBS, for instance (Riddle and Guerry 2016; Halpin et al. 2018).

The sialylated form of LOS is associated not only with GBS development, but also with *C. jejuni* pathogenicity. Kuijf and co-workers showed that the extent of TLR4 signaling and dendritic cell activation was more pronounced upon stimulation with LOS of *C. jejuni* isolates from three GBS patients as compared to the non-sialylated LOS of the corresponding sialyltransferase knockout (CST-II mutant) strains, indicating that LOS sialylation increases the immune responses upon *C. jejuni* LOS (Kuijf et al. 2010). Moreover, *C. jejuni* strains expressing sialylated LOS (classes A, B, and C) have been shown to be significantly more invasive than strains expressing non-sialylated LOS (classes D and E). The inactivation of the LOS sialyltransferase via knockout mutagenesis in three GBS-associated *C. jejuni* strains expressing sialylated LOS (GB2, GB11, and GB19) further resulted in less invasiveness than the respective wild-type counterparts (Louwen et al. 2008). Of clinical relevance, *C. jejuni* ganglioside-like LOS-expressing isolates have been shown to be linked to severe enterocolitis with bloody diarrhea given their strong attachment to the Caco-2 intestinal epithelial cells (Louwen et al. 2012). Taken together, these findings underline that particularly the sialylated form of *C. jejuni* LOS plays a major role in triggering severe forms of human campylobacteriosis which in turn can lead to post-infectious sequelae such as GBS (Dorrell and Wren 2007).

4 Murine Infection Models in Campylobacteriosis Research

Since the discovery of *Campylobacter* species as enteric pathogens, research on campylobacteriosis has been hampered by the lack of reliable experimental in vivo models mainly because chicken and mice are anergic to LPS/LOS and do not exert pro-inflammatory immune responses upon *C. jejuni* colonization. Other vertebrate species such as the highly LPS/LOS susceptible piglets and ferrets have been successfully used for studying numerous aspects of *C. jejuni* pathogenesis, but the limited availability of genetically standardized animals restricted the broad use of these model organisms for campylobacteriosis research (Vitovec et al. 1989; Babakhani et al. 1993; Fox et al. 1987; Stahl et al. 2014). Besides the convenience in generating sufficient numbers of vertebrate animals, the physiological and genetic similarities between mice and humans brought mice into the spotlight as model organism for the study of human biology. Nevertheless, despite their phylogenetic relatedness, a multitude of differences with regard to genetic/epigenetic and to environment evolvement among many others limit final conclusions drawn from experimental procedures (Perlman 2016). As compared to humans, conventional laboratory wild-type mice, for instance, are highly resistant to stable gastrointestinal colonization by *C. jejuni* due to the physiological CR exerted by the murine gut microbiota composition (Heimesaat and Bereswill 2015; Bereswill et al. 2011; Haag et al. 2012c; Masanta et al. 2013; Dorrell and Wren 2007). These limitations led to a ban on murine *C. jejuni* infection models in the late twentieth century (Heimesaat and Bereswill 2015). A limited number of studies used wild-type mice with a defined, restricted, or even absent gut microbiota. Such alterations in the gut microbiota composition led to an efficient colonization of the intestinal tract by *C. jejuni* at high loads, resulting in intestinal pro-inflammatory responses and overt intestinal inflammation (Chang and Miller 2006). Furthermore, genetically modified mice such as MyD88 and NF-κB deficient mice (Fox et al. 2004; Watson et al. 2007) have been successfully used for the study of *C. jejuni* pathogenicity. However, the use of these animals is limited, since even under naïve conditions mice are prone to develop severe clinical conditions due to their compromised immune functions in consequence of respective gene deficiencies (Heimesaat and Bereswill 2015). Throughout the past years, several investigators including ourselves successfully developed various mouse models for campylobacteriosis research. For instance, Mansfield and co-workers (2007) established an up-and-coming murine *C. jejuni* infection model applying IL-10 deficient mice, which displayed severe symptoms upon pathogenic infection as seen in human campylobacteriosis. Hence, novel murine *C. jejuni* infection models, which are based on abrogation of CR and LOS resistance by antibiosis and genetic or dietary modifications, will be further described in detail below.

4.1 Germfree and Secondary Abiotic Mice

Remarkably, investigations in the 1980s demonstrated that *C. jejuni* was able to colonize the intestines of germfree and antibiotic-pretreated mice (Yrios and Balish 1986a, c; Jesudason et al. 1989; Yrios and Balish 1985, 1986b). Several *C. jejuni* strains isolated from human fecal samples caused clinical symptoms such as diarrhea in germfree athymic and euthymic BALB/c mice upon peroral infection (Yrios and Balish 1986c). In line with this, our own studies revealed that the depletion of the murine microbiota by broad-spectrum antibiotic treatment facilitated *C. jejuni* colonization in the gastrointestinal tract of secondary abiotic wild-type mice, resulting in pro-inflammatory immune responses in the colon upon peroral infection (Bereswill et al. 2011; Heimesaat and Bereswill 2015). Even following single antibiotic pretreatment with vancomycin (Stahl et al. 2014), ciprofloxacin, penicillin (Iizumi et al. 2016), or ampicillin (O'Loughlin et al. 2015) conventional wild-type mice could be colonized by *C. jejuni* upon oral challenge. Before infection, vancomycin-treated mice displayed depleted intestinal *Bacteroidetes* and *Clostridia* (Stahl et al., 2014), whereas ampicillin application was accompanied with decreased *Firmicutes* such as *Enterococcus faecalis* in the murine intestinal tract (O'Loughlin et al. 2015). Conversely, in one study *C. jejuni* strain 11168 was able to colonize the intestine of conventionally colonized C57BL/6 mice exhibiting a robust Th1-directed anti-*C. jejuni*-specific antibody response for several weeks (Mansfield et al. 2007). The absence of the CR against *C. jejuni* colonization in conventional mice is most possibly due to differences in the microbiota composition. It was experimentally proven that *C. jejuni* colonize conventional mice in the presence of elevated *E. coli* concentrations in the intestinal microbiota (Haag et al. 2012c). Moreover, successful *C. jejuni* colonization of wild-type mice with such a modified gut microbiota resulted in colonic epithelial apoptosis and both innate and adaptive pro-inflammatory immune responses mimicking key features of initial human campylobacteriosis. However, these mice did not develop typical clinical symptoms such as bloody diarrhea (Bereswill et al. 2011; Haag et al. 2012c; Mousavi et al. 2020a).

Despite their close evolutionary relationship, several *Campylobacter* species display different intestinal colonization properties (Alter et al. 2011; Genger et al. 2020). It has been recently shown, for instance, that *C. coli,* but not *C. jejuni,* could effectively colonize conventional wild-type mice that had been perorally challenged with high loads of the respective strain even without antibiotic pretreatment (Genger et al. 2020; Bereswill et al. 2011). These results further underline that murine CR against *Campylobacter* might be species- or even strain-specific, a phenomenon which needs to be investigated in more detail in the future.

4.2 Human Microbiota-Associated Mice

Given the possibilities in generating sufficient numbers of experimental animals in a relatively short period of time, the convenience in housing and handling as well as the genetical similarities with humans, mice constitute the most commonly used animal models in clinical research. However, there are distinct differences between mice and men by nature (Lundberg 2019). Among these, as already stated earlier, conventionally colonized laboratory wild-type mice display a strong CR against (entero)pathogens including *C. jejuni* determined by their gut microbiota composition. Given the importance of the gut microbiota during initiation and perpetuation of intestinal inflammatory morbidities in mice and men (Heimesaat et al. 2006, 2010, 2007a, 2007b; Erridge et al. 2010; Fiebiger et al. 2016; Haag et al. 2012c; Bereswill et al. 2011), we aimed at generating an experimental model that enabled us to elucidate the triangle relationships, i.e., the "ménage à trois" between the enteropathogen *C. jejuni,* the vertebrate host immunity, and the human commensal gut microbiota. Therefore, secondary abiotic wild-type mice were perorally subjected to a complex human gut microbiota by fecal microbiota transplantation (FMT) derived from healthy human donors. Remarkably, the human gut microbiota could sufficiently establish within the murine host during the first week and remained relatively stable for up to six weeks post-challenge (Bereswill et al. 2011). Following peroral infection, *C. jejuni* could stably colonize alongside the intestinal tract of these human microbiota-associated (hma) mice, which was, however, not the case in a secondary abiotic control group that had obtained a complex murine gut microbiota by FMT. This further provides evidence that it is the host-specific composition of the gut microbiota that renders the vertebrate host resistant against or susceptible toward enteropathogens including *C. jejuni*. Remarkably, *C. jejuni* induced apoptotic changes in colonic epithelia and both innate and adaptive pro-inflammatory immune responses in the infected large intestines that can also be observed in *C. jejuni*-infected patients. However (with respect to the gut microbiota composition), "humanized" mice did not display overt *C. jejuni*-induced clinical signs such as bloody diarrhea, for instance. Previously, we were able to show that treatment of *C. jejuni*-infected secondary abiotic and also of hma wild-type mice by murine FMT could effectively lower intestinal pathogen loads and alleviate induced intestinal and remarkably, even systemic immune responses (Heimesaat et al. 2019a, b). These preclinical murine FMT intervention studies suggest that changes in the gut microbiota composition upon application of pre- or probiotic formulations might be promising future options to treat pathogen shedding by asymptomatic carriers which is critical in food production and immunosuppressed individuals, for instance.

As *C. jejuni* infection model, hma mice further constitute valuable tools to investigate *C. jejuni* virulence factors in vivo. Orally challenged hma mice with *C. jejuni* isogenic mutants deficient in the formate dehydrogenase subunit D (Δ*fdhD*) or in the formic acid receptor (Δ*cj0952c*) harbored comparable bacterial loads in the gastrointestinal tract as compared to the wild-type strain (Bereswill et al. 2011). However, mice infected with either mutant strains displayed less pronounced immunopathological responses as indicated by lower numbers of colonic apoptotic epithelial cells, neutrophilic granulocytes, T lymphocytes, and regulatory T cells, resulting in less distinct colonic secretion of pro-inflammatory cytokines including IL-6 as compared to parental strain-infected mice. Hence, these results point toward the essential role of formic acid metabolism in the pathogenicity of *C. jejuni* in vivo (Bereswill et al. 2011).

4.3 Infant Mice

Despite the exposure of the fetus to the maternal bloodstream, which crosses the placenta and enters the amniotic fluid, the colonization of commensal bacteria starts as early as during birth and the first year of life and has a strong impact on lifelong health (Walker 2017). In fact, the physiological CR against pathogens including *C. jejuni* develops in an age-dependent fashion, with high susceptibility during very early life (Haag et al. 2012a). Distinct differences in TLR4 expressing features between fetal, neonatal, and adult murine intestinal epithelial cells have been demonstrated previously given that fetal, but not adult intestinal epithelial cells displayed susceptibility to bacterial LPS (Lotz et al. 2006).

In early studies, the infant mouse model was commonly used to unravel *C. jejuni* virulence factors (Newell 1986; Newell et al. 1985; Newell and Pearson 1984). Additionally, various so-called competition assays, in which the colonization of a *C. jejuni* mutant strain was compared to a parental counterpart strain, revealed stable intestinal colonization of *C. jejuni* in infant mice (Newell et al. 1985; Diker et al. 1992). Remarkably, cultural analyses of the colonic microbiota composition in infant mice immediately after weaning (i.e., 3 weeks of age) indicated higher intraluminal *E. coli* loads but lower *Lactobacillus* spp. numbers as compared to adult mice (3 months of age) (Haag et al. 2012a). In line with our study showing that exogenous application of *E. coli* subsequently leading to increased commensal enterobacterial loads could abrogate CR against *C. jejuni* (Haag et al. 2012c), the higher *E. coli* loads in infant mice might be one of the factors explaining the susceptibility of infant mice towards *C. jejuni.* Remarkably, as opposed to adult wild-type mice with a modified gut microbiota, *C. jejuni* was able to cause clinical signs upon peroral challenge of infant mice such as diarrhea with pronounced mucus discharge and blood during the first week that was self-limited and resolved within two weeks (Kazmi et al. 1984; Haag et al. 2012a).

Based upon the results obtained from infection experiments in infant mice, the role of distinct immune cells and of inflammatory mediators involved in the initiation and perpetuation of campylobacteriosis was further examined. Studies revealed that IL-23, IL22, and IL-18 are differentially involved in mediating *C. jejuni*-induced immunopathology (Heimesaat et al. 2016). Infant mice lacking IL-18 showed higher susceptibility to *C. jejuni* colonization, increases in numbers of innate as well as adaptive immune cells such as neutrophils, T and B lymphocytes in the colonic mucosa and lamina propria, respectively, that were less distinct when compared to infected wild-type counterparts. In contrast to IL-18 deficient mice, the secretion of pro-inflammatory mediators, including TNF-α, IFN-γ, IL-6, and monocyte chemoattractant protein-1 (MCP-1), was more pronounced in IL-23p19$^{-/-}$ mice upon *C. jejuni* infection (Heimesaat et al. 2016). These results underline the complex *C. jejuni* immunopathogenesis, which can be further investigated using the infant mouse as a promising infection model.

The infant mouse infection model was further used to investigate the impact of the *htrA* gene, encoding for a serine protease, in pathogen–host interactions and induction of immunopathology upon *C. jejuni* infection. To address this, conventionally colonized infant mice were infected with the *C. jejuni* knockout mutant Δ*htrA*. Mutant strain-infected mice were colonized at high loads but were less distinctly suffering from clinical signs such as bloody diarrhea and displayed less pronounced pathogen-induced colonic epithelial apoptotic and intestinal as well as extra-intestinal pro-inflammatory immune responses as compared to wild-type strain-infected controls. These results hence provide evidence for the importance of the serine protease HtrA as a new virulence factor mediating *C. jejuni*-induced disease in vivo (Heimesaat et al. 2014b).

4.4 Zinc-Deficient Mice

Zinc constitutes a nutritionally fundamental trace element and is after iron the second most abundant essential trace metal in the human body (Gammoh and Rink 2017). Many immunological pathways are modulated by zinc, and it is of note that the NF-κB and thus TLR4 signaling pathways are suppressed by zinc (Gammoh and Rink

2017). Furthermore, previous investigations revealed that malnourishment alters the composition of the gut microbiota and in consequence, the nutrient flow and immune responses of the host (Mayneris-Perxachs et al. 2016; Ghosh et al. 2014). Therefore, zinc deficiency is accompanied by compromised immune functions and metabolic flexibility, resulting in higher susceptibility to bacterial infections as observed in zinc-deficient mice generated by dietary zinc deprivation (Corbin et al. 2008). Most importantly, zinc is required for LOS/LPS resistance in mice and thus, zinc deficiency enhances the susceptibility of the animals to inflammation induced by *C. jejuni* LOS during intestinal infection. Remarkably, zinc treatment prevents mice from endotoxin-induced lethality (Snyder and Walker 1976). The deficiency in zinc-dependent antimicrobial peptides such as calprotectin secreted by neutrophils compromises the physiological CR and, hence, facilitates intestinal colonization of mice by enteropathogens such as *C. jejuni* (Corbin et al. 2008). In line with these results, it is noteworthy that decreased serum zinc concentrations were measured in patients suffering from bacterial infections (Vikbladh 1951; Auerbach 1965), indicating a negative correlation between serum zinc levels and susceptibility to infections.

Giallourou and co-workers reported that zinc deficiency had a direct effect on *C. jejuni* colonization and induced inflammatory responses in antibiotic-pretreated mice receiving a zinc-deprived diet (Giallourou et al. 2018). In support of other studies, the authors reported that in addition to feeding the zinc-deficient diet the application of antibiotics was necessary to overcome CR and to facilitate *C. jejuni* colonization in the murine intestinal tract resulting in diarrhea and persistent bloody stool upon *C. jejuni* infection, thus mimicking the "classical" clinical symptoms of human campylobacteriosis (Giallourou et al. 2018). In pretreated mice receiving a zinc-deprived diet, *C. jejuni* induced both intestinal and systemic inflammatory responses, as indicated by pronounced luminal mucus discharge, gut epithelial damage, and immune cells including neutrophilic granulocytes infiltrating the infected intestinal mucosa and lamina propria which can also be observed in human campylobacteriosis (Giallourou et al. 2018). Given that standard laboratory mouse feeds are enriched with pharmacological zinc concentrations, the zinc-depleted murine infection model provides an excellent novel tool to investigate various zinc-dependent aspects of *C. jejuni*-induced immunopathogenesis. Moreover, zinc may be considered as a pharmacological agent against Gram-negative bacterial infections in particular and can reduce infectious burden via its immune-modulatory (i.e., anti-inflammatory) properties (Thambiayya et al. 2012; Rink and Haase 2007; Haase and Rink 2009; Walker and Black 2004). It is further noteworthy that patients suffering from Crohn's disease were more prone to *C. jejuni* infections as compared to individuals without preexisting intestinal morbidities (Sarabi Asiabar et al. 2019) and that the detection of *C. jejuni* in gastrointestinal biopsies correlated with decreased serum zinc concentration in Crohn's disease patients (Vikbladh 1951; Satsangi et al. 1987).

4.5 SIGIRR-Deficient Mice

The single Ig IL-1 receptor-related molecule (SIGIRR) plays a critical role in intestinal inflammation by maintaining the microbial tolerance of the colonic epithelium and inhibiting the Toll/interleukin-1 receptor and MyD88-dependent signaling pathways (Xiao et al. 2007; Wald et al. 2003). Mice deficient in SIGIRR (Sigirr$^{-/-}$) were used to study gastrointestinal infections caused by *Citrobacter rodentium* and *Salmonella enterica* serovar Typhimurium (Sham et al. 2013). Following oral application of either pathogen, Sigirr$^{-/-}$ mice were rapidly infected and developed intestinal inflammation as opposed to wild-type mice. Upon oral *C. jejuni* challenge, however, conventionally colonized Sigirr$^{-/-}$ mice could only sporadically be colonized and displayed, if at all, rather mild signs of intestinal inflammation (Sham et al. 2013). Peroral application of a single antibiotic compound, namely vancomycin, resulted in pronounced gut microbiota shifts in conventionally colonized Sigirr$^{-/-}$ mice characterized by depleted *Bacteroidetes* and *Clostridia,* and conversely, promoted growth of lactobacilli (Russell et al. 2012). These alterations of the murine microbiota composition allowed *C. jejuni* to colonize the intestinal lumen of Sigirr$^{-/-}$ mice at high loads upon oral infection (Stahl et al. 2014) and to induce gross intestinal pathology that was accompanied by higher expression levels of pro-inflammatory mediators such as TNF-α, IFN-γ, and IL-17 in the infected Sigirr$^{-/-}$ as compared to wild-type mice (Stahl et al. 2014). Overall, the induced clinical course of disease in *C. jejuni*-infected Sigirr$^{-/-}$ mice was, however, rather moderate and self-limiting (Stahl et al. 2014; Stahl and Vallance 2015).

The Sigirr$^{-/-}$ mouse model could also be applied to elucidate the role of TLR4 in campylobacteriosis. *C. jejuni* infection experiments revealed that TLR4-deficient Sigirr$^{-/-}$ mice were clinically far less compromised and exhibited less distinct pathogen-induced intestinal as well as extra-intestinal including systemic sequelae as compared to infected Sigirr$^{-/-}$ counterparts (Stahl et al. 2014). Following peroral infection of TLR4$^{-/-}$ Sigirr$^{-/-}$ mice, *C. jejuni* colonized the intestines at high loads, but induced only mild, if any, signs of enteritis. Compared to Sigirr$^{-/-}$ counterparts, TLR4-deficient Sigirr$^{-/-}$ mice displayed less cecal TNF-α and IFN-γ expression upon infection, which was comparable to uninfected control mice (Stahl et al. 2014). These results further underline the essential role of TLR4-dependent LOS signaling in *C. jejuni* pathogenesis. Overall, given the immunopathological similarities between *C. jejuni*-infected Sigirr$^{-/-}$ mice and human campylobacteriosis, these mice can be used as a reliable in vivo model of self-limiting human *C. jejuni* infection (Stahl and Vallance 2015).

4.6 IL-10 Deficient Mice

The so far described murine infection models resemble the self-limiting course of *C. jejuni* infection in humans. The anti-inflammatory cytokine IL-10 plays a central role in the self-limitation of inflammation and is required for murine resistance to LOS and LPS via the inhibition of TLR-4-dependent innate immune responses (Haag et al. 2012b; Warren et al. 2010; 2006, Robertson et al. 2007; Emoto et al. 2003; Moore et al. 2001). In order to generate a robust and reproducible clinical mouse model for severe campylobacteriosis, we took advantage of IL-$10^{-/-}$ mice that had been successfully applied for the study of *C. jejuni* infection in earlier studies (Mansfield et al. 2007; Lippert et al. 2009). Therefore, IL-$10^{-/-}$ mice were pretreated with broad-spectrum antibiotics starting immediately after weaning in order to abolish the physiological CR preventing *C. jejuni* infection and to eliminate potential colitogenic stimuli from the commensal gut microbiota leading to chronic colitis in IL-$10^{-/-}$ mice after 3 months or more (Haag et al. 2012b). *C. jejuni* was able to colonize the gastrointestinal tract of secondary abiotic IL-$10^{-/-}$ mice at high pathogenic loads within one week following peroral infection and caused non-self-limiting acute enterocolitis, characterized by wasting and bloody inflammatory diarrhea as seen in severe human campylobacteriosis (Haag et al. 2012b; 2019, Mousavi et al. 2020b, c). In addition, both *C. jejuni*-induced innate and adaptive pro-inflammatory immune responses were not limited to the intestinal tract, but could also be observed in extra-intestinal, including systemic compartments (Haag et al. 2012b; Heimesaat et al. 2014a). The isolation of *C. jejuni* from mesenteric lymph nodes, liver, kidneys, spleen, and cardiac blood indicated the translocation ability of *C. jejuni* from the inflamed intestines to extra-intestinal and even systemic tissue sites (Heimesaat et al. 2014a). It is of note that commensal *E. coli* lacking any invasive or other pathogenic properties did not induce pathology in this model (Haag et al. 2012b). Overall, the secondary abiotic IL-$10^{-/-}$ mouse model has been proven valuable for investigating the immunopathological impact of defined *C. jejuni* virulence factors involved in campylobacteriosis (Heimesaat and Bereswill 2015; Heimesaat et al. 2014a; Masanta et al. 2013; Fiebiger et al. 2016; Schmidt et al. 2019a, b; Heimesaat et al. 2014c). For instance, following peroral infection with a *C. jejuni* strain lacking Cj0268c, which is involved in cellular adhesion and invasion, mice displayed less severe pathogen-induced immunopathology in the intestinal tract as compared to mice that had been infected with the parental or the complemented strain (Heimesaat et al. 2014c). Furthermore, upon peroral challenge of secondary abiotic IL-$10^{-/-}$ mice with *C. jejuni* Δ*flaA/B* (lacking the flagella genes *flaA* and *flaB*) or *C. jejuni* Δ*cadF* (lacking the adhesin CadF), we could show that *C. jejuni* flagellar motility, but not adhesion exerted by CadF, is required for induced immunopathology in the murine host (Schmidt et al. 2019a). Besides the infant mouse model, also secondary abiotic IL-$10^{-/-}$ mice proved to serve as a reliable model in order to dissect the immunopathological impact of the serine protease HtrA during peracute *C. jejuni*-induced disease in more detail. Whereas 6 days following wild-type strain infection, mice were suffering from severe ulcerative enterocolitis, deletion of *C. jejuni* HtrA (Δ*htrA*) resulted in only mild disease (Fig. 1) (Heimesaat et al. 2014a). Remarkably, upon peroral infection with an isogenic *C. jejuni* strain carrying a protease-inactive HtrA due to a single-point mutation at S197A in the active center (S197A), mice also displayed far less distinct microscopic inflammatory sequelae as compared to wild-type infected counterparts (Fig. 1) (Schmidt et al. 2019b).

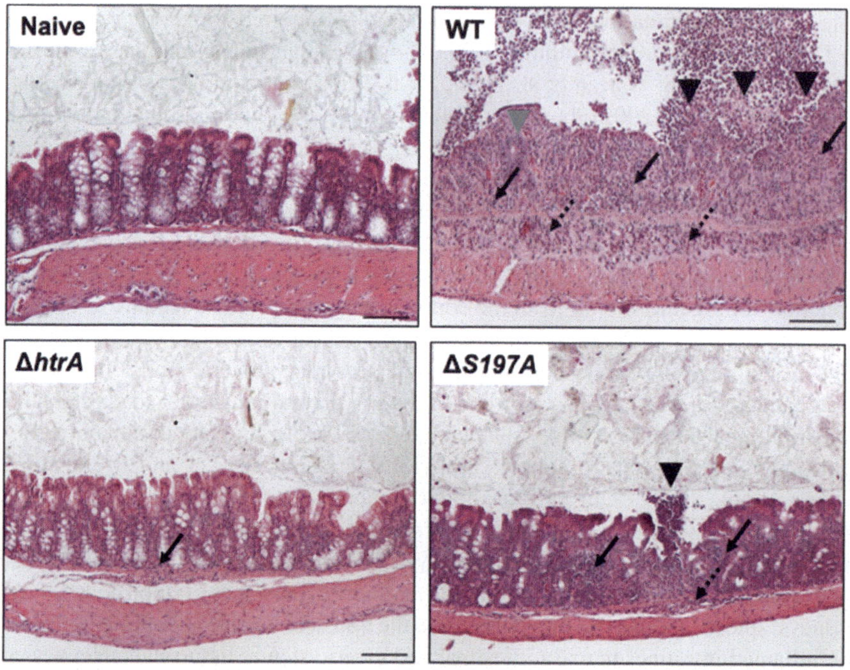

Fig. 1 Colonic histopathological changes following peroral infection of secondary abiotic IL-10 deficient mice with wild-type *C. jejuni* or strains mutated in the serine protease HtrA. Secondary abiotic mice were generated by broad-spectrum antibiotic treatment. Histopathological changes in the colonic mucosa and lamina propria were assessed in hematoxylin and eosin stained paraffin sections of large intestinal *ex vivo* biopsies obtained from uninfected (naïve) mice (upper left), of mice 6 days following peroral infection with the *C. jejuni* wild-type strain (NCTC11168 WT; upper right), with isogenic *C. jejuni* strains either lacking the serine protease HtrA (NCTC11168Δ*htrA*; lower left) or carrying protease-inactive HtrA with a single-point mutation in the active center (NCTC11168Δ*htrA*/*htrA* S197A; lower right) as illustrated in representative photomicrographs out of three representative experiments (100 × magnification; scale bar 100 μm). Solid and dotted arrows indicate mucosal and submucosal infiltrates, respectively; black arrow heads mucosal bleeding due to ulcerations, and gray arrowheads goblet cell loss

In support of the above-discussed findings, IL-10 deficient mice with three different genetic backgrounds (C57BL/6, non-obese diabetic (NOD), and C3H/HeJ lacking TLR4) displayed more pronounced *C. jejuni*-induced colitis than mice without IL-10 gene deficiency (Mansfield et al. 2008). The less distinct intestinal as well as extra-intestinal including systemic sequelae upon *C. jejuni* infection in TLR4-deficient IL-10$^{-/-}$ mice provided in vivo evidence for the major role of LOS/TLR4 signaling as immunopathological mechanisms of *C. jejuni* infection (Haag et al. 2012b; Bereswill et al. 2011). Following *C. jejuni* infection, secondary abiotic NOD2$^{-/-}$ IL-10$^{-/-}$ mice harbored the pathogen at high loads, but developed less severe enterocolitis as compared to IL-10$^{-/-}$ counterparts (Heimesaat et al. 2017) indicating that NOD2 signaling increases severity of campylobacteriosis. Further studies revealed the regulatory role of mammalian target of rapamycin (mTOR) and the phosphatidylinositol 3-kinase-γ (PI3K-γ, upstream of mTOR) in *C. jejuni*-induced colitis (Sun et al. 2012, 2013). The inactivation of mTOR and PI3K-γ via rapamycin affects several innate immune signaling pathways involved in TLR4 activation and IL-10 production (Snyder and Amiel 2018; Brown et al. 2011; Ohtani et al. 2012; Lorne et al. 2009) and thus ameliorated *C. jejuni*-induced intestinal accumulation of neutrophilic granulocytes which correlated with reduced inflammatory responses in the intestinal tract (Sun et al. 2012, 2013; He et al. 2019).

Since antimicrobial resistance in general has turned out to be a major concern for public health and also *Campylobacter* become increasingly resistant to clinically important antibiotics (Luangtongkum et al. 2009), it is of utmost relevance to search for novel antibiotics-independent approaches to combat and/or even to prevent campylobacteriosis. In our preclinical intervention studies, we addressed the potential anti-pathogenic and immunomodulatory properties of the polyphenolic compounds curcumin and resveratrol, of the cresol analogue carvacrol, of vitamin D (25-OH-cholecalciferol), and of vitamin C (ascorbate) in our clinical murine infection model for acute human campylobacteriosis (Mousavi et al. 2019, 2020b, c; Lobo de Sá et al. 2019). Secondary abiotic IL-10$^{-/-}$ mice that were treated with vitamin C or carvacrol, for instance, harbored lower colonic pathogen loads and were suffering from less severe enterocolitis upon *C. jejuni* infection as compared to placebo control animals. Additionally, these treatment regimens dampened apoptotic epithelial and pro-inflammatory immune cell responses in the intestines that were accompanied by less pronounced pro-inflammatory cytokine secretion (Mousavi et al. 2020b, c). Potent immunomodulatory properties during acute *C. jejuni*-induced enterocolitis could also be achieved by peroral application of vitamin D in *C. jejuni* infection (Mousavi et al. 2019). The alleviated *C. jejuni*-induced disease observed in our applied clinical murine model for peracute human campylobacteriosis highlights the application of defined vitamins and plant-derived compounds with antimicrobial and/or anti-inflammatory properties as a promising novel option for both the treatment of campylobacteriosis and the reduction of *C. jejuni* colonization in the intestines of vertebrate livestock animals.

5 Concluding Remarks

In conclusion, series of studies have shown that the colonization resistance against *C. jejuni* caused by the murine gut microbiota composition could be abrogated via antibiotic treatment, thereby opening novel avenues for the development of reliable murine infection models for the analyses of pathogen–host interactions. For example, *C. jejuni* could effectively colonize the gut of antibiotics pretreated mice with deficiencies in IL-10 or SIGIRR and cause clinical signs similar to those observed during human campylobacteriosis. Furthermore, studies with TLR4 IL-10 double-deficient as well as TLR4 SIGIRR double-deficient mice provided further evidence for the essential role of LOS signaling in *C. jejuni* infection. These novel findings underline the pivotal impact of LOS as a major trigger for *C. jejuni*-mediated inflammation and by providing promising options for novel targets in the combat of human campylobacteriosis.

Acknowledgements This work is supported by the German Federal Ministry of Education and Research (BMBF) in frame of the zoonoses research consortium PAC-Campylobacter to S.B. and M.M.H. (project IP7 01KI1725D).

References

Adler HE, DaMassa AJ (1979) Toxicity of endotoxin to chicks. Avian Dis 23(1):174–178

Alter T, Bereswill S, Glünder G, Haag L-M, Hänel I, Heimesaat M, Lugert R, Rautenschlein S, Weber R, Zautner A (2011) Die Campylobacteriose des Menschen. Bundesgesundheitsblatt-Gesundheitsforschung-Gesundheitsschutz 54(6):728–734. https://doi.org/10.1007/s00103-011-1289-y

Atarashi K, Tanoue T, Oshima K, Suda W, Nagano Y, Nishikawa H, Fukuda S, Saito T, Narushima S, Hase K, Kim S, Fritz JV, Wilmes P, Ueha S, Matsushima K, Ohno H, Olle B, Sakaguchi S, Taniguchi T, Morita H, Hattori M, Honda K (2013) Treg induction by a rationally selected mixture of clostridia strains from the human microbiota. Nature 500(7461):232–236. https://doi.org/10.1038/nature12331

Auerbach S (1965) Zinc content of plasma, blood, and erythrocytes in normal subjects and in patients with hodgkin's disease and various hematologic disorders. J Lab Clin Med 65:628–637

Babakhani FK, Bradley GA, Joens LA (1993) Newborn piglet model for campylobacteriosis. Infect Immun 61(8):3466–3475. https://doi.org/10.1128/IAI.61.8.3466-3475.1993

Backert S, Tegtmeyer N, Cróinín TÓ, Boehm M, Heimesaat MM (2017) Chapter 1 - Human campylobacteriosis. In: Klein G (ed) *Campylobacter*. Academic Press, pp 1–25. https://doi.org/10.1016/B978-0-12-803623-5.00001-0

Beery JT, Hugdahl MB, Doyle MP (1988) Colonization of gastrointestinal tracts of chicks by *Campylobacter jejuni*. Appl Environ Microbiol 54(10):2365–2370. https://doi.org/10.1128/AEM.54.10.2365-2370.1988

Bereswill S, Fischer A, Plickert R, Haag LM, Otto B, Kuhl AA, Dasti JI, Zautner AE, Munoz M, Loddenkemper C, Gross U, Gobel UB, Heimesaat MM (2011) Novel murine infection models provide deep insights into the "menage a trois" of *Campylobacter jejuni*, microbiota and host innate immunity. PLoS ONE 6(6):e20953. https://doi.org/10.1371/journal.pone.0020953

Boehm M, Simson D, Escher U, Schmidt AM, Bereswill S, Tegtmeyer N, Backert S, Heimesaat MM. (2018) Function of Serine Protease HtrA in the Lifecycle of the Foodborne Pathogen Campylobacter jejuni. Eur J Microbiol Immunol (Bp). 8(3):70–77. https://doi.org/10.1556/1886.2018.00011

Bolton DJ (2015) Campylobacter virulence and survival factors. Food Microbiol 48:99–108. https://doi.org/10.1016/j.fm.2014.11.017

Brown J, Wang H, Hajishengallis GN, Martin M (2011) TLR-signaling networks: an integration of adaptor molecules, kinases, and cross-talk. J Dent Res 90(4):417–427. https://doi.org/10.1177/0022034510381264

Buffie CG, Pamer EG (2013) Microbiota-mediated colonization resistance against intestinal pathogens. Nature Rev Immunol 13(11):790–801. https://doi.org/10.1038/nri3535

Cario E, Podolsky DK (2000) Differential alteration in intestinal epithelial cell expression of toll-like receptor 3 (TLR3) and TLR4 in inflammatory bowel disease. Infect Immun 68(12):7010. https://doi.org/10.1128/IAI.68.12.7010-7017.2000

Chang C, Miller JF (2006) *Campylobacter jejuni* colonization of mice with limited enteric flora. Infect Immun 74(9):5261–5271. https://doi.org/10.1128/IAI.01094-05
Cochet F, Peri F (2017) The Role of Carbohydrates in the Lipopolysaccharide (LPS)/Toll-Like Receptor 4 (TLR4) Signalling. Int J Molecul Sci 18(11). https://doi.org/10.3390/ijms18112318
Corbin BD, Seeley EH, Raab A, Feldmann J, Miller MR, Torres VJ, Anderson KL, Dattilo BM, Dunman PM, Gerads R, Caprioli RM, Nacken W, Chazin WJ, Skaar EP (2008) Metal chelation and inhibition of bacterial growth in tissue abscesses. Science 319(5865):962–965. https://doi.org/10.1126/science.1152449
Cummings JH, Pomare EW, Branch WJ, Naylor CP, Macfarlane GT (1987) Short chain fatty acids in human large intestine, portal, hepatic and venous blood. Gut 28(10):1221–1227. https://doi.org/10.1136/gut.28.10.1221
Dasti JI, Tareen AM, Lugert R, Zautner AE, Gross U (2010) *Campylobacter jejuni*: a brief overview on pathogenicity-associated factors and disease-mediating mechanisms. Int J Med Microbiol 300(4):205–211. https://doi.org/10.1016/j.ijmm.2009.07.002
de Zoete MR, Keestra AM, Roszczenko P, van Putten JP (2010) Activation of human and chicken toll-like receptors by *Campylobacter* spp. Infect Immun 78(3):1229–1238. https://doi.org/10.1128/iai.00897-09
Dicksved J, Ellstrom P, Engstrand L, Rautelin H (2014) Susceptibility to *Campylobacter* infection is associated with the species composition of the human fecal microbiota. mBio 5(5):e01212–01214. https://doi.org/10.1128/mBio.01212-14
Diker K, Hascelik G, Diker S (1992) Colonization of infant mice with flagellar variants of *Campylobacter jejuni*. Acta microbiologica Hungarica 39(2):133–136
Dorrell N, Wren BW (2007) The second century of *Campylobacter* research: recent advances, new opportunities and old problems. Curr Opin Infect Dis 20(5):514–518. https://doi.org/10.1097/QCO.0b013e3282a56b15
Ducarmon QR, Zwittink RD, Hornung BVH, van Schaik W, Young VB, Kuijper EJ (2019) Gut microbiota and colonization resistance against bacterial enteric infection. Microbiol Molecul Biol Rev MMBR 83(3). https://doi.org/10.1128/mmbr.00007-19
EFSA (2019) European Food Safety Authority. european centre for disease, prevention control,the european union one health 2018 Zoonoses Report. EFSA J 17(12):e05926. https://doi.org/10.2903/j.efsa.2019.5926
Ellström P, Hansson I, Nilsson A, Rautelin H, Engvall EO (2016) Lipooligosaccharide locus classes and putative virulence genes among chicken and human *Campylobacter jejuni* isolates. BMC Microbiol 16(1):116. https://doi.org/10.1186/s12866-016-0740-5
Emoto M, Emoto Y, Brinkmann V, Miyamoto M, Yoshizawa I, Staber M, van Rooijen N, Hamann A, Kaufmann SH (2003) Increased resistance of LFA-1-deficient mice to lipopolysaccharide-induced shock/liver injury in the presence of TNF-alpha and IL-12 is mediated by IL-10: a novel role for LFA-1 in the regulation of the proinflammatory and anti-inflammatory cytokine balance. J Immunol 171(2):584–593. https://doi.org/10.4049/jimmunol.171.2.584
Erridge C, Duncan SH, Bereswill S, Heimesaat MM (2010) The induction of colitis and ileitis in mice is associated with marked increases in intestinal concentrations of stimulants of TLRs 2, 4, and 5. PLoS ONE 5(2):e9125. https://doi.org/10.1371/journal.pone.0009125
Ferrero RL, Lee A (1988) Motility of *Campylobacter jejuni* in a viscous environment: comparison with conventional rod-shaped bacteria. J Gen Microbiol 134(1):53–59. https://doi.org/10.1099/00221287-134-1-53
Fiebiger U, Bereswill S, Heimesaat MM (2016) Dissecting the interplay between intestinal microbiota and host immunity in health and disease: lessons learned from germfree and gnotobiotic animal models. Eur J Microbiol Immunol (Bp) 6(4):253–271. https://doi.org/10.1556/1886.2016.00036
Fox JG, Ackerman JI, Taylor N, Claps M, Murphy JC (1987) *Campylobacter jejuni* infection in the ferret: an animal model of human campylobacteriosis. Am J Veterinary Res 48(1):85–90
Fox JG, Rogers AB, Whary MT, Ge Z, Taylor NS, Xu S, Horwitz BH, Erdman SE (2004) Gastroenteritis in NF-kappaB-deficient mice is produced with wild-type *Camplyobacter jejuni* but not

with *C. jejuni* lacking cytolethal distending toxin despite persistent colonization with both strains. Infect Immun 72(2):1116–1125. https://doi.org/10.1128/iai.72.2.1116-1125.2004
Gammoh NZ, Rink L (2017) Zinc in infection and inflammation. Nutrients 9(6):624. https://doi.org/10.3390/nu9060624
Genger C, Kløve S, Mousavi S, Bereswill S, Heimesaat MM (2020) The conundrum of colonization resistance against *Campylobacter* reloaded: the gut microbota composition in conventional mice does not prevent from *Campylobacter coli* infection. Eur J Microbiol Immunol (Bp). https://doi.org/10.1556/1886.2020.00004
Ghosh TS, Gupta SS, Bhattacharya T, Yadav D, Barik A, Chowdhury A, Das B, Mande SS, Nair GB (2014) Gut microbiomes of Indian children of varying nutritional status. PLoS ONE 9(4):e95547. https://doi.org/10.1371/journal.pone.0095547
Giallourou N, Medlock GL, Bolick DT, Medeiros PH, Ledwaba SE, Kolling GL, Tung K, Guerry P, Swann JR, Guerrant RL (2018) A novel mouse model of *Campylobacter jejuni* enteropathy and diarrhea. PLoS Pathog 14(3):e1007083. https://doi.org/10.1371/journal.ppat.1007083
Gilbert M, Brisson JR, Karwaski MF, Michniewicz J, Cunningham AM, Wu Y, Young NM, Wakarchuk WW (2000) Biosynthesis of ganglioside mimics in *Campylobacter jejuni* OH4384. Identification of the glycosyltransferase genes, enzymatic synthesis of model compounds, and characterization of nanomole amounts by 600-mhz (1)h and (13)c NMR analysis. J Biol Chem 275 (6):3896–3906. https://doi.org/10.1074/jbc.275.6.3896
Gilbert M, Godschalk PC, Karwaski MF, Ang CW, van Belkum A, Li J, Wakarchuk WW, Endtz HP (2004) Evidence for acquisition of the lipooligosaccharide biosynthesis locus in *Campylobacter jejuni* GB11, a strain isolated from a patient with Guillain-Barre syndrome, by horizontal exchange. Infect Immun 72(2):1162–1165. https://doi.org/10.1128/iai.72.2.1162-1165.2004
Gilbert M, Karwaski MF, Bernatchez S, Young NM, Taboada E, Michniewicz J, Cunningham AM, Wakarchuk WW (2002) The genetic bases for the variation in the lipo-oligosaccharide of the mucosal pathogen, *Campylobacter jejuni*. Biosynthesis of sialylated ganglioside mimics in the core oligosaccharide. J Biol Chem 277(1):327–337. https://doi.org/10.1074/jbc.M108452200
Godschalk PC, Kuijf ML, Li J, St Michael F, Ang CW, Jacobs BC, Karwaski MF, Brochu D, Moterassed A, Endtz HP, van Belkum A, Gilbert M (2007) Structural characterization of *Campylobacter jejuni* lipooligosaccharide outer cores associated with Guillain-Barre and Miller Fisher syndromes. Infect Immun 75(3):1245–1254. https://doi.org/10.1128/iai.00872-06
Golec M (2007) Cathelicidin LL-37: LPS-neutralizing, pleiotropic peptide. Ann Agric Environ Med 14(1):1–4
Gradel KO, Nielsen HL, Schonheyder HC, Ejlertsen T, Kristensen B, Nielsen H (2009) Increased short- and long-term risk of inflammatory bowel disease after *salmonella* or *campylobacte*r gastroenteritis. Gastroenterology 137(2):495–501. https://doi.org/10.1053/j.gastro.2009.04.001
Guarner F, Malagelada JR (2003) Gut flora in health and disease. Lancet 361(9356):512–519. https://doi.org/10.1016/s0140-6736(03)12489-0
Guerry P, Szymanski CM, Prendergast MM, Hickey TE, Ewing CP, Pattarini DL, Moran AP (2002) Phase variation of *Campylobacter jejuni* 81-176 lipooligosaccharide affects ganglioside mimicry and invasiveness in vitro. Infect Immun 70(2):787–793. https://doi.org/10.1128/iai.70.2.787-793.2002
Haag LM, Fischer A, Otto B, Grundmann U, Kuhl AA, Gobel UB, Bereswill S, Heimesaat MM (2012a) *Campylobacter jejuni* infection of infant mice: acute enterocolitis is followed by asymptomatic intestinal and extra-intestinal immune responses. Eur J Microbiol Immunol (Bp) 2(1):2–11. https://doi.org/10.1556/EuJMI.2.2012.1.2
Haag LM, Fischer A, Otto B, Plickert R, Kuhl AA, Gobel UB, Bereswill S, Heimesaat MM (2012b) *Campylobacter jejuni* induces acute enterocolitis in gnotobiotic IL-10-/- mice via Toll-like-receptor-2 and -4 signaling. PLoS ONE 7(7):e40761. https://doi.org/10.1371/journal.pone.0040761

Haag LM, Fischer A, Otto B, Plickert R, Kuhl AA, Gobel UB, Bereswill S, Heimesaat MM (2012c) Intestinal microbiota shifts towards elevated commensal *Escherichia coli* loads abrogate colonization resistance against *Campylobacter jejuni* in mice. PLoS ONE 7(5):e35988. https://doi.org/10.1371/journal.pone.0035988

Haase H, Rink L (2009) Functional significance of zinc-related signaling pathways in immune cells. Ann Rev Nutri 29:133–152. https://doi.org/10.1146/annurev-nutr-080508-141119

Habib I, Louwen R, Uyttendaele M, Houf K, Vandenberg O, Nieuwenhuis EE, Miller WG, van Belkum A, De Zutter L (2009) Correlation between genotypic diversity, lipooligosaccharide gene locus class variation, and caco-2 cell invasion potential of *Campylobacter jejuni* isolates from chicken meat and humans: contribution to virulotyping. Appl Environ Microbiol 75(13):4277–4288. https://doi.org/10.1128/aem.02269-08

Hajjar AM, Ernst RK, Tsai JH, Wilson CB, Miller SI (2002) Human Toll-like receptor 4 recognizes host-specific LPS modifications. Nat Immunol 3(4):354–359. https://doi.org/10.1038/ni777

Halpin AL, Gu W, Wise ME, Sejvar JJ, Hoekstra RM, Mahon BE (2018) Post-Campylobacter Guillain Barré Syndrome in the USA: secondary analysis of surveillance data collected during the 2009-2010 novel Influenza A (H1N1) vaccination campaign. Epidemiol Infect 146(13):1740–1745. https://doi.org/10.1017/S0950268818001802

He Z, Gharaibeh RZ, Newsome RC, Pope JL, Dougherty MW, Tomkovich S, Pons B, Mirey G, Vignard J, Hendrixson DR, Jobin C (2019) *Campylobacter jejuni* promotes colorectal tumorigenesis through the action of cytolethal distending toxin. Gut 68(2):289–300. https://doi.org/10.1136/gutjnl-2018-317200

Heimesaat MM, Alutis M, Grundmann U, Fischer A, Tegtmeyer N, Bohm M, Kuhl AA, Gobel UB, Backert S, Bereswill S (2014a) The role of serine protease HtrA in acute ulcerative enterocolitis and extra-intestinal immune responses during *Campylobacter jejuni* infection of gnotobiotic IL-10 deficient mice. Front Cell Infect Microbiol 4:77. https://doi.org/10.3389/fcimb.2014.00077

Heimesaat MM, Alutis ME, Grundmann U, Fischer A, Gobel UB, Bereswill S (2016) The Role of IL-23, IL-22, and IL-18 in *Campylobacter Jejuni* Infection of Conventional Infant Mice. Eur J Microbiol Immunol (Bp) 6(2):124–136. https://doi.org/10.1556/1886.2016.00008

Heimesaat MM, Bereswill S (2015) Murine infection models for the investigation of *Campylobacter jejuni*–host interactions and pathogenicity. Berl Munch Tierarztl Wochenschr 128(3–4):98–103

Heimesaat MM, Bereswill S, Fischer A, Fuchs D, Struck D, Niebergall J, Jahn HK, Dunay IR, Moter A, Gescher DM, Schumann RR, Gobel UB, Liesenfeld O (2006) Gram-negative bacteria aggravate murine small intestinal Th1-type immunopathology following oral infection with *Toxoplasma gondii*. J Immunol 177(12):8785–8795. https://doi.org/10.4049/jimmunol.177.12.8785

Heimesaat MM, Fischer A, Alutis M, Grundmann U, Boehm M, Tegtmeyer N, Gobel UB, Kuhl AA, Bereswill S, Backert S (2014b) The impact of serine protease HtrA in apoptosis, intestinal immune responses and extra-intestinal histopathology during *Campylobacter jejuni* infection of infant mice. Gut Pathog 6:16. https://doi.org/10.1186/1757-4749-6-16

Heimesaat MM, Fischer A, Jahn HK, Niebergall J, Freudenberg M, Blaut M, Liesenfeld O, Schumann RR, Gobel UB, Bereswill S (2007a) Exacerbation of murine ileitis by Toll-like receptor 4 mediated sensing of lipopolysaccharide from commensal *Escherichia coli*. Gut 56(7):941–948. https://doi.org/10.1136/gut.2006.104497

Heimesaat MM, Fischer A, Siegmund B, Kupz A, Niebergall J, Fuchs D, Jahn HK, Freudenberg M, Loddenkemper C, Batra A, Lehr HA, Liesenfeld O, Blaut M, Gobel UB, Schumann RR, Bereswill S (2007b) Shift towards pro-inflammatory intestinal bacteria aggravates acute murine colitis via Toll-like receptors 2 and 4. PLoS ONE 2(7):e662. https://doi.org/10.1371/journal.pone.0000662

Heimesaat MM, Grundmann U, Alutis ME, Fischer A, Bereswill S (2017) Absence of Nucleotide-Oligomerization-Domain-2 Is associated with less distinct disease in *Campylobacter jejuni* infected secondary abiotic IL-10 deficient mice. Front Cell Infect Microbiol 7:322. https://doi.org/10.3389/fcimb.2017.00322

Heimesaat MM, Lugert R, Fischer A, Alutis M, Kuhl AA, Zautner AE, Tareen AM, Gobel UB, Bereswill S (2014c) Impact of *Campylobacter jejuni* cj0268c knockout mutation on intestinal colonization, translocation, and induction of immunopathology in gnotobiotic IL-10 deficient mice. PLoS ONE 9(2):e90148. https://doi.org/10.1371/journal.pone.0090148

Heimesaat MM, Mrazek K, Bereswill S (2019a) Murine Fecal Microbiota Transplantation Alleviates Intestinal and Systemic Immune Responses in *Campylobacter jejuni* Infected Mice Harboring a Human Gut Microbiota. Front Immunol 10:2272. https://doi.org/10.3389/fimmu.2019.02272

Heimesaat MM, Mrazek K, Bereswill S (2019b) Murine fecal microbiota transplantation lowers gastrointestinal pathogen loads and dampens pro-inflammatory immune responses in *Campylobacter jejuni* infected secondary abiotic mice. Sci Rep 9(1):19797. https://doi.org/10.1038/s41598-019-56442-7

Heimesaat MM, Nogai A, Bereswill S, Plickert R, Fischer A, Loddenkemper C, Steinhoff U, Tchaptchet S, Thiel E, Freudenberg MA, Gobel UB, Uharek L (2010) MyD88/TLR9 mediated immunopathology and gut microbiota dynamics in a novel murine model of intestinal graft-versus-host disease. Gut 59(8):1079–1087. https://doi.org/10.1136/gut.2009.197434

Hermans D, Pasmans F, Heyndrickx M, Van Immerseel F, Martel A, Van Deun K, Haesebrouck F (2012) A tolerogenic mucosal immune response leads to persistent *Campylobacter jejuni* colonization in the chicken gut. Crit Rev Microbiol 38(1):17–29. https://doi.org/10.3109/1040841x.2011.615298

HMPC (2012) Human Microbiome Project Consortium 2012. Struct Function Divers Healthy Human Microbiome Nature 486:207–214. https://doi.org/10.1038/nature11234

Hooper LV, Macpherson AJ (2010) Immune adaptations that maintain homeostasis with the intestinal microbiota. Nature Rev Immunol 10(3):159–169. https://doi.org/10.1038/nri2710

Hornung V, Rothenfusser S, Britsch S, Krug A, Jahrsdorfer B, Giese T, Endres S, Hartmann G (2002) Quantitative expression of toll-like receptor 1-10 mRNA in cellular subsets of human peripheral blood mononuclear cells and sensitivity to CpG oligodeoxynucleotides. J Immunol 168(9):4531–4537. https://doi.org/10.4049/jimmunol.168.9.4531

Houliston RS, Vinogradov E, Dzieciatkowska M, Li J, St Michael F, Karwaski M-F, Brochu D, Jarrell HC, Parker CT, Yuki N, Mandrell RE, Gilbert M (2011) Lipooligosaccharide of *Campylobacter jejuni*: similarity with multiple types of mammalian glycans beyond gangliosides. J Biol Chem 286(14):12361–12370. https://doi.org/10.1074/jbc.M110.181750

Iizumi T, Taniguchi T, Yamazaki W, Vilmen G, Alekseyenko AV, Gao Z, Perez Perez GI, Blaser MJ (2016) Effect of antibiotic pre-treatment and pathogen challenge on the intestinal microbiota in mice. Gut Pathog 8:60. https://doi.org/10.1186/s13099-016-0143-z

Imaeda H, Yamamoto H, Takaki A, Fujimiya M (2002) In vivo response of neutrophils and epithelial cells to lipopolysaccharide injected into the monkey ileum. Histochemistry Cell Biol 118(5):381–388. https://doi.org/10.1007/s00418-002-0458-3

Iwami K-i, Matsuguchi T, Masuda A, Kikuchi T, Musikacharoen T, Yoshikai Y (2000) Cutting edge: naturally occurring soluble form of mouse Toll-like receptor 4 inhibits lipopolysaccharide signaling. J Immunol 165(12):6682–6686. https://doi.org/10.4049/jimmunol.165.12.6682

Jandhyala SM, Talukdar R, Subramanyam C, Vuyyuru H, Sasikala M, Nageshwar Reddy D (2015) Role of the normal gut microbiota. World J Gastroenterol 21(29):8787–8803. https://doi.org/10.3748/wjg.v21.i29.8787

Jesudason MV, Hentges DJ, Pongpech P (1989) Colonization of mice by *Campylobacter jejuni*. Infect Immun 57(8):2279–2282. https://doi.org/10.1128/IAI.57.8.2279-2282.1989

Jungi TW, Farhat K, Burgener IA, Werling D (2011) Toll-like receptors in domestic animals. Cell Tissue Res 343(1):107–120. https://doi.org/10.1007/s00441-010-1047-8

Kampmann C, Dicksved J, Engstrand L, Rautelin H (2016) Composition of human faecal microbiota in resistance to *Campylobacter* infection. Clinical Microbiol Infect: Official Publication Europ Soci Clinical Microbiol Infect Dis 22(1):61.e61–61.e68. https://doi.org/10.1016/j.cmi.2015.09.004

Kannaki TR, Reddy MR, Shanmugam M, Verma PC, Sharma RP (2010) Chicken toll-like receptors and their role in immunity. World's Poultry Sci J 66(4):727–738. https://doi.org/10.1017/S0043933910000693

Karlyshev AV, Ketley JM, Wren BW (2005) The Campylobacter jejuni glycome. FEMS Microbiol Rev 29(2):377–390. https://doi.org/10.1016/j.femsre.2005.01.003

Kawasaki K, Akashi S, Shimazu R, Yoshida T, Miyake K, Nishijima M (2000) Mouse toll-like receptor 4· MD-2 complex mediates lipopolysaccharide-mimetic signal transduction by Taxol. J Biol Chem 275(4):2251–2254. https://doi.org/10.1074/jbc.275.4.2251

Kazmi SU, Roberson BS, Stern NJ (1984) Animal-passed, virulence-enhanced *Campylobacter jejuni* causes enteritis in neonatal mice. Current Microbiol 11(3):159–164

Keestra AM, de Zoete MR, Bouwman LI, van Putten JP (2010) Chicken TLR21 is an innate CpG DNA receptor distinct from mammalian TLR9. J Immunol 185(1):460–467. https://doi.org/10.4049/jimmunol.0901921

Ketloy C, Engering A, Srichairatanakul U, Limsalakpetch A, Yongvanitchit K, Pichyangkul S, Ruxrungtham K (2008) Expression and function of Toll-like receptors on dendritic cells and other antigen presenting cells from non-human primates. Vet Immunol Immunopathol 125(1–2):18-30. https://doi.org/10.1016/j.vetimm.2008.05.001

Kim S, Covington A, Pamer EG (2017) The intestinal microbiota: Antibiotics, colonization resistance, and enteric pathogens. Immunol Rev 279(1):90–105. https://doi.org/10.1111/imr.12563

Kist M, Bereswill S (2001) Campylobacter jejuni. Contribut Microbiol 8:150–165. https://doi.org/10.1159/000060405

Kuijf ML, Samsom JN, van Rijs W, Bax M, Huizinga R, Heikema AP, van Doorn PA, van Belkum A, van Kooyk Y, Burgers PC (2010) TLR4-mediated sensing of *Campylobacter jejuni* by dendritic cells is determined by sialylation. J Immunol 185(1):748–755. https://doi.org/10.4049/jimmunol.0903014

Larson CL, Shah DH, Dhillon AS, Call DR, Ahn S, Haldorson GJ, Davitt C, Konkel ME (2008) *Campylobacter jejuni* invade chicken LMH cells inefficiently and stimulate differential expression of the chicken CXCLi1 and CXCLi2 cytokines. Microbiology 154(Pt 12):3835–3847. https://doi.org/10.1099/mic.0.2008/021279-0

Ley RE, Bäckhed F, Turnbaugh P, Lozupone CA, Knight RD, Gordon JI (2005) Obesity alters gut microbial ecology. Proc National Acad Sci United States Amer 102(31):11070–11075. https://doi.org/10.1073/pnas.0504978102

Ley RE, Hamady M, Lozupone C, Turnbaugh PJ, Ramey RR, Bircher JS, Schlegel ML, Tucker TA, Schrenzel MD, Knight R, Gordon JI (2008) Evolution of mammals and their gut microbes. Science 320(5883):1647–1651. https://doi.org/10.1126/science.1155725

Lin J, Sahin O, Michel LO, Zhang Q (2003) Critical role of multidrug efflux pump CmeABC in bile resistance and in vivo colonization of *Campylobacter jejuni*. Infect Immun 71(8):4250–4259. https://doi.org/10.1128/iai.71.8.4250-4259.2003

Lippert E, Karrasch T, Sun X, Allard B, Herfarth HH, Threadgill D, Jobin C (2009) Gnotobiotic IL-10; NF-kappaB mice develop rapid and severe colitis following *Campylobacter jejuni* infection. PLoS ONE 4(10):e7413. https://doi.org/10.1371/journal.pone.0007413

Lobo de Sá FD, Butkevych E, Nattramilarasu PK, Fromm A, Mousavi S, Moos V, Golz JC, Stingl K, Kittler S, Seinige D (2019) Curcumin mitigates immune-induced epithelial barrier dysfunction by *Campylobacter jejuni*. Int J Molecul Sci 20(19):4830. https://doi.org/10.3390/ijms20194830

Lorne E, Zhao X, Zmijewski JW, Liu G, Park YJ, Tsuruta Y, Abraham E (2009) Participation of mammalian target of rapamycin complex 1 in Toll-like receptor 2- and 4-induced neutrophil activation and acute lung injury. Am J Respir Cell Mol Biol 41(2):237–245. https://doi.org/10.1165/rcmb.2008-0290OC

Lotz M, Gütle D, Walther S, Ménard S, Bogdan C, Hornef MW (2006) Postnatal acquisition of endotoxin tolerance in intestinal epithelial cells. J Experim Med 203(4):973–984. https://doi.org/10.1084/jem.20050625

Louwen R, Heikema A, van Belkum A, Ott A, Gilbert M, Ang W, Endtz HP, Bergman MP, Nieuwenhuis EE (2008) The sialylated lipooligosaccharide outer core in *Campylobacter jejuni* is an important determinant for epithelial cell invasion. Infect Immun 76(10):4431–4438. https://doi.org/10.1128/iai.00321-08

Louwen R, Nieuwenhuis EE, van Marrewijk L, Horst-Kreft D, de Ruiter L, Heikema AP, van Wamel WJ, Wagenaar JA, Endtz HP, Samsom J, van Baarlen P, Akhmanova A, van Belkum A (2012) *Campylobacter jejuni* translocation across intestinal epithelial cells is facilitated by ganglioside-like lipooligosaccharide structures. Infect Immun 80(9):3307–3318. https://doi.org/10.1128/iai.06270-11

Luangtongkum T, Jeon B, Han J, Plummer P, Logue CM, Zhang Q (2009) Antibiotic resistance in *Campylobacter*: emergence, transmission and persistence. Future Microbiol 4(2):189–200. https://doi.org/10.2217/17460913.4.2.189

Luethy PM, Huynh S, Ribardo DA, Winter SE, Parker CT, Hendrixson DR (2017) Microbiota-derived short-chain fatty acids modulate expression of *Campylobacter jejuni* determinants required for commensalism and Virulence. MBio 8(3):e00407–00417. https://doi.org/10.1128/mBio.00407-17

Lundberg R (2019) Humanizing the gut microbiota of mice: Opportunities and challenges. Lab Anim 53(3):244–251. https://doi.org/10.1177/0023677218787554

Malik-Kale P, Parker CT, Konkel ME (2008) Culture of *Campylobacter jejuni* with sodium deoxycholate induces virulence gene expression. J Bacteriol 190(7):2286–2297. https://doi.org/10.1128/JB.01736-07

Mansfield LS, Bell JA, Wilson DL, Murphy AJ, Elsheikha HM, Rathinam VA, Fierro BR, Linz JE, Young VB (2007) C57BL/6 and congenic interleukin-10-deficient mice can serve as models of *Campylobacter jejuni* colonization and enteritis. Infect Immun 75(3):1099–1115. https://doi.org/10.1128/IAI.00833-06

Mansfield LS, Patterson JS, Fierro BR, Murphy AJ, Rathinam VA, Kopper JJ, Barbu NI, Onifade TJ, Bell JA (2008) Genetic background of IL-10(-/-) mice alters host-pathogen interactions with *Campylobacter jejuni* and influences disease phenotype. Microb Pathog 45(4):241–257. https://doi.org/10.1016/j.micpath.2008.05.010

Marchesi JR, Adams DH, Fava F, Hermes GD, Hirschfield GM, Hold G, Quraishi MN, Kinross J, Smidt H, Tuohy KM (2016) The gut microbiota and host health: a new clinical frontier. Gut 65(2):330–339. https://doi.org/10.1136/gutjnl-2015-309990

Masanta WO, Heimesaat MM, Bereswill S, Tareen AM, Lugert R, Gross U, Zautner AE (2013) Modification of intestinal microbiota and its consequences for innate immune response in the pathogenesis of campylobacteriosis. Clin Dev Immunol 2013:526860. https://doi.org/10.1155/2013/526860

Mathur H, Rea MC, Cotter PD, Hill C, Ross RP (2015) The sactibiotic subclass of bacteriocins: an update. Current Protein Peptide Sci 16(6):549–558. https://doi.org/10.2174/1389203716666150515124831

Mayneris-Perxachs J, Lima AA, Guerrant RL, Leite AM, Moura AF, Lima NL, Soares AM, Havt A, Moore SR, Pinkerton R, Swann JR (2016) Urinary N-methylnicotinamide and beta-aminoisobutyric acid predict catch-up growth in undernourished Brazilian children. Sci Rep 6:19780. https://doi.org/10.1038/srep19780

Medzhitov R (2007) Recognition of microorganisms and activation of the immune response. Nature 449(7164):819–826. https://doi.org/10.1038/nature06246

Medzhitov R, Janeway C Jr (2000) Innate immune recognition: mechanisms and pathways. Immunol Rev 173:89. https://doi.org/10.1034/j.1600-065x.2000.917309.x

Mogensen TH (2009) Pathogen recognition and inflammatory signaling in innate immune defenses. Clin Microbiol Rev 22 (2):240–273, Table of Contents. https://doi.org/10.1128/cmr.00046-08

Moore KW, de Waal Malefyt R, Coffman RL, O'Garra A (2001) Interleukin-10 and the interleukin-10 receptor. Annu Rev Immunol 19:683–765. https://doi.org/10.1146/annurev.immunol.19.1.683

Mortensen NP, Kuijf ML, Ang CW, Schiellerup P, Krogfelt KA, Jacobs BC, van Belkum A, Endtz HP, Bergman MP (2009) Sialylation of *Campylobacter jejuni* lipo-oligosaccharides is associated with severe gastro-enteritis and reactive arthritis. Microbes Infect 11(12):988–994. https://doi.org/10.1016/j.micinf.2009.07.004

Mousavi S, Bereswill S, Heimesaat MM (2020a) Novel Clinical *Campylobacter jejuni* Infection models based on sensitization of mice to Lipooligosaccharide, a major bacterial factor triggering innate immune responses in human campylobacteriosis. Microorganisms 8(4). https://doi.org/10.3390/microorganisms8040482

Mousavi S, Escher U, Thunhorst E, Kittler S, Kehrenberg C, Bereswill S, Heimesaat MM (2020b) Vitamin C alleviates acute enterocolitis in *Campylobacter jejuni* infected mice. Sci Rep 10(1):2921. https://doi.org/10.1038/s41598-020-59890-8

Mousavi S, Lobo de Sá FD, Schulzke J-D, Bücker R, Bereswill S, Heimesaat MM (2019) Vitamin D in Acute Campylobacteriosis-Results From an Intervention Study Applying a Clinical Campylobacter jejuni Induced Enterocolitis Model. Front Immunol 10:2094–2094. https://doi.org/10.3389/fimmu.2019.02094

Mousavi S, Schmidt A-M, Escher U, Kittler S, Kehrenberg C, Thunhorst E, Bereswill S, Heimesaat MM (2020c) Carvacrol ameliorates acute campylobacteriosis in a clinical murine infection model. Gut Pathog 12:2–2. https://doi.org/10.1186/s13099-019-0343-4

Muller J, Meyer B, Hanel I, Hotzel H (2007) Comparison of lipooligosaccharide biosynthesis genes of *Campylobacter jejuni* strains with varying abilities to colonize the chicken gut and to invade Caco-2 cells. J Med Microbiol 56(Pt 12):1589–1594. https://doi.org/10.1099/jmm.0.47305-0

Muzio M, Bosisio D, Polentarutti N, D'Amico G, Stoppacciaro A, Mancinelli R, van't Veer C, Penton-Rol G, Ruco LP, Allavena P, Mantovani A (2000) Differential expression and regulation of toll-like receptors (TLR) in human leukocytes: selective expression of TLR3 in dendritic cells. J Immunol 164(11):5998–6004. https://doi.org/10.4049/jimmunol.164.11.5998

Naito M, Frirdich E, Fields JA, Pryjma M, Li J, Cameron A, Gilbert M, Thompson SA, Gaynor EC (2010) Effects of sequential Campylobacter jejuni 81-176 lipooligosaccharide core truncations on biofilm formation, stress survival, and pathogenesis. J Bacteriol 192(8):2182–2192. https://doi.org/10.1128/jb.01222-09

Newell DG (1986) Monoclonal antibodies directed against the flagella of *Campylobacter jejuni:* production, characterization and lack of effect on the colonization of infant mice. J Hyg (Lond) 96(2):131–141. https://doi.org/10.1017/s0022172400065906

Newell DG, McBride H, Dolby JM (1985) Investigations on the role of flagella in the colonization of infant mice with *Campylobacter jejuni* and attachment of *Campylobacter jejuni* to human epithelial cell lines. J Hyg (Lond) 95(2):217–227. https://doi.org/10.1017/s0022172400062653

Newell DG, Pearson A (1984) The invasion of epithelial cell lines and the intestinal epithelium of infant mice by *Campylobacter jejuni/coli.* J Diarrhoeal Dis Res 2(1):19–26

Nishida A, Inoue R, Inatomi O, Bamba S, Naito Y, Andoh A (2018) Gut microbiota in the pathogenesis of inflammatory bowel disease. Clinical J Gastroenterol 11(1):1–10. https://doi.org/10.1007/s12328-017-0813-5

Nomura F, Akashi S, Sakao Y, Sato S, Kawai T, Matsumoto M, Nakanishi K, Kimoto M, Miyake K, Takeda K, Akira S (2000) Cutting edge: endotoxin tolerance in mouse peritoneal macrophages correlates with down-regulation of surface toll-like receptor 4 expression. J Immunol 164(7):3476–3479. https://doi.org/10.4049/jimmunol.164.7.3476

O'Hara AM, Shanahan F (2006) The gut flora as a forgotten organ. EMBO Rep 7(7):688–693. https://doi.org/10.1038/sj.embor.7400731

O'Loughlin JL, Samuelson DR, Braundmeier-Fleming AG, White BA, Haldorson GJ, Stone JB, Lessmann JJ, Eucker TP, Konkel ME (2015) The intestinal microbiota influences *Campylobacter jejuni* colonization and extraintestinal dissemination in mice. Appl Environ Microbiol 81(14):4642–4650. https://doi.org/10.1128/AEM.00281-15

O'Neill LA, Fitzgerald KA, Bowie AG (2003) The Toll-IL-1 receptor adaptor family grows to five members. Trends Immunol 24(6):286–290. https://doi.org/10.1016/s1471-4906(03)00115-7

Cróinín OT, Backert S (2012) Host epithelial cell invasion by *Campylobacter jejuni:* trigger or zipper mechanism? Front Cell Infect Microbiol 2:25. https://doi.org/10.3389/fcimb.2012.00025

Oh E, Jeon B (2015) Contribution of surface polysaccharides to the resistance of *Campylobacter jejuni* to antimicrobial phenolic compounds. J Antibiot 68(9):591. https://doi.org/10.1038/ja.2015.26

Ohtani M, Hoshii T, Fujii H, Koyasu S, Hirao A, Matsuda S (2012) Cutting edge: mTORC1 in intestinal CD11c + CD11b + dendritic cells regulates intestinal homeostasis by promoting IL-10 production. J Immunol 188(10):4736–4740. https://doi.org/10.4049/jimmunol.1200069

Oldfield NJ, Moran AP, Millar LA, Prendergast MM, Ketley JM (2002) Characterization of the *Campylobacter jejuni* heptosyltransferase II gene, waaF, provides genetic evidence that extracellular polysaccharide is lipid A core independent. J Bacteriol 184(8):2100–2107. https://doi.org/10.1128/jb.184.8.2100-2107.2002

Pachathundikandi SK, Lind J, Tegtmeyer N, El-Omar EM, Backert S (2015) Interplay of the Gastric Pathogen *Helicobacter pylori* with Toll-Like Receptors. Biomed Res Int. 2015:192420. https://doi.org/10.1155/2015/192420

Pachathundikandi SK, Tegtmeyer N, Arnold IC, Lind J, Neddermann M, Falkeis-Veits C, Chattopadhyay S, Brönstrup M, Tegge W, Hong M, Sticht H, Vieth M, Müller A, Backert S (2019) T4SS-dependent TLR5 activation by Helicobacter pylori infection. Nat Commun. 10(1):5717. https://doi.org/10.1038/s41467-019-13506-6

Palmer C, Bik EM, DiGiulio DB, Relman DA, Brown PO (2007) Development of the human infant intestinal microbiota. PLoS Biol 5(7):e177. https://doi.org/10.1371/journal.pbio.0050177

Peng SL (2005) Signaling in B cells via Toll-like receptors. Current Opin Immunol 17(3):230–236. https://doi.org/10.1016/j.coi.2005.03.003

Perera VN, Nachamkin I, Ung H, Patterson JH, McConville MJ, Coloe PJ, Fry BN (2007) Molecular mimicry in *Campylobacter jejuni*: role of the lipo-oligosaccharide core oligosaccharide in inducing anti-ganglioside antibodies. FEMS Immunol Med Microbiol 50(1):27–36. https://doi.org/10.1111/j.1574-695X.2007.00225.x

Perlman RL (2016) Mouse models of human disease: An evolutionary perspective. Evol Med Public Health 2016(1):170–176. https://doi.org/10.1093/emph/eow014

Poltorak A, He X, Smirnova I, Liu M-Y, Huffel CV, Du X, Birdwell D, Alejos E, Silva M, Galanos C, Freudenberg M, Ricciardi-Castagnoli P, Layton B, Beutler B (1998) Defective LPS signaling in C3H/HeJ and C57BL/10ScCr Mice: Mutations in *Tlr4* Gene. Science 282(5396):2085–2088. https://doi.org/10.1126/science.282.5396.2085

Pomare E, Branch W, Cummings J (1985) Carbohydrate fermentation in the human colon and its relation to acetate concentrations in venous blood. J Clinical Invest 75(5):1448–1454. https://doi.org/10.1172/JCI111847

Purchiaroni F, Tortora A, Gabrielli M, Bertucci F, Gigante G, Ianiro G, Ojetti V, Scarpellini E, Gasbarrini A (2013) The role of intestinal microbiota and the immune system. Eur Rev Med Pharmacol Sci 17(3):323–333

Qureshi ST, Lariviere L, Leveque G, Clermont S, Moore KJ, Gros P, Malo D (1999) Endotoxin-tolerant mice have mutations in Toll-like receptor 4 (Tlr4). J Exp Med 189(4):615–625. https://doi.org/10.1084/jem.189.4.615

Rakoff-Nahoum S, Paglino J, Eslami-Varzaneh F, Edberg S, Medzhitov R (2004) Recognition of commensal microflora by toll-like receptors is required for intestinal homeostasis. Cell 118(2):229–241. https://doi.org/10.1016/j.cell.2004.07.002

Rathinam VAK, Appledorn DM, Hoag KA, Amalfitano A, Mansfield LS (2009) *Campylobacter jejuni*-induced activation of dendritic cells involves cooperative signaling through Toll-like receptor 4 (TLR4)-MyD88 and TLR4-TRIF axes. Infect Immun 77(6):2499–2507. https://doi.org/10.1128/IAI.01562-08

Repaske DR, Adler J (1981) Change in intracellular pH of *Escherichia coli* mediates the chemotactic response to certain attractants and repellents. J Bacteriol 145(3):1196–1208. https://doi.org/10.1128/JB.145.3.1196-1208.1981

Riddle MS, Guerry P (2016) Status of vaccine research and development for *Campylobacter jejuni*. Vaccine 34(26):2903–2906. https://doi.org/10.1016/j.vaccine.2016.02.080

Ridley AM, Toszeghy MJ, Cawthraw SA, Wassenaar TM, Newell DG (2008) Genetic instability is associated with changes in the colonization potential of Campylobacter jejuni in the avian intestine. J Appl Microbiol 105(1):95–104. https://doi.org/10.1111/j.1365-2672.2008.03759.x

Rink L, Haase H (2007) Zinc homeostasis and immunity. Trends in Immunol 28(1):1–4. https://doi.org/10.1016/j.it.2006.11.005
Robertson SA, Care AS, Skinner RJ (2007) Interleukin 10 regulates inflammatory cytokine synthesis to protect against lipopolysaccharide-induced abortion and fetal growth restriction in mice. Biol Reprod 76(5):738–748. https://doi.org/10.1095/biolreprod.106.056143
Robertson SA, Skinner RJ, Care AS (2006) Essential role for IL-10 in resistance to lipopolysaccharide-induced preterm labor in mice. J Immunol 177(7):4888–4896. https://doi.org/10.4049/jimmunol.177.7.4888
Russell SL, Gold MJ, Hartmann M, Willing BP, Thorson L, Wlodarska M, Gill N, Blanchet MR, Mohn WW, McNagny KM, Finlay BB (2012) Early life antibiotic-driven changes in microbiota enhance susceptibility to allergic asthma. EMBO Rep 13(5):440–447. https://doi.org/10.1038/embor.2012.32
Sarabi Asiabar A, Asadzadeh Aghdaei H, Zamani S, Bokaie S, Zali MR, Feizabadi MM (2019) Molecular detection of *Campylobacter jejuni* in patients with Crohn's disease in Iran. Med J Islam Repub Iran 33:76–76. https://doi.org/10.34171/mjiri.33.76
Satsangi J, Wolstencroft RA, Cason J, Ainley CC, Dumonde DC, Thompson RP (1987) Interleukin 1 in Crohn's disease. Clin Exp Immunol 67(3):594–605
Schmidt A-M, Escher U, Mousavi S, Tegtmeyer N, Boehm M, Backert S, Bereswill S, Heimesaat MM (2019a) Immunopathological properties of the *Campylobacter jejuni* flagellins and the adhesin CadF as assessed in a clinical murine infection model. Gut Pathog 11:24–24. https://doi.org/10.1186/s13099-019-0306-9. https://doi.org/10.1186/s13099-019-0306-9
Schmidt AM, Escher U, Mousavi S, Boehm M, Backert S, Bereswill S, Heimesaat MM (2019b) Protease activity of *Campylobacter jejuni* HtrA modulates distinct intestinal and systemic immune responses in infected secondary abiotic IL-10 deficient mice. Front Cell Infect Microbiol 9:1–12. https://doi.org/10.3389/fcimb.2019.00079
Sekirov I, Russell SL, Antunes LCM, Finlay BB (2010) Gut microbiota in health and disease. Physiol Rev 90(3):859–904. https://doi.org/10.1152/physrev.00045.2009
Sham HP, Yu EYS, Gulen MF, Bhinder G, Stahl M, Chan JM, Brewster L, Morampudi V, Gibson DL, Hughes MR (2013) SIGIRR, a negative regulator of TLR/IL-1R signalling promotes Microbiota dependent resistance to colonization by enteric bacterial pathogens. PLoS Pathogens 9(8). https://doi.org/10.1371/journal.ppat.1003539
Snyder JP, Amiel E (2018) Regulation of dendritic cell immune function and metabolism by cellular nutrient sensor mammalian target of rapamycin (mTOR). Front Immunol 9:3145. https://doi.org/10.3389/fimmu.2018.03145
Snyder SL, Walker RI (1976) Inhibition of lethality in endotoxin-challenged mice treated with zinc chloride. Infect Immun 13(3):998. https://doi.org/10.1128/IAI.13.3.998-1000.1976
Stahl M, Frirdich E, Vermeulen J, Badayeva Y, Li X, Vallance BA, Gaynor EC (2016) The Helical Shape of *Campylobacter jejuni* Promotes *In Vivo* pathogenesis by aiding transit through intestinal mucus and colonization of crypts. Infect Immun 84(12):3399–3407. https://doi.org/10.1128/iai.00751-16
Stahl M, Ries J, Vermeulen J, Yang H, Sham HP, Crowley SM, Badayeva Y, Turvey SE, Gaynor EC, Li X, Vallance BA (2014) A novel mouse model of *Campylobacter jejuni* gastroenteritis reveals key pro-inflammatory and tissue protective roles for Toll-like receptor signaling during infection. PLoS Pathog 10(7):e1004264. https://doi.org/10.1371/journal.ppat.1004264
Stahl M, Vallance BA (2015) Insights into *Campylobacter jejuni* colonization of the mammalian intestinal tract using a novel mouse model of infection. Gut Microbes 6(2):143–148. https://doi.org/10.1080/19490976.2015.1016691
Stephenson HN, John CM, Naz N, Gundogdu O, Dorrell N, Wren BW, Jarvis GA, Bajaj-Elliott M (2013) Campylobacter jejuni lipooligosaccharide sialylation, phosphorylation, and amide/ester linkage modifications fine-tune human Toll-like receptor 4 activation. J Biol Chem 288(27):19661–19672. https://doi.org/10.1074/jbc.M113.468298

Sun M, Wu W, Liu Z, Cong Y (2017) Microbiota metabolite short chain fatty acids, GPCR, and inflammatory bowel diseases. J Gastroenterol 52(1):1–8. https://doi.org/10.1007/s00535-016-1242-9

Sun X, Liu B, Sartor RB, Jobin C (2013) Phosphatidylinositol 3-kinase-gamma signaling promotes *Campylobacter jejuni*-induced colitis through neutrophil recruitment in mice. J Immunol 190(1):357–365. https://doi.org/10.4049/jimmunol.1201825

Sun X, Threadgill D, Jobin C (2012) Campylobacter jejuni induces colitis through activation of mammalian target of rapamycin signaling. Gastroenterology 142(1):86–95.e85. https://doi.org/10.1053/j.gastro.2011.09.042

Sun X, Winglee K, Gharaibeh RZ, Gauthier J, He Z, Tripathi P, Avram D, Bruner S, Fodor A, Jobin C (2018) Microbiota-derived metabolic factors reduce campylobacteriosis in mice. Gastroenterology 154(6):1751–1763.e1752. https://doi.org/10.1053/j.gastro.2018.01.042

Szymanski CM, King M, Haardt M, Armstrong GD (1995) *Campylobacter jejuni* motility and invasion of Caco-2 cells. Infect Immun 63(11):4295–4300. https://doi.org/10.1128/IAI.63.11.4295-4300.1995

Thambiayya K, Wasserloos K, Kagan VE, Stoyanovsky D, Pitt BR (2012) A critical role for increased labile zinc in reducing sensitivity of cultured sheep pulmonary artery endothelial cells to LPS-induced apoptosis. Am J Physiol Lung Cell Mol Physiol 302(12):L1287–1295. https://doi.org/10.1152/ajplung.00385.2011

Ulevitch RJ (1999) Endotoxin opens the toll gates to innate immunity. Nature Med 5(2):144–145. https://doi.org/10.1038/5504

Vaara M (1992) Agents that increase the permeability of the outer membrane. Microbiol Rev 56(3):395–411

Van Deun K, Pasmans F, Ducatelle R, Flahou B, Vissenberg K, Martel A, Van den Broeck W, Van Immerseel F, Haesebrouck F (2008) Colonization strategy of *Campylobacter jejuni* results in persistent infection of the chicken gut. Vet Microbiol 130(3–4):285-297. https://doi.org/10.1016/j.vetmic.2007.11.027

van Dijk A, Herrebout M, Tersteeg-Zijderveld MH, Tjeerdsma-van Bokhoven JL, Bleumink-Pluym N, Jansman AJ, Veldhuizen EJ, Haagsman HP (2012) *Campylobacter jejuni* is highly susceptible to killing by chicken host defense peptide cathelicidin-2 and suppresses intestinal cathelicidin-2 expression in young broilers. Vet Microbiol 160(3–4):347–354. https://doi.org/10.1016/j.vetmic.2012.05.034

van Mourik A, Steeghs L, van Laar J, Meiring HD, Hamstra HJ, van Putten JP, Wosten MM (2010) Altered linkage of hydroxyacyl chains in lipid A of *Campylobacter jejuni* reduces TLR4 activation and antimicrobial resistance. J Biol Chem 285(21):15828–15836. https://doi.org/10.1074/jbc.M110.102061

Vašl J, Oblak A, Gioannini TL, Weiss JP, Jerala R (2009) Novel roles of lysines 122, 125, and 58 in functional differences between human and murine MD-2. J Immunol 183(8):5138–5145. https://doi.org/10.4049/jimmunol.0901544

Vaure C, Liu Y (2014) A comparative review of toll-like receptor 4 expression and functionality in different animal species. Front Immunol 5:316. https://doi.org/10.3389/fimmu.2014.00316

Vikbladh I (1951) Studies on zinc in blood II. Scand J Clin Lab Invest 3(Suppl. 2):1–74

Vitovec J, Koudela B, Sterba J, Tomancova I, Matyas Z, Vladik P (1989) The gnotobiotic piglet as a model for the pathogenesis of *Campylobacter jejuni* infection. Zentralblatt fur Bakteriologie: Int J Med Microbiol 271(1):91–103. https://doi.org/10.1016/s0934-8840(89)80058-1

Vogel SN, Madonna GS, Wahl LM, Rick PD (1984) Stimulation of spleen cells and macrophages of C3H/HeJ mice by a lipid A precursor derived from *Salmonella typhimurium*. Clinical Infect Dis 6(4):535–541. https://doi.org/10.1093/clinids/6.4.535

Vollaard E, Clasener H (1994) Colonization resistance. Antimicrobial Agents Chemotherapy 38(3):409. https://doi.org/10.1128/aac.38.3.409

Wald D, Qin J, Zhao Z, Qian Y, Naramura M, Tian L, Towne J, Sims JE, Stark GR, Li X (2003) SIGIRR, a negative regulator of Toll-like receptor–interleukin 1 receptor signaling. Nature Immunol 4(9):920–927. https://doi.org/10.1038/ni968

Walker CF, Black RE (2004) Zinc and the risk for infectious disease. Annu Rev Nutr 24:255–275. https://doi.org/10.1146/annurev.nutr.23.011702.073054

Walker WA (2017) The importance of appropriate initial bacterial colonization of the intestine in newborn, child, and adult health. Pediatr Res 82(3):387–395. https://doi.org/10.1038/pr.2017.111

Warren HS, Fitting C, Hoff E, Adib-Conquy M, Beasley-Topliffe L, Tesini B, Liang X, Valentine C, Hellman J, Hayden D, Cavaillon JM (2010) Resilience to bacterial infection: difference between species could be due to proteins in serum. J Infect Dis 201(2):223–232. https://doi.org/10.1086/649557

Watson RO, Novik V, Hofreuter D, Lara-Tejero M, Galan JE (2007) A MyD88-deficient mouse model reveals a role for Nramp1 in *Campylobacter jejuni* infection. Infect Immun 75(4):1994–2003. https://doi.org/10.1128/iai.01216-06

Weindl G, Naglik JR, Kaesler S, Biedermann T, Hube B, Korting HC, Schaller M (2007) Human epithelial cells establish direct antifungal defense through TLR4-mediated signaling. J. Clinical Investig 117(12):3664–3672. https://doi.org/10.1172/JCI28115

Xiao H, Gulen MF, Qin J, Yao J, Bulek K, Kish D, Altuntas CZ, Wald D, Ma C, Zhou H, Tuohy VK, Fairchild RL, de la Motte C, Cua D, Vallance BA, Li X (2007) The Toll-interleukin-1 receptor member SIGIRR regulates colonic epithelial homeostasis, inflammation, and tumorigenesis. Immunity 26(4):461–475. https://doi.org/10.1016/j.immuni.2007.02.012

Yamamoto M, Sato S, Hemmi H, Uematsu S, Hoshino K, Kaisho T, Takeuchi O, Takeda K, Akira S (2003) TRAM is specifically involved in the Toll-like receptor 4-mediated MyD88-independent signaling pathway. Nat Immunol 4(11):1144–1150. https://doi.org/10.1038/ni986

Young CR, Ziprin RL, Hume ME, Stanker LH (1999) Dose response and organ invasion of day-of-hatch Leghorn chicks by different isolates of *Campylobacter jejuni*. Avian Dis 43(4):763–767

Young KT, Davis LM, Dirita VJ (2007) *Campylobacter jejuni:* molecular biology and pathogenesis. Nat Rev Microbiol 5(9):665–679. https://doi.org/10.1038/nrmicro1718

Yrios JW, Balish E (1985) Colonization and pathogenesis of *Campylobacter* spp. in athymic and euthymic germfree mice. Prog Clin Biol Res 181:199–202

Yrios JW, Balish E (1986a) Colonization and infection of athymic and euthymic germfree mice by *Campylobacter jejuni* and *Campylobacter fetus* subsp. fetus Infect Immun 53(2):378–383. https://doi.org/10.1128/IAI.53.2.378-383.1986

Yrios JW, Balish E (1986b) Immune response of athymic and euthymic germfree mice to *Campylobacter* spp. Infect Immun 54(2):339–346. https://doi.org/10.1128/IAI.54.2.339-346.1986

Yrios JW, Balish E (1986c) Pathogenesis of *Campylobacter spp.* In: athymic and euthymic germfree mice. Infect Immun 53(2):384–392. https://doi.org/10.1128/IAI.53.2.384-392.1986

Zhang Q, Sahin O (2020) Campylobacteriosis. Diseases of poultry 12th edn, pp 754–769

Natural Competence and Horizontal Gene Transfer in *Campylobacter*

Julia Carolin Golz and Kerstin Stingl

Contents

Abstract Thermophilic *Campylobacter*, in particular *Campylobacter jejuni*, *C. coli* and *C. lari* are the main relevant *Campylobacter* species for human infections. Due to their high capacity of genetic exchange by horizontal gene transfer (HGT), rapid adaptation to changing environmental and host conditions contribute to successful spreading and persistence of these foodborne pathogens. However, extensive HGT can exert dangerous side effects for the bacterium, such as the incorporation of gene fragments leading to disturbed gene functions. Here we discuss mechanisms of HGT, notably natural transformation, conjugation and bacteriophage transduction and limiting regulatory strategies of gene transfer. In particular, we summarize the current knowledge on how the DNA macromolecule is exchanged between single cells. Mechanisms to stimulate and to limit HGT obviously coevolved and maintained an optimal balance. Chromosomal rearrangements and incorporation of harmful mutations are risk factors for survival and can result in drastic loss of fitness. In *Campylobacter*, the restricted recognition and preferential uptake of free DNA

J. C. Golz · K. Stingl (✉)
Department of Biological Safety, National Reference Laboratory for Campylobacter, German Federal Institute for Risk Assessment (BfR), Diedersdorfer Weg 1, 12277 Berlin, Germany
e-mail: kerstin.stingl@bfr.bund.de

J. C. Golz
e-mail: julia.golz@bfr.bund.de

S. Backert (ed.), *Fighting Campylobacter Infections*, Current Topics in Microbiology and Immunology 431, https://doi.org/10.1007/978-3-030-65481-8_10

from relatives are mediated by a short methylated DNA pattern and not by a classical DNA uptake sequence as found in other bacteria. A class two CRISPR-Cas system is present but also other DNases and restriction–modification systems appear to be important for *Campylobacter* genome integrity. Several lytic and integrated bacteriophages have been identified, which contribute to genome diversity. Furthermore, we focus on the impact of gene transfer on the spread of antibiotic resistance genes (resistome) and persistence factors. We discuss remaining open questions in the HGT field, supposed to be answered in the future by current technologies like whole-genome sequencing and single-cell approaches.

1 Introduction

Horizontal gene transfer (HGT) is the exchange of genetic material and plays a major role in genetic diversity of pathogens (Lawrence 2005; Daubin and Szollosi 2016). Therefore, HGT in *Campylobacter jejuni* is thought to lead to host adaptation and fitness enhancement. There are three types of HGT, natural transformation, conjugation and phage transduction (Fig. 1). During natural transformation, free environmental DNA is taken up and incorporated into the genome upon homologous recombination or in case of plasmids by plasmid reconstitution and replication. Free DNA might occur in the environment by active secretion from bacterial cells or by cell lysis. Conjugation, however, is limited to DNA exchange between donor and recipient cells being in physical contact with each other. Transduction describes the genetic exchange mediated by bacteriophages. HGT in *Campylobacter* is the main driving force for the outstanding genetic diversity of this pathogen (Wilson et al. 2009; Sheppard et al. 2008). In Sect. 2, we discuss various HGT mechanisms in thermophilic *Campylobacter*, including *C. jejuni*, *C. coli* and *C. lari*, which are the *Campylobacter* species most frequently implicated in human gut disease.

However, genetic changes harbor the risk of harmful mutations or unfavorable chromosomal rearrangements for the bacteria. Therefore, mechanisms for the regulation of DNA entry and recombination into the bacterial chromosome co-evolved. CRISPR-Cas can be considered as the bacterial immune system protecting cells form invading bacteriophages or plasmids (Hille et al. 2018). However, other nucleases including restriction–modification systems play an important role for limiting harmful transfer of genetic material into the foodborne pathogen and are discussed in Sect. 3. Nevertheless, HGT bears the advantage of rapid host adaptation due to fitness enhancement and, e.g., spread of antibiotic resistances. Hence, in Sect. 4, we focus on the current knowledge of interspecies gene transfer and acquisition of novel beneficial genetic traits in thermophilic *Campylobacter* spp.

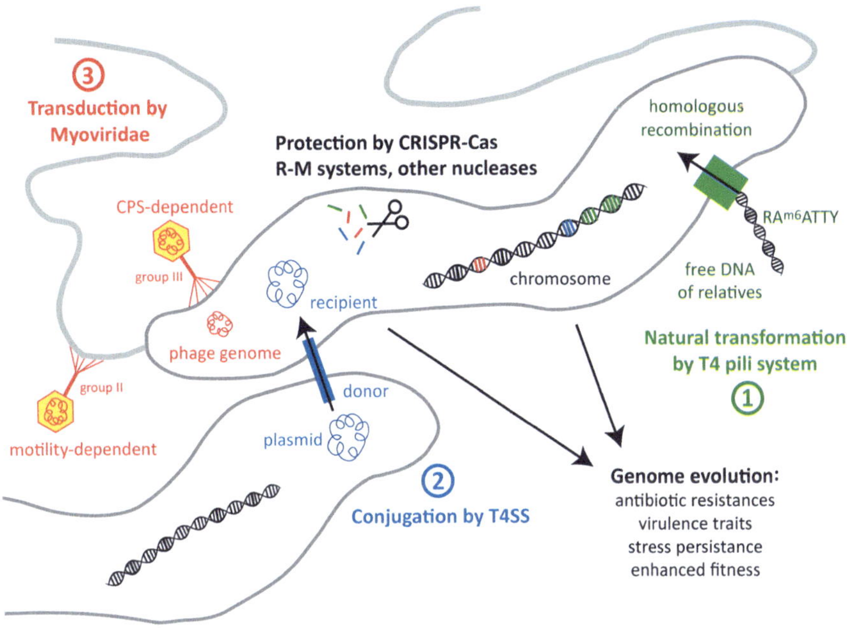

Fig. 1 Overview of horizontal gene transfer (HGT) mechanisms, genetic barriers and impact on pathogen adaptation. The three mechanisms of HGT are depicted for thermophilic *Campylobacter* spp. Natural transformation of free external DNA (in green) occurs via a type II secretion/T4 pili system, which displays homology to the competence machinery of *Neisseria* (detailed in Fig. 2). Transfer of plasmids via conjugation (in blue) is mediated by type IV secretion systems (T4SS) in direct cell–cell contact. Two main classes of bacteriophages of the family Myoviridae mediate genetic diversity by transduction (in red). The Fletcherviruses, group III, CP8-like phages depend on capsular polysaccharide (CPS) modifications as receptor for host entry. The Firehammer, group II, CP220-like phages need motile bacteria for infection. *Campylobacter* limits natural transformation by selection of DNA from relatives harboring a methylated RA^{m6}ATTY profile, provided by activity of the CtsM methylase. Periplasmic nucleases and cytoplasmic restriction–modification systems as well as the CRISPR-Cas type II-C system provide additional barriers for incoming DNA. HGT leads to frequent homologous and rare non-homologous recombination of genetic material, acquisition of plasmids and/or rearrangements of chromosomal loci, which in turn shapes genome evolution. Hence, *Campylobacter* populations genetically diversify providing preadaptation to changing environments, such as presence of antimicrobials, switch of hosts and environmental stress, thereby enhancing the bacterium's overall fitness for survival and transmission

2 Mechanisms of Horizontal Gene Transfer

Most studies are based on classical approaches, in which HGT is followed using a selective marker and phenotypic characterization of resulting bacterial colonies after incorporation of the transferred marker gene (Table 1, first column). The enormous capacity of transfer of a chloramphenicol selection marker by natural transformation in *C. jejuni* was impressively shown by establishing a plate DNA transformation assays for screening mutants (Wiesner et al. 2003). The principle of the assay was the

Table 1 Overview on experimental approaches to decipher HGT mechanisms in bacteria and evolutionary impact

	Classical approaches	Single-cell assays	Whole-genome sequencing
Principle	Transfer of (antibiotic) markers for phenotypic selection	Uptake of (fluorescently) labeled DNA and epifluorescent or electron microscopy (bacteriophage, conjugation)	Bioinformatic analysis of whole genome before/after gene transfer
Methods	Mixture of donor/recipient cells, addition of free DNA or bacteriophages	Addition of fluorescently labeled DNA, donor/recipient cells or bacteriophages	DNA extraction from field strains or after *in vitro* gene transfer events; Fragment library preparation
Analysis	Selection with antimicrobial agent; phenotypic characterization	Analysis with fluorescence microscope or by TEM with/without antibodies	MiSeq, NextSeq, PacBio; Analysis: K-mer, SNP or cgMLST analysis for phylogeny; annotation tools
Readout	Number of CFU with marker vs. total CFU; frequency of gene transfer	Microscopic photo for localization of gene transfer; detection and quantification of gene transfer in single cells	Genome diversity; recombination frequency; chromosomal rearrangements; virulence/antibiotic resistance determinants
Power/advantage of approach	Detection and quantification of the final result of gene transfer processes	Localization of components for gene transfer visible at single-cell level; direct monitoring of different steps of gene transfer possible; parameters for induction of gene transfer can be identified	High throughput method; identification of multiple gene transfer events and impact of gene transfer on whole bacterium/bacterial population
Drawback/disadvantage of approach	Readout does not distinguish between different steps of gene transfer; CFU is biased due to fastidious nature of *Campylobacter*	Accessibility/visibility of DNA during transfer limited; detection and quantification need differential approaches for visualization	Mechanisms of gene transfer are only indirectly visible

TEM, transmission electron microscopy; MiSeq and NextSeq, middle-scale and large-scale short-read, massive parallel whole-genome sequencing methods of Illumina; PacBio, single-molecule real-time sequencing technique of Pacific Biosciences; SNP, single nucleotide polymorphism; cgMLST, core genome multi-locus sequence typing; CFU, colony-forming units

spreading of a countable number of *C. jejuni* cells on an agar plate, which had been overlaid by 2.5 μg of transforming DNA, harboring a given selection marker. After two days of growth, bacterial colonies were patched on agar plates with and without antibiotic. Intriguingly, the authors observed that nearly all colonies comprised transformed cells. Assuming growth from single cells to visible colonies of around 10^6-10^7 cells, natural transformation occurred within approximately 20–25 generations. Two days of incubation were sufficient to generate a bacterial population with adequate capacity of adaptive survival based on a former single cell. The final result of HGT is monitored in classical approaches but the readout cannot distinguish between different steps of gene transfer. Furthermore, in some settings in which the cells are exposed to stress conditions, the parameter colony-forming units (CFU) can be biased due to the fastidious nature of *Campylobacter* spp. and might not reflect full capacity of gene transfer.

Single-cell approaches have the advantage of dissecting different steps of HGT and to localize DNA uptake/transfer complexes (Table 1, second column). The detection and quantification of gene transfer are feasible at the level of single cells, displaying phenotypic heterogeneity. Parameters for induction of gene transfer can more directly be identified, since the assay does not depend on the complete process including the incorporation and expression of a marker gene. However, accessibility and, thus, visibility of DNA during the transfer event are limited. For example, covalently labeled DNA can only be followed into the periplasm and transfer of DNA into the cytoplasm is only indirectly monitored by disappearance of fluorescence of non-covalently labeled DNA (Stingl et al. 2010). For conjugation and transduction, DNA is steadily protected within biological compartments, and the detection by antibodies using transmission electron microscopy (TEM) is a stochastic event. Thus, differential approaches combined with the construction and characterization of mutants are necessary for complete monitoring and quantification of DNA transfer.

A recent approach focusses on whole-genome analysis in order to monitor the overall effects of HGT on population dynamics (Table 1, third column). Different platforms for whole-genome sequencing are used and quality as well as interpretation parameters are currently harmonized in order to optimally compare datasets of different laboratories. Ideally, all three approaches are combined to reveal the complete process and impact of HGT in the foodborne pathogen.

2.1 Natural Transformation and Uptake of Free DNA

Natural transformation was first discovered almost one century ago in *Streptococcus pneumoniae*, when phenotypic changes upon addition of heat-inactivated virulent bacteria to a recipient non-virulent culture were observed (Griffith 1928). Avery and colleagues (1944) pinpointed the transforming agent as DNA. The term "competence" depicts the state, in which cells are able to take up free DNA and naturally transform, i.e., integrate genetic material into their genome or replicate epichromosomal elements autonomously. Potential benefits of natural transformation include the repair of mutations by incoming homologous DNA and the acquisition of new genes and, therefore, new functions, e.g., antibiotic resistance genes or virulence factors. In addition, DNA might serve as nutrient supply by offering a reservoir for recycling of nucleotides. Besides extracellular DNA might serve as a matrix for the formation of biofilms and can enhance persistence of the pathogen outside the host (Feng et al. 2018; Svensson et al. 2014). For comprehensive reviews on natural transformation in other bacteria refer e. g. to Dubnau and Blokesch (2019) and Bakkali and colleagues (2013). Since uptake of foreign DNA might represent a danger of acquiring harmful mutations, competence development is usually a highly regulated process (Johnston et al. 2014). Only few information is available on parameters controlling competence development in *Campylobacter* spp. *C. jejuni* seems to show the highest transformation levels under optimal growth conditions, but transformation also occurred, when growth was restricted at higher pH (Vegge et al. 2012). However, it is unclear, if already expressed DNA uptake complexes still functioned under growth limiting conditions or if competence development still occurred. Prolonged incubation times in the presence of DNA were performed in this study, which do not allow distinguishing activity of DNA uptake complexes from transcriptional regulation of competence genes. Wilson and colleagues (2003) suggested that lower CO_2 levels led to decreased competence in *C. jejuni* strains, although also here pH effects cannot be ruled out.

Since *Campylobacter* are Gram-negative bacteria, free DNA for natural transformation has to be transported i) over the outer membrane into the periplasm and ii) across the inner membrane into the cytoplasm. *Campylobacter* harbors gene homologues of a type II secretion/type IV pilus system that were shown to be essential for DNA uptake in other organisms (Table 2, Fig. 2) (Parkhill et al. 2000; Gundogdu et al. 2007). An at least 1000-fold reduction in transformation frequency was observed by Wiesner and colleagues (2003) using a transposon-based mutagenesis approach in eleven genes, nine of them were named *Campylobacter* transformation system (*cts*) genes (Table 2). Six of the genes are located in an operon, *ctsF-ctsE-ctsX-ctsP-ctsD-ctsR*. The remaining three *cts* genes, *ctsG, ctsT* and *ctsW* are separately located on the chromosome. CtsP and CtsE harbor nucleotide-binding sites (Walker A and B boxes) and are proper candidates for empowering uptake of the DNA macromolecule and/or assembly of a (pseudo-)pilus, like ComGA or PilF/T in *B. subtilis* or *Neisseria*, respectively (Beauchamp et al. 2015). CtsP physically interacts with the unique CtsX protein, both located in the membrane, while CtsE seems to be located in the soluble fraction (Beauchamp et al. 2015). *Campylobacter* recognizes DNA from relatives by using the methylated RAATTY site (see Sect. 3). In *N. gonorrhoeae*, PilQ constitutes the outer membrane pore (Drake and Koomey 1995), mediating entry of external DNA into the periplasm. *C. jejuni* harbors the *pilQ* homolog *ctsD* (Wiesner et al. 2003), which might have similar function as outer membrane DNA pore in *C. jejuni.* The genes *ctsF*, *ctsG* and *ctsT* have homology to *comGB*, *comGC* (*pilE* in Gram-negative bacteria) and *comGD* of *B. subtilis*, playing putative roles in function and assembly of the type IV (pseudo-)pilus system. In particular, it was suggested that ComGB displays an integral membrane protein forming the base for pilus assembly, with ComGC as major and ComGD as minor pilins (Chen et al. 2006). Retraction of DNA bound to type IV competence pili in *Vibrio* was recently demonstrated (Ellison et al. 2018). It remains to be shown, if a similar mechanism for "grabbing" external DNA is present in *Campylobacter* or if a pseudopilus is sufficient for DNA uptake as shown for *Neisseria* (Obergfell and Seifert 2016).

In addition, transposon insertion in *ceuB,* which is located in an operon structure with *ceuC, ceuD, ceuE,* encoding the enterochelin uptake system important for iron acquisition, resulted in impaired natural transformation. Furthermore, also *ctsW*, *proC* and the downstream region of *ansA* led to reduced transformation rates (Wiesner

Table 2 Genes implicated in natural transformation by *C. jejuni*

Gene name[1]	Putative function in natural transformation	Cc	Cl	Reference
comEC (Cj1211)	Competence family protein; predicted integral membrane channel for transport of DNA into cytoplasm	✓	✓	Jeon et al. 2008
comE (Cj0011c)	Periplasmic DNA-binding protein; generates force for pulling DNA macromolecule over the outer membrane in other bacteria; role unclear in *Campylobacter* spp.	×	×	Jeon and Zhang 2007; Meric et al. 2014
ctsD/pilQ (Cj1474c)	Type II secretion/T4 pilus system; potential outer membrane pore/secretin for transport of DNA into periplasm	✓	✓	Wiesner et al. 2003
ctsP (Cj1473c)	Type II secretion/T4 pilus system; ATP/GTP-binding protein with Walker A and B boxes; peripheral membrane protein	✓	×	Wiesner et al. 2003; Beauchamp et al. 2015
ctsX (Cj1472c)	Unique membrane protein; transformation system protein with unknown function; interacts with CtsP	✓	✓	Wiesner et al. 2003; Beauchamp et al. 2015
ctsE/pilF/comGA (Cj1471c)	Type II secretion/T4 pilus system; ATP/GTP-binding protein with Walker A and B boxes; present in soluble fraction	✓	✓	Wiesner et al. 2003; Beauchamp et al. 2015
ctsF/pilG/comGB (Cj1470c)	Type II secretion/T4 pilus system, membrane protein, putatively constitutes platform for pilus/pseudopilus assembly	✓	✓	Wiesner et al. 2003
ctsG/pilE/comGC (Cj1343c)	Type II secretion/T4 pilus system, periplasmic protein; major (?) pre-pilin	✓	✓	Wiesner et al. 2003
ctsT/comGD (Cj1077)	Type II secretion/T4 pilus system; Periplasmic protein; pre-pilin	✓	×	Wiesner et al. 2003

(continued)

Table 2 (continued)

Gene name[1]	Putative function in natural transformation	Cc	Cl	Reference
Cj1078*	Type II secretion/T4 pilus system; Periplasmic protein; prepilin; not yet been demonstrated to function in natural transformation in *C. jejuni*	✓	✗	Wiesner et al. 2003
Cj0825*/*pilD*/*comC*	Putative prepilin peptidase with transmembrane helices	✓	✓	Wiesner et al. 2003
ctsW (Cj1028c)	Not involved in DNA uptake (maybe role in cytoplasmic transport or recombination); purine/pyrimidine phosphoribosyltransferase	✓	✓	Wiesner et al. 2003
ctsM (Cj0208)	DNA modification methylase of RAATTY motif; important for recognition of free DNA	✓	✓	Beauchamp et al. 2017

[1] Gene names as annotated for *C. jejuni* strain NCTC11168 with homologs in *Neisseria* or *Bacillus subtilis* are depicted with putative function in natural transformation
*, evidence only by BLAST homology; BLAST searches based on the translated nucleotide database using a protein query; accession 13.04.2020; Cc, *C. coli*, Cl, *C. lari*
✓, presence and ✗, absence of homologous genes is depicted

et al. 2003). The role of these genes in natural transformation is still unknown. Fry and colleagues reported that a *galE* mutant, with defects in lipopolysaccharide (LPS) synthesis, showed a mild 20-fold reduction in DNA uptake and chromosomal integration, which might be indicative of LPS influencing the function of the DNA uptake machinery (Fry et al. 2000).

The periplasmic DNA-binding competence protein (Com) ComE was shown to facilitate DNA entry into the periplasm in *Neisseria* and *Vibrio* by generation of a pulling force on double-stranded (ds) DNA upon binding (Hepp and Maier 2016; Seitz et al. 2014). The role of the *C. jejuni comE* homolog (Cj0011c) has not been unraveled completely. Transformation rates in Cj0011c knockout mutants were only decreased by 10- to 50-fold (Jeon and Zhang 2007). Interestingly, *C. coli*, which displays similar gene homologs of a type II secretion/type IV pilus system as *C. jejuni* (Table 2), lacks a *comE* homolog (Meric et al. 2014). *comE* is also missing in C. *lari* (Table 2). Hence, in *C. coli* and *C. lari* a role of ComE in natural transformation can be ruled out.

Once the DNA reaches the periplasm, dsDNA has to be unzipped for import of single-stranded (ss) DNA into the cytoplasm mediated by the inner membrane channel ComEC in all so far known competent bacteria (Dubnau and Blokesch 2019). Absence of transformation activity in *C. jejuni comEC* (Cj1211) insertional mutants was demonstrated, whereas binding and uptake of radiolabeled DNA were not impaired (Jeon et al. 2008). Depending on homology, incoming DNA will be

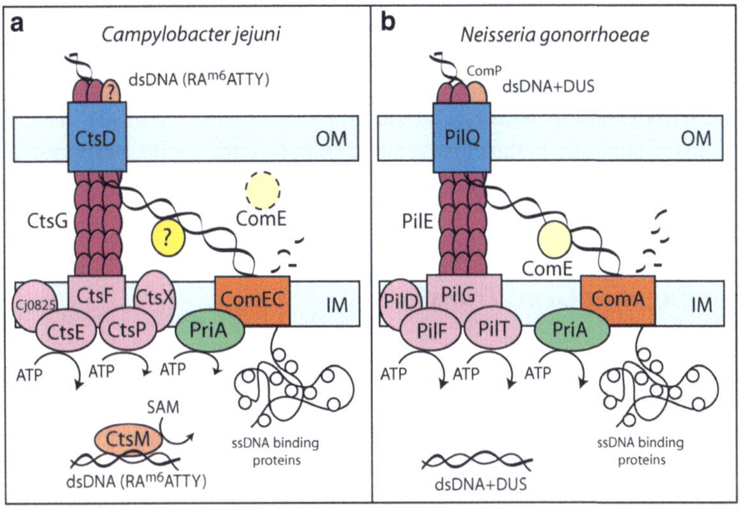

Fig. 2 Working model of the DNA uptake complex for natural transformation in *Campylobacter jejuni* A compared to the system in *Neisseria gonorrhoeae* B Uptake of free external DNA probably occurs in two steps. Transport into the periplasm is mediated by a type II secretion/T4 pili system. Homology analysis suggests CtsD as the outer membrane porin. CtsG might be the major pilin but further pilin proteins CtsT and Cj1078 were identified in *C. jejuni* (see also Table 2). CtsF might form the basis for pilus/pseudopilus. CtsE and CtsP were proposed as ATPases, eventually empowering the DNA uptake process and/or pili assembly. The unique membrane protein CtsX was shown to interact with CtsP. The role of ComE as DNA-binding protein in the periplasm is enigmatic, since a homolog is lacking in *C. coli* and *C. lari*. ComEC appears to be the inner membrane channel as proposed for all competent bacteria, leading to import of single-stranded DNA into the cytoplasm, eventually empowered by PriA. In *Neisseria,* the minor pilin ComP recognizes a specific DNA uptake sequence (DUS) for selective uptake of DNA from relatives. In *C. jejuni,* the methylated RA^{m6}ATTY motif is recognized by a yet unknown receptor. Methylation is mediated by the CtsM methylase. OM, outer membrane; IM, inner membrane; SAM, S-adenosylmethionine; ATP, adenosine triphosphate; dsDNA, double-stranded DNA; ssDNA, single-stranded DNA

incorporated into the genome by homologous recombination. Site-directed homologous recombination of transformed plasmid DNA was observed with homologous regions of at least 286 bp, whereas 125-270 bp homology only led to a random and rare non-homologous insertion into the chromosome (Richardson and Park 1997). However, although non-homologous integration of DNA is infrequent, this mechanism guarantees the incorporation of completely novel genes.

C. jejuni strain 81–176 carries the plasmid pVir, encoding homologs to *Helicobacter pylori cag* pathogenicity island as well as homologs to type IV secretion systems (Bacon et al. 2002; Fischer et al. 2020). The role of pVir in natural transformation is not completely understood. Nevertheless, Bacon and colleagues (2000) showed an 80% reduction in transformation frequency in a *comB3* mutant, whereas *virB11* inactivation did not show a reduced transformation activity. Mutation of one of the glycosylation sites in the glycoprotein VirB10 or deletion of *virB10* showed a mild ~ tenfold decrease in transformation efficiency (Larsen et al. 2004). Knockout of the

N-linked protein glycosylation system (*pgl*), e.g., by deletion of *pglB* or *pglE* led to a drastic 10,000-fold decreased transformation rate (Larsen et al. 2004), suggesting that glycosylation of proteins is essential for natural transformation. Mutations in *virD4* and *comB1* led to wild-type transformation activity (Wiesner et al. 2003), thus, unlike the situation in the close relative *H. pylori*, the VirB/ComB system does not seem to play a major role for DNA uptake in *C. jejuni*.

2.2 Conjugative Gene Transfer

In conjugation processes, DNA is transferred from a donor to a recipient cell through cell-to-cell contact (Lederberg and Tatum 1946). To date 177 plasmid sequences from *Campylobacter* spp. are released at NCBI (https://www.ncbi.nlm.nih.gov/genome/browse#!/plasmids/campylobacter; accession on 22.09.2020). The size of currently identified *C. jejuni* and *C. coli* plasmids ranges from 1.3 to 190 kb. The presence of megaplasmids was shown in various strains from retail (Marasini and Fakhr 2016, 2014; Ghatak et al. 2017; Gunther et al. 2016). The transfer of plasmid-encoded antibiotic resistances in *Campylobacter* has frequently been reported (Taylor et al. 1981; Velazquez et al. 1995; Gibreel et al. 2004; Batchelor et al. 2004; Pratt and Korolik 2005; Zeng et al. 2015; Tang et al. 2017). Type-1 plasmids (pTet) harboring *tetO* are most prevalent in *C. jejuni* and *C. coli* (Schmidt-Ott et al. 2005; Marasini et al. 2018). Although only the *tetO* gene is representative for all pTet plasmids, most of them carry a VirB-type IV secretions system for conjugation. Type-2 plasmids were primarily found in *C. coli* strains, which are characterized to have a size between 24 and 32 kb and bear several *trb* genes for conjugative transfer as well as *virD4*, *traI* and *traQ* (Marasini et al. 2018).

The pVir plasmid of *C. jejuni* strain 81-176 mentioned above is categorized as the "prototype" of the type-3 plasmids. Small plasmids < 6 kb were categorized as type-4 plasmids despite absence of homologous genes shared between them. They contain genes with hypothetical function and replication initiator genes and await further investigations.

Pratt and Korolik (2005) showed that conjugation frequencies of a plasmids encoding *tetO* from donor strains to a recipient strain varied between ~10^{-8} and 10^{-6} within 6 h of mating. Interestingly, also co-transfer of a smaller plasmid was observed together with a larger plasmid conferring resistance to tetracycline (Pratt and Korolik 2005). Absence of conjugation in some strains was observed, indicative of barriers, e.g., restriction–modification systems and/or inability of plasmids to replicate in specific strains. Strain dependency of conjugation rates was identified in a different study, showing variations from 10^{-8} to 10^{-3} (Zeng et al. 2015). Hence, it might be concluded that natural transformation with transformation rates of ~ 10^{-3}–10^{-2} is a more efficient way of HGT in *Campylobacter* spp. However, further studies are needed to collect more data on different field strains and to correlate in vitro with in vivo HGT frequencies.

Conjugation efficiency was induced 100–1000-fold in strains with low-frequency conjugation (LFC) upon 30 min heat shock at around 50 °C (Zeng et al. 2015). Recently, Zeng and colleagues (2018) identified the restriction–modification enzyme CjeI (Cj1051c) as crucial factor for reduced conjugation rate in the LFC strain NCTC11168. In high conjugation frequency (HCF) strains, 1000-fold reduced conjugation frequency was observed upon chromosomal complementation with *cjeI*. The *cjeI* mutants showed enhanced conjugation efficiency, which was nearly independent of heat shock, suggesting that CjeI was the heat-inactivated limiting factor of successful conjugational transfer of plasmids in LFC strains. It was previously observed that CjeI also restricted incoming DNA during natural transformation (Holt et al. 2012). Restriction barriers are discussed in more detail in Sect. 3.

Interestingly, unidirectional DNaseI-resistant conjugation-like transfer of a chromosomal resistance gene was observed from *H. pylori* to *C. jejuni* (Oyarzabal et al. 2007), demonstrating the potential of bacteria of the class Campylobacterales for genetic exchange (Fernandez-Gonzalez et al. 2014).

2.3 Phage Transduction and Genomic Rearrangements

Campylobacter bacteriophages have been isolated from diverse matrices, including food, animals and environments (for a recent review on isolation methods, see Jäckel et al. 2019), indicating that the pathogen is constantly exposed to phages in its natural habitat. *Campylobacter* bacteriophages were first reported in 1968 in *C. fetus* (formerly *Vibrio fetus*) upon induction of lytic phase by the bactericidal agent mitomycin C (Firehammer and Border 1968). For details on the application of bacteriophages for *Campylobacter* infection control, the reader should refer to Chap. 6 of this book.

Most sequenced *Campylobacter* phages (CP) belong to the family Myoviridae, displaying long contractile tails (Javed et al. 2014; NCBI Taxonomy Browser, accession 22.09.2020). They are categorized into two main groups, the Firehammervirus, group II, CP220-like and the Fletcherviruses, group III, CP8-like phages. Group I phages with large genomes of ~320 kb are, however, rare. DNA from *Campylobacter* phages was observed to be refractory to digestion by several restriction enzymes (Sails et al. 1998), which was recently attributed to complete exchange of deoxyguanosine (dG) by modified bases in phage DNA (Crippen et al. 2019).

In general, bacteriophage predation was shown to lead to chromosomal rearrangements in bacteria and, therefore, phages could also contribute to *Campylobacter* genomic evolution. For example, up to 590 kb in *C. jejuni* were inverted due to inversions caused by Mu-like phages (Scott et al. 2007). Interestingly, *C. jejuni* carrying the bacteriophage in its chromosome were resistant to infections by other bacteriophages but revealed an inefficient colonization of the chicken. Besides, the integration of phage-like elements into the genome can lead to genomic changes, visible by altered pulsed-field gel electrophoresis (PFGE) patterns of cleaved chromosomal DNA (Barton et al. 2007).

The process of phage transduction can be divided into several steps. Initially, the bacteriophage has to interact with a receptor on the bacterial cell. Principally, *Campylobacter* phage infection was shown to be either dependent on modifications of the capsular polysaccharides (Sorensen et al. 2011) or on motile flagella (Baldvinsson et al. 2014). Receptor-type dependency could be correlated with phage genus (Sorensen et al. 2015). While CP81-like Fletcherviruses were dependent on capsular polysaccharide (CPS), thereby unable to infect acapsular (Δ*kpsM*) mutants, the CP220-like Firehammerviruses were deficient of infecting non-motile (Δ*motA*) *C. jejuni* strains. The receptor in *C. jejuni* NCTC 11168 for the Myoviridae phage F336 was shown to be an *O*-methyl phosphoramidate attached to 2-acetamido-2-deoxy-D-galactofuranose (GalfNAc) on the capsular polysaccharide (Sorensen et al. 2011). A frameshift in the phase variable homopolymeric G tract of gene Cj1421 resulting in a non-functional O-methyl phosphoramidate (MeOPN) transferase conferred resistance against phage F336. This is because the receptor is unavailable due to lack of receptor attachment to CPS. In addition, Cj1422, another phase variable gene, was shown to attach MeOPN to a heptose in CPS in *C. jejuni*, which confers resistance to F336 (Holst Sorensen et al. 2012; Aidley et al. 2017). The existence of further CPS receptors independent of MeOPN in CPS-dependent phages was suggested recently (Gencay et al. 2018).

In addition, it was shown that a conserved glycan-specific phage protein, Gp047 renamed FlaGrab, recognizes 7-acetamidino-modified pseudaminic acid residues on *Campylobacter* flagella, inhibiting bacterial growth (Javed et al. 2015). In particular, FlaGrab exposure led *C. jejuni* cells to downregulate expression of energy metabolism genes, which was dependent on a functional flagellar motor and was host strain-dependent, irrespective of the level of motility (Sacher et al. 2020). However, FlaGrab is also present in CPS-dependent phages, but is not part of the phage capsule. Thus, it was speculated that FlaGrab is not involved in phage entry, but presents an important protein in the phage lifecycle. It may either function as extracellular effector molecule upon phage-induced cell lysis, improving new infection by reduction of host motility or intracellularly during phage infection (Javed et al. 2015).

The transcriptional bacterial response upon infection of a CP8-like type-III phage NCTC 12673 revealed regulation of an unknown operon with some homology to T4 phage superinfection exclusion and antitoxin genes, as well as multidrug efflux pumps and oxidative stress defense genes (Sacher et al. 2018). Mutants of the *cmeABC* efflux pump were more susceptible for phage infections, while loss of catalase and superoxide dismutase genes led to enhanced phage resistance (Sacher et al. 2018). Thus, it seems that phage infection modulates the capacity of the host to resist antimicrobial treatment and oxidative stress, probably as part of phage–host dynamics. Interestingly, RidA, previously shown to play a role in flagella–flagella interactions due to regulation of flagellar glycan modification and motility (Reuter et al. 2015), was observed to also function in bacteriophage infectivity (Irons et al. 2019). However, the exact molecular mechanism is not yet clear. Taken together, more studies are clearly needed to fully understand *Campylobacter* phage lifecycle and the complex interaction with their host.

The ganglioside-like structures GM1 and GD1 generated by the *cts-II*-encoded sialyltransferase play a role in resistance against bacteriophages (Louwen et al. 2013). This was first suggested by the observation that isolates involved in Guillain-Barré syndrome induction showed lower susceptibility to a panel of 29 bacteriophages. Furthermore, a *ΔctsII* mutant showed increased susceptibility to bacteriophage than the wild-type bacteria. Bioinformatic screening revealed a correlation between the presence of *ctsII* and a degenerated CRISPR-Cas system (see also Sect. 3) in *C. jejuni* strains, indicating that virulence-associated ganglioside-like structures might serve as bacteriophage defense system.

While above we have discussed the current knowledge on lytic phages, also chromosomally integrated prophages have been described in various *Campylobacter* strains. For example, *C. jejuni* strain RM1221 carries four so-called *Campylobacter jejuni*-integrated elements (CJIEs), three of which (CJIE1, 2 and 4) seem to originate from phages (Barton et al. 2007) and the fourth (CJIE3) putatively from an integrated plasmid (Fouts et al. 2005; Parker et al. 2006). The Mu-like phage CJIE1 is integrated upstream of the *argC* gene, encoding an N-acetyl-γ-glutamyl-phosphate reductase. CJIE2 and CJIE4 are integrated at the 3'end of arginyl- and methionyl-tRNA genes. CJIE3 is integrated into the 3' end of an arginyl-tRNA. CJIE1 encodes typical Mu and Mu-like phage proteins. CJIE2 and CJIE4 potentially encode methylases, endonucleases and repressors. CJIE1 was present among ~ 1/7 of *Campylobacter* isolates obtained from surveillance programs in Canada (Clark 2011). Most of these isolates were *C. jejuni* but CJIE1 was also present in *C. coli* and *C. upsaliensis*. The sequence and structure of the integrated CJIE1 varied, leading to protein alterations (Clark and Ng 2008). Furthermore, integration loci varied in different *C. jejuni* strains (Parker et al. 2006). Similarly to CJIE1, also CJIE2 and CJIE4 were inserted at different loci in the *Campylobacter* chromosome in different strains (Clark and Ng 2008). However, until now, induction of these CJIE prophages to lytic phase was unsuccessful (Clark and Ng 2008).

3 Barriers to Horizontal Gene Transfer

While HGT is crucial for the acquisition of novel genetic material and beneficial adaptation to changing environments, introgression of foreign DNA in bacterial genomes can also lead to tremendous fitness loss. The fact that *Campylobacter* is well protected by HGT barriers becomes obvious, since genetic manipulation of *Campylobacter* is hampered using constructs amplified in cloning strains of *Escherichia coli* (Gardner and Olson 2012). In the following, we address the aspect of barriers to HGT and focus on the CRISPR-Cas system and on other nucleases, including restriction–modification systems protecting *Campylobacter* against incoming foreign DNA. Furthermore, we address the question how *C. jejuni* can select for DNA of relatives, without using a classical DNA uptake sequence as demonstrated for other bacteria.

3.1 CRISPR-Cas and Nucleases

"Bacterial immunity" based on clustered regularly interspaces short palindromic repeats (CRISPR) and CRISPR-associated (Cas) proteins might be a powerful mechanism for restriction of horizontal gene transfer in *Campylobacter*. These systems are present in ~ 40% of complete bacterial and ~ 85% of archaeal genomes (Makarova et al. 2020). The principle is that incoming foreign DNA is memorized by incorporation of small fragments in CRISPR regions. Upon repeated entry, complementary CRISPR rRNAs (crRNA) in complex with Cas proteins target invading DNA for degradation. CRISPR-Cas systems are classified into two classes, six types and at least 33 subtypes (Makarova et al. 2020). While class 1 systems use multiple Cas proteins building up the effector complex, the class 2 system uses a single-protein effector, e.g., Cas9 in case of *Campylobacter* spp. Class 2 systems currently include three types and 17 subtypes. *C. jejuni* harbors a class 2, type II-C CRISPR-Cas system. It consists of the genes *cas1*, *cas2* and *cas9* as well as a trans-activating CRISPR RNA (TracrRNA). Cas1 and Cas2 are suggested to acquire and integrate new protospacers (Yosef et al. 2012). Cas9 participates in spacer acquisition (Heler et al. 2015; Wei et al. 2015). CRISPR loci are transcribed as a single pre-crRNA precursor, which is processed to crRNAs by the bacterial non-Cas RNase III in type II systems. In turn, crRNAs in complex with Cas9 silence invading plasmid or phage DNA, which bear sequence homology to the integrated spacer sequences. DNA strand breaks at stalled replication forks induce RecBCD-dependent spacer acquisition. In order to avoid autoimmunity, chromosomal loci were protected against spacer acquisition by relatively abundant Chi sites in *E. coli*, at which dsDNA break repair is stimulated in bacteria (Levy et al. 2015). However, self-DNA might be integrated into the CRISPR loci at very low frequency (Stern et al. 2010). Cas4-like proteins in *Campylobacter* bacteriophages were suggested to modify spacer element acquisition in favor of phage evasion due to preferential integration of host sequences in CRISPR loci (Hooton and Connerton 2014). Thus, coevolution of phages with the host leads to continuous modulation of genome dynamics.

The optimal size of the bacterial memory is dependent on the diversity of threats, i.e., phages. Since the effectiveness of response is dependent on the number/concentration of crRNA-Cas complexes with matching specificity, the depth of memory was proposed to be limited to 10–100 spacers in bacteria (Bradde et al. 2020). Based on the current database called CRISPRCasFinder, hosted at the University of Paris-Saclay, the numbers of predicted CRISPR loci in *C. jejuni* and *C. coli* range from 0 to 11 (median = 1; nCj = 207, nCc = 37, accession 22.09.2020 (at https://crisprcas.i2bc.paris-saclay.fr/MainDb/StrainList), harboring each one to multiple spacers (Grissa et al. 2007; Couvin et al. 2018). However, low transcription of crRNAs and TracrRNA was observed in *C. jejuni* RM1221 due to a stop-mutation in *cas9* (Dugar et al. 2013). Thus, the authors suggested that absence of CRISPR loci or truncation of *cas9* enabled acquisition of prophages or plasmids and that active CRISPR and mobile elements are mutually exclusive. Although Cas9 nucleases usually target dsDNA, a recent study demonstrated that in *C. jejuni* also endogenous ssRNA was targeted by CjCas9 (Dugar et al. 2018). Hence, it was proposed by the authors that CjCas9 may also serve to target RNA viruses or even regulate endogenous gene expression, which should be investigated in the future.

Apart from the CRISPR-Cas system, periplasmic nucleases were reported to degrade incoming genomic DNA, thereby inhibiting natural transformation. The periplasmic DNase, encoded by the *dns* gene (CJE0256) from Mu-like prophage CJIEI, inhibits natural transformation in RM1221 (Gaasbeek et al. 2009). Transformability of field strains correlated with presence or absence of *dns*. Homologs of DNA/RNA non-specific endonucleases were subsequently also detected on the prophages CJIE2 and CJIE4 and inhibition of natural transformation levels by around 30–40-fold were demonstrated (Gaasbeek et al. 2010).

3.2 Methylation-Dependent DNA Recognition

It has been reported long time ago that *C. jejuni* preferentially takes up DNA from siblings, although the mechanisms were completely unknown (Wang and Taylor 1990). However, *C. jejuni* does not have a typical DNA uptake sequence (DUS), like it was demonstrated for *N. gonorrhoeae* (Goodman and Scocca 1988), with the minor pilin protein ComP identified as specific receptor (Cehovin et al. 2013). Nevertheless, *C. jejuni* selects DNA of relatives and discriminates against foreign DNA. By single-molecule real-time sequencing (SMRT), a high degree of methylation of chromosomal DNA became apparent, with the RAATTY site being the only methylation site shared between *C. coli* BfR-CA-09557 and other *C. jejuni* strains (Zautner et al. 2015). By deletion of the respective methylase gene *ctsM* (named as *Campylobacter* transformation system methyltransferase), it was shown that *C. jejuni* recognizes the adenine N6 (exocyclic NH_2-group at the sixth position of the purine ring) methylated RAATTY site of free external DNA as first step of natural transformation (Beauchamp et al. 2017). *ctsM* mutants were not impaired in DNA uptake, indicating that CtsM itself is not involved in recognition and/or transport of methylated DNA. The authors also demonstrated that *E. coli* plasmids could successfully be transformed into *C. jejuni* after methylation with the *E. coli* *Eco*RI methylase. In this case, one of the four RAATTY-sites, namely GAATTC is methylated, which was sufficient for DNA uptake in *Campylobacter*. This study presented a major advantage for future genetic manipulation of *Campylobacter* spp., since researchers can substantially improve genetic manipulation by methylation of plasmid constructs via commercially available *Eco*RI methylase prior to transformation in the respective *Campylobacter* host. The native *Eco*RI system is restricted to a special strain of *E. coli* and not ubiquitously found in this species, explaining that DNA from *E. coli* does not present a substrate for natural transformation of *Campylobacter* spp. It remains to be investigated, which components of the DNA uptake complex recognize the methylated RAATTY motif, in order to decipher the mechanism of selective DNA uptake in the foodborne pathogen.

Apart from CtsM as methylase, other restriction–modification systems are thought to constitute a genetic barrier for incoming DNA. The restriction–modification type IIG enzyme Cj1051c was shown to lower transformation efficiency using a *C. jejuni* derived plasmid by 1000-fold (Holt et al. 2012). Cj1051c was shown to also drastically reduce conjugation efficiency among *C. jejuni* strains (Zeng et al. 2018). Since *Campylobacter* genomes harbor diverse methylation profiles and various restriction-methylation genes, that are also strain-dependent (O'Loughlin et al. 2015; Zautner et al. 2015), it is expected that several other restriction–modification systems play crucial roles in establishing genetic barriers, even against relatives, favoring clonal spreading.

4 Impact of Gene Transfer on *Campylobacter* Fitness

As discussed above, during evolution *Campylobacter* spp. have developed powerful means for HGT and coevolved with incoming genetic material in order to balance the acquisition of novel material and putative detrimental effects. In the following, we address the beneficial impact of gene transfer and report on fitness advantages due to enormous genetic plasticity of the foodborne pathogen.

4.1 Spread of Resistomes and Persistence Factors

Human infections by *Campylobacters* are commonly caused by consumption and handling of raw poultry meat (for more details see Chap. 1 of this book). While most human campylobacteriosis cases are self-limiting, antibiotic treatment, in particular the use of macrolides or fluoroquinolones, was reported in around one-third of the patients (Rosner et al. 2017). The spread of antibiotic resistances by HGT enables preadaptation to changing environments and leads to diversification of the bacterial population. The observation of different resistances shared between *C. jejuni* and *C. coli* strains isolated from livestock, sewage and human disease indicated frequent spread of plasmids and multidrug-resistant genomic islands (MDRGIs) by HGT (Mourkas et al. 2019). Spread of antibiotic resistance between *C. jejuni* strains by natural transformation was reinforced in biofilms versus planktonic environments (Bae et al. 2014). For details on biofilm formation and quorum sensing, the reader should refer to Chap. 11 of this book. Biofilms contain extracellular DNA and are thought to convey enhanced persistence of host-associated pathogens in the environment. Their role in the dissemination of antibiotic resistances remains to be studied in more detail.

One of the well-known and the most prevailing resistance mechanisms of *Campylobacter* against macrolides in European strains is the point mutation A2075G in the 23S rRNA gene. This mutation is associated with a substantial decrease in bacterial fitness (Wang et al. 2014; Luangtongkum et al. 2012), probably leading to the currently observed low rates of macrolide resistance in *Campylobacter* spp. from livestock (EFSA 2020). Recently, *C. jejuni* and *C. coli* strains were isolated carrying the gene *ermB*, encoding an rRNA methylase, conferring resistance against macrolides in Asia (Qin et al. 2014; Du et al. 2018; Cheng et al. 2020; Liu et al. 2017), Europe (Florez-Cuadrado et al. 2016) and USA (Chen et al. 2018). Up to now, nine types of *ermB*-carrying MDRGI have been identified in *Campylobacter* spp. Besides *ermB*, these islands include resistances against aminoglycosides, such as gentamicin, kanamycin, streptomycin, spectinomycin or streptothricin, as well as ampicillins and tetracyclines (Wang et al. 2014; Florez-Cuadrado et al. 2016; Chen et al. 2018). *C. coli* strains were also identified, harboring *ermB* on different plasmids (Wang et al. 2014). The published NCBI sequences of *Campylobacter ermB* present four different allele variants. Comparative genome analysis revealed identical *ermB* sequences in *Campylobacter*, *Streptococcus suis*, *Enterococcus faecium* and *Clostridium difficile* isolates from different matrices (Florez-Cuadrado et al. 2017), suggesting multiple HGT events among different species. Especially the spread of macrolide resistance is of great danger, since macrolides are often drugs of choice to treat campylobacteriosis in humans (Rosner et al. 2017).

A variant of the multidrug efflux pump RE-CmeABC (for resistance-enhancing *Campylobacter* multidrug efflux system ABC), displaying sequence variation and enhanced expression due to a mutation in the promotor region, was shown to be spread via natural transformation and homologous recombination (Yao et al. 2016). This "super" pump conveys increased minimal inhibitory concentrations (MICs) against antimicrobials, such as ciprofloxacin, erythromycin, phenicols and tetracycline.

As long as the acquired antibiotic resistance determinant does not lead to fitness decrease, it can stably remain in a strain and is readily spread to other strains. For example, the transfer of *tetO* in *C. jejuni* was demonstrated to occur in vivo in chicken even without selection pressure (Avrain et al. 2004). This is especially important because it demonstrates that in case a long-term antimicrobial is discontinued, the resistance might persist and even spread to other strains. It is further stated that tetracyline resistance determinant, *tetO*, originated from Gram-positive cocci (Sougakoff et al. 1987) and kanamycin resistance seems to originate from Gram-positive cocci or from *Enterobacteriaceae* (Ouellette et al. 1987; Gibreel and Skold 1998). Apart from tetracycline resistance, resistance against (fluoro-)quinolones was observed to be another example of resistance determinant, not necessarily vanishing upon cease of antibiotic use. The resistance is conferred by the C257T point mutation in the gene encoding gyrase subunit A (*gyrA*). This point mutation was shown to even exert a fitness advantage on certain *Campylobacter* strains in the in vivo chicken gut environment (Luo et al. 2005). The spread of fluoroquinolone resistance in distinct clonal lineages might at least partially be explained by this fitness enhancement, although an additional selective pressure by antibiotic usage cannot be ruled out (Kovac et al. 2015; Leekitcharoenphon et al. 2018).

Not only the dissemination of antibiotic resistances bears risks for human health, but also the spread of bacterial persistence factors can increase the adaptive potential favoring the pathogens' survival and transmission. However, since the function of gene variations is mostly unknown and it is expected that multiple gene exchanges synergistically lead to a beneficial adaptation, reports are scarce on the acquisition of novel traits other than antibiotic resistances. Mosaic sequence exchange in the highly similar flagellin genes *flaA* and *flaB* was observed on the intra- and intergenomic level (Wassenaar et al. 1995; Harrington et al. 1997). *Campylobacter* virulence is dependent on motility and, thus, a functional flagellar system. Hence, variations of the involved structural genes lead to variants, putatively evading host immune response.

Phongsisay and colleagues (2006) showed that human ganglioside-like structures, such as GM1, were readily transformable to strains not associated with Guillain-Barré syndrome induction in humans. The resulting transformants had acquired large DNA fragments and presented a high degree of genetic and phenotypic variation, corroborating the enormous potential of *C. jejuni* for genome plasticity upon natural transformation. Another interesting study highlighted successful HGT of genes with metabolic functions (Vorwerk et al. 2015). In particular, most *Campylobacter* strains are not capable of catabolizing glucose. Nevertheless, some *C. coli* strains harbor a genomic island, which allows using glucose as an energy source through the metabolic Entner–Doudoroff pathway. This locus was transferred between *C. coli* strains as well as between *C. coli* and *C. jejuni*, conferring glycolytic activity (Vorwerk et al. 2015), suggesting that this metabolic trait was acquired in order to optimize energy supply in distinct niches.

4.2 Interspecies Gene Transfer

As reported above, *C. jejuni* differentiates DNA of relatives by recognition of the methylated RAATTY profile, mediated by the N-adenine specific methylase CstM (Beauchamp et al. 2017). *ctsM* homologs are present in thermophilic *Campylobacter* spp., suggesting that gene transfer is enabled between different species. The manifestation of incoming DNA is further dependent on the degree of homology and of strain-specific restriction–modification systems as well as nucleases, which function as genetic barriers (see Sect. 3).

Genetic exchanges can also be analyzed using genome analysis of bacterial populations. The population structure of *C. jejuni* is different from *C. coli* even though their core genomes show a nucleotide sequence identity of ~ 85%, and they colonize similar habitats (Dingle et al. 2005). From nearly 3,000 MLST types, 11% of *C. coli* sequence types showed *C. jejuni* origin, vice versa this was only estimated for 0.6% of the *C. jejuni* types (Sheppard et al. 2008). This indicated a considerable but asymmetric gene flow between the two major thermophilic *Campylobacter* species. *C. jejuni* has a very diverse structure, with over 40 clonal complexes. In *C. coli,* only three different clades were identified (Sheppard et al. 2012). Clade 1 is predominantly found in clinical and animal farm samples and comprises the majority of all isolated and sequenced *C. coli* strains, whereas clade 2 and 3 were found in waterfowl and riparian environment. A genetic exchange between *C. jejuni* and *C. coli* of clade 1 was observed previously (Sheppard et al. 2011), while clade 2 and 3 were unaffected by *C. jejuni* introgression, probably due to separated niches and lack of contact with *C. jejuni*. A separation of individual clones with rare or no contact to others and a host tropism can explain why some strains isolated from the same host in different geographic location are more related than strains from different hosts (Sheppard and Maiden 2015). The study by Epping and colleagues (2020) analyzing whole-genome sequences of more than 490 *C. jejuni* strains obtained from Germany and Canada showed a strong host association and enables to further study host adaptation on the level of subsets of variant genes. For more details, the reader should refer to Chap. 3 of this book. Frequent HGT events might also give rise to a population of *Campylobacter* strains that are called "generalists," able to colonize multiple hosts.

Introgression can occur as mosaic recombination of gene alleles. Consistent with asymmetric gene flow between the two species, the exchange from *C. jejuni* into *C. coli* was 17 times more frequently observed than from *C. coli* to *C. jejuni* (Sheppard et al. 2011). However, based on frequent genetic exchange, a convergence between the species *C. coli* and *C. jejuni* was postulated (Sheppard et al. 2008). For *C. jejuni* and *C. coli*, there are 44 clonal complexes and 11,111 sequence types defined (https://pubmlst.org, accession on 22.09.2020). Interestingly, nearly 40% of sequence types are not assigned to a clonal complex, demonstrating the diverse genome structure of these two major species. However, *C. jejuni* diversity is much greater than *C. coli* as defined by core genome phylogeny (Golz et al. 2020), with yet unknown reason.

We have identified *C. coli* strains as a fraction of clade 1, which have undergone recent ongoing extended introgression by *C. jejuni* sequences (Golz et al. 2020). These strains were particularly isolated from chicken eggs, i.e., from fecal contamination on egg shells. K-mer analysis on whole-genome sequences revealed that these "hybrid" strains had incorporated up to 15% of genomic sequences from *C. jejuni* along the whole genome. However, a more in-depth analysis showed that recombination events were not random but followed a common pattern. In particular, *C. jejuni* introgression occurred in a common set of genes, implicated in stress defense. Hence, this genome alteration might represent a functional adaptation to survival in a harsh environment and confirms the enormous potential of natural transformation in shaping *Campylobacter* genomes.

5 Concluding Remarks

Due to high levels of genetic exchange by natural transformation, conjugation or transduction, *Campylobacter* shows an enormous genome diversity. This widens the pathogens adaptive potential and enables colonization of multiple hosts and successful survival in the environment, although the microaerobic bacterium is generally stress-sensitive and fastidious. Also, spread of antibiotic resistances endangers therapy options for treatment of campylobacteriosis (Oyarzabal and Backert 2012). The mechanisms of HGT in *Campylobacter* are yet poorly understood, and there is an urgent need to understand more in detail how the pathogen adapts by gene acquisition and/or gene variation. For example, open questions remain of how HGT is regulated in the pathogen, i.e., under which conditions gene transfer is most active and efficient. Once parameters are revealed that inhibit competence development and/or the function of HGT mechanisms, those critical elements could serve as target for the development of HGT inhibition. Especially in the context of control strategies such as chemical decontamination, bacteriophage treatment or vaccine development, it will be crucial to have a second-line strategy for prevention of pathogen adaptation. Therefore, the inhibition of HGT in *Campylobacter* is a promising approach in combating *Campylobacter*.

Acknowledgements This work is supported by the German Federal Ministry of Education and Research (BMBF) in frame of the zoonoses research consortium PAC-CAMPY to K.S. (project IP3/01KI1725B).

References

Aidley J, Sorensen MCH, Bayliss CD, Brondsted L (2017) Phage exposure causes dynamic shifts in the expression states of specific phase-variable genes of *Campylobacter jejuni*. Microbiology 163(6):911–919. https://doi.org/10.1099/mic.0.000470

Avery OT, Macleod CM, McCarty M (1944) Studies on the chemical nature of the substance inducing transformation of pneumococcal types: induction of transformation by a desoxyribonucleic acid fraction isolated from pneumococcus type III. J Exp Med 79(2):137–158. https://doi.org/10.1084/jem.79.2.137

Avrain L, Vernozy-Rozand C, Kempf I (2004) Evidence for natural horizontal transfer of *tetO* gene between *Campylobacter jejuni* strains in chickens. J Appl Microbiol 97(1):134–140. https://doi.org/10.1111/j.1365-2672.2004.02306.x

Bacon DJ, Alm RA, Burr DH, Hu L, Kopecko DJ, Ewing CP, Trust TJ, Guerry P (2000) Involvement of a plasmid in virulence of *Campylobacter jejuni* 81-176. Infect Immun 68(8):4384–4390. https://doi.org/10.1128/iai.68.8.4384-4390.2000

Bacon DJ, Alm RA, Hu L, Hickey TE, Ewing CP, Batchelor RA, Trust TJ, Guerry P (2002) DNA sequence and mutational analyses of the pVir plasmid of *Campylobacter jejuni* 81-176. Infect Immun 70(11):6242–6250. https://doi.org/10.1128/iai.70.11.6242-6250.2002

Bae J, Oh E, Jeon B (2014) Enhanced transmission of antibiotic resistance in *Campylobacter jejuni* biofilms by natural transformation. Antimicrob Agents Chemother 58(12):7573–7575. https://doi.org/10.1128/AAC.04066-14

Bakkali M (2013) Could DNA uptake be a side effect of bacterial adhesion and twitching motility? Arch Microbiol 195(4):279–289. https://doi.org/10.1007/s00203-013-0870-1

Baldvinsson SB, Sorensen MC, Vegge CS, Clokie MR, Brondsted L (2014) *Campylobacter jejuni* motility is required for infection of the flagellotropic bacteriophage F341. Appl Environ Microbiol 80(22):7096–7106. https://doi.org/10.1128/AEM.02057-14

Barton C, Ng L-K, Tyler SD, Clark CG (2007) Temperate bacteriophages affect pulsed-field gel electrophoresis patterns of *Campylobacter jejuni*. J Clin Microbiol 45(2):386–391. https://doi.org/10.1128/JCM.01513-06

Batchelor RA, Pearson BM, Friis LM, Guerry P, Wells JM (2004) Nucleotide sequences and comparison of two large conjugative plasmids from different *Campylobacter* species. Microbiology 150(Pt 10):3507–3517. https://doi.org/10.1099/mic.0.27112-0

Beauchamp JM, Erfurt RS, DiRita VJ (2015) Characterization and localization of the *Campylobacter jejuni* transformation system proteins CtsE, CtsP, and CtsX. J Bacteriol 197(3):636–645. https://doi.org/10.1128/JB.02434-14

Beauchamp JM, Leveque RM, Dawid S, DiRita VJ (2017) Methylation-dependent DNA discrimination in natural transformation of *Campylobacter jejuni*. Proc Natl Acad Sci U S A 114(38):E8053–E8061. https://doi.org/10.1073/pnas.1703331114

Bradde S, Nourmohammad A, Goyal S, Balasubramanian V (2020) The size of the immune repertoire of bacteria. Proc Natl Acad Sci U S A 117(10):5144–5151. https://doi.org/10.1073/pnas.1903666117

Cehovin A, Simpson PJ, McDowell MA, Brown DR, Noschese R, Pallett M, Brady J, Baldwin GS, Lea SM, Matthews SJ, Pelicic V (2013) Specific DNA recognition mediated by a type IV pilin. Proc Natl Acad Sci U S A 110(8):3065–3070. https://doi.org/10.1073/pnas.1218832110

Chen I, Provvedi R, Dubnau D (2006) A macromolecular complex formed by a pilin-like protein in competent *Bacillus subtilis*. J Biol Chem 281(31):21720–21727. https://doi.org/10.1074/jbc.M604071200

Chen JC, Tagg KA, Joung YJ, Bennett C, Francois Watkins L, Eikmeier D, Folster JP (2018) Report of *erm*(B)(+) *Campylobacter jejuni* in the United States. Antimicrobial agents and chemotherapy 62(6). https://doi.org/10.1128/AAC.02615-17

Cheng Y, Zhang W, Lu Q, Wen G, Zhao Z, Luo Q, Shao H, Zhang T (2020) Point deletion or insertion in CmeR-Box, A2075G substitution in 23S rRNA, and presence of *erm*(B): are key factors of erythromycin resistance in *Campylobacter jejuni* and *Campylobacter coli* isolated from central China. Front Microbiol 11:203. https://doi.org/10.3389/fmicb.2020.00203

Clark CG (2011) Sequencing of CJIE1 prophages from *Campylobacter jejuni* isolates reveals the presence of inserted and (or) deleted genes. Can J Microbiol 57(10):795–808. https://doi.org/10.1139/w11-069

Clark CG, Ng L-K (2008) Sequence variability of *Campylobacter* temperate bacteriophages. BMC Microbiol 8(1):49. https://doi.org/10.1186/1471-2180-8-49

Couvin D, Bernheim A, Toffano-Nioche C, Touchon M, Michalik J, Neron B, Rocha EPC, Vergnaud G, Gautheret D, Pourcel C (2018) CRISPRCasFinder, an update of CRISRFinder, includes a portable version, enhanced performance and integrates search for Cas proteins. Nucleic Acids Res 46(W1):W246–W251. https://doi.org/10.1093/nar/gky425

Crippen CS, Lee YJ, Hutinet G, Shajahan A, Sacher JC, Azadi P, de Crecy-Lagard V, Weigele PR, Szymanski CM (2019) Deoxyinosine and 7-Deaza-2-Deoxyguanosine as carriers of genetic information in the DNA of *Campylobacter* Viruses. J Virol 93(23). https://doi.org/10.1128/JVI.01111-19

Daubin V, Szollosi GJ (2016) Horizontal Gene transfer and the history of life. Cold Spring Harb Perspect Biol 8(4):a018036. https://doi.org/10.1101/cshperspect.a018036

Dingle KE, Colles FM, Falush D, Maiden MC (2005) Sequence typing and comparison of population biology of *Campylobacter coli* and *Campylobacter jejuni*. J Clin Microbiol 43(1):340–347. https://doi.org/10.1128/JCM.43.1.340-347.2005

Drake SL, Koomey M (1995) The product of the *pilQ* gene is essential for the biogenesis of type IV pili in *Neisseria gonorrhoeae*. Mol Microbiol 18(5):975–986. https://doi.org/10.1111/j.1365-2958.1995.18050975.x

Du Y, Wang C, Ye Y, Liu Y, Wang A, Li Y, Zhou X, Pan H, Zhang J, Xu X (2018) Molecular identification of multidrug-resistant *Campylobacter* species from diarrheal patients and poultry meat in Shanghai China. Front Microbiol 9:1642. https://doi.org/10.3389/fmicb.2018.01642

Dubnau D, Blokesch M (2019) Mechanisms of DNA uptake by naturally competent bacteria. Annu Rev Genet 53:217–237. https://doi.org/10.1146/annurev-genet-112618-043641

Dugar G, Herbig A, Forstner KU, Heidrich N, Reinhardt R, Nieselt K, Sharma CM (2013) High-resolution transcriptome maps reveal strain-specific regulatory features of multiple *Campylobacter jejuni* isolates. PLoS Genet 9(5):e1003495. https://doi.org/10.1371/journal.pgen.1003495

Dugar G, Leenay RT, Eisenbart SK, Bischler T, Aul BU, Beisel CL, Sharma CM (2018) CRISPR RNA-dependent binding and cleavage of endogenous RNAs by the *Campylobacter jejuni* Cas9. Mol Cell 69(5):893–905 e897. https://doi.org/10.1016/j.molcel.2018.01.032

EFSA (2020) The European union summary report on antimicrobial resistance in zoonotic and indicator bacteria from humans, animals and food in 2017/2018. EFSA Journal 18(3):e06007. https://doi.org/10.2903/j.efsa.2020.6007

Ellison CK, Dalia TN, Vidal Ceballos A, Wang JC, Biais N, Brun YV, Dalia AB (2018) Retraction of DNA-bound type IV competence pili initiates DNA uptake during natural transformation in *Vibrio cholerae*. Nat Microbiol 3(7):773–780. https://doi.org/10.1038/s41564-018-0174-y

Epping L, Piro RM, Knüver MT, Borowiak M, Huber C, Thürmer A, Malorny B, Stingl K, Fruth A, Wieler LH, Semmler T (2020) Genome-wide identification of Host-Specificity of *Campylobacter jejuni* in Germany based on Whole Genome Data. In: 6th Joint conference of DGHM and VAAM, Leipzig, Germany

Feng J, Ma L, Nie J, Konkel ME, Lu X (2018) Environmental stress-induced bacterial lysis and extracellular DNA release contribute to *Campylobacter jejuni* biofilm formation. Appl Environ Microbiol 84(5). https://doi.org/10.1128/AEM.02068-17

Fernandez-Gonzalez E, Backert S (2014) DNA transfer in the gastric pathogen *Helicobacter pylori*. J Gastroenterol. 49(4):594–604. https://doi.org/10.1007/s00535-014-0938-y

Firehammer B, Border M (1968) Isolation of temperate bacteriophages from *Vibrio fetus*. Am J Vet Res 29(11):2229–2235

Fischer W, Tegtmeyer N, Stingl K, Backert S (2020) Four chromosomal type IV secretion systems in *Helicobacter pylori*: composition, structure and function. Front Microbiol 11:1592. https://doi.org/10.3389/fmicb.2020.01592

Florez-Cuadrado D, Ugarte-Ruiz M, Meric G, Quesada A, Porrero MC, Pascoe B, Saez-Llorente JL, Orozco GL, Dominguez L, Sheppard SK (2017) Genome comparison of erythromycin resistant *Campylobacter* from turkeys identifies hosts and pathways for horizontal spread of *erm*(B) Genes. Front Microbiol 8:2240. https://doi.org/10.3389/fmicb.2017.02240

Florez-Cuadrado D, Ugarte-Ruiz M, Quesada A, Palomo G, Dominguez L, Porrero MC (2016) Description of an *erm*(B)-carrying *Campylobacter coli* isolate in Europe. J Antimicrob Chemother 71(3):841–843. https://doi.org/10.1093/jac/dkv383

Fouts DE, Mongodin EF, Mandrell RE, Miller WG, Rasko DA, Ravel J, Brinkac LM, DeBoy RT, Parker CT, Daugherty SC, Dodson RJ, Durkin AS, Madupu R, Sullivan SA, Shetty JU, Ayodeji MA, Shvartsbeyn A, Schatz MC, Badger JH, Fraser CM, Nelson KE (2005) Major structural differences and novel potential virulence mechanisms from the genomes of multiple *Campylobacter* species. PLoS Biol 3(1):e15. https://doi.org/10.1371/journal.pbio.0030015

Fry BN, Feng S, Chen YY, Newell DG, Coloe PJ, Korolik V (2000) The *galE* gene of *Campylobacter jejuni* is involved in lipopolysaccharide synthesis and virulence. Infect Immun 68(5):2594–2601. https://doi.org/10.1128/iai.68.5.2594-2601.2000

Gaasbeek EJ, Wagenaar JA, Guilhabert MR, van Putten JP, Parker CT, van der Wal FJ (2010) Nucleases encoded by the integrated elements CJIE2 and CJIE4 inhibit natural transformation of *Campylobacter jejuni*. J Bacteriol 192(4):936–941. https://doi.org/10.1128/jb.00867-09

Gaasbeek EJ, Wagenaar JA, Guilhabert MR, Wosten MM, van Putten JP, van der Graaf-van Bloois L, Parker CT, van der Wal FJ (2009) A DNase encoded by integrated element CJIE1 inhibits natural transformation of *Campylobacter jejuni*. J Bacteriol 191(7):2296–2306. https://doi.org/10.1128/jb.01430-08

Gardner SP, Olson JW (2012) Barriers to horizontal gene transfer in *Campylobacter jejuni*. Adv Appl Microbiol 79:19–42. https://doi.org/10.1016/B978-0-12-394318-7.00002-4

Gencay YE, Sorensen MCH, Wenzel CQ, Szymanski CM, Brondsted L (2018) Phase variable expression of a single phage receptor in *Campylobacter jejuni* NCTC12662 influences sensitivity toward several diverse CPS-dependent phages. Front Microbiol 9:82. https://doi.org/10.3389/fmicb.2018.00082

Ghatak S, He Y, Reed S, Strobaugh T Jr, Irwin P (2017) Whole genome sequencing and analysis of *Campylobacter coli* YH502 from retail chicken reveals a plasmid-borne type VI secretion system. Genom Data 11:128–131. https://doi.org/10.1016/j.gdata.2017.02.005
Gibreel A, Skold O (1998) High-level resistance to trimethoprim in clinical isolates of *Campylobacter jejuni* by acquisition of foreign genes (*dfr1* and *dfr9*) expressing drug-insensitive dihydrofolate reductases. Antimicrob Agents Chemother 42(12):3059–3064. https://doi.org/10.1128/AAC.42.12.3059
Gibreel A, Skold O, Taylor DE (2004) Characterization of plasmid-mediated *aphA-3* kanamycin resistance in *Campylobacter jejuni*. Microb Drug Resist 10(2):98–105. https://doi.org/10.1089/1076629041310127
Golz JC, Epping L, Knuver MT, Borowiak M, Hartkopf F, Deneke C, Malorny B, Semmler T, Stingl K (2020) Whole genome sequencing reveals extended natural transformation in *Campylobacter* impacting diagnostics and the pathogens adaptive potential. Sci Rep 10(1):3686. https://doi.org/10.1038/s41598-020-60320-y
Goodman SD, Scocca JJ (1988) Identification and arrangement of the DNA sequence recognized in specific transformation of *Neisseria gonorrhoeae*. Proc Natl Acad Sci U S A 85(18):6982–6986. https://doi.org/10.1073/pnas.85.18.6982
Griffith F (1928) The significance of pneumococcal types. J Hyg (Lond) 27(2):113–159. https://doi.org/10.1017/s0022172400031879
Grissa I, Vergnaud G, Pourcel C (2007) The CRISPRdb database and tools to display CRISPRs and to generate dictionaries of spacers and repeats. BMC Bioinformatics 8:172. https://doi.org/10.1186/1471-2105-8-172
Gundogdu O, Bentley SD, Holden MT, Parkhill J, Dorrell N, Wren BW (2007) Re-annotation and re-analysis of the *Campylobacter jejuni* NCTC11168 genome sequence. BMC Genom 8:162. https://doi.org/10.1186/1471-2164-8-162
Gunther NWt, Reichenberger ER, Bono JL (2016) Complete genome sequence of UV-resistant *Campylobacter jejuni* RM3194, Including an 81.08-Kilobase Plasmid. Genome Announc 4(2). https://doi.org/10.1128/genomeA.00305-16
Harrington CS, Thomson-Carter FM, Carter PE (1997) Evidence for recombination in the flagellin locus of *Campylobacter jejuni*: implications for the flagellin gene typing scheme. J Clin Microbiol 35(9):2386–2392. https://doi.org/10.1128/JCM.35.9.2386-2392.1997
Heler R, Samai P, Modell JW, Weiner C, Goldberg GW, Bikard D, Marraffini LA (2015) Cas9 specifies functional viral targets during CRISPR-Cas adaptation. Nature 519(7542):199–202. https://doi.org/10.1038/nature14245
Hepp C, Maier B (2016) Kinetics of DNA uptake during transformation provide evidence for a translocation ratchet mechanism. Proc Natl Acad Sci U S A 113(44):12467–12472. https://doi.org/10.1073/pnas.1608110113
Hille F, Richter H, Wong SP, Bratovic M, Ressel S, Charpentier E (2018) The biology of CRISPR-Cas: backward and forward. Cell 172(6):1239–59. https://doi.org/10.1016/j.cell.2017.11.032
Holst Sorensen MC, van Alphen LB, Fodor C, Crowley SM, Christensen BB, Szymanski CM, Brondsted L (2012) Phase variable expression of capsular polysaccharide modifications allows *Campylobacter jejuni* to avoid bacteriophage infection in chickens. Front Cell Infect Microbiol 2:11. https://doi.org/10.3389/fcimb.2012.00011
Holt JP, Grant AJ, Coward C, Maskell DJ, Quinlan JJ (2012) Identification of Cj1051c as a major determinant for the restriction barrier of *Campylobacter jejuni* strain NCTC11168. Appl Environ Microbiol 78(22):7841–7848. https://doi.org/10.1128/AEM.01799-12
Hooton SP, Connerton IF (2014) *Campylobacter jejuni* acquire new host-derived CRISPR spacers when in association with bacteriophages harboring a CRISPR-like Cas4 protein. Front Microbiol 5:744. https://doi.org/10.3389/fmicb.2014.00744
Irons J, Sacher JC, Szymanski CM, Downs DM (2019) Cj1388 is a RidA homolog and is required for flagella biosynthesis and/or function in *Campylobacter jejuni*. Front Microbiol 10:2058. https://doi.org/10.3389/fmicb.2019.02058

Jäckel C, Hammerl JA, Hertwig S (2019) *Campylobacter* phage isolation and characterization: what we have learned so far. Methods Protoc 2(1). https://doi.org/10.3390/mps2010018

Javed MA, Ackermann HW, Azeredo J, Carvalho CM, Connerton I, Evoy S, Hammerl JA, Hertwig S, Lavigne R, Singh A, Szymanski CM, Timms A, Kropinski AM (2014) A suggested classification for two groups of *Campylobacter* myoviruses. Arch Virol 159(1):181–190. https://doi.org/10.1007/s00705-013-1788-2

Javed MA, Sacher JC, van Alphen LB, Patry RT, Szymanski CM (2015) A flagellar glycan-specific protein encoded by *Campylobacter* phages inhibits host cell growth. Viruses 7(12):6661–6674. https://doi.org/10.3390/v7122964

Jeon B, Muraoka W, Sahin O, Zhang Q (2008) Role of Cj1211 in natural transformation and transfer of antibiotic resistance determinants in *Campylobacter jejuni*. Antimicrob Agents Chemother 52(8):2699–2708. https://doi.org/10.1128/aac.01607-07

Jeon B, Zhang Q (2007) Cj0011c, a periplasmic single- and double-stranded DNA-binding protein, contributes to natural transformation in *Campylobacter jejuni*. J Bacteriol 189(20):7399–7407. https://doi.org/10.1128/jb.01012-07

Johnston C, Martin B, Fichant G, Polard P, Claverys JP (2014) Bacterial transformation: distribution, shared mechanisms and divergent control. Nat Rev Microbiol 12(3):181–196. https://doi.org/10.1038/nrmicro3199

Kovac J, Cadez N, Stessl B, Stingl K, Gruntar I, Ocepek M, Trkov M, Wagner M, Smole Mozina S (2015) High genetic similarity of ciprofloxacin-resistant *Campylobacter jejuni* in central Europe. Front Microbiol 6:1169. https://doi.org/10.3389/fmicb.2015.01169

Larsen JC, Szymanski C, Guerry P (2004) N-linked protein glycosylation is required for full competence in *Campylobacter jejuni* 81-176. J Bacteriol 186(19):6508–6514. https://doi.org/10.1128/jb.186.19.6508-6514.2004

Lawrence JG (2005) Horizontal and vertical gene transfer: the life history of pathogens. Contrib Microbiol 12:255–271. https://doi.org/10.1159/000081699

Lederberg J, Tatum EL (1946) Gene recombination in *Escherichia coli*. Nature 158(4016):558. https://doi.org/10.1038/158558a0

Leekitcharoenphon P, Garcia-Graells C, Botteldoorn N, Dierick K, Kempf I, Olkkola S, Rossi M, Nykäsenoja S, Malorny B, Stingl K (2018) Comparative genomics of quinolone-resistant and susceptible Campylobacter jejuni of poultry origin from major poultry producing European countries (GENCAMP). EFSA Support Publ 15(5):1398E. https://doi.org/10.2903/sp.efsa.2018.EN-1398

Levy A, Goren MG, Yosef I, Auster O, Manor M, Amitai G, Edgar R, Qimron U, Sorek R (2015) CRISPR adaptation biases explain preference for acquisition of foreign DNA. Nature 520(7548):505–510. https://doi.org/10.1038/nature14302

Liu D, Deng F, Gao Y, Yao H, Shen Z, Wu C, Wang Y, Shen J (2017) Dissemination of *erm*(B) and its associated multidrug-resistance genomic islands in *Campylobacter* from 2013 to 2015. Vet Microbiol 204:20–24. https://doi.org/10.1016/j.vetmic.2017.02.022

Louwen R, Horst-Kreft D, de Boer AG, van der Graaf L, de Knegt G, Hamersma M, Heikema AP, Timms AR, Jacobs BC, Wagenaar JA, Endtz HP, van der Oost J, Wells JM, Nieuwenhuis EE, van Vliet AH, Willemsen PT, van Baarlen P, van Belkum A (2013) A novel link between *Campylobacter jejuni* bacteriophage defence, virulence and Guillain-Barre syndrome. Eur J Clin Microbiol Infect Dis 32(2):207–226. https://doi.org/10.1007/s10096-012-1733-4

Luangtongkum T, Shen Z, Seng VW, Sahin O, Jeon B, Liu P, Zhang Q (2012) Impaired fitness and transmission of macrolide-resistant *Campylobacter jejuni* in its natural host. Antimicrob Agents Chemother 56(3):1300–1308. https://doi.org/10.1128/AAC.05516-11

Luo N, Pereira S, Sahin O, Lin J, Huang S, Michel L, Zhang Q (2005) Enhanced *in vivo* fitness of fluoroquinolone-resistant *Campylobacter jejuni* in the absence of antibiotic selection pressure. Proc Natl Acad Sci U S A 102(3):541–546. https://doi.org/10.1073/pnas.0408966102

Makarova KS, Wolf YI, Iranzo J, Shmakov SA, Alkhnbashi OS, Brouns SJJ, Charpentier E, Cheng D, Haft DH, Horvath P, Moineau S, Mojica FJM, Scott D, Shah SA, Siksnys V, Terns MP, Venclovas C, White MF, Yakunin AF, Yan W, Zhang F, Garrett RA, Backofen R, van der Oost

J, Barrangou R, Koonin EV (2020) Evolutionary classification of CRISPR-Cas systems: a burst of class 2 and derived variants. Nat Rev Microbiol 18(2):67–83. https://doi.org/10.1038/s41579-019-0299-x

Marasini D, Fakhr MK (2014) Exploring PFGE for detecting large plasmids in *Campylobacter jejuni* and *Campylobacter coli* isolated from various retail meats. Pathogens 3(4):833–844. https://doi.org/10.3390/pathogens3040833

Marasini D, Fakhr MK (2016) Whole-genome sequencing of a *Campylobacter jejuni* strain isolated from retail chicken meat reveals the presence of a megaplasmid with Mu-like prophage and multidrug resistance genes. Genome Announc 4(3). https://doi.org/10.1128/genomeA.00460-16

Marasini D, Karki AB, Buchheim MA, Fakhr MK (2018) Phylogenetic relatedness among plasmids harbored by *Campylobacter jejuni* and *Campylobacter coli* isolated from retail meats. Front Microbiol 9:2167. https://doi.org/10.3389/fmicb.2018.02167

Meric G, Yahara K, Mageiros L, Pascoe B, Maiden MC, Jolley KA, Sheppard SK (2014) A reference pan-genome approach to comparative bacterial genomics: identification of novel epidemiological markers in pathogenic *Campylobacter*. PLoS ONE 9(3):e92798. https://doi.org/10.1371/journal.pone.0092798

Mourkas E, Florez-Cuadrado D, Pascoe B, Calland JK, Bayliss SC, Mageiros L, Meric G, Hitchings MD, Quesada A, Porrero C, Ugarte-Ruiz M, Gutierrez-Fernandez J, Dominguez L, Sheppard SK (2019) Gene pool transmission of multidrug resistance among *Campylobacter* from livestock, sewage and human disease. Environ Microbiol 21(12):4597–4613. https://doi.org/10.1111/1462-2920.14760

O'Loughlin JL, Eucker TP, Chavez JD, Samuelson DR, Neal-McKinney J, Gourley CR, Bruce JE, Konkel ME (2015) Analysis of the *Campylobacter jejuni* genome by SMRT DNA sequencing identifies restriction-modification motifs. PLoS ONE 10(2):e0118533. https://doi.org/10.1371/journal.pone.0118533

Obergfell KP, Seifert HS (2016) The pilin N-terminal domain maintains *Neisseria gonorrhoeae* transformation competence during pilus phase variation. PLoS Genet 12(5):e1006069. https://doi.org/10.1371/journal.pgen.1006069

Ouellette M, Gerbaud G, Lambert T, Courvalin P (1987) Acquisition by a *Campylobacter*-like strain of aphA-1, a kanamycin resistance determinant from members of the family *Enterobacteriaceae*. Antimicrob Agents Chemother 31(7):1021–1026. https://doi.org/10.1128/aac.31.7.1021

Oyarzabal OA, Backert S (2012) Microbial Food Safety: An Introduction (Food Science Text Series). Springer, Heidelberg

Oyarzabal OA, Rad R, Backert S (2007) Conjugative transfer of chromosomally encoded antibiotic resistance from *Helicobacter pylori* to *Campylobacter jejuni*. J Clin Microbiol 45(2):402–408. https://doi.org/10.1128/JCM.01456-06

Parker CT, Quiñones B, Miller WG, Horn ST, Mandrell RE (2006) Comparative genomic analysis of *Campylobacter jejuni* strains reveals diversity due to genomic elements similar to those present in *C. jejuni* strain RM1221. J Clinical Microbiol 44(11):4125–4135. https://doi.org/10.1128/JCM.01231-06

Parkhill J, Wren BW, Mungall K, Ketley JM, Churcher C, Basham D, Chillingworth T, Davies RM, Feltwell T, Holroyd S, Jagels K, Karlyshev AV, Moule S, Pallen MJ, Penn CW, Quail MA, Rajandream MA, Rutherford KM, van Vliet AH, Whitehead S, Barrell BG (2000) The genome sequence of the food-borne pathogen *Campylobacter jejuni* reveals hypervariable sequences. Nature 403(6770):665–668. https://doi.org/10.1038/35001088

Phongsisay V, Perera VN, Fry BN (2006) Exchange of lipooligosaccharide synthesis genes creates potential Guillain-Barre syndrome-inducible strains of *Campylobacter jejuni*. Infect Immun 74(2):1368–1372. https://doi.org/10.1128/IAI.74.2.1368-1372.2006

Pratt A, Korolik V (2005) Tetracycline resistance of Australian *Campylobacter jejuni* and *Campylobacter coli* isolates. J Antimicrob Chemother 55(4):452–460. https://doi.org/10.1093/jac/dki040

Qin S, Wang Y, Zhang Q, Zhang M, Deng F, Shen Z, Wu C, Wang S, Zhang J, Shen J (2014) Report of ribosomal RNA methylase gene *erm*(B) in multidrug-resistant *Campylobacter coli*. J Antimicrob Chemother 69(4):964–968. https://doi.org/10.1093/jac/dkt492

Reuter M, Periago PM, Mulholland F, Brown HL, van Vliet AH (2015) A PAS domain-containing regulator controls flagella-flagella interactions in *Campylobacter jejuni*. Front Microbiol 6:770. https://doi.org/10.3389/fmicb.2015.00770

Richardson PT, Park SF (1997) Integration of heterologous plasmid DNA into multiple sites on the genome of *Campylobacter coli* following natural transformation. J Bacteriol 179(5):1809–1812. https://doi.org/10.1128/jb.179.5.1809-1812.1997

Rosner BM, Schielke A, Didelot X, Kops F, Breidenbach J, Willrich N, Golz G, Alter T, Stingl K, Josenhans C, Suerbaum S, Stark K (2017) A combined case-control and molecular source attribution study of human *Campylobacter* infections in Germany, 2011-2014. Sci Rep 7(1):5139. https://doi.org/10.1038/s41598-017-05227-x

Sacher JC, Flint A, Butcher J, Blasdel B, Reynolds HM, Lavigne R, Stintzi A, Szymanski CM (2018) Transcriptomic analysis of the *Campylobacter jejuni* response to T4-Like phage NCTC 12673 infection. Viruses 10(6). https://doi.org/10.3390/v10060332

Sacher JC, Shajahan A, Butcher J, Patry RT, Flint A, Hendrixson DR, Stintzi A, Azadi P, Szymanski CM (2020) Binding of phage-encoded FlaGrab to motile *Campylobacter jejuni* flagella inhibits growth, downregulates energy metabolism, and requires specific flagellar glycans. Front Microbiol 11:397. https://doi.org/10.3389/fmicb.2020.00397

Sails A, Wareing D, Bolton F, Fox A, Curry A (1998) Characterisation of 16 *Campylobacter jejuni* and *C. coli* typing bacteriophages. J Med Microbiol 47(2):123–128. https://doi.org/10.1099/00222615-47-2-123

Schmidt-Ott R, Pohl S, Burghard S, Weig M, Gross U (2005) Identification and characterization of a major subgroup of conjugative *Campylobacter jejuni* plasmids. J Infect 50(1):12–21. https://doi.org/10.1016/j.jinf.2004.02.013

Scott AE, Timms AR, Connerton PL, Carrillo CL, Radzum KA, Connerton IF (2007) Genome dynamics of *Campylobacter jejuni* in response to bacteriophage predation. PLoS Pathog 3(8):e119. https://doi.org/10.1371/journal.ppat.0030119

Seitz P, Pezeshgi Modarres H, Borgeaud S, Bulushev RD, Steinbock LJ, Radenovic A, Dal Peraro M, Blokesch M (2014) ComEA is essential for the transfer of external DNA into the periplasm in naturally transformable *Vibrio cholerae* cells. PLoS Genet 10(1):e1004066. https://doi.org/10.1371/journal.pgen.1004066

Sheppard SK, Jolley KA, Maiden MC (2012) A Gene-By-Gene Approach to Bacterial Population Genomics: Whole Genome MLST of *Campylobacter*. Genes (Basel) 3(2):261–277. https://doi.org/10.3390/genes3020261

Sheppard SK, Maiden MC (2015) The evolution of *Campylobacter jejuni* and *Campylobacter coli*. Cold Spring Harb Perspect Biol 7(8):a018119. https://doi.org/10.1101/cshperspect.a018119

Sheppard SK, McCarthy ND, Falush D, Maiden MC (2008) Convergence of *Campylobacter* species: implications for bacterial evolution. Science 320(5873):237–239. https://doi.org/10.1126/science.1155532

Sheppard SK, McCarthy ND, Jolley KA, Maiden MC (2011) Introgression in the genus *Campylobacter*: generation and spread of mosaic alleles. Microbiology 157(Pt 4):1066–1074. https://doi.org/10.1099/mic.0.045153-0

Sorensen MC, Gencay YE, Birk T, Baldvinsson SB, Jackel C, Hammerl JA, Vegge CS, Neve H, Brondsted L (2015) Primary isolation strain determines both phage type and receptors recognised by *Campylobacter jejuni* bacteriophages. PLoS ONE 10(1):e0116287. https://doi.org/10.1371/journal.pone.0116287

Sorensen MC, van Alphen LB, Harboe A, Li J, Christensen BB, Szymanski CM, Brondsted L (2011) Bacteriophage F336 recognizes the capsular phosphoramidate modification of *Campylobacter jejuni* NCTC11168. J Bacteriol 193(23):6742–6749. https://doi.org/10.1128/JB.05276-11

Sougakoff W, Papadopoulou B, Nordmann P, Courvalin P (1987) Nucleotide sequence and distribution of gene *tetO* encoding tetracycline resistance in *Campylobacter coli*. FEMS Microbiol Lett 44(1):153–159. https://doi.org/10.1111/j.1574-6968.1987.tb02260.x

Stern A, Keren L, Wurtzel O, Amitai G, Sorek R (2010) Self-targeting by CRISPR: gene regulation or autoimmunity? Trends Genet 26(8):335–340. https://doi.org/10.1016/j.tig.2010.05.008

Stingl K, Muller S, Scheidgen-Kleyboldt G, Clausen M, Maier B (2010) Composite system mediates two-step DNA uptake into *Helicobacter pylori*. Proc Natl Acad Sci U S A 107(3):1184–1189. https://doi.org/10.1073/pnas.0909955107

Svensson SL, Pryjma M, Gaynor EC (2014) Flagella-mediated adhesion and extracellular DNA release contribute to biofilm formation and stress tolerance of *Campylobacter jejuni*. PLoS ONE 9(8):e106063. https://doi.org/10.1371/journal.pone.0106063

Tang Y, Dai L, Sahin O, Wu Z, Liu M, Zhang Q (2017) Emergence of a plasmid-borne multidrug resistance gene *cfr(C)* in foodborne pathogen *Campylobacter*. J Antimicrob Chemother 72(6):1581–1588. https://doi.org/10.1093/jac/dkx023

Taylor DE, De Grandis SA, Karmali MA, Fleming PC (1981) Transmissible plasmids from *Campylobacter jejuni*. Antimicrob Agents Chemother 19(5):831–835. https://doi.org/10.1128/aac.19.5.831

Vegge CS, Brondsted L, Ligowska-Marzeta M, Ingmer H (2012) Natural transformation of *Campylobacter jejuni* occurs beyond limits of growth. PLoS ONE 7(9):e45467. https://doi.org/10.1371/journal.pone.0045467

Velazquez JB, Jimenez A, Chomon B, Villa TG (1995) Incidence and transmission of antibiotic resistance in *Campylobacter jejuni* and *Campylobacter coli*. J Antimicrob Chemother 35(1):173–178. https://doi.org/10.1093/jac/35.1.173

Vorwerk H, Huber C, Mohr J, Bunk B, Bhuju S, Wensel O, Sproer C, Fruth A, Flieger A, Schmidt-Hohagen K, Schomburg D, Eisenreich W, Hofreuter D (2015) A transferable plasticity region in *Campylobacter coli* allows isolates of an otherwise non-glycolytic food-borne pathogen to catabolize glucose. Mol Microbiol 98(5):809–830. https://doi.org/10.1111/mmi.13159

Wang Y, Taylor DE (1990) Natural transformation in *Campylobacter* species. J Bacteriol 172(2):949–955. https://doi.org/10.1128/jb.172.2.949-955.1990

Wang Y, Zhang M, Deng F, Shen Z, Wu C, Zhang J, Zhang Q, Shen J (2014) Emergence of multidrug-resistant *Campylobacter* species isolates with a horizontally acquired rRNA methylase. Antimicrob Agents Chemother 58(9):5405–5412. https://doi.org/10.1128/AAC.03039-14

Wassenaar TM, Fry BN, van der Zeijst BA (1995) Variation of the flagellin gene locus of *Campylobacter jejuni* by recombination and horizontal gene transfer. Microbiology 141(Pt 1):95–101. https://doi.org/10.1099/00221287-141-1-95

Wei Y, Terns RM, Terns MP (2015) Cas9 function and host genome sampling in Type II-A CRISPR-Cas adaptation. Genes Dev 29(4):356–361. https://doi.org/10.1101/gad.257550.114

Wiesner RS, Hendrixson DR, DiRita VJ (2003) Natural Transformation of *Campylobacter jejuni* Requires Components of a Type II Secretion System. J Bacteriol 185(18):5408–5418. https://doi.org/10.1128/jb.185.18.5408-5418.2003

Wilson DJ, Gabriel E, Leatherbarrow AJ, Cheesbrough J, Gee S, Bolton E, Fox A, Hart CA, Diggle PJ, Fearnhead P (2009) Rapid evolution and the importance of recombination to the gastroenteric pathogen *Campylobacter jejuni*. Mol Biol Evol 26(2):385–397. https://doi.org/10.1093/molbev/msn264

Wilson DL, Bell JA, Young VB, Wilder SR, Mansfield LS, Linz JE (2003) Variation of the natural transformation frequency of *Campylobacter jejuni* in liquid shake culture. Microbiology 149(Pt 12):3603–3615. https://doi.org/10.1099/mic.0.26531-0

Yao H, Shen Z, Wang Y, Deng F, Liu D, Naren G, Dai L, Su CC, Wang B, Wang S, Wu C, Yu EW, Zhang Q, Shen J (2016) Emergence of a potent multidrug efflux pump variant that enhances *Campylobacter* resistance to multiple antibiotics. mBio 7(5). https://doi.org/10.1128/mBio.01543-16

Yosef I, Goren MG, Qimron U (2012) Proteins and DNA elements essential for the CRISPR adaptation process in *Escherichia coli*. Nucleic Acids Res 40(12):5569–5576. https://doi.org/10.1093/nar/gks216

Zautner AE, Goldschmidt AM, Thurmer A, Schuldes J, Bader O, Lugert R, Gross U, Stingl K, Salinas G, Lingner T (2015) SMRT sequencing of the *Campylobacter coli* BfR-CA-9557 genome sequence reveals unique methylation motifs. BMC Genom 16:1088. https://doi.org/10.1186/s12864-015-2317-3

Zeng X, Ardeshna D, Lin J (2015) Heat shock-enhanced conjugation efficiency in standard *Campylobacter jejuni* strains. Appl Environ Microbiol 81(13):4546–4552. https://doi.org/10.1128/aem.00346-15

Zeng X, Wu Z, Zhang Q, Lin J (2018) A cotransformation method to identify a restriction-modification enzyme that reduces conjugation efficiency in *Campylobacter jejuni*. Appl Environ Microbiol 84(23). https://doi.org/10.1128/AEM.02004-18

Molecular Mechanisms of *Campylobacter* Biofilm Formation and Quorum Sensing

Christoph Püning, Yulan Su, Xiaonan Lu, and Greta Gölz

Contents

Abstract Even though *Campylobacter* spp. are known to be fastidious organisms, they can survive within the natural environment. One mechanism to withstand unfavourable conditions is the formation of biofilms, a multicellular structure composed of different bacterial and other microbial species which are embedded in an extracellular matrix. High oxygen levels, low substrate concentrations and the presence of external DNA stimulate the biofilm formation by *C. jejuni*. These

C. Püning · Y. Su · G. Gölz (✉)
Institute of Food Safety and Food Hygiene, Free University Berlin, Koenigsweg 69D, 14163 Berlin, Germany
e-mail: greta.goelz@fu-berlin.de

C. Püning
e-mail: Christoph.puening@fu-berlin.de

Y. Su
e-mail: yulan.su@fu-berlin.de

X. Lu
Food, Nutrition, and Health Program, Faculty of Land and Food Systems, The University of British Columbia, 2205 East Mall, Vancouver, BC V6T 1Z4, Canada
e-mail: xiaonan.lu@ubc.ca

S. Backert (ed.), *Fighting Campylobacter Infections*, Current Topics in Microbiology and Immunology 431, https://doi.org/10.1007/978-3-030-65481-8_11

external factors trigger internal adaptation processes, e.g. via regulating the expression of genes encoding proteins required for surface structure formation, as well as motility, stress response and antimicrobial resistance. Known genes impacting biofilm formation will be summarized in this review. The formation of biofilms as well as the expression of virulence genes is often regulated in a cell density depending manner by quorum sensing, which is mediated via small signalling molecules termed autoinducers. Even though quorum sensing mechanisms of other bacteria are well understood, knowledge on the role of these mechanisms in *C. jejuni* biofilm formation is still scarce. The LuxS enzyme involved in generation of autoinducer-2 is present in *C. jejuni,* but autoinducer receptors have not been identified so far. Phenotypes of *C. jejuni* strains lacking a functional *luxS* like reduced growth, motility, oxygen stress tolerance, biofilm formation, adhesion, invasion and colonization are also summarized within this chapter. However, these phenotypes are highly variable in distinct *C. jejuni* strains and depend on the culture conditions applied.

1 Introduction

Compared to other food-borne pathogenic bacteria, *Campylobacter* spp. are susceptible to various stressors including elevated ambient oxygen concentrations, dehydration and UV-light, which are present in the natural environments and in food processing plants. Nevertheless, *Campylobacter* spp. are widespread in the environment and persist in the food production chain indicating that these bacteria are capable to survive these unfavourable conditions (Boronowsky et al. 2014; Golz et al. 2018; Hansson et al. 2018; Tram et al. 2020a). However, how they regulate their stress responses and environmental adaptation is still not fully understood as campylobacters are lacking several classical regulatory factors. One microbial strategy to survive within hostile surroundings is the formation of biofilms. Biofilms are organized aggregates of microorganisms encased by an extracellular matrix. This extracellular matrix structures the biofilm and also protects microorganisms from stressful conditions present outside of the biofilm (Kostakiotis et al. 2013). The process of biofilm formation as well as the expression of virulence factors is often coordinated at a multicellular stage, which depends on the detection of the cell density via quorum sensing (QS) systems which are present in many bacteria, fungi and parasites (Mukherjee and Bassler 2019). Within this article, we summarize information on external factors and genes involved in biofilm formation and QS of *C. jejuni.*

2 Microbial Biofilm Formation

Bacteria can switch from a planktonic single-cell lifestyle to a multicellular lifestyle, e.g. in biofilms, and back to planktonic style. In biofilms, bacterial species live in close contact with communities which can also contain fungi, algae, protists and archaea

(Flemming et al. 2016). These biofilms can be found either attached to a surface or as free-floating aggregates, which are both surrounded by a matrix of extracellular polymeric substances (EPS) (Joshua et al. 2006; Roy et al. 2018). Depending on the microorganisms within the biofilm, the EPS consists of proteins, nucleic acids, polysaccharides, lipids and other compounds which form part of highly viscous watery solutions (Flemming et al. 2016). Within these biofilms, microorganisms are protected from several external stressors—such as dehydration, and exposure to oxygen radicals, disinfectants or antimicrobial substances—and grow much more slowly compared to planktonic cells, thereby facilitating survival under unfavourable conditions in diverse environmental niches. Furthermore, microorganisms within biofilms can support each other by exchanging of substrates or by degradation of toxigenic substances (Flemming et al. 2016). The ability to form biofilms and to colonize preformed biofilms as well as the specific architecture of biofilms depends on the microbial composition, the genetic background of the individual strains involved and the environmental conditions.

2.1 Building and Dispersion of Microbial Biofilms

Biofilm formation takes place in three major steps: In the first two steps, the microorganisms build up microcolonies by attachment to surfaces and/or to each other, and the production of EPS establishes the biofilm structure, which matures the microcolonies into a three dimensional architecture. In the third phase, the microorganisms actively or passively detach from the biofilm and are released back to the planktonic lifestyle. In bacterial biofilms, surface or cell-to-cell attachment is mediated by extracellular adhesive appendages, like flagella, pili or outer membrane proteins, secreted adhesins as well as by the molecular structure and adhesive properties of the abiotic surfaces (Kostakioti et al. 2013). Once the microcolonies are built, multiple regulatory networks translate signals to concerted gene expression changes, which lead to building of the extracellular matrix and mediate the spatial and temporal reorganization of the microbial cells within the final biofilm (Petrova and Sauer 2016). The biofilm matures into a well-organized architecture, with intervening water channels for nutrient and waste exchange which is embedded in a viscous EPS matrix (Coughlan et al. 2016). In the final state, biofilms represent highly dynamic structures, in which the bacteria could disperse passively or actively. Passive dispersal is due to external shear forces or abrasion when the biofilm structure grows (Kaplan 2010). Active dispersion of biofilms is triggered by beneficial conditions outside the biofilm or detrimental conditions inside the biofilm. These include scarcity of substrates including carbon and energy sources, accumulation of signalling molecules and in case of *Campylobacter* also elevated oxygen levels (Kostakioti et al. 2013, Petrova and Sauer 2016). The release of microorganisms from biofilms is supported by increased motility. Active dispersion of biofilms can be mediated by bacterial secretion of EPS-degrading enzymes including glycosidases, lipases, proteases and deoxyribonuleases, as well as by production of surfactants (Kaplan 2010).

2.2 *Methods to Analyse Biofilms*

Analysis of biofilms is focussed on the quantification and successful measurement of several multiple parameters including the biomass and architecture of biofilms, the bacterial viability, attachment and motility within biofilms and the composition of the EPS (reviewed by Azeredo et al. 2017). Briefly, the total amount of the biofilm-mass is commonly quantified by indirect staining methods, e.g. by the Crystal Violet Assay, while the viable cell count can be determined by, e.g. direct plating, flow cytometry or live/dead staining combined with confocal laser scanning microscopy (CLSM). CLSM is further applied to study the spatial structure of biofilms. The metabolic activity of bacteria in biofilms can be measured by colorimetric determination of the conversion of tetrazolium salts to formazan by a spectrophotometer. The amount of initially attached bacteria can be quantified by direct plating or microscopic methods.

2.3 *Environmental Conditions Influencing* Campylobacter jejuni *Biofilm Formation*

Investigations focused on the biofilm formation capacity of *C. jejuni* were mostly conducted under laboratory conditions with well-defined reference strains of the pathogen as monospecies biofilms, which do not reflect the situation outside the laboratory (Lamas et al. 2018; Teh et al. 2014). The resulting data demonstrate that *C. jejuni* bacteria are able to form biofilms on glass, polystyrene and stainless steel surfaces (Joshua et al. 2006; Li et al. 2017; Oh et al. 2016; Teh et al. 2016; Wagle et al. 2019). However, the whole biofilm formation process of *C. jejuni* is modulated by many extrinsic and intrinsic factors which will be discussed in more detail.

2.3.1 Substrate Availability and Oxygen

Distinct external stress conditions which all depend on the specific metabolic properties of *C. jejuni* have been found to regulate the biofilm formation and lifestyle of the pathogen (Fig. 1). Corresponding results confirmed that biofilm formation enables *C. jejuni* to survive hostile environmental conditions. Nutrient availability is a key factor in the regulation of biofilm formation by *C. jejuni*. Starvation induces biofilm formation by *C. jejuni* which was indicated by significantly higher biofilm production by bacteria grown in less nutrient-rich Mueller–Hinton medium as compared to bacteria grown in Brucella or Bolton broth (Reeser et al. 2007). Similarly, addition of fucose

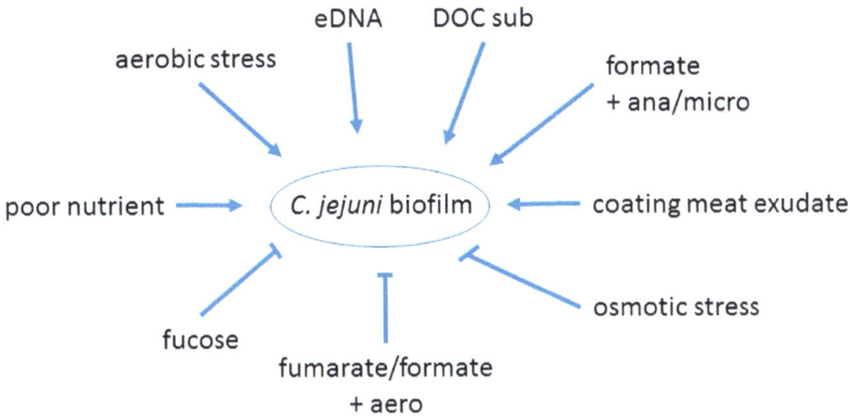

Fig. 1 Impact of environmental conditions relevant to formation and survival of *C. jejuni* biofilms. The biofilm formation by *C. jejuni* is enhanced by starvation, aerobic stress, extracellular DNA (eDNA), sublethal bile salt (desoxycholate, DOC sub) concentrations, formate at microaerobic (micro) and anaerobic (ana) conditions, as well as surface coating with meat exudates. Osmotic stress (induced by NaCl, glucose and sucrose), fumarate and formate at aerobic conditions and fucose decreased biofilm formation

inhibited biofilm formation of *C. jejuni* strains encoding enzymes required for fucose utilization (Dwivedi et al. 2016). In contrast, meat exudate significantly enhanced biofilm formation of *C. jejuni* grown on surfaces or in liquid media. However, this might be rather due to enhanced attachment than to active biofilm formation (Brown et al. 2014; Li et al. 2017; Wagle et al. 2019). In further support of the role of nutrients in biofilm formation, a recent study demonstrated that addition of energy sources such as fumarate and formate enhanced biofilm formation in a microaerobic atmosphere, but reduced biofilm formation under aerobic conditions (Kassem et al. 2017). Supplementation of growth media with formate additionally enhanced biofilm formation under anaerobic conditions (Kassem et al. 2017). In most studies, aerobic atmosphere enhanced the biofilm formation of several *C. jejuni* strains (Feng et al. 2018; Pascoe et al. 2015; Reuter et al. 2010; Stetsenko et al. 2019; Turonova et al. 2015; Zhong et al. 2020). Results from a recent study revealed that extracellular DNA (eDNA) enhances biofilm formation by *C. jejuni* (Feng et al. 2018). Interestingly, release of eDNA was induced by exposure of *C. jejuni* to aerobic conditions. In other studies, however, biofilm formation was similar or even lower if *C. jejuni* were incubated under aerobic conditions (Kassem et al. 2017; Reeser et al. 2007; Teh et al. 2017). These conflicting results might be explained by the different strains and methods used (see also Section Genetic Background and Genes Impacting Biofilm Formation of *C. jejuni*). Besides that, also small genomic variations within clones of one strain might influence investigated phenotypes, as recently shown for several clones of the reference strain NCTC11168 by Pascoe and co-workers (2019). Furthermore, the application of the bile salt deoxycholate in sublethal concentrations enhanced the biofilm formation of *C. jejuni*, while no differences in biofilm formation have been observed by the addition of sublethal concentrations of other detergents, such as Triton X-100, Tween-20 or sodium dodecyl sulphate (Svensson et al. 2009). In contrast, osmotic stress generated by NaCl, glucose or sucrose inhibited biofilm formation of *C. jejuni* (Reeser et al. 2007). The knowledge about the influence of temperature on *C. jejuni* biofilm production is still scarce. In two studies, biofilm production was higher if *C. jejuni* was incubated at 37 °C as compared to 25 °C or 20 °C, respectively (Reeser et al. 2007; Wagle et al. 2019). Taken together, biofilm formation of *C. jejuni* is influenced by multiple factors. Under laboratory conditions, biofilm formation was induced by nutrient starvation and oxygen stress, while osmotic stress rather reduced the biofilm formation. However, as the results obtained by the studies described above were generated in artificial systems, the transferability of these results to the real world is only limited. The multitude of conflicting results obtained in this highly innovative field of research underlines the urgent need for standardization and better control of future studies on factors influencing *C. jejuni* biofilm formation as a major mechanism to survive outside the vertebrate hosts.

2.3.2 Other Bacterial Species in Multispecies Biofilms

Even though *C. jejuni* forms biofilms in monocultures, the biomass of these monospecies biofilms is much lower as compared to biofilms formed by monocultures of *Pseudomonas aeruginosa* or *Escherichia coli*. If *C. jejuni* were co-cultivated in biofilms with *E. coli*, *P. aeruginosa*, *Enterococcus faecalis*, *Salmonella enterica* or *Staphylococcus simulans*, the survival of *C. jejuni* was prolonged as compared to monocultured cells, and the biofilm-mass was increased to levels produced by the co-cultured species (Feng et al. 2016; Indikova et al. 2015; Teh et al. 2019, 2010; Zhong et al. 2020). Furthermore, it has been demonstrated that in *C. jejuni-Salmonella* dual-species biofilms *C. jejuni* is located at the bottom of the biofilms in areas with high eDNA concentrations, while *Salmonella* is located at the top of the biofilm and in areas where less eDNA is present (Feng et al. 2018). It was assumed that other bacteria in co-cultures establish a more favourable environment, e.g. by lowering the oxygen level, providing CO_2 and alteration of metabolite concentrations (Zhong et al. 2020). Taken together, these results indicate that *C. jejuni* is able to colonize multispecies biofilms but the use of multispecies biofilms as a target for pathogen control via biosafety measures awaits further investigations.

2.3.3 Antimicrobial Substances

Within biofilms, microorganisms are protected against the antimicrobial activities of various substances including well-established antibiotics (Sharma et al. 2019). The molecular mechanisms by which biofilms protect bacteria from antimicrobial activity are multifactorial. The EPS structure hampers penetration of distinct antibiotics and can contain enzymes which actively inactivate antibiotics by molecular modifications (Hall and Mah 2017). In addition, the dormant state of bacteria in biofilms may passively enhance the tolerance to antimicrobial substances (Petrova and Sauer 2016). On the other hand, the close cell proximity within biofilms and the eDNA in the EPS structure support horizontal gene transfer. In accordance, *C. jejuni* transfers chromosomally encoded antibiotic resistance genes more frequently in biofilms as compared to bacteria in the planktonic lifestyle (Bae et al. 2014). Furthermore, antibiotic resistance is also influencing the biofilm formation ability of *C. jejuni* strains. Of 206 *C. jejuni* and *C. coli* strains isolated from poultry products, biofilm-producing strains possessed a significantly higher resistance to ampicillin, neomycin, sulfamethoxazole, amikacin, clindamycin and erythromycin as compared to strains unable to form biofilms (Zhang et al. 2017). Another study reported that fluoroquinolone resistance of *C. jejuni* is associated with an increased ability to form biofilms in oxygen-rich environments (Whelan et al. 2019). These aspects of enhanced antimicrobial resistance gene transfer within biofilms and higher biofilm formation in antibiotic resistant strains indicate the necessity to control and reduce *C. jejuni* biofilms.

2.4 *Genetic Background and Genes Impacting Biofilm Formation of* C. jejuni

The transition from planktonic lifestyle to the embedding of bacterial cells in the biofilm matrices goes along with substantial alterations in gene expression, which result in the production of adhesive surface molecules and in a comprehensive metabolic reprogramming (Kostakiotis et al. 2013). Recently, it has been reported that the expression of approx. 600 genes was differentially regulated during the biofilm formation of *C. jejuni*, with increased expression of genes involved in iron metabolism and acquisition, cell division, glycan production and attachment and reduced expression of genes involved in energy metabolism, amino acid catabolism and chemotaxis (Tram et al. 2020b). However, which of these changes are responsible for biofilm formation itself or which are going along with altered lifestyle in the established biofilm have to be determined. Nevertheless, several genes, impacting the biofilm formation capacity of *C. jejuni*, are summarized in Table 1, and their putative involvement in the biofilm formation process is described in more detail below.

2.4.1 Genetic Background of Individual *C. jejuni* Strains

The composition of genes differentially regulated during biofilm formation and genes directly involved in the synthesis of biofilm matrix molecules is highly variable in genomes of individual *C. jejuni* strains. These differences are suspected to be responsible for the fact that some strains of the pathogen form only weak or nearly no biofilm-mass and others produce biofilm-mass in large amounts (Bronnec et al.

Table 1 *C. jejuni* genes impacting the biofilm formation

Gene	Function	Mutation[a]	Biofilm formation	Reference
Stress response				
ahpC	Alkyl hydroperoxide reductase	lack	increased	Oh and Jeon (2014)
		over	reduced	Oh and Jeon (2014)
katA	Catalase A	lack	reduced	Oh and Jeon (2014)
perR	Peroxide stress response regulator	lack	reduced	Oh and Jeon (2014)
cosR	*Campylobacter* oxidative stress regulator	over	reduced	Oh and Jeon (2014)
cprS	*Campylobacter* planktonic growth regulation sensor	lack	increased	Svensson et al. (2009)
csrA	Carbon-starvation regulator	lack	reduced	Fields and Thompson (2008)
pta	Polyphosphate acetyltransferase Pta	lack	reduced	Joshua et al. (2006)
dps	Iron-binding protein	lack	reduced	Theoret et al. (2012)
spoT	Guanosine-3′,5′-bis(Diphosphate) 3′-pyrophosphohydrolase	lack	increased	Svensson et al. (2009)
recA	Recombinase A	lack	increased	Feng et al. (2018)
ppk-1	Polyphosphate kinase	lack	increased	Drozd et al. (2014)
ppk-2	Polyphosphate kinase	lack	increased	Drozd et al. (2014)
phoX	Alkaline phosphatase	lack	increased	Drozd et al. (2014)
Surface structures				
peb-4	Adhesion	lack	reduced	Asakura et al. (2007)
		lack	increased	Rathbun et al. (2009)
pglB	Oligosaccharyltransferase	lack	increased	Cain et al. (2019)
eptC	Phosphoethanolamine transferase	lack	reduced	Lim & Kim (2017)
waaF	Heptosyltransferase II	lack	increased	Naito et al. (2010)
lgtF	Glycosyltransferase	lack	increased	Naito et al. (2010)

(continued)

Table 1 (continued)

Gene	Function	Mutation[a]	Biofilm formation	Reference
Flagella				
flhA	Flagellar biosynthesis protein	lack	reduced	Kalmokoff et al. (2006)
fliA	Sigma factor 28	lack	reduced	Kalmokoff et al. (2006)
flaA	Major flagellin A	lack	reduced	Li et al. (2017)
flaB	Minor flagellin B	lack	reduced	Li et al. (2017)
flaC	Secreted flagellin	lack	reduced	Kalmokoff et al. (2006)
flaG	Flagellar filament length control	lack	reduced	Kalmokoff et al. (2006)
flgA	Flagella basal body *p*-ring formation protein	lack	reduced	Kim et al. (2015)
fliS	Flagellar secretion chaperon	lack	reduced	Joshua et al. (2006)
pflA	Paralyzed flagellum protein	lack	reduced	Svensson et al. (2014)
cj1324	Flagellar glycosylation protein	lack	reduced	Howard et al. (2009)
Chemotaxis				
tlp3	Transducer-like protein-3	lack	increased	Rahman et al. (2014)
tlp8	Transducer-like protein-8	lack	reduced	Chandrashekhar et al. (2015)
cheA	Histidine kinase sensor	lack	reduced	Reuter et al. (2020)
cheY	Cytoplasmic response regulator	lack	reduced	Reuter et al. (2020)
		lack	increased	Tram et al. (2020b)
cheW	Phosphotransferase	lack	reduced	Reuter et al. (2020)
		lack	increased	Tram et al. (2020b)
cheV	Phosphotransferase	lack	reduced	Reuter et al. (2020)
Others				
cje1441	Extracellular DNase	lack	increased	Brown et al. (2015)

[a]Lack: lack of function, over: overexpression

2016; Feng et al. 2018; Joshua et al. 2006; Melo et al. 2017). For example, *C. jejuni* strains encoding for extracellular DNases, mostly located on the mobile elements CJIE1, CJIE2 and CJIE4, are unable or only poor biofilm producer and are further able to remove pre-established biofilms of other *C. jejuni* strains (Brown et al. 2015).

Moreover, biofilm formation capacities of individual *C. jejuni* isolates are significantly associated with distinct multilocus sequence types (MLST) and with several clonal complexes, which display specific features concerning host adaptation, termed host-generalists and host-specialists, respectively (see also Pascoe et al. 2015 and Chapters "Population Biology and Comparative Genomics of *Campylobacter* Species" and "Emission Sources of *Campylobacter* from Agricultural Farms, Impact on Environmental Contamination and Intervention Strategies" in this book). It is of note that a strong biofilm formation capacity of *C. jejuni* isolates is correlated with the absence of specific host adaptation, leading to the fact that the host-generalist group of *C. jejuni* isolates displays an enhanced capacity for biofilm formation. Furthermore, nearly 2/3 of the *C. jejuni* isolates belonging to the chicken-specialists belonged to the group of weak biofilm producers (Pascoe et al. 2015). Even though genes with a robust association to biofilm formation differed between the isolates of the host-generalist group, most of these genes are involved in adhesion, motility, glycosylation, capsular polysaccharides and oxidative stress response (Pascoe et al. 2015). Taken together, these findings provide evidence that the genomic repertoire necessary for biofilm formation is highly variable within *C. jejuni* isolates and that biofilm formation is more important for isolates that are not adapted to specific vertebrate hosts.

2.4.2 Flagella-Associated Genes and Motility

Besides the involvement in motility and chemotaxis, the flagella of *C. jejuni* is also crucial for secretion of proteins, autoagglutination, microcolony formation and avoidance of the innate immune response (Guerry 2007), indicating that mutation of the flagella might have multifactorial effects. Generally, motility mediated by flagella is essential for the biofilm formation capacity of *C. jejuni*. Loss of motility caused by targeted mutation of flagella-associated *C. jejuni* genes *flhA*, *fliA*, *flaA*, *flaB*, *flaC*, *flaG*, *flgA* and *fliS*, resulted in impaired biofilm formation (Feng et al. 2018; Joshua et al. 2006; Kalmokoff et al. 2006; Kim et al. 2015; Li et al. 2017; Reuter et al, 2010; Turonova et al. 2015). Besides the fact that flagella-associated motility is essential to reach substrates where biofilms can be formed, also flagella-associated attachment seems to impact *C. jejuni* biofilm formation. This was supported by the observation that aflagellated *C. jejuni* mutants (mutation of *flhA*) formed less biofilm-mass as compared to *pflA* mutants with paralyzed flagella only (Svensson et al. 2014). Furthermore, the biofilm formation capacity of *C. jejuni* depends on flagellar *O*-linked glycan modifications. This was shown by targeted deletion of the *cj1324* gene, which resulted in the loss of flagellar sugar modifications and reduced biofilm formation but does not alter the motility (Howard et al. 2009). Additionally, the reduced biofilm-mass formation of a *flaA/flaB* mutant could be restored by addition of chicken meat exudate (Li et al. 2017). Taken together, these findings indicate that surface attachment mediated by the flagella is essential for *C. jejuni* biofilm formation.

2.4.3 Chemotaxis-Associated Genes

Directed movement of bacteria is interactively controlled and directed by the sensing of attractants or repellents by transducer-like proteins (Tlp). The activation of Tlp results in a signalling cascade mediated by the Che proteins, which modulate flagellar rotation (Tram et al. 2020b). Deletion of *cheY* and *cheW* genes in *C. jejuni* enhanced the formation of biofilm-mass, even though motility of both mutants was significantly reduced in the planktonic state (Tram et al. 2020b). The authors suggested that the enhanced biofilm-mass production could be due to the higher autoagglutination displayed by these mutants. In contrast, defects in robust biofilm formation at the air-media interface were reported for *C. jejuni* mutants lacking functional *cheA*, *cheY*, *cheW* or *cheV* genes (Reuter et al. 2020). The authors concluded that the chemotaxis signalling system is rather necessary for organized biofilm formation at the air-media interface than for biofilm formation per se. The contradicting findings described in these studies might also be due to differences in the experimental conditions or biofilm detection assays applied. Moreover, deletion of the chemoreceptor Tlp3 resulted in enhanced biofilm formation, while deletion of Tlp8 resulted in reduced biofilm formation rates by respective *C. jejuni* mutants (Chandrashekhar et al. 2015; Rahman et al. 2014). These data indicate that distinct chemotactic compounds as well as chemotaxis signalling pathway are essentially involved in biofilm formation by *C. jejuni*.

2.4.4 Stress Response-Associated Genes

The influence of oxidative stress on the biofilm formation capacity of *C. jejuni* has been intensively investigated at the molecular level. Deletion of alkyl hydroperoxide reductase (*ahpC*) and catalase A (*katA*) genes increased biofilm formation by the respective mutant strains (Oh and Jeon 2014). Results from confocal laser scanning microscopy support the assumption that AhpC is involved in the development of *C. jejuni* microcolonies at the early stages of biofilm formation. This role of *ahpC* was further confirmed elegantly by genetic manipulation of *perR* and *cosR* genes encoding positive and negative regulators of *ahpC*, respectively (Oh and Jeon 2014; Turonova et al. 2015). The important role of oxygen stress responses in biofilm formation of *C. jejuni* was further confirmed by the finding that deletion of the sensor for the *Campylobacter* planktonic growth regulation system (*cprS*) reduced oxidative stress resistance, but enhanced biofilm formation in respective mutants (Svensson et al. 2009). However, deletion of the gene encoding the major carbon-starvation regulator *csrA* also rendered *C. jejuni* more prone to aerobic stress but reduced the biofilm formation capacity, which is in contrast to many other bacteria in which *csrA* represses the biofilm formation (Fields and Thompson 2008). However, given that the translation of more than 100 genes is dysregulated in a *csrA* mutant, it is difficult to determine which of them are responsible for the observed phenotype (Fields et al. 2016; El Abbar et al. 2019). The role of *csrA* in biofilm formation of *C. jejuni* is further supported by the fact that deletion of the gene encoding polyphosphate acetyltransferase Pta (Cj0688), also under post-transcriptional control of *csrA*, resulted in reduced biofilm formation (Joshua et al. 2006). Additionally, a *C. jejuni* mutant lacking the gene for the iron-binding protein Dps displayed increased susceptibility to H_2O_2 but reduced biofilm formation (Theoret et al. 2012). Deletion of *spoT* (involved in the stringent stress response) and recombinase A (*recA*) enhanced biofilm formation especially at aerobic conditions (Feng et al. 2018; Svensson et al. 2009). In addition, results from both studies demonstrated that the lack of *spoT* and *recA* enhanced lysis of the bacteria thereby releasing high molecular DNA, which is one of the prerequisites for bacterial biofilm production.

Finally, the *C. jejuni* biofilm production is linked to intracellular levels of inorganic polyphosphates, which play crucial roles in stress tolerance and virulence of the pathogen (Kumar et al. 2016). Deletion of genes coding for both polyphosphate kinases Pkk 1 and Pkk 2 as well as for the alkaline phosphatase PhoX (Cj0145) resulted in enhanced *C. jejuni* biofilm production and surface attachment. (Drozd et al. 2014; Gangaiah et al. 2009, 2010). Taken together these data demonstrate that various stressors induce biofilm formation of *C. jejuni* via activation of the major stress response regulons known to date.

2.4.5 Surface Structure-Associated Genes

The production of the peptidyl prolyl cis–trans isomerase Peb4, involved in folding of integral outer membrane proteins, is increased in *C. jejuni* cells living in biofilms (Kalmokoff et al. 2006). Mutational analysis of the corresponding gene revealed that Peb4 is required for both adhesion and attachment of *C. jejuni* to host cells in vitro and for biofilm-mass formation (Asakura et al. 2007). In contrast, deletion of this gene in another *C. jejuni* strain resulted in enhanced biofilm-mass formation (Rathbun et al. 2009). These conflicting results might be due to strain-specific variations in the genetic background or by polar effects of the mutation strategy, but this awaits further evaluation. In addition, protein glycosylation is essentially involved in *C. jejuni* biofilm formation. Mutational analysis of the *pglB* gene by targeted deletion revealed that *N*-linked protein glycosylation reduces the biofilm formation capacity of *C. jejuni*, is required for resistance to heat and salt but decreases the resistance to peroxide (Cain et al. 2019). In contrast, *N*-linked protein glycosylation mediated by EptC enhances biofilm formation, indicating that the modulation of biofilm formation by *N*-linked glycosylation is highly dependent on the glycosylated proteins involved (Cullen et al. 2013; Lim and Kim 2017; Scott et al. 2012). Finally, *C. jejuni* lipooligosaccharide (LOS) structures influence the biofilm formation capacity as indicated by enhanced biofilm formation in *C. jejuni waaF* or *lgtF* deletion mutants with truncated LOS. However, mutational analysis by targeted deletion revealed that LOS modifications by GalT or CstII enzymes did not influence the biofilm-mass, which was comparable in deletion mutants and the wild-type strain (Naito et al. 2010). Besides the LOS surface structure, *C. jejuni* has the ability to coat its surface with a polysaccharide capsule (CPS), being the major serodeterminant of the Penner scheme (Karlyshev et al. 2000). Given that polysaccharides are a common component in the EPS, the knowledge about the impact of CPS on the biofilm formation capacity of *C. jejuni* is still scarce. Deletion of the gene *kpsM*, involved in the transport of capsular polysaccharides across the inner membrane, resulted in enhanced biofilm formation of this uncapsulated *C. jejuni* mutant (Joshua et al. 2006). However, the mechanisms responsible for this phenotype have to be elucidated in future studies. In conclusion, these observations indicate that glycosylation state of surface molecules is essentially involved in *C. jejuni* biofilm formation.

2.5 Control Strategies Targeting C. jejuni *Biofilms*

Given that the EPS structure of biofilms protects the microorganisms from physical, chemical and environmental stresses, disruption of the EPS structure is a favoured strategy to combat bacterial pathogens in biofilms (Devaraj et al. 2019). Since eDNA is an essential component of the EPS produced by many bacteria, DNase treatment is a promising measure for inhibition of biofilm formation and for the degradation of established biofilms which has been also successfully proven for *C. jejuni* biofilms (Brown et al. 2015; Feng et al. 2018; Sharma and Pagedar Singh 2018; Svensson et al. 2014). In addition, treatment of *C. jejuni* with the phytochemicals trans-cinnamonaldehyde, eugenol and carvacrol before and after biofilm formation reduced the biofilm-mass (Wagle et al. 2019). Application of all three substances at bactericidal concentrations killed the majority of bacterial cells also in mature biofilms within 10 min (Wagle et al. 2019). Notably, sublethal concentrations of these phytochemicals downregulated periplasmic nitrate reductase NapA involved in energy generation and the chaperon DnaK involved in stress responses by *C. jejuni* cells in the biofilms (Wagle et al. 2019). While the mechanisms by which phytochemicals reduce *C. jejuni* biofilm formation capacity await further investigation, it seems noteworthy that citrus extracts reduced the biofilm-mass of *C. jejuni* (Castillo et al. 2014), most likely by reduction of AI-2 activity (as described in Section Phenotypes of *C. jejuni luxS* Mutants). Finally, biofilm-mass formation by *C. jejuni* in mono- and multispecies cultures was significantly inhibited by zinc oxide nanoparticles, which are small and have a high oxidative potential (Melo et al. 2017; Zhong et al. 2020). In summary, even though several strategies to inhibit *C. jejuni* biofilm formation or to eliminate *C. jejuni* in mature biofilms have been developed, their efficacy as hygiene measures under practical conditions still needs to be investigated in detail.

3 Quorum Sensing

Bacteria adapt their metabolism according to the surrounding environment not only within single cells but also at a multicellular level (Miller et al. 2002). Several processes such as biofilm formation, expression of virulence factors, competence for DNA-uptake or bioluminescence are of particular benefit in multicellular communities (Mukherjee and Bassler 2019). To collectively regulate these processes, bacteria use a cell-to-cell communication system known as quorum sensing (QS). QS is mediated by small signalling molecules, termed autoinducers (AIs), which accumulate in the environment in a cell density dependent manner. The AIs bind to specific bacterial receptors and induce the expression of distinct target genes. Depending on the signalling molecule produced and the presence of appropriate receptors, bacteria can communicate on intra-species, inter-species, inter-genera as well as inter-kingdom levels. The regulation by QS is assumed to be a highly complex process since many QS processes involve more than one signal-receptor combination, exerting their functions in a hierarchical cascade (Abisado et al. 2018). For example, four different QS-pathways are known in *P. aeruginosa*, namely the Las-, Rhl-, Pqs- and IQS-systems. Expression of virulence genes is regulated by AI-RhlR complex, and for the induction of RhlR-system, one of the other three QS-pathways is required (Papenfort and Bassler 2016). Furthermore, it has been described that some bacteria might only sense an AI without the ability to produce it. This is also true for *P. aeruginosa*, which does not produce AI-2, whereas AI-2 molecules generated by other bacteria alter the gene expression in this pathogen (Duan et al. 2003).

3.1 Quorum Sensing Signalling Mechanisms

Three major categories of signalling molecules, namely AI-1, AI oligopeptides (AIP) and AI-2, have been described. AI-1 are used by Gram-negative bacteria, while AIP serve as signalling molecules in Gram-positive bacteria. Both Gram-positive and Gram-negative bacteria utilize AI-2 (a furanone) as signalling molecules. To date, additional AI molecules were identified such as the *Pseudomonas* quinolone signal, diffusible signal factors and AI-3. It is reasonable to postulate that additional AI molecules exist (LaSarre and Federle 2013; Papenfort and Bassler 2016).

AI-1 molecules are acylated homoserine lactones (AHL) composed of an invariant homoserine lactone ring attached to an acyl chain, which can vary in the length of carbon atoms, in saturation and in the oxidation state (LaSarre and Federle 2013). These AHLs are synthesized from S-adenosylmethionine (SAM) by concerted action of the LuxI enzyme family members and acylated acyl carrier proteins. Notably, AI-2 is a by-product of the activated methyl cycle (AMC). Within the AMC, LuxS catalyzes the cleavage of S-ribosylhomocysteine (SRH) to homocysteine and 4,5-dihydroxyl-2,3-pentanedion (DPD), which spontaneously cyclize into AI-2 (Winzer et al. 2002). While *Vibrio harveyi* recognizes the borated form of AI-2, *E. coli* and other *Enterobacteriaceae* sense the borate-free form of AI-2 (Chen et al. 2002; Miller et al. 2002). Even though the knowledge about QS mechanisms in other bacterial species is constantly growing, information regarding QS in bacteria of the genus *Campylobacter* is rather limited. In 2002, the presence of a *luxS* gene homolog and active production of AI-2 by *C. jejuni* was reported for the first time (Elvers and Park 2002). Whereas several other *Campylobacter* species also produce AI-2, no AI-2 production could be determined in *C. lari*, *C. insulanigrae* and *C. peloridis* (Golz et al. 2012; Tazumi et al. 2011). So far, no AI-1 synthase has been identified in the *C. jejuni* genome. Only one publication described the production of a putative AI-1 molecule (cjA) by *C. jejuni* (Moorhead and Griffiths 2011). The structure of cjA could not be determined, but it was demonstrated that addition of exogenous AI-1 compounds induced the expression of the *C. jejuni* virulence genes *cadF*, *ciaB*, *cdtB* and *flaA* and supported the transition of the pathogen to the dormant—so-called viable but not culturable (VBNC)—state. To date, no additional *C. jejuni* QS signalling molecules have been identified. While most AI-1 molecules can diffuse freely across bacterial membrane, several AI-1 as well as hydrophilic AI-2 molecules might require active transport across the cell membrane (LaSarre and Federle 2013; Pereira et al. 2013). In *E. coli*, AI-2 export is mediated by YdgG, a transmembrane protein belonging to the large group of the so-called AI-2 exporter superfamily (Herzberg et al. 2006; Rettner and Saier 2010). So far, no further AI export systems have been described. However, AI-2 export in *C. jejuni* is modulated by a small non-coding RNA (CjNC110). Mutational analysis by targeted deletion of the CjNC110 sequence revealed decreased extracellular AI-2 levels but increased intracellular levels of AI-2, suggesting that CjNC110 is required for modulation of the AI-2 transport to the extracellular space (Kreuder et al. 2020).

Gram-negative bacteria commonly sense AI-1 molecules by cytoplasmic LuxR-Type receptors, which act as transcription factors or by two-component membrane-bound histidine kinases (Papenfort and Bassler 2016). For detection of AI-2, different receptor types have been described so far. *Vibrionaceae* sense AI-2 by a transmembrane receptor, thereby inducing an intracellular signalling cascade. In contrast, AI-2 is imported and phosphorylated via ABC-transporters by several *Enterobacteriaceae, Bacillaceae* and *Rhizobiaceae* (Pereira et al. 2013). The phosphorylated AI-2 stabilizes transcription factors, which in turn enable the regulation of target gene expression. For *E. coli* and *Helicobacter pylori*, chemoreceptors have been identified sensing AI-2 as chemoattractant and chemorepellent, respectively (Hegde et al. 2011; Rader et al. 2011). However, the low sequence homologies of the AI-2 receptors led to the postulation that additional receptor types may exist (Papenfort and Bassler 2016; Pereira et al. 2013). No AI-2 receptor homolog has been identified in *Campylobacter* yet. However, the results obtained from an AI-2 uptake assay prompted us to speculate that *C. jejuni* might perceive AI-2 by a two-component regulatory system rather than by an ABC-transporter system (Adler et al. 2015).

3.2 *Phenotypes of* C. jejuni luxS *Mutants*

It is still under debate whether *C. jejuni* is using AI-2 to regulate their behaviour as mostly conflicting results were reported. Whether these conflicting results depend on strain variation, culture conditions, methods and/or mutation strategies applied has to be elucidated in the future. Nevertheless, we tried to summarize the findings on putative QS-related *C. jejuni* phenotypes published so far. Since no specific AI-2 receptor of *Campylobacter* is known so far, AI-2-dependent phenotypes have primarily been investigated using *luxS* mutants of various *C. jejuni* strains. Given that LuxS is required for the AMC, it is necessary to complement all experimental assays including a *luxS* mutant by the addition of exogenous AI-2 to determine whether the phenotypes observed are due to interrupted metabolism or lack of AI-2. Recent investigations confirmed that homocysteine and SHR concentrations were significantly reduced or enhanced in a *C. jejuni luxS* mutant compared to the parental strain, respectively (Mou and Plummer 2016). However, reduction of the methionine and SAM concentrations as a result of the *luxS* deletion was less pronounced as expected. Furthermore, the methylome profile of this *luxS* mutant was comparable to that of the wild-type (Mou et al. 2014), indicating that the observed phenotypes of *luxS* mutants are not due to a complete lack of methionine or SAM metabolites (Mou and Plummer 2016). Furthermore, no morphological changes in cell shape or flagella morphology have been determined for *luxS* mutants of *C. jejuni* strains 81116, NCTC11168 or IA3902 (Jeon et al. 2003; Mou and Plummer 2016). The phenotypes of *C. jejuni luxS* mutants are summarized in Fig. 2.

Despite that fact that besides AI-2, also the disruption of the AMC may influence bacterial growth, the multiplication of *C. jejuni luxS* mutants has been extensively

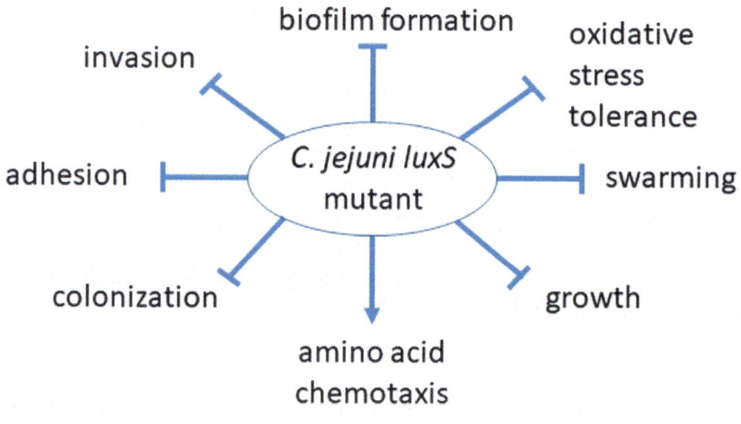

Fig. 2 Overview of *C. jejuni luxS* mutant phenotypes. Besides enhanced chemotaxis towards amino acids, reduced colonization, adhesion, invasion, biofilm formation and swarming abilities as well as reduced oxidative stress tolerance and growth kinetics have been described for *C. jejuni luxS* mutants. However, several phenotypes were only observed for different *C. jejuni luxS* mutants or under specific culture conditions (for details see text)

investigated. Remarkably, reduced growth rates were reported for the *C. jejuni* strain 81–176 with inactivated *luxS* gene, but not for the strains NCTC11168, 81116 and M129 (Elvers and Park 2002; He et al. 2008; Holmes et al. 2009; Jeon et al. 2003; Mou and Plummer 2016; Plummer 2012; Quinones et al. 2009; Reeser et al. 2007). Strain-specific differences in growth-related phenotypes of *C. jejuni luxS* mutants were confirmed by a detailed analysis of various strains grown under different conditions (Adler et al. 2014). These results indicated that the NCTC11168Δ*luxS* mutant in which the *luxS* gene is replaced by an antibiotic resistance gene showed reduced growth in comparison with the wild-type strain both under substrate limited and nutrient-rich conditions. In contrast, two different *luxS* mutants of *C. jejuni* strain 81–176 exhibited growth defects under substrate limited conditions only. Genetic complementation restored the growth kinetics of both mutants of strain 81–176, while the chemical complementation by AI-2 only partially restored growth of the Δ*luxS* mutant of the *C. jejuni* NCTC11168 strain. These data indicate that *C. jejuni* growth might be influenced by *luxS* and AI-2 but in a strain-dependent manner and under certain nutritional conditions only.

Results from a majority of studies showed that motility of *C. jejuni luxS* mutants on swarming plates is strongly reduced, which was independent of strain background or culture conditions (Adler et al. 2014; Elvers and Park 2002; Holmes et al. 2009; Jeon et al. 2003; Plummer et al. 2011; Quinones et al. 2009; Simunovic et al. 2020). However, for the 81–176Δ*luxS* mutant constructed by He and colleagues (2008), reduced motility was only detected if bacteria were incubated on Mueller–Hinton medium-based swarming plates at 37 °C. In Brucella broth, however, the motility of this mutant was neither reduced at 37 °C nor at 42 °C (Adler et al. 2014; He et al. 2008). In contrast, the motility of the 81–176::*luxS* mutant (insertion of antibiotic resistance cassette within the *luxS* gene) constructed by Quinones and co-workers (2009) was reduced in both media and at both temperatures (Adler et al. 2014; Quinones et al. 2009). These results suggest that differences in some strain-specific phenotypic properties of *C. jejuni luxS* mutants are indeed caused by polar effects generated by the genetic manipulations applied. Even though the motility of the NCTC11168Δ*luxS* mutant was not restored by the addition of exogenous AI-2 in the study of Holmes and colleagues (2009), the motility of other *C. jejuni luxS* mutants was at least partially restored by genetic complementation or upon the addition of exogenous AI-2 (Adler et al. 2014; Plummer et al. 2011; Quinones et al. 2009). The latter studies revealed that AI-2 influences the motility of *C. jejuni* on swarming agar. So far, the mechanisms of AI-2-dependent regulation are not understood. Several studies investigating gene expression patterns of *luxS* mutants revealed conflicting results. While reduced *flaA* gene expression was reported for a 81116*luxS* mutant, no difference on protein level nor in flagellar morphology was observed (Jeon et al. 2003). Several flagellar assembly/regulation genes were differentially expressed in 81–176Δ*luxS* cultivated at 42 °C, even though under these conditions the swarming ability of the mutant was comparable to the wild-type (He et al. 2008). Furthermore, Holmes and colleagues (2009) determined the downregulation of several flagellar-associated genes and subsequently reduced swarming capabilities, but the authors could neither restore the gene expression pattern nor the phenotype by adding exogenous AI-2. Therefore, the confusing and in part contradicting results obtained by mutational analysis of the *C. jejuni luxS* locus indicate that further work under standardized and better controlled experimental conditions is essential for the investigation of the complex mechanisms underlying the interactions of *C. jejuni* LuxS and/or AI-2 with motility.

Whether AI-2 exhibits a direct chemotactic function in *C. jejuni* has not been determined yet. Nevertheless, when compared to the wild-type strain, a Δ*luxS* mutant of the 81–176 strain displayed enhanced chemotactic behaviour towards amino acids (Quinones et al. 2009). Holmes and co-workers (2009) reported reduced mRNA-levels of the genes encoding *cheA* and the chemoreceptors Tlp1, -2 and -4 (Cj1506, Cj0144, Cj0262) in a NCTC11168Δ*luxS* mutant. However, no significantly different regulation in expression of the *cheA, cheB, cheR, cheV* and *cheW* genes has been observed for the Δ*luxS* mutant of *C. jejuni* strain 81–176 (He et al. 2008).

Molecular mechanisms related to host-interactions like adhesion, invasion, cytotoxicity and intestinal colonization are basic to *C. jejuni* pathogenicity (see Chapter "*Campylobacter* Virulence Factors and Molecular Host–pathogen Interactions" of this book). Expression of *Pseudomonas*, *Vibrio cholerae* and *E. coli* virulence factors is regulated by QS (Furniss and Clements 2018; Jiang et al. 2019; Papenfort and Bassler 2016). However, studies investigating the AI-2-dependent regulation of pathogenicity in *Campylobacter* are still scarce. *C. jejuni* LuxS was essential for adhesion of *C. jejuni* 81–176 as demonstrated with a *luxS* mutant of this strain and cultured LMH cells in vitro (Quinones et al. 2009). In contrast, deletion of *luxS* did not alter adhesion of *C. jejuni* strain NCTC11168 on INT-407 cells, while the invasion rate of the Δ*luxS* mutant used in this study was reduced (Simunovic et al. 2020). Interestingly, the invasion rate of a NCTC11168Δ*luxS* mutant was only slightly reduced in Caco-2 cells (Elvers and Park 2002). Whether these highly varying and confusing observations were due to different properties of *luxS* mutants or of the different cell lines remains open. However, complementation with exogenous AI-2 is needed to prove that all these phenotypes were caused by the lack of AI-2 and did not result from disruption of the AMC.

Additional contradicting results concerning the influence of *luxS* on *C. jejuni* colonization capacity were obtained by the analysis of *C. jejuni luxS* mutant strains in animal models in vivo. While the *luxS* mutant of *C. jejuni* strain IA3902 displayed a loss of chicken colonization, this ability was only reduced in a *luxS* mutant of *C. jejuni* strain 81–176, whereas the NCTC11168 *luxS* mutant colonized chickens with similar rates compared to the wild-type strain (Plummer et al. 2012; Quinones et al. 2009). It is not clear yet if these contradicting findings are the result of real strain-dependent differences or are caused by the mutational strategy applied. No general conclusion regarding the impact of AI-2 on the *C. jejuni* colonization capabilities could be drawn. Given that AI-2 produced by commensal microbiota could also have an impact on the phenotype of *C. jejuni luxS* mutants, and despite the difficulties in determining whether altered phenotypes were due to lack of AI-2 or disrupted AMC, the results summarized here should be interpreted with caution.

The NCTC11168 *luxS* mutant constructed by Elvers and Park (2002) did not show altered hydrogen peroxide or paraquat susceptibility, while the 81–176 *luxS* mutant was less resistant to cumene hydroperoxide and hydrogen peroxide as compared to the wild-type (He et al. 2008). Gene expression analysis revealed that expression of the peroxide stress defence-related genes *ahpC* (encoding an alkyl hydroxide reductase) and *tpx* (encoding a thiol peroxidase) was reduced in the 81–176 *luxS* mutant after oxidative stress treatment as compared to the wild-type strain (He et al. 2008). The important role of *LuxS* in the *C. jejuni* oxidative stress response is further underlined by the fact that the *Campylobacter* oxidative stress regulator (CosR) negatively regulates the expression of *luxS* (Hwang et al. 2011). Furthermore, a *C. jejuni* NCTC11168 *luxS* mutant displayed lower survival rates as compared to the wild-type strain at cold stress (Ligowska et al. 2011). However, whether these stress responses are directly modulated via AI-2-dependant QS remains to be elucidated.

The biofilm-mass developed by a Δ*luxS* mutant of *C. jejuni* strain M129 was significantly reduced compared to the wild-type and could be partially restored by the addition of cell free supernatants of the wild-type strain (Reeser et al. 2007). Furthermore, reduced adhesion on polystyrene surfaces was reported for a NCTC11168 *luxS* mutant (Simunovic et al. 2020). In contrast, the attachment on stainless steel coupons was comparable for both the NCTC11168 *luxS* mutant and the wild-type strain (Bezek et al. 2016). In addition, biofilm formation of the *C. jejuni* strain 81–176 was reduced by the application of AI-1 molecule cjA (Moorhead and Griffiths 2011). Taken together, these data suggest that the process of *C. jejuni* biofilm formation is regulated by concerted action of several QS systems like in other bacteria (Paluch et al. 2020; Papenfort and Bassler 2016).

3.3 Quorum Quenching

The inhibition of QS, also termed quorum quenching (QQ), has raised much attention in recent years (Paluch et al. 2020). QQ could be implemented as a preventive or therapeutic approach to combat pathogenic bacteria and could be achieved at different stages, e.g. by inhibition of signalling molecule production, degradation of signalling molecules or blockage of the receptor. This is underlined by increasing numbers of patents and applications for QQ compounds and their functions (reviewed by Chen et al. 2018). Even though the exact role of AI-2-mediated QS has not been elucidated for *C. jejuni*, several authors investigated putative agents that can interrupt QS mechanisms. For example, the application of citrus extract nearly eliminated AI-2 activity in cell-free supernatant of several *C. jejuni* strains and further reduced their motility, biofilm formation, adhesion and invasion of HeLa cells as well as the expression of *cadF* and *ciaB* virulence genes (Castillo et al. 2014, 2015). In support, nearly all the 20 natural plant extracts investigated by Simunovic and co-workers (2020) altered several phenotypes of *C. jejuni*. The ethanolic extract of *Rhodiola rosea* (roseroot) had the greatest potential to inhibit AI-2 production, motility, adhesion to polystyrene surface and invasion into INT-407 cells by strain NCTC11168, which were comparably to phenotypes observed for the *C. jejuni* Δ*luxS* mutant. Furthermore, none of the tested compounds exerted a synergistic effect on the phenotype of the *C. jejuni*

Δ*luxS* mutant. These data implicate that these compounds could inhibit AI-2 QS circuit and thereby alter the behaviour of *C. jejuni*. Furthermore, quinolinone alkaloid mixture extracts from the tree *Euodia ruticarpa* reduced AI-2 production and number of attached bacteria on a polystyrene surface by *C. jejuni* NCTC11168 as well as by the Δ*luxS* and Δ*cmeB* mutants (Bezek et al. 2016). However, as attachment of the parental strain and the Δ*luxS* mutant were comparable, the authors concluded that the effects of the extract are not related to AI-2-dependent QQ.

4 Concluding Remarks

The comprehensive review of the literature documented in this book chapter indicates that some aspects of QS and biofilm formation by *C. jejuni* have been investigated to date, but both processes are still not well understood at the molecular level. Obvious shortcomings in these important fields of research are caused by lack of precise genetic analysis of the biological systems involved and by the extensive genetic variation of *C. jejuni* at the isolate level. These limitations should be overcome in the future by standardized and complete genetic analysis including the mutation strategies applied and by whole genome analysis of *C. jejuni* at the strain level. The facts that *C. jejuni* produces AI-2 and that some phenotypes of *luxS* mutants could be partially restored by exogenous AI-2 point towards regulatory functions of AI-2 in *C. jejuni*. Therefore, regarding the QS system, it seems highly recommended that future research should focus on the identification and biochemical characterization of a possible AI-2 receptor including complementation of manipulated pathways by exogenous AI molecules to finally prove the QS-dependent phenotypes of artificially generated *C. jejuni* mutants. The promising results obtained by AI-2-dependent QS-signalling should strengthen intensive research on potential additional AI molecules and their regulatory functions in *C. jejuni* and other *Campylobacter* species.

The manifold environmental and intrinsic conditions affecting *C. jejuni* biofilm formation provide strong evidence that *C. jejuni* actively produces biofilms to survive unfavourable conditions outside vertebrate hosts. Therefore, a deeper understanding of *C. jejuni* biofilm formation is a key to direct future research for improvement of biosafety and hygiene in slaughter and food processing lines. Altogether, biochemical properties of *C. jejuni* QS and biofilms will guide the development of innovative and novel strategies to diminish the entry or cross-contamination of *Campylobacter* in livestock and the food processing chain.

Acknowledgements This work is supported by the German Federal Ministry of Education and Research (BMBF) in frame of the zoonoses research consortium PAC-*Campylobacter* to C.P. and G.G. (project IP2/01KI1725A).

References

Abisado RG, Benomar S, Klaus JR, Dandekar AA, Chandler JR (2018) Bacterial quorum sensing and microbial community interactions. mBio 9. https://doi.org/10.1128/mBio.02331-17

Adler L, Alter T, Sharbati S, Golz G (2014) Phenotypes of *Campylobacter jejuni luxS* mutants are depending on strain background, kind of mutation and experimental conditions. PLoS ONE 9:e104399. https://doi.org/10.1371/journal.pone.0104399

Adler L, Alter T, Sharbati S, Gölz G (2015) The signalling molecule autoinducer-2 is not internalised in *Campylobacter jejuni*. Berl Munch Tierarztl 128:6

Asakura H, Yamasaki M, Yamamoto S, Igimi S (2007) Deletion of *peb4* gene impairs cell adhesion and biofilm formation in *Campylobacter jejuni*. FEMS Microbiol Lett 275:278–285. https://doi.org/10.1111/j.1574-6968.2007.00893.x

Azeredo J, Azevedo NF, Briandet R, Cerca N, Coenye T, Costa AR, Desvaux M, Di Bonaventura G, Hébraud M, Jaglic Z, Kačániová M, Knøchel S, Lourenço A, Mergulhão F, Meyer RL, Nychas G, Simões M, Tresse O, Sternberg C (2017) Critical review on biofilm methods. Crit Rev Microbiol 43(3):313–351. https://doi.org/10.1080/1040841X.2016.1208146

Bae J, Oh E, Jeon B (2014) Enhanced transmission of antibiotic resistance in *Campylobacter jejuni* biofilms by natural transformation. Antimicrob Agents Chemother 58(12):7573–5. https://doi.org/10.1128/AAC.04066-14

Bezek K, Kurincic M, Knauder E, Klancnik A, Raspor P, Bucar F, Smole Mozina S (2016) Attenuation of adhesion, biofilm formation and quorum sensing of *Campylobacter jejuni* by *Euodia ruticarpa*. Phytother Res 30:1527–1532. https://doi.org/10.1002/ptr.5658

Bronowski C, James CE, Winstanley C (2014) Role of environmental survival in transmission of *Campylobacter jejuni*. FEMS Microbiol Lett 356:8–19. https://doi.org/10.1111/1574-6968.12488

Bronnec V, Turonova H, Bouju A, Cruveiller S, Rodrigues R, Demnerova K, Tresse O, Haddad N, Zagorec M (2016) Adhesion, biofilm formation, and genomic features of *Campylobacter jejuni* Bf, an atypical strain able to grow under aerobic conditions. Front Microbiol 7:1002. https://doi.org/10.3389/fmicb.2016.01002

Brown HL, Reuter M, Hanman K, Betts RP, van Vliet AH (2015) Prevention of biofilm formation and removal of existing biofilms by extracellular DNases of *Campylobacter jejuni*. PLoS ONE 10:e0121680. https://doi.org/10.1371/journal.pone.0121680

Brown HL, Reuter M, Salt LJ, Cross KL, Betts RP, van Vliet AH (2014) Chicken juice enhances surface attachment and biofilm formation of *Campylobacter jejuni*. Appl Environ Microbiol 80:7053–7060. https://doi.org/10.1128/AEM.02614-14

Cain JA, Dale AL, Niewold P, Klare WP, Man L, White MY, Scott NE, Cordwell SJ (2019) Proteomics reveals multiple phenotypes associated with n-linked glycosylation in *Campylobacter jejuni*. Mol Cell Proteomics 18:715–734. https://doi.org/10.1074/mcp.RA118.001199

Castillo S, Heredia N, Arechiga-Carvajal E, Garcia S (2014) Citrus extracts as inhibitors of quorum sensing, biofilm formation and motility of *Campylobacter jejuni*. Food Biotechnol 28:106–122. https://doi.org/10.1080/08905436.2014.895947

Castillo S, Heredia N, Garcia S (2015) 2(5H)-Furanone, epigallocatechin gallate, and a citric-based disinfectant disturb quorum-sensing activity and reduce motility and biofilm formation of *Campylobacter jejuni*. Folia Microbiol (Praha) 60:89–95. https://doi.org/10.1007/s12223-014-0344-0

Chandrashekhar K, Gangaiah D, Pina-Mimbela R, Kassem II, Jeon BH, Rajashekara G (2015) Transducer like proteins of *Campylobacter jejuni* 81–176: role in chemotaxis and colonization of the chicken gastrointestinal tract. Front Cell Infect Microbiol 5:46. https://doi.org/10.3389/fcimb.2015.00046

Chen X, Schauder S, Potier N, Van Dorsselaer A, Pelczer I, Bassler BL, Hughson FM (2002) Structural identification of a bacterial quorum-sensing signal containing boron. Nature 415:545–549. https://doi.org/10.1038/415545a

Chen X, Zhang L, Zhang M, Liu H, Lu P, Lin K (2018) Quorum sensing inhibitors: a patent review (2014–2018). Expert Opin Ther Pat 28:849–865. https://doi.org/10.1080/13543776.2018.1541174

Coughlan LM, Cotter PD, Hill C, Alvarez-Ordonez A (2016) New weapons to fight old enemies: novel strategies for the (bio)control of bacterial biofilms in the food industry. Front Microbiol 7:1641. https://doi.org/10.3389/fmicb.2016.01641

Cullen TW, O'Brien JP, Hendrixson DR, Giles DK, Hobb RI, Thompson SA, Brodbelt JS, Trent MS (2013) EptC of *Campylobacter jejuni* mediates phenotypes involved in host interactions and virulence. Infect Immun 81:430–440. https://doi.org/10.1128/IAI.01046-12

Devaraj A et al (2019) The extracellular DNA lattice of bacterial biofilms is structurally related to Holliday junction recombination intermediates. Proc Natl Acad Sci USA 116(50):25068–25077. https://doi.org/10.1073/pnas.1909017116

Drozd M, Chandrashekhar K, Rajashekara G (2014) Polyphosphate-mediated modulation of *Campylobacter jejuni* biofilm growth and stability. Virulence 5:680–690. https://doi.org/10.4161/viru.34348

Duan K, Dammel C, Stein J, Rabin H, Surette MG (2003) Modulation of *Pseudomonas aeruginosa* gene expression by host microflora through interspecies communication. Mol Microbiol 50:1477–1491. https://doi.org/10.1046/j.1365-2958.2003.03803.x

Dwivedi R, Nothaft H, Garber J, Xin Kin L, Stahl M, Flint A, van Vliet AH, Stintzi A, Szymanski CM (2016) L-fucose influences chemotaxis and biofilm formation in *Campylobacter jejuni*. Mol Microbiol 101:575–589. https://doi.org/10.1111/mmi.13409

El Abbar FM, Li J, Owen HC, Daugherty CL, Fulmer CA, Bogacz M, Thompson SA (2019) RNA binding by the *Campylobacter jejuni* post-transcriptional regulator *csrA*. Front Microbiol 10:1776. https://doi.org/10.3389/fmicb.2019.01776

Elvers KT, Park SF (2002) Quorum sensing in *Campylobacter jejuni*: detection of a *luxS* encoded signalling molecule. Microbiology 148:1475–1481. https://doi.org/10.1099/00221287-148-5-1475

Feng J, Lamour G, Xue R, Mirvakliki MN, Hatzikiriakos SG, Xu J, Li H, Wang S, Lu X (2016) Chemical, physical and morphological properties of bacterial biofilms affect survival of encased *Campylobacter jejuni* F38011 under aerobic stress. Int J Food Microbiol 238:172–182. https://doi.org/10.1016/j.ijfoodmicro.2016.09.008

Feng J, Ma L, Nie J, Konkel ME, Lu X (2018) Environmental stress-induced bacterial lysis and extracellular DNA release contribute to *Campylobacter jejuni* biofilm formation. Appl Environ Microbiol 84. https://doi.org/10.1128/AEM.02068-17

Fields JA, Thompson SA (2008) *Campylobacter jejuni csrA* mediates oxidative stress responses, biofilm formation, and host cell invasion. J Bacteriol 190:3411–3416. https://doi.org/10.1128/JB.01928-07

Fields JA, Li J, Gulbronson CJ, Hendrixson DR, Thompson SA (2016) *Campylobacter jejuni csrA* regulates metabolic and virulence associated proteins and is necessary for mouse colonization. PLoS ONE 11:e0156932. https://doi.org/10.1371/journal.pone.0156932

Flemming HC, Wingender J, Szewzyk U, Steinberg P, Rice SA, Kjelleberg S (2016) Biofilms: an emergent form of bacterial life. Nat Rev Microbiol 14:563–575. https://doi.org/10.1038/nrmicro.2016.94

Furniss RCD, Clements A (2018) Regulation of the locus of enterocyte effacement in attaching and effacing pathogens. J Bacteriol 200. https://doi.org/10.1128/JB.00336-17

Gangaiah D, Kassem II, Liu Z, Rajashekara G (2009) Importance of polyphosphate kinase 1 for *Campylobacter jejuni* viable-but-nonculturable cell formation, natural transformation, and antimicrobial resistance. Appl Environ Microbiol 75:7838–7849. https://doi.org/10.1128/AEM.01603-09

Gangaiah D, Liu Z, Arcos J, Kassem II, Sanad Y, Torrelles JB, Rajashekara G (2010) Polyphosphate kinase 2: a novel determinant of stress responses and pathogenesis in *Campylobacter jejuni*. PLoS ONE 5:e12142. https://doi.org/10.1371/journal.pone.0012142

Golz G, Adler L, Huehn S, Alter T (2012) *LuxS* distribution and AI-2 activity of *Campylobacter* spp. J Appl Microbiol 112:571–578. https://doi.org/10.1111/j.1365-2672.2011.05221.x

Golz G, Kittler S, Malakauskas M, Alter T (2018) Survival of *Campylobacter* in the food chain and the environment. Curr Clin Microbiol 5:126–134. https://doi.org/10.1007/s40588-018-0092-z

Guerry P (2007) *Campylobacter* flagella: not just for motility. Trends Microbiol 15:456–461. https://doi.org/10.1016/j.tim.2007.09.006

Hall CW, Mah TF (2017) Molecular mechanisms of biofilm-based antibiotic resistance and tolerance in pathogenic bacteria. FEMS Microbiol Rev 41(3):276–301. https://doi.org/10.1093/femsre/fux010

Hansson I, Sandberg M, Habib I, Lowman R, Engvall EO (2018) Knowledge gaps in control of *Campylobacter* for prevention of campylobacteriosis. Transbound Emerg Dis. 65:30–48. https://doi.org/10.1111/tbed.12870

He Y, Frye JG, Strobaugh TP, Chen CY (2008) Analysis of AI-2/*luxS*-dependent transcription in *Campylobacter jejuni* strain 81–176. Foodborne Pathog Dis 5:399–415. https://doi.org/10.1089/fpd.2008.0106

Hegde M, Englert DL, Schrock S, Cohn WB, Vogt C, Wood TK, Manson MD, Jayaraman A (2011) Chemotaxis to the quorum-sensing signal AI-2 requires the Tsr chemoreceptor and the periplasmic LsrB AI-2-binding protein. J Bacteriol 193:768–773. https://doi.org/10.1128/JB.01196-10

Herzberg M, Kaye IK, Peti W, Wood TK (2006) YdgG (TqsA) controls biofilm formation in *Escherichia coli* K-12 through autoinducer 2 transport. J Bacteriol 188:587–598. https://doi.org/10.1128/JB.188.2.587-598.2006

Holmes K, Tavender TJ, Winzer K, Wells JM, Hardie KR (2009) AI-2 does not function as a quorum sensing molecule in *Campylobacter jejuni* during exponential growth in vitro. BMC Microbiol 9:214. https://doi.org/10.1186/1471-2180-9-214

Howard SL, Jagannathan A, Soo EC, Hui JP, Aubry AJ, Ahmed I, Karlyshev A, Kelly JF, Jones MA, Stevens MP, Logan SM, Wren BW (2009) *Campylobacter jejuni* glycosylation island important in cell charge, legionaminic acid biosynthesis, and colonization of chickens. Infect Immun 77:2544–2556. https://doi.org/10.1128/IAI.01425-08

Hwang S, Kim M, Ryu S, Jeon B (2011) Regulation of oxidative stress response by CosR, an essential response regulator in *Campylobacter jejuni*. PLoS ONE 6:e22300. https://doi.org/10.1371/journal.pone.0022300

Indikova I, Humphrey TJ, Hilbert F (2015) Survival with a helping hand: *Campylobacter* and microbiota. Front Microbiol 6:1266. https://doi.org/10.3389/fmicb.2015.01266

Jeon B, Itoh K, Misawa N, Ryu S (2003) Effects of quorum sensing on *flaA* transcription and autoagglutination in *Campylobacter jejuni*. Microbiol Immunol 47:833–839. https://doi.org/10.1111/j.1348-0421.2003.tb03449.x

Jiang Q, Chen J, Yang C, Yin Y, Yao K (2019) Quorum sensing: a prospective therapeutic target for bacterial diseases. Biomed Res Int 2019:2015978. https://doi.org/10.1155/2019/2015978

Joshua GW, Guthrie-Irons C, Karlyshev AV, Wren BW (2006) Biofilm formation in *Campylobacter jejuni*. Microbiology 152:387–396. https://doi.org/10.1099/mic.0.28358-0

Kalmokoff M, Lanthier P, Tremblay TL, Foss M, Lau PC, Sanders G, Austin J, Kelly J, Szymanski CM (2006) Proteomic analysis of *Campylobacter jejuni* 11168 biofilms reveals a role for the motility complex in biofilm formation. J Bacteriol 188:4312–4320. https://doi.org/10.1128/JB.01975-05

Kaplan JB (2010) Biofilm dispersal: mechanisms, clinical implications, and potential therapeutic uses. J Dent Res 89(3):205–18. https://doi.org/10.1177/0022034509359403

Karlyshev AV, Linton D, Gregson NA, Lastovica AJ, Wren BW (2000) Genetic and biochemical evidence of a *Campylobacter jejuni* capsular polysaccharide that accounts for Penner serotype specificity. Mol Microbiol 35(3):529–541. https://doi.org/10.1046/j.1365-2958.2000.01717.x

Kassem II, Candelero-Rueda RA, Esseili KA, Rajashekara G (2017) Formate simultaneously reduces oxidase activity and enhances respiration in *Campylobacter jejuni*. Sci Rep 7:40117. https://doi.org/10.1038/srep40117

Kim JS, Park C, Kim YJ (2015) Role of *flgA* for flagellar biosynthesis and biofilm formation of *Campylobacter jejuni* NCTC11168. J Microbiol Biotechn 25:1871–1879. https://doi.org/10.4014/jmb.1504.04080

Kostakioti M, Hadjifrangiskou M, Hultgren SJ (2013) Bacterial biofilms: development, dispersal, and therapeutic strategies in the dawn of the postantibiotic era. Cold Spring Harb Perspect Med 3:a010306. https://doi.org/10.1101/cshperspect.a010306

Kreuder AJ, Ruddell B, Mou K, Hassall A, Zhang Q, Plummer PJ (2020) Small non-coding RNA CjNC110 influences motility, autoagglutination, AI-2 localization, hydrogen peroxide sensitivity and chicken colonization in *Campylobacter jejuni*. Infect Immun. https://doi.org/10.1128/IAI.00245-20

Kumar A, Gangaiah D, Torrelles JB, Rajashekara G (2016) Polyphosphate and associated enzymes as global regulators of stress response and virulence in *Campylobacter jejuni*. World J Gastroenterol 22:7402–7414. https://doi.org/10.3748/wjg.v22.i33.7402

Lamas A, Regal P, Vazquez B, Miranda JM, Cepeda A, Franco CM (2018) *Salmonella* and *Campylobacter* biofilm formation: a comparative assessment from farm to fork. J Sci Food Agr 98:4014–4032. https://doi.org/10.1002/jsfa.8945

LaSarre B, Federle MJ (2013) Exploiting quorum sensing to confuse bacterial pathogens. Microbiol Mol Biol Rev 77:73–111. https://doi.org/10.1128/MMBR.00046-12

Li J, Feng J, Ma L, de la Fuente NC, Golz G, Lu X (2017) Effects of meat juice on biofilm formation of *Campylobacter* and *Salmonella*. Int J Food Microbiol 253:20–28. https://doi.org/10.1016/j.ijfoodmicro.2017.04.013

Ligowska M, Cohn MT, Stabler RA, Wren BW, Brondsted L (2011) Effect of chicken meat environment on gene expression of *Campylobacter jejuni* and its relevance to survival in food. Int J Food Microbiol 145(Suppl 1):S111–S115. https://doi.org/10.1016/j.ijfoodmicro.2010.08.027

Lim ES, Kim JS (2017) Role of *eptC* in Biofilm Formation by *Campylobacter jejuni* NCTC11168 on Polystyrene and Glass Surfaces. J Microbiol Biotechn 27:1609–1616. https://doi.org/10.4014/jmb.1610.10046

Melo RT, Mendonca EP, Monteiro GP, Siqueira MC, Pereira CB, Peres PABM, Fernandez H, Rossi DA (2017) Intrinsic and extrinsic aspects on *Campylobacter jejuni* biofilms. Front Microbiol 8:1332. https://doi.org/10.3389/fmicb.2017.01332

Miller MB, Skorupski K, Lenz DH, Taylor RK, Bassler BL (2002) Parallel quorum sensing systems converge to regulate virulence in *Vibrio cholerae*. Cell 110:303–314. https://doi.org/10.1016/s0092-8674(02)00829-2

Moorhead SM, Griffiths MW (2011) Expression and characterization of cell-signalling molecules in *Campylobacter jejuni*. J Appl Microbiol 110:786–800. https://doi.org/10.1111/j.1365-2672.2010.04934.x

Mou KT, Muppirala UK, Severin AJ, Clark TA, Boitano M, Plummer PJ (2014) A comparative analysis of methylome profiles of *Campylobacter jejuni* sheep abortion isolate and gastroenteric strains using PacBio data. Front Microbiol 5:782. https://doi.org/10.3389/fmicb.2014.00782

Mou KT, Plummer PJ (2016) The impact of the *luxS* mutation on phenotypic expression of factors critical for *Campylobacter jejuni* colonization. Vet Microbiol 192:43–51. https://doi.org/10.1016/j.vetmic.2016.06.011

Mukherjee S, Bassler BL (2019) Bacterial quorum sensing in complex and dynamically changing environments. Nat Rev Microbiol 17:371–382. https://doi.org/10.1038/s41579-019-0186-5

Naito M, Frirdich E, Fields JA, Pryjma M, Li J, Cameron A, Gilbert M, Thompson SA, Gaynor EC (2010) Effects of sequential *Campylobacter jejuni* 81–176 lipooligosaccharide core truncations on biofilm formation, stress survival, and pathogenesis. J Bacteriol 192:2182–2192. https://doi.org/10.1128/JB.01222-09

Oh E, Jeon B (2014) Role of alkyl hydroperoxide reductase (AhpC) in the biofilm formation of *Campylobacter jejuni*. PLoS ONE 9:e87312. https://doi.org/10.1371/journal.pone.0087312

Oh E, Kim JC, Jeon B (2016) Stimulation of biofilm formation by oxidative stress in *Campylobacter jejuni* under aerobic conditions. Virulence 7:846–851. https://doi.org/10.1080/21505594.2016.1197471

Paluch E, Rewak-Soroczynska J, Jedrusik I, Mazurkiewicz E, Jermakow K (2020) Prevention of biofilm formation by quorum quenching. Appl Microbiol Biotechnol 104:1871–1881. https://doi.org/10.1007/s00253-020-10349-w

Papenfort K, Bassler BL (2016) Quorum sensing signal-response systems in Gram-negative bacteria. Nat Rev Microbiol 14:576–588. https://doi.org/10.1038/nrmicro.2016.89

Pascoe B, Meric G, Murray S, Yahara K, Mageiros L, Bowen R, Jones NH, Jeeves RE, Lappin-Scott HM, Asakura H, Sheppard SK (2015) Enhanced biofilm formation and multi-host transmission evolve from divergent genetic backgrounds in *Campylobacter jejuni*. Environ Microbiol 17:4779–4789. https://doi.org/10.1111/1462-2920.13051

Pascoe B, Williams LK, Calland JK, Meric G, Hitchings MD, Dyer M, Ryder J, Shaw S, Lopes BS, Chintoan-Uta C, Allan E, Vidal A, Fearnley C, Everest P, Pachebat JA, Cogan TA, Stevens MP, Humphrey TJ, Wilkinson TS, Cody AJ, Colles FM, Jolley KA, Maiden MCJ, Strachan N, Pearson BM, Linton D, Wren BW, Parkhill J, Kelly DJ, van Vliet AHM, Forbes KJ, Sheppard SK (2019) Domestication of *Campylobacter jejuni* NCTC11168. Microb Genom 5. https://doi.org/10.1099/mgen.0.000279

Pereira CS, Thompson JA, Xavier KB (2013) AI-2-mediated signalling in bacteria. FEMS Microbiol Rev 37:156–181. https://doi.org/10.1111/j.1574-6976.2012.00345.x

Petrova OE, Sauer K (2016) Escaping the biofilm in more than one way: desorption, detachment or dispersion. Curr Opin Microbiol 30:67–78. https://doi.org/10.1016/j.mib.2016.01.004.

Plummer P, Sahin O, Burrough E, Sippy R, Mou K, Rabenold J, Yaeger M, Zhang Q (2012) Critical role of *luxS* in the virulence of *Campylobacter jejuni* in a guinea pig model of abortion. Infect Immun 80:585–593. https://doi.org/10.1128/IAI.05766-11

Plummer P, Zhu J, Akiba M, Pei D, Zhang Q (2011) Identification of a key amino acid of *luxS* involved in AI-2 production in *Campylobacter jejuni*. PLoS ONE 6:e15876. https://doi.org/10.1371/journal.pone.0015876

Plummer PJ (2012) LuxS and quorum-sensing in *Campylobacter*. Front Cell Infect Microbiol 2:22. https://doi.org/10.3389/fcimb.2012.00022

Quinones B, Miller WG, Bates AH, Mandrell RE (2009) Autoinducer-2 production in *Campylobacter jejuni* contributes to chicken colonization. Appl Environ Microbiol 75:281–285. https://doi.org/10.1128/AEM.01803-08

Rader BA, Wreden C, Hicks KG, Sweeney EG, Ottemann KM, Guillemin K (2011) *Helicobacter pylori* perceives the quorum-sensing molecule AI-2 as a chemorepellent via the chemoreceptor TlpB. Microbiology 157:2445–2455. https://doi.org/10.1099/mic.0.049353-0

Rahman H, King RM, Shewell LK, Semchenko EA, Hartley-Tassell LE, Wilson JC, Day CJ, Korolik V (2014) Characterisation of a multi-ligand binding chemoreceptor CcmL (Tlp3) of *Campylobacter jejuni*. PLoS Pathog 10:e1003822. https://doi.org/10.1371/journal.ppat.1003822

Rathbun KM, Hall JE, Thompson SA (2009) Cj0596 is a periplasmic peptidyl prolyl cis-trans isomerase involved in *Campylobacter jejuni* motility, invasion, and colonization. BMC Microbiol 9:160. https://doi.org/10.1186/1471-2180-9-160

Reeser RJ, Medler RT, Billington SJ, Jost BH, Joens LA (2007) Characterization of *Campylobacter jejuni* biofilms under defined growth conditions. Appl Environ Microbiol 73:1908–1913. https://doi.org/10.1128/AEM.00740-06

Rettner RE, Saier MH Jr (2010) The autoinducer-2 exporter superfamily. J Mol Microbiol Biotechnol 18:195–205. https://doi.org/10.1159/000316420

Reuter M, Mallett A, Pearson BM, van Vliet AH (2010) Biofilm formation by *Campylobacter jejuni* is increased under aerobic conditions. Appl Environ Microbiol 76:2122–2128. https://doi.org/10.1128/AEM.01878-09

Reuter M, Ultee E, Toseafa Y, Tan A, van Vliet AHM (2020) Inactivation of the core *cheVAWY* chemotaxis genes disrupts chemotactic motility and organised biofilm formation in *Campylobacter jejuni*. FEMS Microbiol Lett https://doi.org/10.1093/femsle/fnaa198

Roy R, Tiwari M, Donelli G, Tiwari V (2018) Strategies for combating bacterial biofilms: A focus on anti-biofilm agents and their mechanisms of action. Virulence 9:522–554. https://doi.org/10.1080/21505594.2017.1313372

Scott NE, Nothaft H, Edwards AV, Labbate M, Djordjevic SP, Larsen MR, Szymanski CM, Cordwell SJ (2012) Modification of the *Campylobacter jejuni* N-linked glycan by EptC protein-mediated addition of phosphoethanolamine. J Biol Chem 287:29384–29396. https://doi.org/10.1074/jbc.M112.380212

Sharma K, Pagedar Singh A (2018) Antibiofilm effect of DNase against single and mixed species biofilm. Foods 7(3):42. https://doi.org/10.3390/foods7030042

Sharma D, Misba L, Khan AU (2019) Antibiotics versus biofilm: an emerging battleground in microbial communities. Antimicrob Resist Infect Control 8:76. https://doi.org/10.1186/s13756-019-0533-3

Simunovic K, Ramic D, Xu C, Smole Mozina S (2020) Modulation of *Campylobacter jejuni* motility, adhesion to polystyrene surfaces, and invasion of INT407 cells by quorum-sensing inhibition. Microorganisms 8. https://doi.org/10.3390/microorganisms8010104

Stetsenko VV, Efimochkina NR, Pichugina TV (2019) Growth and persistence of *Campylobacter jejuni* in foodstuffs. Bull Exp Biol Med 166:759–765. https://doi.org/10.1007/s10517-019-04435-x

Svensson SL, Davis LM, MacKichan JK, Allan BJ, Pajaniappan M, Thompson SA, Gaynor EC (2009) The CprS sensor kinase of the zoonotic pathogen *Campylobacter jejuni* influences biofilm formation and is required for optimal chick colonization. Mol Microbiol 71:253–272. https://doi.org/10.1111/j.1365-2958.2008.06534.x

Svensson SL, Pryjma M, Gaynor EC (2014) Flagella-mediated adhesion and extracellular DNA release contribute to biofilm formation and stress tolerance of *Campylobacter jejuni*. PLoS ONE 9:e106063. https://doi.org/10.1371/journal.pone.0106063

Tazumi A, Negoro M, Tomiyama Y, Misawa N, Itoh K, Moore JE, Millar BC, Matsuda M (2011) Uneven distribution of the *luxS* gene within the genus *Campylobacter*. Br J Biomed Sci 68:19–22. https://doi.org/10.1080/09674845.2011.11732836

Teh AH, Lee SM, Dykes GA (2014) Does *Campylobacter jejuni* form biofilms in food-related environments? Appl Environ Microbiol 80(17):5154–5160. https://doi.org/10.1128/AEM.01493-14

Teh AH, Lee SM, Dykes GA (2016) The influence of prior modes of growth, temperature, medium, and substrate surface on biofilm formation by antibiotic-resistant *Campylobacter jejuni*. Curr Microbiol 73:859–866. https://doi.org/10.1007/s00284-016-1134-5

Teh AHT, Lee SM, Dykes GA (2017) Identification of potential *Campylobacter jejuni* genes involved in biofilm formation by EZ-Tn5 Transposome mutagenesis. BMC Res Notes 10:182. https://doi.org/10.1186/s13104-017-2504-1

Teh AHT, Lee SM, Dykes GA (2019) Association of some *Campylobacter jejuni* with *Pseudomonas aeruginosa* biofilms increases attachment under conditions mimicking those in the environment. PLoS ONE 14:e0215275. https://doi.org/10.1371/journal.pone.0215275

Teh KH, Flint S, French N (2010) Biofilm formation by *Campylobacter jejuni* in controlled mixed-microbial populations. Int J Food Microbiol 143:118–124. https://doi.org/10.1016/j.ijfoodmicro.2010.07.037

Theoret JR, Cooper KK, Zekarias B, Roland KL, Law BF, Curtiss R, Joens LA (2012) The *Campylobacter jejuni* Dps homologue is important for in vitro biofilm formation and cecal colonization of poultry and may serve as a protective antigen for vaccination. Clin Vaccine Immunol 19:1426–1431. https://doi.org/10.1128/CVI.00151-12

Tram G, Day CJ, Korolik V (2020a) Bridging the Gap: A Role for *Campylobacter jejuni* Biofilms. Microorganisms 8(3):452. https://doi.org/10.3390/microorganisms8030452

Tram G, Klare WP, Cain JA, Mourad B, Cordwell SJ, Day CJ, Korolik V (2020b) Assigning a role for chemosensory signal transduction in *Campylobacter jejuni* biofilms using a combined omics approach. Sci Rep 10:6829. https://doi.org/10.1038/s41598-020-63569-5

Turonova H, Briandet R, Rodrigues R, Hernould M, Hayek N, Stintzi A, Pazlarova J, Tresse O (2015) Biofilm spatial organization by the emerging pathogen *Campylobacter jejuni*: comparison between NCTC11168 and 81–176 strains under microaerobic and oxygen-enriched conditions. Front Microbiol 6:709. https://doi.org/10.3389/fmicb.2015.00709

Wagle BR, Upadhyay A, Upadhyaya I, Shrestha S, Arsi K, Liyanage R, Venkitanarayanan K, Donoghue DJ, Donoghue AM (2019) Trans-cinnamaldehyde, eugenol and carvacrol reduce *Campylobacter jejuni* biofilms and modulate expression of select genes and proteins. Front Microbiol 10:1837. https://doi.org/10.3389/fmicb.2019.01837

Whelan MVX, Ardill L, Koide K, Nakajima C, Suzuki Y, Simpson JC, Cróinín TÓ (2019) Acquisition of fluoroquinolone resistance leads to increased biofilm formation and pathogenicity in *Campylobacter jejuni*. Sci Rep 9:18216. https://doi.org/10.1038/s41598-019-54620-1

Winzer K, Hardie KR, Burgess N, Doherty N, Kirke D, Holden MT, Linforth R, Cornell KA, Taylor AJ, Hill PJ, Williams P (2002) LuxS: its role in central metabolism and the in vitro synthesis of 4-hydroxy-5-methyl-3(2H)-furanone. Microbiology 148:909–922. https://doi.org/10.1099/00221287-148-4-909

Zhang T, Dong J, Cheng Y, Lu Q, Luo Q, Wen G, Liu G, Shao H (2017) Genotypic diversity, antimicrobial resistance and biofilm-forming abilities of *Campylobacter* isolated from chicken in Central China. Gut Pathog 9:62. https://doi.org/10.1186/s13099-017-0209-6

Zhong X, Wu Q, Zhang J, Ma Z, Wang J, Nie X, Ding Y, Xue L, Chen M, Wu S, Wei X, Zhang Y (2020) *Campylobacter jejuni* biofilm formation under aerobic conditions and inhibition by ZnO nanoparticles. Front Microbiol 11:207. https://doi.org/10.3389/fmicb.2020.00207

Correction to: Fighting *Campylobacter* Infections

Steffen Backert

Correction to:
S. Backert (ed.),
Fighting Campylobacter Infections, Current Topics in Microbiology and Immunology 431, https://doi.org/10.1007/978-3-030-65481-8

In the original version of the chapter "The Data Behind Risk Analysis of *Campylobacter Jejuni* and *Campylobacter Coli* Infections" by mistake the family name of the author Racem Ben Romdhane had been indicated as Romdhane instead of Ben Romdhane and his given name had been indicated as Racam Ben instead of Racem. This has now been corrected.

And

The original version of the chapter "Population Biology and Comparative Genomics of *Campylobacter* Species" was published with the copyright holder "The Author(s), under exclusive license to Springer Nature Switzerland AG" and without open access. This has now been changed to the copyright holder "The Author(s)" and open access under the terms of the Creative Commons Attribution 4.0 International License (https://creativecommons.org/licenses/by/4.0/).

For further details see license information in the chapter.

The chapters and the book have been updated with the changes.

The updated version of these chapters can be found at
https://doi.org/10.1007/978-3-030-65481-8_2
https://doi.org/10.1007/978-3-030-65481-8_3

S. Backert (ed.), *Fighting Campylobacter Infections*, Current Topics in Microbiology and Immunology 431, https://doi.org/10.1007/978-3-030-65481-8_12

The manufacturer's authorised representative in the EU is Springer Nature Customer Service Centre GmbH, Europaplatz 3, 69115 Heidelberg, Germany. If you have any concerns regarding our products, please contact ProductSafety@springernature.com

Printed and bound by CPI Group (UK) Ltd, Croydon, CR0 4YY
23/09/2026
02221544-0009